Large-Scale Mammalian Cell Culture Technology

Bioprocess Technology

Series Editor

W. Courtney McGregor

Xoma Corporation
Berkeley, California

Large-Scale Mammalian Cell Culture Technology

edited by

ANTHONY S. LUBINIECKI

SmithKline Beecham Pharmaceuticals
King of Prussia, Pennsylvania

MARCEL DEKKER, INC.   New York and Basel

Library of Congress Cataloging-in-Publication Data

Large-scale mammalian cell culture technology/edited by Anthony S.
  Lubiniecki.
      p.   cm. -- (Bioprocess technology; v. 10)
    Includes bibliographical references.
    Includes index.
    ISBN 0-8247-8327-1 (alk. paper)
    1. Cell culture. 2. Biotechnology. 3 Mammals--Cytology-
-Technique. I. Lubiniecki, Anthony S. II. Series.
    [DNLM: 1. Biological Products. 2. Cells, Cultured. 3.Mam-
mals. W1 BI88U v. 10/QS 525 L322]
    QH585.L364 1990
    660'.6--dc20
    DNLM/DLC
    for Library of Congress   90-3897
                                                                  CIP

MARCEL DEKKER, INC.
270 Madison Avenue, New York, New York   10016

Current printing (last digit):
10 9 8 7 6 5 4 3 2 1

PRINTED IN THE UNITED STATES OF AMERICA

To Robin and Greg, whose support and patience
enabled this work to be completed

Series Introduction

Bioprocess technology encompasses all of the basic and applied sciences as well as the engineering required to fully exploit living systems and bring their products to the marketplace. The technology that develops is eventually expressed in various methodologies and types of equipment and instruments built up along a bioprocess stream. Typically in commercial production, the stream begins at the bioreactor, which can be a classical fermentor, a cell culture perfusion system, or an enzyme bioreactor. Then comes separation of the product from the living systems and/or their components followed by an appropriate number of purification steps. The stream ends with bioproduct finishing, formulation, and packaging. A given bioprocess stream may have some tributaries or outlets and may be overlaid with a variety of monitoring devices and control systems. As with any stream, it will both shape and be shaped with time. Documenting the evolutionary shaping of bioprocess technology is the purpose of this series.

Now that several products from recombinant DNA and cell fusion techniques are on the market, the new era of bioprocess technology is well established and validated. Books of this series represent developments in various segments of bioprocessing that have paralleled progress in the life sciences. For obvious proprietary reasons, some developments in industry, although validated, may be published only later, if at all. Therefore, our continuing series will follow the growth of this field as it is available from both academia and industry.

W. Courtney McGregor

Preface

Large-scale cell culture is a hybrid technology that combines aspects of many different disciplines, including cell biology, biochemistry, fermentation, virology and microbiology, and protein chemistry. This technology has been employed historically to prepare virus vaccines, but more recently to produce protein pharmaceuticals. The need for these products to be safe, effective, and economical has provided the impetus for developmental efforts. These efforts have led to improvements in the technology and are occurring regularly. This growth, which has been especially noticeable over the past 7 years, has prompted the present work to be written.

This book critically reviews the current state of the field. The aim of the book is to offer its readers a view of what is happening and how the different disciplines interact to produce safe products. To understand that process it is necessary to examine each specialty and to describe the benefits, shortcomings, and interactions of each.

From the genesis of this work, the chapter authors and I felt this book should be of value in stimulating the current generation of graduate students to enter the field of cell culture technology. We also wish to encourage professionals of related fields to diversify into cell culture technology. At this time, more trained workers are needed to satisfy the current industrial demand and to increase the rate at which new discoveries can be made.

Accordingly, we have prepared a manuscript that strives to be comprehensive, readable, and critical. The chapters include basic theory as well as current applications and problems. Wherever possible, interrelationships of diverse concepts are illustrated. We hope that this interdisciplinary approach to cell culture methodology will stimulate readers to start or continue a career in this complex, but rewarding, area of technology.

Anthony S. Lubiniecki

Contents

Contributors

Robert Arathoon   Genentech, Inc., South San Francisco, California

Donald G. Bergmann   SmithKline Beecham Pharmaceuticals, King of Prussia, Pennsylvania

John R. Birch   Celltech Ltd., Slough, England

Stuart E. Builder   Genentech, Inc., South San Francisco, California

Edward T. Cox   Genentech, Inc., South San Francisco, California

David E. DeLucia   Verax Corporation, Lebanon, New Hampshire

Arthur Y. Elliott   Merck Pharmaceutical Manufacturing Division of Merck & Co., Inc., West Point, Pennsylvania

N. B. Finter   Wellcome Biotechnology Ltd., Beckenham, England

A. J. M. Garland   Wellcome Biotechnology Ltd., Beckenham, England

Robert L. Garnick   Genentech, Inc., South San Francisco, California

Bryan Griffiths   Public Health Laboratory Service, Centre for Applied Microbiology and Research, Salisbury, England

Edward G. Hayman*   Verax Corporation, Lebanon, New Hampshire

Wei-Shou Hu   Department of Chemical Engineering and Materials Science, University of Minnesota, Minneapolis, Minnesota

Andrew J. S. Jones   Genentech, Inc., South San Francisco, California

Randal J. Kaufman   Genetics Institute, Cambridge, Massachusetts

Anthony S. Lubiniecki   SmithKline Beecham Pharmaceuticals, King of Prussia, Pennsylvania

Jennie P. Mather   Genentech, Inc., South San Francisco, California

Laurie H. May   Genentech, Inc., South San Francisco, California

Mark G. Oberg[†]   Department of Chemical Engineering and Materials Science, University of Minnesota, Minneapolis, Minnesota

John R. Ogez   Genentech, Inc., South San Francisco, California

Melvin S. Oka   SmithKline Beecham Pharmaceuticals, King of Prussia, Pennsylvania

Judith A. Poiley   Hazleton Laboratories America, Inc., Kensington, Maryland

Nitya G. Ray   Verax Corporation, Lebanon, New Hampshire

Shaul Reuveny   Israel Institute for Biological Research, Ness-Ziona, Israel

Peter W. Runstadler, Jr.   Verax Corporation, Lebanon, New Hampshire

Randall G. Rupp   Somatogen Corporation, Broomfield, Colorado

Johanna v. G. Sample   Verax Corporation, Lebanon, New Hampshire

Mary B. Sliwkowski   Genentech, Inc., South San Francisco, California

William R. Srigley   Invitron Corporation, St. Louis, Missouri

---

Present affiliations:
*T-Cell Sciences, Cambridge, Massachusetts.
[†]Exxon Chemical—Baytown Olefins Plant, Baytown, Texas.

R. C. Telling   Wellcome Biotechnology Ltd., Beckenham, England

James N. Thomas   Immunex Corporation, Seattle, Washington

Mary Tsao   Genentech, Inc., South San Francisco, California

Amar S. Tung   Verax Corporation, Lebanon, New Hampshire

Joseph E. Tyler   Abbott Biotech, Inc., Needham Heights, Massachusetts

James L. Vaughn   Insect Pathology Laboratory, Plant Sciences Institute, Agricultural Research Service, U.S. Department of Agriculture, Beltsville, Maryland

Stefan A. Weiss   GIBCO–Life Technologies, Inc., Grand Island, New York

Michael E. Wiebe   Genentech, Inc., South San Francisco, California

Large-Scale Mammalian Cell Culture Technology

1

Large-Scale Mammalian Cell Culture:
A Perspective

N. B. FINTER, A. J. M. GARLAND, and R. C. TELLING
Wellcome Biotechnology Ltd., Beckenham, England

I. INTRODUCTION

While the large-scale culture of mammalian cells may seem a comparatively recent development, both human (1) and hamster cells (2,3) had already been grown in 1000-liter culture tanks some 20 years ago. Work on such a scale was at first carried out in only a very few institutions whose special needs made it worthwhile to tackle the initially formidable technological and other problems.
One such need was to produce foot-and-mouth disease (FMD) virus vaccines on a commercial scale for veterinary use; another was to make large amounts of human interferon (IFN) for clinical evaluation. Elsewhere, large-scale cell culture was generally thought to be a difficult, clumsy, and uneconomical way of manufacturing, with the further disadvantage that its products might in the end prove to be unacceptable for use in terms of safety.

Concerns about safety stemmed from studies in the 1950s and 1960s with cultures in magnetically stirred "spinner" flasks. These showed that the cells which grew readily in suspension culture and which could be subcultured without apparent limit were all "transformed" and indeed could be distinguished by these characteristics from "normal" cells grown directly from human or animal tissues, and diploid fibroblasts. Most transformed cells were derived from tumors, were aneuploid with at least some chromosomal abnormalities, and when injected into neonatal or immunosuppressed animals led to the formation of tumors. Since little was known at that time about the mechanisms responsible for the transformed state, the idea of using such cells as a source of medicinal products met with little favor: it was feared that the products might be contaminated with some virus or oncogenic factor derived from the cells with potentially disastrous consequences for the recipients.

Today, attitudes have changed as the result of developments in the scale and reliability of mammalian cell culture and the proven safety and efficacy of proteins derived from them. In many countries there are now regulations governing pharmaceutical manufacture from suspension cell cultures and a number of products made in this way have been licensed for use. Indeed, mammalian cell cultures are now the preferred source of a number of important proteins for use in human and animal medicine, especially those which are relatively large, complex, or glycosylated. This chapter will trace the developments which have influenced present views, with special reference to the important precedents which FMD virus vaccines and interferon have provided.

II. PRODUCTION OF FOOT-AND-MOUTH DISEASE VIRUS VACCINE FROM BHK 21 CELLS

Foot-and-mouth disease (FMD) is enzootic in cattle, sheep, goats, and pigs in many parts of the world and, unless controlled, has major economic consequences through its effects on animal productivity and international trade. In 1947, Frenkel (4) showed that FMD virus could be grown in surviving fragments of fresh bovine tongue epithelium incubated in a suitable medium. Today such cultures, handled in tanks on an industrial scale, still provide virus which is used to make vaccines. However, this system has the disadvantages common to all primary tissue culture systems and cannot provide enough vaccine to satisfy the worldwide demand.

In 1962, Mowat and Chapman (5), working at the U.K. government Animal Virus Research Institute (AVRI) at Pirbright, succeeded in growing FMD viruses in the anchorage-dependent cell line, BHK21, clone 13 (6). In further work, Capstick and coworkers at the AVRI adapted BHK cells to grow and replicate FMD viruses while in suspension culture (7). Further studies by these workers at the AVRI and subsequently in the adjoining Wellcome FMD laboratories defined the environmental conditions needed for optimal cell growth and production of viruses, and showed how the process could be scaled up for industrial use. These various studies identified both the similarities and the differences between mammalian cell culture and the systems already established for microbial culture; they will be described in some detail since many of the developments which turned the initial laboratory experimental system into a robust industrial process have proved to have general application in large-scale mammalian cell culture work.

Initial experiments at AVRI in 200- and 800-ml glass vessels (8,9) were soon scaled up into stainless steel vessels of 30- and 100-liter capacity, based on those available for the submerged culture of bacteria (10–13). They were made of an austenitic chromium–nickel alloy, with internal welds ground smooth and the internal finish a dull polish. The vessel ends were dished and the lid was attached to the body with a full-faced flanged gasket. To reduce the number of connections into the tank and consequent sterilization problems, pressure relief safety valves were not fitted; instead, the vessels were designed for a working pressure of 60 psi, and air, steam, and other services were supplied at maximum pressures well below this limit. Valves in contact with the culture media were of the diaphragm type or were stainless steel ball-plug valves in which leakage via the spindle is prevented by a self-adjusting polytetrafluoroethylene diaphragm seal (14).

The cells were stirred with a single-vaned disk impeller mounted on a stirrer shaft which passed through a double mechanical seal. Culture temperature, pH, and stirrer speed were automatically controlled (3,11,13). Temperature was found not to be critical over the range 33–37°C. The bicarbonate-CO_2 buffering system proved inadequate when the cell number exceeded about a million cells/ml and the pH fell below 7.0. Therefore, to improve pH control, air was automatically injected below the impeller when the pH fell below the desired value, and CO_2 similarly when the pH was too high (11,12). This method controlled pH to within ±0.05 pH unit. At a temperature of 35°C and a pH of 7.4, a fivefold increase in BHK cells to about 2.5×10^6/ml was consistently obtained within 48 hr (2,3,13,14). The cells grew best at a pO_2 level of 80 mm Hg, but in-place oxygen electrodes proved unstable and subject to fouling during prolonged culture (15,16). Since Tengerdy (17) had shown that the redox potential difference between aerated and deoxygenated culture medium was linearly proportional to the logarithm of the dissolved oxygen concentration, redox electrodes were installed: these controlled the automatic addition of sparge air whenever the redox potential fell below a desired value, approximately equivalent to a pO_2 level of 40–80 mm Hg. With this indirect method, the rate of growth and cell densities matched those resulting when

direct aeration measurements were used for control (Telling, unpublished observations).

Master cell banks were fully characterized according to U.S.
federal requirements (18) and the cells were stored frozen at $-65°C$
or in liquid nitrogen. All cells used in production were revived
from such a bank.

An inexpensive medium based on Eagle's formula, supplemented
with 5% adult bovine serum and tryptose phosphate broth, was developed which gave good productivity in the tank cultures. The
complete medium was sterilized by filtration through asbestos-
cellulose sheets (19) held in filter holders permanently connected to
the culture tanks. These and the connecting pipework were sterilized by steam injection; as a precaution, the pressure within the
system was kept positive throughout the cycle of sterilization, cool-
ing, and filtration.

The integrity of the vessels was improved by the development
of magnetic stirring (20): a driven magnet immediately adjacent to
the outside of the upper dished end of the tank induces rotation of
a follower magnet mounted immediately opposite within the vessel on
an axle shaft bearing the impeller. This elimination of a continuous
drive shaft with its attendant mechanical seals reduced the chances
of accidental microbial contamination. For the same reason, hardpiped steam-sterilizable sample and addition points were added.
These changes in plant design together with attention to details in
plant operation improved the long-term maintenance of sterility with
the closed system of tanks and pipes, and enabled the use of a
"solera" system of culture which has proved indispensable for larger
scale culture work. In this system, cells are grown for months at a
time in continuous culture in a particular tank; cells are drawn off
as required to feed the virus production vessels, leaving a volume
of cell suspension in the solera tank which is diluted with fresh
growth medium, so that fresh cycles of cell growth take place.
This major difference from the practices used in microbial fermentation reflects the very much slower rate of growth of animal cells;
several weeks may be needed from the time cells are revived from
the master cell bank until there are sufficient cells to seed the
largest vessels.

Experimental work confirmed that reproducible growth of the
BHK 21 cells and production of FMD viruses could be obtained and
that successive batch cultures could be carried out in the same
tank over extended periods of time without problems from microbial
contamination.

FMD virus strains grown in the system described were the
source of inactivated vaccines which were shown to be potent and
to protect animals of the target species against challenge infection

(20,21). As already pointed out, there were initial concerns that a vaccine of such an origin might prove to be tumorigenic. BHK cells contain C-type virus particles; they also become aneuploid and have an increasing oncogenic potential in the species of origin — the hamster — with successive passages in suspension culture (22, 23). However, a product for veterinary use enables direct safety tests to be carried out in the target species; accordingly, high concentrations of cells and cell extracts containing nucleic acids were injected into animals of various species, which were observed for periods of up to 2 or 3 years (individual animals and groups have been followed for periods up to 16 years; Garland, unpublished). All these studies confirmed that tumors only developed in one species, the hamster, and then only if intact cells were injected (22,23); the vaccine production process ensures that all cells are destroyed or removed during the course of manufacture.

The technology thus developed on the pilot plant scale has been used since 1967 for the design and operation of the industrial scale FMD virus vaccine production plants of the Wellcome (now Cooper Animal Health) group, located in eight countries in Europe, Africa, and South America. These now utilize vessels of up to 5000-liter capacity to produce the particular FMD virus serotypes of local importance. By 1976 more than 1 million liters of vaccine had been issued annually from these plants, and by 1983 approximately 2 million liters of vaccine, equivalent to 350 million doses of monovalent vaccine (20,24). The same basic technology has been used by Wellcome to make inactivated rabies virus vaccine for veterinary use from BHK 21 cells (25,26).

III. MASS CULTURE OF LYMPHOBLASTOID CELLS AT THE ROSWELL PARK MEMORIAL INSTITUTE

While FMD virus vaccines were being developed in the United Kingdom, a pioneer study was under way at the Roswell Park Memorial Institute in upper New York State. George Moore and his colleagues constructed a plant for growing human lymphoblastoid cells in suspension in a 1000-liter culture tank (1), at the time the largest scale used for mammalian cells. Their study showed that work at this scale was feasible and led to the development of a number of media, particularly RPMI 1640, which are still used for the culture of lymphoblastoid cells. However, no uses could then be found for the cells which were produced in large quantities; financial support ceased and the plant was abandoned.

This project was ahead of its time, and with hindsight it seems very likely that the harvests from some of these lymphoblastoid cell cultures contained various lymphokines in quantities sufficient to permit their isolation and purification, had this been realized.

IV. DEVELOPMENT OF LYMPHOBLASTOID
       INTERFERON AT WELLCOME

The clinical evaluation of the interferons (IFN) was greatly impeded
by a supply shortage for more than two decades after the original
discovery in 1957 (27). Leukocyte IFN preparations were made from
human white blood cells obtained from transfusion blood (28), and
these enabled clinical studies to start in the early 1970s, but only
relatively small amounts could be made in this way.

At the Wellcome Research Laboratories in Beckenham, studies
with IFN started in 1959. In 1974, Finter and colleagues decided to
explore alternatives to human blood cells as a source of IFN in order
to make substantially larger amounts available for clinical trials. By
then, Wellcome staff already had considerable experience with the
culture of BHK cells in tanks, and so it was logical to see if IFN
could be made in a similar system. After many human cell lines had
been screened (29), the Namalwa line of lymphoblastoid cells was
selected for further study, as these cells grew well and formed large
amounts of IFN. The program led to the large-scale production of a
highly purified human lymphoblastoid IFN preparation, which is now
the commercial product, Wellferon. After preliminary scale-up work
in a 50-liter tank, a 1000-liter pilot plant was commissioned early in
1978; this was followed by production from plants based on 2000-
liter vessels in 1980, and on multiple vessels of 8000- or 10,000-
liter capacity from 1983 onward.

Experience soon showed that while the equipment and technology
needed to make FMD vaccines and IFN are basically similar, the
latter process is technically much more demanding, reflecting the
more fastidious cell, the more complex production system, and the
more stringent requirements for a product intended for human use.

Namalwa cells grow relatively slowly, with a doubling time of
about 35 hr, and are therefore cultured in a continuous solera system. In larger plants, solera vessels in a series of increasing size
generate the cells needed for IFN production and provide a source
for rapid reseeding if one of the series is lost for whatever reason.
IFN is made in a batch process in which cells are transferred to a
production tank, treated with butyrate, and subsequently induced
by the addition of Sendai virus (30). The production vessels are
sterilized between each batch of cells, but the solera tanks are
maintained in continuous culture for as long as possible. With attention to detail in design and operation, the average duration of the
solera cultures has been progressively extended from a few weeks on
the pilot plant to 10 months or more on the present plants, with
great benefits to the efficiency and economy of production. Factors

contributing to this improvement have been a progressive increase
in the use of welded joints in the pipework and of diaphragm valves
rather than ball valves, together with changes in the types of filter
and housings used to supply sterile medium to the vessels. Instrumentation and control systems have also steadily improved in reliability, and increasingly digital recorders/controllers are used for
pH, temperature, and stirrer speed measurements and for gas
additions.

The medium used is RPMI 1640, supplemented with tryptone and
3.5% adult bovine serum. The serum is processed in-house by
Wellcome, tested for compliance with its quality control specification,
sterilized by passage through asbestos-free cartridge filters, and
stored frozen in convenient volumes. Before use, the serum is irradiated at 3.5 MRads, an amount more than sufficient to destroy
any contaminants likely to be present in the crude serum pool.

In the manufacture of an antibiotic or synthetic drug for medicinal use, more importance is attached to the quality and composition
of the final product than on in-process controls. Conversely, in
the production of a conventional live attenuated virus vaccine, control of the starting materials and in-process quality is critical,
since only limited tests on the composition of the final product are
possible. Wellferon in many ways resembles the former category of
products more closely than the latter; nevertheless, the quality of
the starting materials is checked as part of normal good manufacturing practice and there are extensive in-process quality checks at
all stages. Only cells derived from the tested master cell bank are
used in production. Medium is made from chemicals which have
been tested, bar-coded, and released from quarantine by quality
assurance. A computer-controlled manufacturing recipe, linked to a
bar-code reader and weighing system, ensures that the correct
chemicals and quantities are used. Medium concentrates are made in
large quantities and stored, if necessary frozen at $-20°C$. The
requisite quantity of each concentrate is combined to make a batch
of complete medium, thus providing uniformity in composition over a
considerable period of time even when more than 8000 liters are used
each day.

Satisfactory primary production has been obtained with water
that has been deionized and then distilled; or deionized followed by
reverse osmosis; or merely deionized. Thus, and depending on
the quality of the local water supply, the water used in the medium
can be prepared by whatever system is most economical; in later
stages of production and purification, water complying with the
pharmacopoeial standards of water for injection must be used.

V. ACCEPTABILITY FOR CLINICAL USE OF PRODUCTS MADE FROM MAMMALIAN CELL CULTURES

For the reasons already discussed, there was until quite recently a general reluctance to use cultures of transformed mammalian cells as the source of medicinal products. However, virus vaccines are needed for the routine immunization of children, and these consist of a complex mixture of active vaccine virus with nucleic acids and proteins derived from the host cell and the medium. The primary cultures of monkey kidney cells formerly used as the source of live attenuated poliovirus vaccines were frequently contaminated with simian viruses with an unknown potential for causing disease in man (indeed, many doses of vaccine contaminated with SV40 virus were issued before the existence of this virus was discovered; subsequent epidemiological studies have shown that, fortunately, the recipients seem to have suffered no ill effects). In the face of this problem, there was a well-recognized and urgent need to find alternative manufacturing substrates. As a result, the regulatory authorities in various countries, including the United Kingdom and subsequently the United States, agreed that human diploid cell lines such as WI-38 and MRC-5 could be used to make polio, rubella, and other live attenuated virus vaccines, subject to detailed controls to ensure that the cells were free from microbial and virus contaminants and retained their "normal" karyological status and growth characteristics during the permitted subculture life of 30−40 population doublings.

Wellcome staff discussed the principle of using Namalwa cells for the manufacture of interferon for clinical use with representatives of the U.K. and U.S. regulatory authorities from 1975 onward. They found a receptive attitude and, indeed, active encouragement, for compared with the use of buffy coat cells as the substrate for production, Namalwa cells, even though transformed and of tumor origin, could be controlled to a much greater degree. It was clearly understood that the final product would have to be highly purified and free from contamination with potentially noxious agents before it could be used in man. Fortunately, in the context of a pioneering exercise, the human alpha-interferons are very stable proteins which withstand exposure to highly acidic conditions and to some organic solvents without loss of biological activity: such treatments are involved in the four-stage purification sequence used in the production of Wellferon.

No viruses or other microbiological contaminants could be detected in Namalwa cells from the master cell bank, apart from Epstein-Barr (EB) virus: this is present in every cell in the form of a single copy of the complete virus genome, but under no circumstances is any infectious virus formed. Since nearly all human beings are infected with EB virus at some time in their life and most have antibodies to it, this finding did not seem of serious import.

Although Namalwa cells seem free from contamination with any objectionable agent, theoretically they could still contain agents for which the requisite cultural conditions have not yet been developed (e.g., the human T-cell leukemia viruses were only discovered recently). Steps were therefore taken to validate the purification system in terms of its ability to remove or inactivate viruses of every possible type. For this purpose, high-titer preparations of various human and animal viruses were added, one at a time, to preparations of crude Namalwa interferon which were processed in the routine way. Nearly all of the viruses tested were destroyed or eliminated during the first stage of purification, and the few which survived disappeared during the further stages. Because it was technically very difficult to detect very small amounts of these viruses in the presence of high concentrations of human IFN, tests were also carried out with viruses added to the medium rather than to the IFN product; in addition, and on the advice of the U.S. Bureau of Biologics, Food and Drug Administration, further similar studies were undertaken with bacterial viruses, since these grow to even higher titers than most animal viruses and are not affected by IFN. All the bacteriophages tested, whether large, small, DNA-containing or RNA-containing, and representative of every known class of virus, were completely inactivated during the stages of Wellferon purification. This work was extended to demonstrate the elimination of the scrapie agent, included as a representative of the class of slow viruses (31).

Direct tests showed that only the occasional batch of crude Namalwa IFN contained any detectable DNA; the greatest quantity ever measured was 20 μg/ml in a batch in which, due to a technical failure, most of the cells died and very little IFN was formed. No EB virus DNA could be detected. When radiolabeled Namalwa cell DNA preparations, both intact and of oncogene size, were added at different stages during the purification, nearly all counts were eliminated from the purified product. From the maximum recorded level of DNA in a crude product and the cumulative degree of elimination, the final level of any DNA contamination of Wellferon was calculated to be far below the level of biological significance (32,33): direct tests on every batch confirm that less than 10 pg of DNA (the smallest amount which can be detected) is present in a dose.

Wellferon was first used in government-approved clinical trials in the United Kingdom in 1979 and in the United States in 1980. It has now been administered to many thousands of patients and has been licensed for sale in a number of countries in Europe, North America, and Asia for treatment of patients with various clinical conditions including virus-associated tumors caused by human papillomaviruses, hairy cell leukemia, and renal cancer. Further trials are in progress in a number of other important clinical conditions, including chronic active hepatitis and chronic myeloid leukemia.

In the past 10 years, there have been a number of international
meetings at which the issues and choices involved in the decision to
use a particular cell substrate as a source for the manufacture of
medicinal products have been debated (34,35). There was at first
considerable opposition to the idea of using transformed cells, but
gradually views have changed as the issues have been discussed at
succeeding meetings and as guidelines for controlling the safety and
efficacy of products derived from such cultures have evolved. The
evidence accumulated during the development of Wellferon, the first
such product, played an important role in these debates.

VI. FUTURE DEVELOPMENTS

The general acceptance of transformed cells grown in mass culture
as a manufacturing source for therapeutic proteins was undoubtedly
accelerated by the perceived need for a more satisfactory source of
human interferon at a time when recombinant DNA technology had
not yet provided an alternative. Even now, a single recombinant
IFN such as $\alpha 2$ is not the same as the mixture of 22 different alpha-
interferons (36) present in Wellferon. This precedent has helped
to establish the use of large-scale cell cultures as the source of
many other proteins needed for therapeutic or diagnostic use.

Emphasis has been given in this chapter to the use of conventional stirred tanks for growing mammalian cells on a large scale,
both because they are the ones most commonly used and for historical reasons. Experience at Wellcome has shown that such tanks
are robust, versatile, and efficient, and plants based on them can
produce very large amounts of product at an economical price. Although this basic technology is now very well developed, there will
undoubtedly be refinements in the future. Attempts are already
being made to achieve greater cell densities by perfusion or re-
feeding techniques, and to improve control and productivity through
the use of new sensors, microprocessors, and the application of
artificial intelligence techniques. Serum-free medium formulations
have been developed which greatly simplify downstream purification,
an area of major importance in a complete production system, while
at the same time progress has been made with purification and
analytical techniques.

The revival of interest in cultured cells has also led to the re-
examination and refinement of a number of other cell culture options.
Airlift fermenters of up to 2000-liter capacity have been developed
and used, in particular for the culture of hybridoma cells (37).
Methods have been devised for the large-scale culture of anchorage-
dependent cells, including their culture on the surface of buoyant
beads, the microcarrier technique introduced by van Wezel (38).
Various systems involving hollow fibers or a ceramic matrix have

also been used, especially for the growth of hybridoma cells producing monoclonal antibodies. Commercially available equipment embodying such methods may include arrangements for perfusion or automatic harvesting under microprocessor control. These and other technical developments will increase the ease, convenience, and economy of culturing cells on a large scale. Thus, while there are alternative substrates for the expression of proteins, such as bacterial yeast or insect cells, and other approaches to the manufacture of therapeutic proteins, such as total chemical synthesis, it seems likely that mammalian cells will retain their importance as a manufacturing source for many years to come.

VII. CONCLUSION

Mammalian cells have now been cultured in suspension in large tanks for more than 20 years. The technology was first developed for industrial purposes at the Wellcome laboratories, initially in order to make foot-and-mouth disease virus vaccines for veterinary use from BHK cells, and later to make alpha-interferons for medical use from Namalwa lymphoblastoid cells. These two products served as important precedents. They established the feasibility of carrying out large-scale culture of mammalian cells for manufacturing purposes. They also showed that products acceptable for medical use can be made from transformed cells.

Along with stirred tanks, many other culture systems have now been devised for growing large numbers of mammalian cells and using them to make gram or kilogram quantities of proteins of medical importance, including tPA, factor VIII, and various monoclonal antibodies. Mammalian cell culture is no longer just a laboratory tool; it is a well-recognized and full-fledged manufacturing technology that is likely to find increased use in years to come.

REFERENCES

1. Moore, G. E., Hasenbusch, P., Garner, R. E., and Burn, A. Pilot plant for mammalian cell cultures. *Biotechnol. Bioeng.* 10: 625 – 640 (1968).
2. Telling, R. C. and Radlett, P. J. Submerged cultures of BHK21 cells up to pilot plant scale. Report of the Research Group, European Comm. for the Control of Foot-and-Mouth Disease, Brescia. Appendix X, pp. 148 – 155 (1969).
3. Telling, R. C. and Radlett, P. J. Large scale cultivation of mammalian cells. *Adv. Appl. Microbiol.* 13:91 – 119 (1970).
4. Frenkel, H. S. The culture of foot-and-mouth disease virus on cattle tongue epithelium. *Bull. Off. Int. Epiz.* 28:155 – 160 (1947).

5. Mowat, G. N. and Chapman, W. G. Growth of foot-and-mouth disease virus in fibroblastic cell line. *Nature* 194:253–255 (1962).
6. Macpherson, I. A. and Stoker, M. G. P. Polyoma transformation of hamster cell clones. An investigation of genetic factors affecting cell competence. *Virology* 16:147–151 (1962).
7. Capstick, P. B., Telling, R. C., Chapman, W. G., and Stewart, D. L. Growth of a cloned strain of hamster kidney cells in suspended culture and their susceptibility to the virus of foot-and-mouth disease. *Nature* 195:1163–1164 (1962).
8. Smith, H. M. and Burrows, T. M. Laboratory apparatus for the suspended culture of tissue cells. *Lab. Pract.* 12:451–453 (1963).
9. Capstick, P. B., Garland, A. J. M., Chapman, W. G., and Masters, R. C. Factors affecting the production of foot-and-mouth disease virus in deep suspension cultures of BHK21 Cl 13 cells. *J. Hyg. (Camb.)* 645:273–280 (1967).
10. Elsworth, R., Capel, G. H., and Telling, R. C. Improvements in the design of a laboratory culture vessel. *J. Appl. Bacteriol.* 21:80–85 (1958).
11. Telling, R. C. and Elsworth, R. Submerged culture of hamster kidney cells in a stainless steel vessel. *Biotechnol. Bioeng.* 7:417–434 (1965).
12. Telling, R. C. and Stone, C. J. A method of automatic pH control of a bicarbonate-CO_2 buffer system for the submerged culture of hamster kidney cells. *Biotechnol. Bioeng.* 6:147–158 (1964).
13. Capstick, P. B. Growth of baby hamster kidney cells in suspension. *Proc. Roy. Soc. Med.* 56:1062–1064 (1963).
14. Radlett, P. J., Telling, R. C., Stone, C. J., and Whiteside, J. P. Improvements in the growth of BHK21 cells in submerged culture. *Appl. Microbiol.* 22:534–537 (1971).
15. Radlett, P. J., Braeme, A. J., and Telling, R. C. A simple steam sterilizable dissolved oxygen electrode. *Lab. Pract.* 21:811–814 (1972).
16. Radlett, P. J., Telling, R. C., Whiteside, J. P., and Maskell, M. A. The supply of oxygen to submerged cultures of BHK21 cells. *Biotechnol. Bioeng.* 14:437–445 (1972).
17. Tengerdy, R. P. Redox potential changes in the 2-keto-L-glucuronic acid fermentation. 1. Correlation between redox potential and dissolved oxygen concentration. *J. Biochem. Microbiol. Tech. Eng.* 3:241–253 (1961).
18. Code of Federal Regulations, Title 9, Chapter III, Part 112.52. Published by the Office of the Federal Register, U.S. National Archives and Records Administration (1983).

19. Telling, R. C., Stone, C. J., and Maskell, M. A. Sterilization of tissue culture medium on a large laboratory scale.
Biotechnol. Bioeng. 8:153−165 (1966).

20. Radlett, P. J. The use of BHK suspension cells for the production of foot-and-mouth disease vaccines. In *Advances in
Biochem. Eng. Biotechnol.*, Vol. 34, Springer-Verlag, Berlin,
pp. 129−146 (1987).

21. Capstick, P. B. and Telling, R. C. Production of FMD
Vaccine in BHK21 Cells. Report of the Research Group,
European Comm. for the Control of Foot-and-Mouth Disease,
Pirbright, VII, pp. 108−113 (1966).

22. Capstick, P. B., Garland, A. J. M., Masters, R. C., and
Chapman, W. G. Some functional and morphological alterations
occurring during and after the adaptation of BHK21 Clone 13
cells to suspension culture. *Exp. Cell Res.* 44:119−128 (1966).

23. Capstick, P. B., Telling, R. C., and Garland, A. J. M.
Utilisation and control of BHK cells in inactivated foot and
mouth disease vaccine production. *Prog. Immunobiol. Stand.*
3:131−135 (1969).

24. Radlett, P. J., Pay, T. W. F., and Garland, A. J. M. The
use of BHK suspension cells for the commercial production of
foot-and-mouth disease vaccines over a twenty year period.
Dev. Biol. Standard. 60:163−170 (1985).

25. Pay, T. W. F., Boge, A., and Menard, F. J. R. R. The
application of foot-and-mouth disease vaccine technology to the
production of inactivated rabies vaccines for use in animals.
In *Rabies in the Tropics* (E. Kilwert, C. Merieux, H. Koprowski,
and K. Bogel, eds.), Springer-Verlag, Berlin, pp. 698−702
(1985).

26. Pay, T. W. F., Boge, A., Menard, F. J. R. R., and Radlett,
P. J. Production of rabies vaccine by an industrial scale
BHK21 suspension cell culture process. *Dev. Biol. Standard.*
60:171−174 (1985).

27. Isaacs, A. and Lindenmann, J. Virus interference. *Proc. Roy.
Soc. London B*, 147:258−267 (1957).

28. Strander, H. and Cantell, K. Production of interferon by
human leukocytes in vitro. *Ann. Med. Exp. Biol. Fenn.* 44:
265−273 (1966).

29. Christofinis, G. J., Steel, C. M., and Finter, N. B. Interferon production by human lymphoblastoid cells of different
origins. *J. Gen. Virol.* 52:169−171 (1981).

30. Johnston, M. D. Sources of interferon for clinical use: Beta-
interferons from human lymphoblastoid cells. *Interferon 4:
In Vivo and Clinical Studies* (N. B. Finter, ed.), Elsevier,
Amsterdam, pp. 81−87 (1985).

31. Finter, N. B. and Fantes, K. H. The purity and safety of interferons prepared for clinical use: The case for lymphoblastoid interferon. *Interferon*, Vol. 2 (I. Gresser, ed.), Academic Press, London, pp. 65–80 (1980).
32. Lowy, D. R. Potential hazards from contaminating DNA that contains oncogenes. In *Abnormal Cells, New Products and Risk*. In vitro monograph No. 6 (H. E. Hopps and J. C. Petricianni, eds.), Tissue Culture Association, Gaithersburg, pp. 36–40 (1985).
33. Finter, N. B., Fantes, K. H., Lockyer, M. J., Lewis, W. G., and Ball, G. D. The DNA content of crude and purified human lymphoblastoid interferon (Namalwa cell) interferon prepared by Wellcome Biotechnology Ltd. In *Abnormal Cells: New Products and Risk*. In vitro monograph No. 6 (H. E. Hopps and J. C. Petricianni, eds.), Tissue Culture Association, Gaithersburg, pp. 125–128 (1985).
34. Petricianni, J. C., Hopps, H. E., and Chapple, P. J. (eds.). *Cell Substrates: Their Use in the Production of Vaccines and Other Biologicals*, Plenum Press, New York, pp. 1–220 (1979).
35. Hopps, H. E. and Petricianni, J. C. (eds.). *Abnormal Cells: New Products and Risk*. In vitro monograph No. 6, Tissue Culture Association, Gaithersburg, pp. 1–180 (1985).
36. Zoon, K. C., zur Nedden, D. L., Enterline, J. C., Manischewitz, J. F., Dyer, D. R., Boykins, R. A., Bekisz, J., and Gerrard, T. L. Chemical and biological characterisation of natural human lymphoblastoid interferon alphas. In *The Biology of the Interferon System* (K. Cantell and H. Schellekens, eds.), Martinius Hijhoff, Amsterdam (in press).
37. Birch, J. R., Lambert, K., Thompson, T. W., Kenney, A. C., and Wood, L. A. Antibody production with airlift fermenters. In *Large Scale Cell Culture Technology* (B. Lydersen, ed.), Hanser, Munich, pp. 1–20 (1987).
38. van Wezel, A. L. Growth of cell-strains and primary cells on microcarriers is in homogenous culture. *Nature* 216:64–65 (1967).

2

Use of Recombinant DNA Technology for Engineering Mammalian Cells to Produce Proteins

RANDAL J. KAUFMAN
Genetics Institute, Cambridge, Massachusetts

I. INTRODUCTION OF GENETIC MATERIAL INTO CELLS

The techniques of gene isolation, gene modification, and gene transfer into appropriate host cells have provided a means to understand protein structure and function by enabling the production of large amounts of proteins that previously could be isolated in only minute quantities, and by allowing the generation of proteins with specific, designed alterations. Although a variety of eukaryotic and prokaryotic host systems are available for the expression of a particular gene, the focus of this chapter is on the use of recombinant DNA technology in the engineering of mammalian cells to express high levels of proteins from heterologous genes.

The advantages of using mammalian cells as a host for the expression of a gene obtained from a higher eukaryote stem from experience that the signals for synthesis, processing, and secretion of these proteins are usually properly and efficiently recognized in mammalian cells. Experience with heterologous gene expression in mammalian cells has demonstrated that (1) proteins can be readily synthesized and secreted into the growth medium; (2) protein folding and disulfide bond formation are usually like that of the natural

protein and therefore the proteins are usually produced in a functional form resistant to degradation; (3) glycosylation, both N- and
O-linked, often occurs at normal positions; (4) other posttranslational modifications can occur, e.g., the proteolytic processing of
propeptides, the gamma-carboxylation of glutamic acid residues,
hydroxylation of aspartic acid and proline residues, sulfation of
tyrosine residues, phosphorylation, and fatty acid addition; and
(5) multimeric proteins of single or multiple subunits can be correctly assembled. Although a number of different host-vector systems have been described for the high-level expression of heterologous genes in mammalian cells, significant success has been obtained by cotransfection of Chinese hamster ovary (CHO) cells with
the desired gene and an amplifiable selectable genetic marker such
as dihydrofolate reductase.

This chapter will first address the various methods for transfer
of genetic material into mammalian cells and methods to select for
stable transformants followed by selection for gene amplification.
The reader is referred elsewhere for reviews on different expression systems (1). Important components in the design of expression vectors will be described, such as potential for gene amplification, transcriptional controls, mRNA processing, and translation
signals. Potential limitations which result from the requirement for
specific posttranslational modifications and the influence of the host
cell in obtaining appropriate expression and secretion will be discussed. The latter portion of this chapter will illustrate some of
these concepts by summarizing the difficulties encountered in obtaining high-level expression of the human clotting factor VIII.

II. DNA-MEDIATED GENE TRANSFER

A variety of methods are available to transfer DNA to mammalian
cells. The most common procedure to obtain stable transformants is
to add DNA directly to cells in the form of a $CaPO_4$ precipitate (2).
A number of modifications of the original procedure which include
lowering the pH of the $CaPO_4$ (3) or adding polyethylene glycol (4),
glycerol (5,6), dimethyl sulfoxide (7), sodium butyrate (8), or
chloroquine (9) after the transfection procedure have been reported
to increase transfection efficiency (frequencies range from 10^{-6} to
10^{-3}/cell depending on the cell line). Diethylaminoethyl (DEAE)-
dextran-mediated DNA transfer is a convenient method for transfection of a high percentage of the cells transiently (10,11) but has
not been reported as useful for obtaining stable transformants. In
these two methods DNA uptake is through endocytosis and DNA
enters the cell via an endocytic vesicle. Consequently, this process

frequently results in DNA rearrangements, possibly as a result of passage through cellular compartments of low pH or containing endonucleases.

In contrast to the above approaches, there are a number of methods in which DNA is directly introduced into the cytoplasm of the cell. One common method is that of polyethylene glycol (PEG)-induced fusion of bacterial protoplasts with mammalian cells (12). This method can be very efficient ($10^{-4}-10^{-2}$/cell) and has been used with cells that are difficult to transfect by the $CaPO_4$ procedure. Protoplast fusion frequently yields multiple copies of the plasmid DNA tandemly integrated into the host chromosome (13). In a similar manner, phospholipid vesicles have been used to deliver plasmid DNA to a wide variety of cells (14). Microinjection of DNA directly into the cytoplasm or nucleus has been used as a highly efficient method per cell (approaches 100%), although a limited number of cells can be injected at one time (15). Electroporation has been used to transfer DNA to cells in which other procedures have been unsuccessful (16-18). Cells to be electroporated are resuspended in phosphate-buffered saline or a similar electrolyte buffer, mixed with the DNA, and then subjected to a high-voltage electric discharge. When cells are exposed to this high voltage, their membranes transiently form small pores through which DNA can pass. Most efficient electroporation occurs at approximately 100-200 V for 1-2 msec. The frequency of stable transformation increases with increasing amounts of DNA and may reach 10% of the electroporated cells.

III. STABLE AND TRANSIENT TRANSFECTION

With most of the methods described above, 5-50% of the cells in the population acquire DNA. These cells transiently express the DNA over a period of several days to several weeks until the DNA is eventually lost from the population. This transient expression of desired genes can be used instead of the more laborious procedure of isolating and characterizing stably transfected cell lines. Transient expression experiments obviate the effects of integration sites on expression and the possibility of selecting cells which harbor mutations in the transfected DNA. Most importantly, the rapid identification of an expression plasmid to direct the synthesis of the desired protein ensures that the expression plasmid is competent before development of stably transfected cell lines.

If a selection procedure for the transfected DNA is applied after DNA transfection, it is possible to isolate cells that have stably integrated the foreign DNA into their genome. The limiting factor for obtaining stable transformants is the frequency of DNA integration, not the frequency of DNA uptake. Different cell lines exhibit different frequencies of stable transformation and in their capacity incorporate foreign DNA. The ability to select cells for the incorporation and expression of two independent genetic markers present on different vectors has been termed cotransformation (19). Different cell lines and transfection methods yield varying frequencies of cotransformation. For example, the frequency of cotransformation in CHO cells is lower than that observed in mouse L cells because mouse L cells can incorporate more DNA into their genome than CHO cells. Cotransformation by $CaPO_4$-mediated DNA transfection is very efficient (19) whereas cotransformation by protoplast fusion of two independent plasmids is very rare. In cotransformation, the transfected DNA becomes ligated together inside the cell and subsequently integrates via nonhomologous recombination into host chromosomal DNA as a unit. However, recent findings suggest that homologous recombination can occur at frequencies 1% that of nonhomologous integration (20). The ability to select for such homologous recombination events makes feasible the ability to replace genes as well as to inactivate specific genes.

IV. SELECTION OF STABLE TRANSFORMANTS

Resistance to cytotoxic drugs is the characteristic most widely used to select for stable transformants. Drug resistance can be recessive or dominant. Genes conferring recessive drug resistance require a particular host which is deficient in the activity under selection. Genes conferring dominant drug resistance can be used independent of the host. Many of the recessive genetic selectable markers are involved in the salvage pathway for purine and pyrimidine biosynthesis. When de novo biosynthesis of purines or pyrimidines is inhibited, the cell can utilize purine and pyrimidine salvage pathways, providing the enzymes (i.e., thymidine kinase, hypoxanthine-guanine phosphoribosyltransferase, adenine phosphoribosyltransferase, or adenosine kinase) necessary for conversion of the nucleoside precursors to the corresponding nucleotides are present (see Fig. 1). These salvage enzymes are not required for cell growth when de novo purine or pyrimidine biosynthesis is functional. Cells deficient for a particular salvage pathway enzyme are viable under normal growth conditions. However, addition of drugs which inhibit the de novo biosynthesis of purines or pyrimidines results in the death of deficient cells because the salvage pathway becomes essential to provide purines and pyrimidines. Cells which acquire the capability of expressing the deficient activity via gene

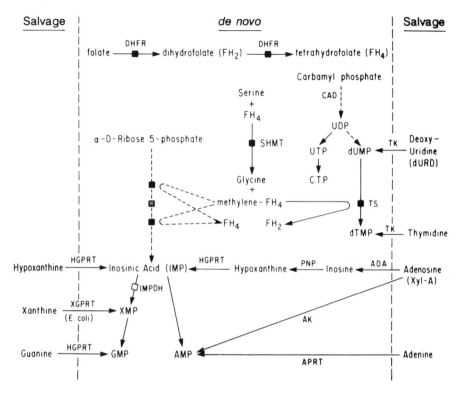

Figure 1 De novo and salvage biosynthetic pathways for purines and pyrimidines involving available selectable markers. Abbreviations for de novo enzymes: DHFR, dihydrofolate reductase; CAD, carbamyl-P synthetase, asparate transcarbamylase, and dihydroorotase (only the latter two activities are indicated in the figure); SHMT, serine hydroxymethyltransferase; TS, thymidylate synthetase; and IMPDH, inosine monophosphate dehydrogenase. Abbreviations for salvage enzymes: TK, thymidine kinase; ADA, adenosine deaminase; PNP, purine nucleoside phosphorylase; AK, adenosine kinase; APRT, adenine phosphoribosyltransferase; HGPRT, hypoxanthine-guanine phosphoribosyltransferase; and XGPRT, *E. coli* xanthine-guanine phosphoribosyltransferase. Solid arrows indicate single reactions. Dashed arrows indicate multiple reactions. The solid square indicates reactions that are inhibited by the folate analog methotrexate and aminopterin. The hatched square indicates the principle reaction inhibited by azaserine. The open square indicates the reaction inhibited by mycophenolic acid. (Reprinted from Ref. 225 with permission.)

transfer can be selected for growth under these conditions. This
has become a common selection technique and is the basis for the
thymidine kinase (TK), adenine phosphoribosyltransferase (APRT),
and hypoxanthine-guanine phosphoribosyltransferase (HGPRT)
selection.

A. Thymidine Kinase (TK)

TK converts deoxyuridine (dUrd) to deoxyuridine monophosphate
(dUMP), and thymidine (dThy) to thymidine monophosphate (dTMP).
It has become a common recessive selection marker because it is
possible to readily select TK-deficient cells by propagation in bromodeoxyuridine (BUdR). BUdR-resistant cells cannot grow in HAT
(hypoxanthine, aminopterin, and thymidine) medium unless they acquire a functional TK gene (21). The folate analog aminopterin
blocks de novo pyrimidine and purine biosynthesis. The addition
of hypoxanthine allows AMP and GMP production through the HGPRT
salvage pathway. Exogenously added thymidine can be converted
to dTMP only if a functional TK is present. Thus, for cells to
grow in HAT medium they require functional TK and HPRT. Introduction of a herpes virus TK gene with the TK promoter into mouse
L TK⁻ cells is conveniently used to select cells containing exogenous
DNA (22). Transfection of a TK gene fragment without the TK
promoter, in order to express suboptimal levels of TK, and stringent selection of HAT medium containing limiting amounts of thymidine
has yielded cells that contain a 50-fold amplification of the transfected DNA (23). Thus, by modifying the selection conditions and
crippling the TK gene, it has been possible to subsequently select
for cells that contain amplified copies of the transfected DNA. By
limiting thymidine concentration and by including methotrexate, a
folate analog, it has been reported that TK can be used as a
selectable marker in cells which contain TK (24). Thus, this approach may prove useful as a method to use TK as a dominant
selectable marker.

B. Xanthine-Guanine Phosphoribosyltransferase (XGPRT)

The *E. coli* enzyme XGPRT converts xanthine to xanthine monophosphate (XMP). This reaction is carried out poorly by the mammalian
enzyme and so it is possible to use bacterial XGPRT as a dominant
selection marker by forcing cells to utilize xanthine (25). In the
presence of HGPRT, hypoxanthine is converted to GMP and AMP.
In order to use XGPRT as a dominant selection marker, the conversion of IMP to XMP by IMPdehydrogenase needs to be blocked by
mycophenolic acid. In this way the only GMP produced is via

xanthine conversion to XMP, carried out by XGPRT. Thus, the components required for XGPRT as a dominant selectable marker are hypoxanthine, xanthine, and mycophenolic acid. It has been possible to select for cells that have amplified the XGPRT gene up to 10-fold by propagation in limiting amounts of xanthine (26).

C. Dihydrofolate Reductase (DHFR)

DHFR catalyzes the conversion of folate to tetrahydrofolate (FH4). FH4 is required for the biosynthesis of glycine from serine, for the biosynthesis of thymidine monophosphate from deoxyuridine monophosphate, and for purine biosynthesis. CHO cells that are deficient in DHFR were isolated after ethyl methanesulfonate and UV irradiation-induced mutagenesis and selection in high specific activity [^{3}H]dUrd (27). Cells containing functional DHFR convert [^{3}H]dUrd to [^{3}H]dTMP, which is incorporated into DNA. This incorporation into DNA is lethal due to disruption of the DNA by radioactive damage. DHFR-deficient cells survive this selection procedure. The DHFR-deficient cells require the addition of thymidine, glycine, and hypoxanthine to the medium and do not grow in the absence of added nucleosides unless they acquire a functional DHFR gene.

Methotrexate, a folate analog, binds and inhibits DHFR, resulting in cell death. When cells are selected for growth in sequentially increasing concentrations of methotrexate, the surviving population contains increased levels of DHFR, which result from an amplification of the DHFR gene (28−31). Highly methotrexate-resistant cells have been obtained that contain several thousand copies of the DHFR gene and express several thousand-fold elevated levels of DHFR. DHFR is frequently used as a cotransformation selection marker in DHFR-deficient CHO cells, making it possible to coamplify and obtain high-level expression of heterologous genes by methotrexate selection.

Although most frequently used as a recessive marker in the DHFR-deficient CHO cells, there have been several adaptations which have made DHFR a useful dominant selectable and amplifiable genetic marker. One approach has been to use a strong promoter for transcription of the foreign DHFR gene. Transfected cells then have the ability to grow in moderate concentrations of methotrexate as the result of high-level expression of the foreign DHFR gene and not from amplification of the endogenous DHFR gene (32). Another approach has been to cotransfect with another dominant selectable marker (e.g., Neo resistance) and to pool the resultant transformants and select in moderate levels of methotrexate. Most frequently the foreign DHFR gene is preferentially amplified over the endogenous DHFR gene (33). Finally, a methotrexate-resistant

DHFR has been cloned which contains a single amino acid change from a leucine to an arginine at position 22. This altered enzyme has a 270-fold elevated K_i relative to the wild-type enzyme. Thus, it is possible to directly select for the presence of this enzyme by growth in moderate levels of methotrexate (34). In addition, the transfected DNA can be amplified by further selection in increasing concentrations of methotrexate. However, the large change in K_i makes it difficult to obtain high degrees of amplification with the mutant DHFR gene.

D. Carbamyl Phosphate Synthetase – Asparate Transcarbamylase – Dihydroorotase (CAD)

CAD is a multifunctional protein that catalyzes the first three steps in UMPbiosynthesis. The complex includes carbamyl-phosphate synthetase, aspartate transcarbamylase, and dihydroorotase. PALA (N-phosphonacetyl-L-aspartic acid) is a transition state analog inhibitor of the aspartate transcarbamylase which causes cell death as a result of UMP depletion. Selection for growth in sequentially increasing concentrations of PALA has been used to select for amplification of CAD gene (35). In practice the use of this selection system has been limited to CAD-deficient CHO cells (36). The full-length cDNA for the mammalian gene has not been isolated intact due to its large size. A cloned copy of the genomic CAD gene has been introduced into CAD-deficient CHO mutants (Urd-A, 37) by selection for growth in the absence of uridine (38). Individual transformants selected for growth in increasing concentrations of PALA exhibit amplification of the introduced CAD gene (36). DNA transfer experiments have also demonstrated that the E. coli aspartate transcarbamylase gene (PyrB) in a eukaryotic expression plasmid is expressed and amplified up to fourfold by PALA selection in aspartate transcarbamylase-deficient CHO cells (39). However, the utility of either the eukaryotic or prokaryotic CAD gene as a dominant selectable marker has not been demonstrated.

E. Adenosine Deaminase (ADA)

ADA has been demonstrated to be a useful dominant selectable and highly amplifiable genetic marker for mammalian cells (40). Although ADA is present in virtually all mammalian cells, it is not an essential enzyme for cell growth. However, under certain conditions, cells require ADA (41–44). Since ADA catalyzes the irreversible conversion of cytotoxic adenine nucleosides to their respective nontoxic inosine analogs, cells propagated in the presence of cytotoxic concentrations of adenosine or cytotoxic adenosine analogs such as 9-B-D-xylofuranosyladenine (Xyl-A) require ADA to detoxify the cytotoxic agent. Once functional ADA is required for cell growth, then 2'-deoxycoformycin (dCF), a tight-binding transition state analog inhibitor of ADA, can be used to select for amplification of the ADA gene. As a result of one selection protocol, ADA was overproduced 11,400-fold and represented 75% of the soluble protein synthesis of the cell (45). ADA can function as a dominant selectable and amplifiable genetic marker (40) because most cells synthesize minute quantities of it. Thus, it is possible to select for introduction of a heterologous ADA gene carried in an efficient expression vector.

F. Multidrug Resistance (mdr)

Cell lines have been obtained which are cross-resistant to a variety of chemotherapeutic agents such as adriamycin, colchicine, and vincristine. The mechanism for resistance is overproduction of a plasma membrane glycoprotein of 170 kDa as a result of gene amplification (46). This membrane protein is involved in transport of these drugs out of the cell (47 – 49). The cloned cDNA can function as a useful dominant and amplifiable selection marker (50). The degree of gene amplification with mdr may be limited since cells exhibiting high levels of drug resistance contain only 50 copies of the mdr gene (50).

G. Ornithine Decarboxylase (ODC)

ODC is the first enzyme in the synthesis of polyamines and is an essential enzyme for cell growth (51). An expression plasmid encoding the ODC cDNA serves as a dominant selectable and amplifiable genetic marker in CHO cells (52). Selection in increasing concentrations of the ODC inhibitor difluoromethylornithine (DFMO) has resulted in a 300-fold amplification of the transfected gene. ODC can also be used to complement the growth of "ODC"-deficient CHO cells (53) in medium lacking putrescine (52).

H. Asparagine Synthetase (AS)

AS is a housekeeping enzyme responsible for the biosynthesis of asparagine from glutamine and aspartic acid. Selection for expression of heterologous AS is possible by selection for growth in glutamine-free medium with the addition of the appropriate concentration of either albizzin, a glutamine analog, or B-hydroxyl aspartamate, an aspartic acid analog, to inhibit the endogenous ASA (54).

The *E. coli* asparagine synthetase gene has also been used as a dominant selectable and amplifiable marker. Since the *E. coli*

enzyme uses ammonia as a nitrogen source, it is resistant to albizzin. Dominant selection for the bacterial AS is possible by
growth in asparagine-free medium containing albizzin. Subsequently, the transfected DNA can be amplified up to 100-fold by growth
in increasing concentrations of B-aspartyl hydroxamate (55).

I. Other Strategies for Transformant Selection

Transformants can be selected on the basis of expression of surface
antigens, e.g., the T-cell antigen leu-2 (56) and the transferrin
receptor (57), by cell sorting using a fluorescent activated cell
sorter and specific antibodies. In addition, sequential selection for
positive fluorescence yields cells that have amplified the gene for
the surface antigen (58). Other fluorescently labeled compounds
that can be used to select cells include a fluorescent conjugate of
methotrexate to select for elevated DHFR levels (59), or aryl hydro-
carbons to select for cells exhibiting elevated levels of aryl hydro-
carbon hydroxylase (60).

V. OPTIMIZING EXPRESSION OF HETEROLOGOUS GENES

The level of protein expression from heterologous genes introduced
into mammalian cells depends on multiple factors including DNA
copy number, efficiency of transcription, mRNA processing, mRNA
transport, mRNA stability and translational efficiency, and protein
processing, secretion, and stability. The rate-limiting step for
high-level expression may be different for different genes. Controls at each one of these levels will be discussed in turn.

A. Gene Amplification

Gene amplification is ubiquitous in nature and many, if not all,
genes become amplified at some frequency (approximately $1/10^4$ although this number can vary extensively) (see 30, 31 for reviews).
Upon appropriate selection conditions, where the growth of cells
containing amplification of a particular gene is favored, a population of cells that contain the amplified gene will outgrow the general population. In the absence of drug selection, the amplified
gene is most frequently lost. The stability of amplified DHFR
genes can be monitored by the ability to stain cells quantitatively
for DHFR with a fluorescent derivative of methotrexate (59).
Analysis by fluorescence-activated cell sorting yields information on
the DHFR content of the cells in a population. The pattern of the
histogram can be used to predict the stability of the cell population

(61). In addition, it is possible to isolate individual viable cells on the basis of their DHFR content and follow them upon continued passage in order to monitor changes in their DNA copy number (61−63).

The degree of stability of the amplified genes frequently correlates with the chromosomal location of the amplified genes and the complexity of the associated karyotypic alterations. Increased instability can be correlated with more complex chromosomal alterations. When amplified genes are on double-minute chromosomes, they are rapidly lost upon propagation in the absence of selection (61). Double-minute chromosomes are small paired chromosomal elements that lack centromeric function and thus segregate randomly at mitosis. When the amplified genes are chromosomally located they are associated with expanded chromosomal regions termed homogeneously staining regions (HSRs) (64). This term comes from the lack of banding in these regions after trypsin-giemsa banding procedures. Typically, when mouse fibroblasts are selected for methotrexate resistance, the endogenous amplified genes are associated with double-minute chromosomes (65). When heterologous DHFR genes are selected for and amplified in mouse fibroblasts the genes are also associated with double-minute chromosomes (32). In contrast, when CHO cells are selected for methotrexate resistance, the endogenous amplified genes are associated with expanded chromosomal regions (64). Similarly, when heterologous DHFR genes are selected for and amplified in CHO cells, they are localized to expanded chromosomal regions (66). In a few cases, the amplified transfected DNA may be present with little cytological perturbation (66). This may reflect a smaller size of the amplified unit within a chromosome. Frequently, cells selected for amplified sequences become tetraploid, a characteristic possibly associated with increased instability. There may also be qualitative differences in the composition of the DNA surrounding the amplified sequences in the resulting HSRs (66).

The mechanism of amplification of a transfected CAD gene upon PALA selection has been extensively studied (67,68). The integrated CAD sequence is first deleted and then undergoes replication as an extrachromosomal element. The extrachromosomal intermediate may increase in size by forming multiples of itself to eventually create double-minute chromosomes. The extrachromosomal sequence can eventually integrate to form a homogeneously staining chromosomal region.

CHO cells selected for amplification of heterologous DHFR genes have generally proven to be very stable upon propagation in the absence of methotrexate (69). Frequently, transfected and amplified DNA is found in HSRs associated with dicentric chromosomes. This finding has led to the proposal that bridge-breakage-fusion cycles (66,70,71) may be involved in some gene amplifications. The loss of telomeric function with subsequent bridge-breakage-fusion cycles could be important in generating chromosome aberrations and perhaps duplications and instability in DNA transfer experiments. Recombination of exogenous DNA with a chromosome might displace or disturb the function of a telomere, thus rendering the chromosome susceptible to instability. After chromosome replication both sister chromatids would contain abnormal termini which might spontaneously fuse to form chromosomal loops or circles and, eventually, dicentric chromosomes. Repeated cycles would generate duplicated inverted chromosomal regions and stabilization could result from acquisition of telomere function, perhaps by chromosome translocation. Chromosomal inversions and translocations have been observed in DNA transfection experiments (66,72−74).

The amplification of DNA in different chromosomal locations has been compared. Results demonstrate that DNA sequences in different chromosomal locations have dramatically different potentials for amplification (36,75). It is not known if this reflects the proximity of the integration site to some specific sequence or to its general location in the chromosome. Transfected sequences which are more unstably associated with the host DNA appear to be amplified preferentially to those which are stably associated (75). For any one transformant, either a subset or all of the transfected sequences present may be amplified (69,75). These results are consistent with a hypothesis that a gradient of DNA amplification exists (76). Transfected sequences at the center of the gradient may be amplified frequently whereas those more distal would be less likely to be amplified. It is not known what sequences determine the center of the gradient, but they might function as an origin of replication.

In order to coamplify a particular DNA sequence, it is important that the DNA sequence or its products not interfere with amplification or be toxic to the cell. Since a minor subset of cells from the total population survives during each step of the selection process, it is possible to select for mutations in the coamplified gene which allow cell viability. Thus, the protein expressed at a high level may be different from the native protein. One example of a mutation occurring upon selection for amplification is the coamplification of the SV40 small t antigen in CHO cells (75). Highly methotrexate-resistant cells were selected that expressed SV40 small t antigen at 15% of the total protein synthesis. However, the t antigen migrated at a molecular weight 2000 Da less than the wild-type t antigen from cells at a lower expression level. Although the basis for this change is not understood, it demonstrates the potential difficulty in assuring the fidelity of proteins produced after selection for gene amplification.

Coamplification of heterologous genes with DHFR has been successfully used to obtain a number of cell lines that express high
levels of a protein from heterologous genes. Some examples include
human tissue plasminogen activater (69), human gamma-interferon
(77,78), human beta-interferon (79), human factor IX (80), human
factor VIII (81), the herpes simplex glycoprotein D (82), and the
alpha and beta subunits of bovine luteinizing hormone (83).

B. Vectors for Expression of cDNA Genes

1. Constitutive Promoters

The best characterized promoter systems are those derived from
viruses (SV40 and adenovirus), primarily due to analysis of mutations generated by natural variation of the virus (for a review,
see 84). Generally, the best studied promoters are constitutive,
i.e., they are always transcriptionally active and there is no simple
way to regulate their activity. There are three important identified
sequence elements that control transcription initiation: the TATAA
box (often at approximately -30 bp with respect to the mRNA initiation site as +1 (85), which probably functions by designating the
start site for RNA polymerase II transcription; the CAAT box (at
approximately -80 bp from the mRNA initiation site) (86), where
factors bind to facilitate transcription initiation; and transcriptional
enhancers, which encompass a variety of core sequences that act to
increase transcription from a promoter in an orientation- and
distance-independent manner (87).

The transcriptional enhancer appears to be a primary regulator
of transcriptional activity. Some enhancers located in viruses,
e.g., polyoma (88) or Moloney murine sarcoma virus (89), show a
host cell preference and thus contribute to the host range of the
virus. Others, like the SV40 enhancer (90), the Rous sarcoma
virus (RSV) LTR (91), and the human cytomegalovirus (CMV) enhancer (92), are very active in a wide variety of cell types from
many species. Enhancers with strict cell type specificity have been
observed in many cellular genes, most notably the immunoglobulin
genes (93–95) and the insulin gene (96), and are likely primary
cis-acting determinants in tissue specificity of gene transcription.
The addition of a strong enhancer can increase transcriptional
activity by 10- to 100-fold. Thus, most expression vectors include a strong enhancer, frequently derived from SV40, RSV, or
CMV.

2. Inducible Promoters

In order to express a protein which is potentially cytotoxic, e.g., regulatory proteins such as the c-myc gene product, it is desirable to use an inducible expression system which is regulated by an external stimulus. The sequences required for induced transcription from such promoters generally function as enhancer elements in that they can act independently of orientation and position with respect to the mRNA transcription initiation site. The inducible promoters which have demonstrated utility for expression of heterologous genes are described below.

   Beta-Interferon Promoter. The beta-interferon gene is highly inducible in fibroblasts by virus infection or by the presence of double-stranded RNA [poly(rI)-poly(rC)] (for review, see 97). The sequences responsible for this induction are found between bases from -77 to -36 from the start site for mRNA transcription. This element contains a 5' segment which functions as a strong constitutive transcription element and a 3' segment which functions as a negative element that prevents transcription without induction (98). This system has been used to obtain a 200-fold induction of beta-interferon expression in CHO cells after cointroduction and coamplification of beta-interferon genes with a DHFR gene (79). These experiments demonstrated that the normal inducibility of the beta-interferon gene is maintained in cells containing 25-fold amplified copies of the interferon gene. It is possible that higher levels of amplification would titrate out the negative factor which is required to decrease the basal level of transcription (98). One potential problem with this induction system is that, when studied, only a small percentage of murine fibroblasts (5–10%) exhibit beta-interferon gene induction in response to poly(rL.rC.) (99). The same studies demonstrate poor activation of the beta-interferon gene in human HeLa cells but also demonstrate that pretreatment with interferon results in high inducibility. Thus, an interferon-inducible factor appears to be required for activation of the beta-interferon gene in the nonresponsive cells.

   Heat Shock Promoter. Heat shock genes are transcriptionally activated when cells are exposed to hyperthermia or a variety of other stresses (100). The molecular basis for this gene activation has been extensively studied using the *Drosophila* heat shock protein 70 gene (*hsp*70) promoter, which is efficiently expressed and tightly regulated in *Drosophila* and mammalian cells. The *Drosophila* *hsp*70 promoter contains a 15-bp sequence upstream from the TATA box which is responsible for activation of the promoter in response to heat and which can activate a heterologous thymidine kinase (TK) gene when placed upstream from the TK TATA box (101). Most heat shock genes that have been studied contain a similar sequence in an equivalent position and a consensus sequence (C — GAA — TTC — G) has been derived (102). This sequence is the binding site for a specific heat shock transcription factor that is activated during heat shock (103,104).

This inducible system has been used to obtain a high level of expression of the c-*myc* protein in CHO cells after coamplification of a *hsp*70 promoter — c-*myc* gene fusion with DHFR (105). Highly amplified cell lines were obtained that contain 2000 copies of the introduced c-*myc* fusion gene, and undetectable levels of c-*myc* mRNA. Incubation of these cells at 43°C resulted in at least a 100-fold induction of c-*myc* mRNA. Translation only occurred when the cells were returned to 37°C. After 3–4 hr, the c-*myc* protein levels reached approximately 1 mg/10^9 cells. These results demonstrate that even at a level of 2000 copies, the *hsp*70 promoter retains inducibility. This suggests that any negative- or positive-acting factor involved in *hsp*70 promoter activity is not limiting. One potential problem with this system is that heat shock is detrimental to the cells and may severely affect the protein secretion machinery of some cell lines (106).

Metallothionein Promoter. Metallothioneins are small, cysteine-rich proteins which play an important role in detoxification of heavy metals and in heavy-metal hemostasis (107). Metallothionein (MT) gene transcription is induced by the presence of heavy bivalent metal ions, such as cadmium and zinc (108), by glucocorticoid hormones (109), and by interferon (110). The mouse MT-1 promoter region contains a metal regulatory element between bases −59 and −46, with respect to the start site for mRNA transcription as +1, which may aact as an inducible enhancer (111,112). Homology in this region between human and mouse MT-I and MT-II genes has identified the sequence CCTTTGCGCCCG as the regulatory region (113,114). Insertion of multiple copies of this sequence can confer zinc "inducibility" on a heterologous promoter (115). Using in vivo competition assays, Sequin et al. (114) showed that there is a cellular factor(s) that in the presence of cadmium interacts with the heavy-metal-responsive element to induce transcription. This promoter has frequently been used successfully in bovine papilloma-based vector systems (116). However, the basal level of expression is generally high, and the induction ratio is poor, generally never greater than 5- to 10-fold. High levels of cadmium are cytotoxic; however, zinc may be used as a less toxic inducing agent.

Glucorticoid Induction. Glucorticoids and other steroid hormones mediate physiological responses in target cells via a specific, high-affinity interaction with cytoplasmic receptors. Hormone binding to receptors results in a conformational change, thereby allowing the hormone-receptor complex to bind specific sites on DNA (117). Hormone-receptor complex binding to DNA results in transcriptional activation of the hormone-responsive gene. The best studied glucocorticoid-inducible promoter is the mouse mammary tumor virus (MMTV) LTR. The glucocorticoid-responsive element (GRE) has been localized between -100 and -200 of the MMTV LTR. These GREs behave like enhancers in that they stimulate transcription of heterologous promoters in an orientation- and position-independent manner (118–121) and contain a consensus hexanucleotide sequence, TGTTG (120). The purified glucocorticoid receptor binds to the GRE sequence in vitro (121–124). This promoter has been used to express a heterologous DHFR in CHO cells (125). However, the induction ratio with this promoter upon addition of dexamethasone, a synthetic glucocorticoid, was only fivefold and the basal, uninduced level was high. Similarly, when the MMTV LTR promoter was used to express the *E. coli* XGPRT gene, a 10-fold gene copy number amplification was achieved in mouse 3T6 fibroblasts by selection in limiting concentrations of xanthine, and the dexamethasone induction of XGPRT expression ranged from 6- to 15-fold (126). This induction ratio is considerably lower than the 50- to 600-fold seen with the MMTV LTR in rat hepatocytes but is consistent with the 10-fold induction of MMTV RNA seen in mouse mammary tumor cells (127). Klessig et al. (128) reported success in expression of the adenovirus DNA-binding protein under control of the MMTV LTR in HeLa cells. Addition of dexamethasone resulted in a 50- to 200-fold induction and after 8 hr the level of expression, from a single integrated copy, reached 15% of that observed in adenovirus-infected cells. One potential limiting feature for glucocorticoid induction is the level of glucocorticoid receptor. Cells which express higher levels of glucocorticoid receptor show a greater induction of the MMTV promoter (129). Cotransfection of a plasmid encoding the glucocorticoid receptor with a plasmid containing the MMTV LTR into monkey CV-1 cells yielded a 500-fold induction of mRNA from the MMTV LTR upon addition of dexamethasone (130). With the availability of cDNA clones for the glucocorticoid (131), estrogen (132), progesterone (133), and other steroid receptors, it should be possible to generate cell lines that express high levels of a specific receptor allowing highly inducible expression.

As genes for other transactivators are cloned and characterized for the structural features required to activate gene transcription, it will be possible to engineer novel potent transactivators to elicit induction of specific promoters. This was recently demonstrated by mixing DNA binding and transcriptional activation domains from yeast, bacteria, and mammalian cells to elicit transactivation of specific genes (134–136).

3. RNA Processing

Splicing. Most higher eukaryotic genes contain introns which are processed from the precursor RNA in the nucleus to generate the mature mRNA. The sequences and factors responsible for RNA splicing were recently reviewed (137). Although many genes do not require introns for mRNA formation when introduced into mammalian cells, there are several examples of genes that have strong requirements for the presence of an intron (138,139). It was reported that optimal expression of the human factor VIII cDNA gene exhibits a requirement for an intron in the cDNA expression vector (140). Whether this results from the extremely large size of the factor VIII cDNA (7.5 kb) or from some other unique feature of the expression vector used in these experiments is not clear. In general, it is probably better to contain introns in cDNA expression vectors for mammalian cells. In contrast, there is a profound requirement for the presence of introns for the expression of genes introduced into transgenic animals (141).

Transcription Termination and Polyadenylation. Transcription termination, 3'-end cleavage, and polyadenylation of precursor mRNAs are essential steps for the biogenesis of mRNA (for review, see 142). In higher eukaryotes, RNA polymerase II transcribes across the polyadenylation site(s) and terminates downstream of the mature 3' termini for the mRNA. To date the DNA sequences which function as transcription termination signals have only been mapped to various restriction fragments at the 3' end of several genes (143–145). Structural features within some of these regions have been compared to identify a sequence which represents a conserved element within the termination region of a number of different genes (146). Further studies are required to identify whether this sequence has any functional significance.

The possibility that an upstream promoter may occlude transcription from a downstream promoter (147) suggests that insertion of a transcription termination signal upstream from the second transcription unit in an expression vector may potentiate expression from that promoter by preventing transcription from extending through the upstream sequences. Indeed, this has been directly

demonstrated by the insertion of a histone H2A gene termination
region or a mouse beta-globin gene termination region between two
tandem alpha-globin transcription units to elicit an approximately
sevenfold increase in expression from the downstream transcription
unit (148). Transcription termination signals would also provide an
approach to minimize transcription from the opposite strand of DNA
which can suppress gene expression through the formation of antisense mRNA (149,150).

The 3' end of eukaryotic mRNA is formed by polyandenylation,
which involves cleavage of the precursor mRNA at a specific site
and then polymerization of about 200 adenylate residues, poly(A),
to the newly generated 3' end (151). Removal of the polyadenylation site decreases expression up to 10-fold (152). Two sequences
important for polyadenylation have been identified. The first is a
highly conserved hexanucleotide AAUAAA, present 11 − 30 nucleotides
upstream of most polyadenylation sites, which forms the recognition
sequence for the cleavage and polyadenylation reaction. Deletions
or point mutations in this sequence prevent the appearance of properly polyadenylated mRNA in vivo (153 − 156). There is also a requirement for a sequence downstream of the poly(A) site for efficient cleavage and polyadenylation (157,158). A loose consensus
sequence for this potential second element was identified as either a
U-rich or a G + U-rich tract (159). However, removal of this sequence in some instances has no effect on the efficiency of polyadenylation (160) but may influence the position of 3' processing
(161). This sequence appears to be required for the formation of
a precleavage complex (162).

Translation of mRNAs. The primary event limiting translation
of mRNAs is at the initiation step. Polypeptide chain synthesis is
initiated when eIF-2 forms a ternary complex with GTP and initiator
Met-tRNA (for a review, see 163). This complex binds a 40S ribosomal subunit. This 40S ribosomal subunit complex binds to the 5'
end of the mRNA and migrates in the 3' direction until it encounters
an AUG triplet which can efficiently serve as the initiator codon.
Subsequently, a 60S ribosomal subunit is added to form an 80S
initiation complex, with concomitant hydrolysis of GTP to GDP. In
order to reinitiate, GDP bond to eIF-2 must be replaced by GTP, a reaction catalyzed by the guanine nucleotide exchange factor (GEF).
The phosphorylation of the alpha subunit of eIF-2 results in stabilization of the GEF-eIF-2-GDP complex, thereby inhibiting recycling of eIF-2. This presumably depletes the amount of functional eIF-2 competent for translation initiation. Two protein kinases have been shown to regulate protein synthesis initiation by

phosphorylation of the alpha subunit of eIF-2. The hemin-controlled inhibitor (HDI) is a protein kinase that has been studied in reticulocytes and is activated by various stimuli including hemin deprivation and heat treatment. The double-stranded RNA-activated inhibitor (DAI) is a protein kinase induced by interferon and its activity is dependent on the presence of double-stranded RNA. Several approaches to the development of expression vectors optimized for protein synthesis initiation are discussed below.

mRNA and Consensus Sequence Requirements. M. Kozak systematically analyzed the sequence requirements for efficient mRNA translation and concluded that the sequence

$$5'-CC_{G}^{A}CCATGG-3'$$

is most efficient for translation initiation (164). Of greatest importance is the purine in the -3 position and then the G in the $+4$ position. Alterations to produce this consensus sequence around the initiator methionine codon may result in little to as much as a 10-fold increase in translation initiation.

In addition to the consensus sequence requirements, there are other features that influence mRNA translation. The presence of upstream initiation codons is detrimental to initiation at the downstream site, particularly if they are not followed by in-frame termination codons before the start site for translation initiation of the desired coding region (165,166). In addition, the presence of significant secondary structure (-50 kcal/mol) near the AUG on the mRNA can be detrimental to translation (167–169).

Translational Efficiency Mediated by eIF-2. Translation of mRNA derived from plasmid DNA after transient transfection of COS-1 cells is very inefficient. This deficiency can be corrected by the introduction of the adenovirus-associated (VA) RNA gene into the COS-1 cell expression vector (170). VA RNA is a small RNA synthesized by RNA polymerase III which has been proposed to block the double-stranded RNA-activated protein kinase (DAI) (171,172). It has been proposed that transcription from both strands of the episomal DNA template in transfected COS-1 cells can generate double-stranded RNA leading to activation of DAI kinase and translation arrest. VA RNA stimulates translation by preventing the phosphorylation of eIF-2 (170,173,174). To date there has been no demonstration of an effect of VA RNA on the translation of mRNA derived from transcription units integrated in the host chromosome.

Polycistronic Expression Vectors. It is now known that polycistronic mRNAs are translated in mammalian cells (175). The most
likely mechanism is one of translation termination and ribosome
scanning with reinitiation at downstream AUGs (176). The utilization of polycistronic mRNA expression vectors containing the desired
gene in the 5' open reading frame and a selectable gene in the 3'
position has been used to select cells that express high levels of
the gene product in the 5' position (177). Since the translation of
the selectable gene in the 3' end is inefficient, polycistronic expression vectors exhibit preferential translation of the 5' desired
gene.

VI. SECRETION AND POSTTRANSLATIONAL
MODIFICATIONS OF EXPRESSED
PROTEINS

Two general pathways of secretion exist in higher cells: the constitutive pathway and the regulated pathway (178). All cells carry
out constitutive secretion and secrete proteins with a variety of
posttranslational modifications. Some cells have the capacity to
store proteins in specialized granules and secrete them upon the
appropriate stimulus. This regulated pathway of secretion was
shown to be required for the correct processing of proinsulin to
insulin (179) and proopiomelanocortin to ACTH (180). In these
cases only the secretory granules contain the enzymes that can
process these specialized proteins.

The highly compartmentalized structure of the eukaryotic cell
requires a mechanism of protein localization to ensure that proteins
are directed to the appropriate organelle such as mitochondria,
peroxisomes, or nucleus, and for the distribution of membrane-
associated or secreted proteins through the endoplasmic reticulum
(ER) and the Golgi complex to their final destination at the cell
surface, in secretory granules, or in lysosomes (for reviews, see
181, 182). The precursor forms of proteins transported through
the ER contain a hydrophobic signal sequence at or near the amino-
terminus. Translation of the mRNAs of secretory and membrane-
bound proteins is arrested at an early stage when the signal recognition particle (SRP) binds to this signal sequence on the nascent
polypeptide chain. This arrested complex interacts with an SRP
receptor (docking protein) on the rough endoplasmic reticulum (ER)
after which protein synthesis continues. In this way the protein is
cotranslationally translocated into the lumen of the ER via the docking protein interaction. The signal sequence is usually cleaved by
the signal peptidase during translocation. Asparagine-linked glycosylation occurs at appropriate recognition sites (asn-X-ser/thr) as

the protein enters the ER. After translocation and high-mannose
core oligosaccharide addition, the protein is transported through the
ER where some trimming of terminal glucose and mannose residues
can occur. Also in the ER, fatty acid addition to the polypeptide
backbone can occur. The protein moves to the Golgi complex where
further asparagine-linked carbohydrate modifications, such as addition of galactose and sialic acid, occur. Also in the Golgi complex,
sulfation of tyrosine and carbohydrate residues, protein phosphorylation, serine and threonine O-linked glycosylation, processing of propeptides, and modifications such as gamma-carboxylation of glutamic
acid residues and beta-hydroxylation of aspartic acid residues occur.
The processing which occurs as a protein transits through the secretory pathway have been reviewed (183,184). All newly synthesized proteins pass through the Golgi complex in a similar manner
until they reach the last compartment where there exist three different pathways which can be taken (185): (1) membrane bound
proteins and proteins which are constitutively secreted are transported to the cell surface; (2) regulated secretory proteins are
transported and stored in secretory granules which are released
upon appropriate stimulus (178); and (3) lysosomal enzymes are
targeted to lysosomal structures (186). To date results have demonstrated that a variety of cell lines have the capacity to appropriately synthesize, glycosylated, and secrete large quantities of
heterologous proteins. These data argue that the primary structure
of a protein must contain the information that determines its final
destination as well as the complex nature of the posttranslational
modifications that occur.

A. Protein Translocation, Signal Cleavage, N-Linked Glycosylation, and Transport Through the ER

To date, most proteins which are expressed at high levels in mammalian cells have been shown to exhibit correct signal peptide
cleavage, appropriate N-linked glycosylation, and transport through
the ER (69,80,177). In one instance, the signal sequence of the
AIDS virus (HIV) envelope glycoprotein gp120 was shown to be
inefficiently processed in CHO cells. This glycoprotein contains a
novel signal sequence which contains an unusually long hydrophobic domain at the amino terminus of the protein which is preceded by a highly charged region. Replacement of the gp120 signal
peptide with a herpes virus glycoprotein D signal peptide resulted
in a dramatic increase in expression (187). However, this protein
was retained within one of the intracellular compartments with little
transport to the cell surface. In order to obtain a secreted form of
gp120, the entire putative transmembrane domain was deleted. The
alteration resulted in efficient secretion of the gp120. This approach

approach was previously used to obtain secreted forms of vesicular stomatitis virus G glycoprotein (188) and the influenza hemagglutinin (188). The secreted gp120 was shown to elicit AIDS retrovirus-neutralizing antibodies (187). These results demonstrate the ability to engineer a protein to both increase its processing efficiency and to redirect a membrane-bound protein for secretion.

In general, the rate of transit through the ER is dependent on the polypeptide backbone of the protein (190). For membrane-bound proteins, the cytoplasmic domain plays the most critical determinant in the rate of transit. Hybrid proteins between the vesicular stomatitis virus and various cytoplasmic domains are transported through the ER at the rate characteristic of the cytoplasmic domain (181). Secreted proteins may vary to great degrees in their rate of transport and it has not been determined as to which protein domains control the individual rate of transit. However, when genetic alterations have been introduced into cloned genes, it has frequently been observed that the resultant proteins are not efficiently transported out of the ER. For example, substitution, addition, or deletions of amino acids within the amino terminus, the transmembrane domain, or the cytoplasmic domain of influenza virus hemagglutinin result in blocked or delayed transport of the molecule from the ER into the Golgi complex (189,191,192). These experiments suggest that altered protein structures may produce a protein which is not recognized by cellular transport machinery or which is improperly folded, to produce aggregated forms within the ER. In the case of immunoglobulins (193) and influenza virus hemagglutinin (194), it has been shown that proper folding and assembly of the heavy and light chains of the immunoglobulin or assembly of the influenza hemagglutinin trimers are essential for their efficient transport out of the ER. Particularly intriguing is the nature of the cellular protein in the ER, which is complexed with the incompletely assembled molecules. In both situations, a protein of 78 kDa is associated with the improperly folded proteins within the ER. This protein has been identified as immunoglobulin-binding protein (BiP) (193). BiP has been shown to be similar, if not identical to GRP78, which is a glucose-regulated protein related to the heat shock family (hsp70) (195). The identification of the role which this protein plays in secretion may provide fundamental insight into mechanisms of transport and secretion (196,197).

B. Posttranslational Modifications and Transport Through the Golgi Complex

A variety of posttranslational modifications can occur in the Golgi apparatus. Surprisingly, proteins synthesized in heterologous cells frequently exhibit the same modifications which occur on the native protein. One example is the expression of the human coagulation factor IX, which is a serine protease zymogen synthesized in the liver and which contains 12 gamma-carboxyglutamic acid residues within the amino terminus of the mature protein. These carboxyl groups are transferred by a vitamin K-dependent reaction and their presence is required for factor IX activity. Expression of factor IX in CHO cells has resulted in high levels of factor IX antigen but negligible activity. Upon addition of vitamin K to the conditioned medium, active factor IX was produced, although a portion of the total factor IX expressed still lacked biological activity. When the active portion was purified, it had a specific activity and levels of gamma-carboxyglutamic acid very similar to plasma-derived human factor IX (80).

VII. PRODUCTION OF FACTOR VIII THROUGH GENETIC MANIPULATION OF MAMMALIAN CELLS

Blood vessel injury is associated with the sequential activation of a series of plasma proteases that leads to the localized generation of thrombin, which converts fibrinogen to fibrin. The hemostatic importance of this cascade is revealed by the existence of bleeding disorders resulting from altered proteolytic activities (198,199). One such disorder is the deficiency of factor VIII in classic hemophilia A. The factor VIII complex has two distinct biological functions: coagulant activity and a role in primary hemostasis. The factor VIII procoagulant protein (factor VIII, antihemophilic factor) and the factor VIII-related antigen (von Willebrand factor, vWF) are under separate genetic control, have distinct biochemical and immunological properties, and have unique and essential physiological functions.

vWF is a large, adhesive, multifunctional glycoprotein that is essential for normal hemostasis (for review, see 198). It is synthesized by endothelial cells and is stored within specialized organelles called Weibel-Palade bodies where it is released upon stimulation. vWF circulates as a series of disulfide-bonded multimers that extend to molecular weights greater than 10^7. The high molecular weight multimers mediate the adhesion of platelets to subendothelium upon vascular injury. A variety of abnormalities in vWF can result in a condition known as von Willebrand's disease.

Factor VIII is an important regulatory molecule in the blood coagulation cascade (198,199). After activation by thrombin, it accelerates the rate of factor X activation by factor IXa, eventually leading to the formation of the fibrin clot. Deficiency of factor VIII is an X-linked chromosomal disorder that has been a major source of

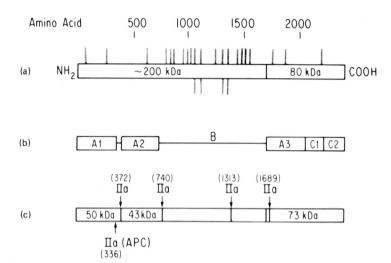

Figure 2   Structural features of factor VIII.   (a) Relative positions
of the two-chain form with the vertical bars representing sites on
potential N-linked glycosylation.   (b) Relative positions of the domains composing factor VIII.   (c) Position of thrombin (IIa) and
activated protein C (APC) cleavage sites.

hemorrhagic morbidity and mortality in affected males.   In the past,
treatment consisted of frequent transfusions of plasma-derived preparations of factor VIII that carried the risk of virus contamination.
The ability to produce an inexpensive, pure preparation of factor
VIII from a non-plasma-derived source would significantly improve
the quality of life for afflicted individuals.

The cloning of the factor VIII cDNA provided the primary structure of the factor VIII protein as shown in Fig. 2 (200–202).   It is
synthesized as a 2351-amino-acid single-chain precursor from which
a 19-amino-acid signal peptide is cleaved upon translocation into the
lumen of the ER.   The precursor is proteolytically processed to
generate an amino terminal-derived heavy chain ranging in size from
90 to 200 kDa and a carboxy terminal-derived light chain of 80 kDa.
Factor VIII is composed of three domains which occur in the order
A1–A2–B–A3–C1–C2:   the A domain occurs twice in the heavy
chain and once in the light chain and has homology to ceruloplasmin,
a copper-binding protein in plasma; a single unique B domain; and
two C domains in the light chain that have homology to phospholipid-
binding proteins (203).   The protein has 25 potential N-linked

glycosylation sites, 19 of which are located in the B domain. The
B domain is proteolytically released upon activation by thrombin
and is not required for procoagulant activity in vitro or in vivo
(204–206). In the plasma factor VIII exists as a metal ion complex
of the 200-kDa heavy chain and the 80-kDa light chain. The complex is noncovalently bound to von Willebrand factor (vWF) at a
ratio of one molecule of factor VIII per 100 molecules of vWF (207,
208). The presence of vWF is required to stabilize factor VIII in
the plasma (209–213).

To date there are no known cell lines that produce factor VIII.
Most evidence has suggested that the hepatocyte is the site of synthesis of factor VIII in vivo; however, the factor VIII mRNA has
been detected in many other tissues besides liver (214). The isolation of a full-length cDNA encoding factor VIII provided the
capability of deriving mammalian cell lines which express factor VIII
in order to study the biosynthesis of factor VIII in a way that was
never before possible (201,202). In addition, the ability to express
biologically active factor VIII in mammalian cell culture and the
availability of the cDNA in which to engineer specific mutations
within the factor VIII molecule have led to a greater understanding
of the structural requirements for factor VIII cofactor activity (215).

A. Expression of Factor VIII in CHO Cells

Dihydrofolate reductase cotransfection and coamplification was used
to derive CHO cells that express human factor VIII. The factor
VIII cDNA was introduced into an expression vector and transfected
with a selectable DHFR gene into DHFR-deficient CHO cells. Upon
selection for the DHFR-positive phenotype by growth in nucleoside-
free medium, and further selection for DHFR gene amplification by
growth in sequentially increasing concentrations of methotrexate,
cells were obtained that harbored 1000 copies of the DHFR selection
marker and the factor VIII cDNA gene. However, expression of
active factor VIII procoagulant activity was 2–3 orders of magnitude
lower than that observed with other genes using similar vectors and
approaches in CHO cells (69,77,78,80,83). Presently, we have
identified three reasons for the low level of factor VIII expression
(81). First, the protein is exquisitely sensitive to proteases and is
degraded upon secretion from the cell. Second, the majority of
factor VIII synthesized is never transported from the ER to the
Golgi. Finally, the factor VIII mRNA level is low as a result of a
posttranscriptional event. To improve factor VIII expression, we
have solved the difficulty in factor VIII stability and have begun
to understand the mechanisms for limited transport from the ER to
Golgi. The factor VIII produced has been purified to apparent

homogeneity and displays properties strikingly similar to those of human plasma-derived factor VIII.

B. vWF Stabilizes Factor VIII by Promoting Association of the Heavy and Light Chains upon Secretion from the Cell

Our initial observation was that the presence of serum in the medium improved the stable accumulation of factor VIII activity. Previous studies had demonstrated that vWF is required in vivo for the survival of factor VIII in plasma (209−213). The requirement for vWF to stabilize factor VIII expressed from CHO cells was tested by addition of wWF at 2.5 µg/ml to serum-free growth medium. This resulted in factor VIII accumulation similar to that observed when cells are grown in medium containing 10% fetal calf serum. These results demonstrated that vWF is the primary serum component required for stable factor VIII expression (81).

The role that vWF provided in promoting stable factor VIII accumulation was analyzed by an [^{35}S]methionine pulse-labeling and chase experiment to study the synthesis and secretion of factor VIII in the presence and absence of serum. These experiments demonstrated that serum increases the rate of secretion of the light chain and promotes a stable association of the heavy and light chains of factor VIII. In the absence of serum, the heavy chain exhibits decreased association with light chain and an increased susceptibility to degradation (81).

C. Coexpression of Factor VIII and vWF Alleviates the Requirement for Exogenous vWF

An approach to alleviate the requirement for exogenous vWF for stabilization of factor VIII is to express vWF in the same cells that express factor VIII. The feasibility of this approach was suggested by the ability of recombinant-derived vWF to stabilize factor VIII when added to serum-free conditioned medium. The approach to deriving cells that express both factor VIII and vWF is outlined in Fig. 3. The factor VIII expression plasmid was introduced with a DHFR selectable marker into DHFR-deficient CHO cells. The cells were selected for increasing resistance to MTX and a cell line obtained that expresses elevated amounts of factor VIII. These cells were then transfected with a vWF expression plasmid and an adenosine deaminase (ADA) selectable marker (40). Selection of the transformants for increasing resistance to the ADA inhibitor, 2'-deoxycoformycin, yielded cells which had amplified the ADA and vWF genes. The resulting cell line 10A13-a expresses factor VIII, vWF, DHFR, and ADA.

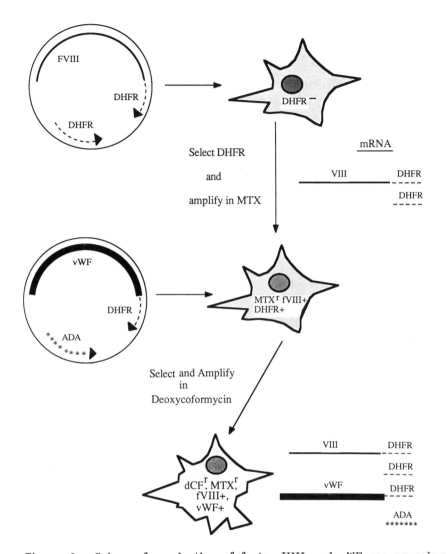

Figure 3   Scheme for selection of factor VIII and vWF coexpressing CHO cells.   A plasmid encoding DHFR and factor VIII was transfected into DHFR-deficient CHO cells and transformants selected by growth in nucleoside-free medium.   Cells were then selected in increasing concentrations of methotrexate in order to select for cells which had amplified the DHFR and factor VIII genes.   Then these cells were transfected with a vWF expression plasmid containing an adenosine deaminase gene and selected for growth in the presence of cytotoxic concentrations of adenosine with increasing amounts of 2'-deoxycoformycin.   The final cell expresses DHFR, ADA, vWF, and factor VIII.   The mRNA structures are depicted below.   The factor VIII, vWF, and DHFR mRNA species should all hybridize to a DHFR probe.

41

   The synthesis and processing of vWF was studied in the coexpressing cell line 10A1-3αA by [35]methionine labeling, immunopre-
cipitation of the cell extracts and conditioned medium with an anti-
vWF specific antiserum, and electrophoresis on SDA-polyacrylamide
gels under reducing and nonreducing conditions. The results demonstrated that the majority of vWF expressed is properly processed
by cleavage of the 100-kDa propetide to generate the mature 230-
kDa form. Some precursor vWF is also secreted. The ratio of the
secreted precursor and mature forms is similar to that observed in
vWF secreted constitutively from human endothelial cells (216).
The degree of vWF multimer formation was analyzed by electrophoresis of the samples on nonreducing gels and exhibited a ladder very
similar to that obtained from vWF constitutively secreted from human
endothelial cells. These results demonstrate that CHO cells can
process and multimerize vWF in a manner very similar to that observed in human endothelial cells.
   The ability of the factor VIII and vWF coexpressing cells to accumulate factor VIII over 3 days of growth in the presence or absence of fetal calf serum (FCS) is shown in Fig. 4. In contrast to
the factor VIII alone expressing cell line 10A1, which does not accumulate factor VIII, the coexpressing cells accumulate factor VIII
activity linearly over 3 days when propagated either in the presence
or in the absence of serum. As described below, coexpression of
vWF did not significantly alter the synthesis, processing, or secretion of factor VIII from these cells. The major difference is that
coexpression of vWF results in the recovery of greater amounts of
factor VIII protein and activity in the conditioned medium. The results suggest that after 24 hr the amount of vWF present in 10%
serum is limiting for the continued accumulation of factor VIII activity and that the coexpressing cells probably supplement the vWF
present in the serum, allowing for high levels of active factor VIII
accumulation.
   The synthesis and processing of factor VIII and vWF were
studied in the coexpressing cell line. These cells express factor
VIII and vWF from similar transcription units each derived from the
adenovirus major late promoter and SV40 enhancer at the 5' end and
a DHFR sequence and the SV40 early polyadenylation site at the 3'
end. Southern blot hybridization analysis of BGlII-digested genomic
DNA using a probe that would identify both genes demonstrated
that the factor VIII transcription unit is present at 20-fold greater
copy number than the vWF transcription unit (Fig. 5). Since the
factor VIII gene was coamplified with the cotransfected DHFR gene,
the cotransfected DHFR gene is also present at a 20-fold greater
level compared to the vWF gene.
   A similar analysis by Northern blot hybridization to identify
factor VIII and vWF mRNA species demonstrated three distinct mRNA

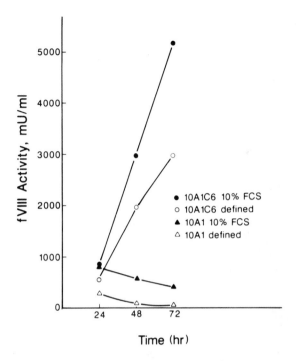

Time (hr)

Figure 4 Accumulation of factor VIII in the presence and absence of vWF coexpression. The factor VIII-producing CHO cell line (10A1) and a factor VIII/vWF-coexpressing cell line (10A1C6) were seeded and grown in medium either containing or lacking 10% fetal calf serum. After 24, 48, and 72 hr, samples were taken and measured for factor VIII activity assay by the Kabi Coatest method.

species hybridizing to a DHFR probe which represent the vWF mRNA, the factor VIII mRNA, and the DHFR mRNA (Fig. 5). The abundance of factor VIII and vWF mRNA are very similar. In contrast, the DHFR mRNA is present in much greater abundance, which is consistent with the high gene copy number. These results demonstrate that factor VIII mRNA is inefficiently expressed per gene copy when compared to the vWF or DHFR mRNAs. Analysis of in vitro nuclear transcription to determine the RNA polymerase density on these transcription units has shown that the difference is not due to different transcription rates. Thus it appears that a post-transcriptional event is responsible for the disproportionately low level of factor VIII mRNA.

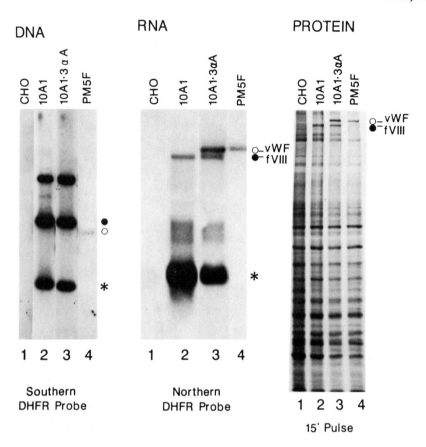

Figure 5   Analysis of DNA copy number, mRNA abundance, and initial translation products in factor VIII and vWF coexpressing cells.   DNA and RNA were isolated from the original CHO cell line (CHO), the factor VIII-producing cell line (10A1), the coexpressing cell line (10A1-3αA), and a vWF only expressing cell line (PM5F), and analyzed by Southern blot and Northern blot hybridization analysis to a DHFR probe.   Cells were also labeled for 15 min with [^{35}S]methionine and cell extracts prepared for analysis by SDA-PAGE and autoradiography.   The open circles represent the vWF-specific genes, mRNA, and protein.   The closed circles represent the factor VIII genes, mRNA, and protein.   The asterisk represents the DHFR genes and mRNA.   For this analysis, the DHFR protein species has run off the bottom of the gel.

The primary translation products in the 10A1 and 10A1-3αA cells were analyzed by a 15-min pulse label with [^{35}S]methionine and SDS-PAGE of the total cell extract (Fig. 5). Two protein species migrating with molecular weights greater than 200 kDa can be identified in 10A1-3αA cells. The upper band represents vWF while the lower band represents factor VIII. The relative amount of primary translation product is proportional to the mRNA abundance for each of these proteins. However, the level of factor VIII antigen in the conditioned medium is approximately 100–200 ng/ml compared to 1–2 μg/ml for vWF. This demonstrates that while both vWF and factor VIII mRNAs are translated with similar efficiency, factor VIII protein is inefficiently processed and secreted from CHO cells in comparison to another large complex glycoprotein, vWF. The results of these studies demonstrate that factor VIII expression is limited due to (1) a low mRNA level and (2) inefficient processing and/or secretion of the protein.

D. Factor VIII Is Inefficiently Transported From the ER to the Golgi

Since a naturally occurring cell line producing factor VIII has never been obtained, little is known of the intracellular processing events required to generate the mature forms of factor VIII found in plasma. Initial [^{35}S]methionine pulse-labeling and chase experiments demonstrated that only a fraction of the factor VIII which is synthesized is ever secreted from CHO cells. The factor VIII that is not secreted is observed in the ER in a complex with the immunoglobulin-binding protein or BiP (196). BiP is identical to the glucose-regulated protein of 78 kDa (GRP78) and is homologous to members of the hsp70 protein family (217). Although the function of BiP remains unknown, it binds proteins to a greater degree when normal protein folding or N-linked glycosylation is blocked and may prevent thrir transport and secretion (194,195,217). It is not known whether BiP acts to detect and remove improperly folded proteins that are secretion-incompetent or whether it actually facilitates protein folding and/or transit through the ER (197).

The importance of normal N-linked glycosylation in secretion efficiency and BiP association was demonstrated by studying the secretion of tissue plasminogen activator (tPA) from CHO cells (196). Inhibition of N-linked glycosylation was obtained by either tunicamycin treatment or by expression of mutant tPA genes generated by site-directed mutagenesis of three asparagine residues to glutamine residues at the sites of N-linked glycosylation. The results demonstrated that disruption of N-linked glycosylation promotes stable BiP association and blocks secretion of tPA. Further insight into the importance of glycosylation was suggested by the analysis of a factor VIII deletion molecule which has the majority of the B-domain and 18 glycosylation sites removed (196). This deleted factor VIII is more efficiently secreted and exhibits a lesser, transient association with BiP. However, the inefficient secretion and stable BiP association of full-length factor VIII appears not to result from a general inability of CHO cells to efficiently glycosylate heavily glycosylated proteins. An analysis of the processing of vWF, which contains 17 N-linked glycosylation sites, demonstrated that this molecule displayed very little BiP association and was efficiently secreted (196). Thus the ability to glycosylate proteins is not limiting in CHO cells. This result has led us to propose that the clustered N-linked glycosylation sites within the B domain of factor VIII may be inefficiently glycosylated, resulting in a protein which fails to achieve native conformation. Such molecules of factor VIII may become bound to BiP and retained in the cell (196).

E. Asparagine–Linked Glycosylation Pattern of Factor VIII Expressed in CHO Cells Is Similar to Human Plasma-Derived Factor VIII

The utilization of N-linked glycosylation sites in factor VIII derived from CHO cells and from human plasma were compared by measuring susceptibility of the N-linked carbohydrate to specific endoglycosidases. Endoglycosidase H (endo-H) specifically removes N-linked residues that are of the high-mannose type (218) and N-glycanse removes all N-linked carbohydrate independent of complexity (219). Factor VIII was purified from normal plasma and from the CHO cell-conditioned medium and treated with thrombin to generate specific cleavage products with molecular masses of 50 kDa, 43 kDa, and a heavily glycosylated B domain of approximately 180 kDa from the heavy chain as well as a 73-kDa doublet from the light chain. Endo-H treatment of the 73-kDa doublet decreases its molecular weight to a similar degree for plasma- and CHO-derived factor VIII (Fig. 6, lanes 1–4). N-glycanse digestion produces a similar shift of the 73-kDa doublet as that obtained by endo-H digestion (Fig. 6, lanes 5 and 6). Thus, the utilization of N-linked sites on the light chain appears to be similar between the recombinant and plasma-derived factor VIII. The pattern of digestion is consistent with the interpretation that N-linked glycosylation of the light chain is of the high-mannose type, although a complex N-glycanase-resistant site on the light chain cannot be ruled out.

Direct carbohydrate analysis of the heavy chain has demonstrated that the majority of the N-linked sites are occupied within the B domain (R. Steinbrink, Genetics Institute, Cambridge, MA, unpublished observations). Analysis of this heavily glycosylated region derived from the heavy chain demonstrates similar resistance to endo-H and sensitivity to *N*-glycanase for both the plasma-derived and CHO-derived factor VIII (Fig. 6, indicated by B). This is consistent with the majority of N-linked carbohydrate being of complex or hybrid type on the B domain. The CHO-derived factor VIII appears to have a slightly higher molecular weight, which may result from increased branching of the N-linked chains.

No mobility change occurs in the 43-kDa peptide after digestion with endo-H or N-glycanase, indicating that the single potential N-linked site is not utilized on either CHO- or plasma-derived factor VIII. The 50-kDa peptide has two potential N-linked glycosylation sites. *N*-glycanse digestion of CHO or plasma-derived factor VIII generates similar 45-kDa species, which result from removal of N-linked oligosaccharides from the 50-kDa thrombin fragment of the heavy chain (Fig. 6, lanes 5 and 6). Endo-H digestion increases the mobility of approximately 50% of the 50-kDa species to an intermediate molecular weight between undigested and *N*-glycanase digested species in both the CHO- and plasma-derived factor VIII (Fig. 6, lanes 3 and 4). This result suggests that a subpopulation of 50-kDa species contains a high mannose type of oligosaccharide and a complex type, while the rest of the molecules contain only complex or hybrid type. The subset of the 50-kDa species of recombinant factor VIII which is completely endo-H-resistant migrates with slightly greater heterogeneity and molecular weight than that of the plasma-derived factor VIII. This may be due to a greater branching of complex- or hybrid-type N-linked carbohydrate present on the 50-kDa thrombin fragment of CHO-derived factor VIII compared to plasma-derived factor VIII. This analysis indicates that CHO cells appropriately recognize and process the multiple N-linked sites of the secretion competent factor VIII similarly to the natural source of plasma factor VIII. This similarity is consistent with the hypothesis that the peptide backbone is the primary determinant in dictating the utilization and nature of N-linked carbohydrate sites (220).

F. Summary of the Synthesis and Processing of Factor VIII

Our current understanding of the steps in the biosynthetic pathway for factor VIII are summarized in Fig. 7. Factor VIII is contrnslationally translocated into the lumen of the ER during which a single peptide of 19 amino acids is cleaved. Since the factor VIII that appears in the conditioned medium has the same amino terminus as that derived from plasma (221–223), the signal cleavage site in CHO cells is probably the same as that which is utilized in vivo. The addition of high-mannose oligosaccharide units occurs in the

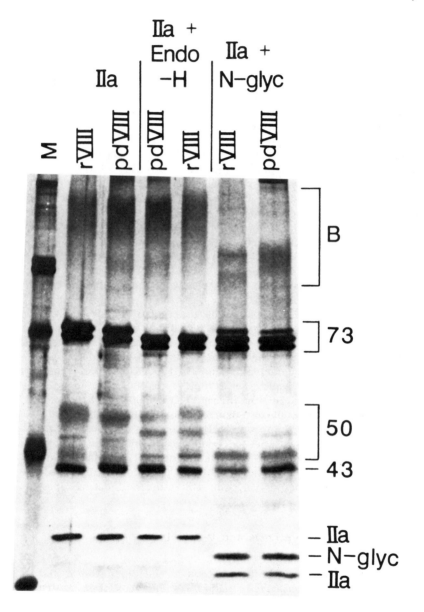

Figure 6   Endoglycosidase sensitivity of CHO-derived and plasma-derived factor VIII. Purified factor VIII from conditioned medium of CHO cells which express factor VIII (rVIII) and from normal human pooled plasma (pdVIII) was treated with thrombin and subsequently digested with endoglycosidase H (Endo-H) or N-glycanase

ER. A significant proportion of the factor VIII synthesized in CHO cells is apparently never secreted but rather is retained in a complex with BiP within the ER (196). The secretion-competent factor VIII transits to the Golgi apparatus where the majority of the N-linked glycosylation sites on the heavy chain are modified to hybrid- or complex-type structures. The N-linked carbohydrate at the two sites within the light chain remain of the high mannose type. Addition of O-linked sugars occurs on the heavy and light chains (R. Steinbrink, Genetics Institute, Cambridge, MA, unpublished observations). Sulfation occurs at tyrosine residues within acidic regions in the heavy chain between domains A1 and A2 and in the amino terminal-derived thrombin cleavage fragment of the light chain (224). Also in the Golgi, factor VIII is cleaved to generate the heavy and light chains. In the presence of vWF, the heavy and light chains assemble into a stable complex joined by a metal ion bridge. In the absence of vWF, the heavy and light chains are secreted unassociated or are rapidly dissociated. Unassociated factor VIII heavy chain is subsequently degraded.

The ability to engineer CHO cells to synthesize factor VIII demonstrates the capability of this host cell to synthesize and properly process a very complex glycoprotein. Factor VIII expression is very low compared to other proteins studied in this host. We have identified several reasons for inefficient factor VIII expression. First, the factor VIII mRNA is poorly expressed. The development of more efficient expression vectors should overcome this limitation. Second, the factor VIII secreted is rapidly degraded in the absence of vWF. Cells that coexpress factor VIII and vWF are able to accumulate factor VIII in serum-free medium. Finally, a proportion of the synthesized factor VIII is never transported out of the ER and is observed in a complex with BiP. We recently derived CHO cells which have reduced levels of BiP. These cells with reduced BiP

(N-glyc). Samples were analyzed by reducing SDS-PAGE and subsequently stained using the silver nitrate protocol. Indicated are the specific thrombin cleavage products of factor VIII: B, the heavily glycosylated B domain; 73, the light-chain doublet; 43, the 43-kDa fragment from the heavy chain; 50, the amino terminal 50-kDa fragment from the heavy chain. Also indicated is the location of thrombin before and after N-glyc digestion (IIa).

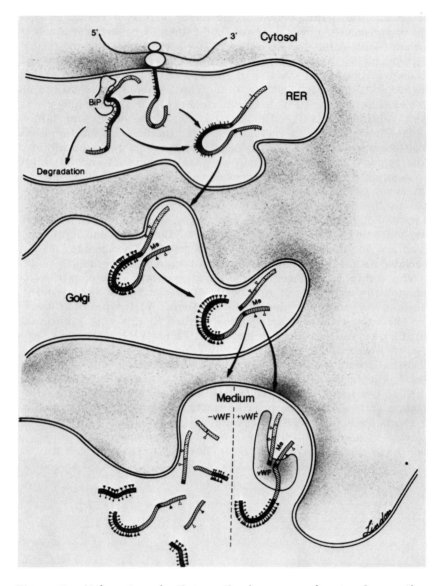

Figure 7 Major steps in the synthesis, processing, and secretion
of factor VIII in mammalian cells. The factor VIII primary translation product is translocated into the lumen of the endoplasmic
reticulum (ER) where N-linked glycosylation occurs (I). A
majority of factor VIII binds tightly to BiP and is probably destined
for destruction. A proportion of the molecules attains proper conformation for transport into the Golgi where complex carbohydrate

exhibit improved section of proteins which avidly bind to BiP
(226). These experiments suggest that proper engineering should
improve the secretion potential of CHO cells.

VIII. SUMMARY

The recent advances in molecular biology have merged with somatic
cell genetics and cell biology to allow mammalian cells to be extremely useful for the expression of foreign genes. This chapter has
focused primarily on the approaches and potential limitations to high-
level expression of proteins in mammalian cells. Future developments will involve the modification of mammalian cells in order to increase the efficiency of the various steps in protein processing and
secretion. The ability to genetically engineer mammalian cells to
produce high levels of desired proteins is presently complemented
by advances in biochemical engineering which involve the ability to
grow mammalian cells in very large volumes or at very high densities with reduced serum requirements. As a result, the cost for
production of gram quantities of a protein from a mammalian host
cell are approaching the cost of proteins similarly derived from
microbial systems with all the advantages that mammalian systems
afford.

ACKNOWLEDGMENTS

I thank Louise Wasley, Monique Davies, Debra Pittman, Patricia
Murtha, David Israel, and Andrew Dorner for their experiments,
which were important in formulating many of the ideas presented.
I also thank Andrew Dorner for reading this chapter critically and
Michelle Wright for assisting in its preparation.

---

modification on N-linked sites (▼), addition of carbohydrate to
serine and theonine residues (●), and sulfation of several tyrosine
residues in highly acidic regions of factor VIII occur. In the
Golgi the protein is cleaved to its mature heavy and light chains.
The presence of vWF in the medium promotes the secretion of heavy
and light chains in a stable metal ion complex (Me). In the absence of vWF, the individual chains are secreted and subsequently
degraded. The A domains (cross-hatched), B domain (solid black),
and the C domains (dotted) are indicated. The two acidic regions
adjacent to the A domains are also indicated (solid black).

REFERENCES

1. Y. Gluzman (ed.). *Eukaryotic Viral Vectors*, Cold Spring Harbor Lab., New York (in press).
2. Graham, F. L. and van der Eb, A. J. A new technique for the assay of infectivity of human adenovirus 5 DNA. *Virology* 52:456−467 (1973).
3. Chen, C. and Okayama, H. High-efficiency transformation of mammalian cells by plasmid DNA. *Mol. Cell. Biol.* 7:2745−2752 (1987).
4. Shen, Y., Hirschhorn, R. R., Mercer, W. E., Surmacz, E., Tsutsui, Y., Soprano, K. J., and Baserga, R. Gene transfer: DNA microinjection compared with DNA transfection with a very high efficiency. *Mol. Cell. Biol.* 2:1145−1154 (1982).
5. Frost, E. and William, J. Mapping temperature-sensitive and host-range mutations of advenovirus type 5 by marker rescue. *Virology* 91:39−50 (1978).
6. Gorman, C., Padmanabhan, R., and Howard, B. H. High efficiency DNA-mediated transformation of primate cell. *Science* 221:551−553 (1983).
7. Stow, N. D. and Wilkie, N. M. An improved techniqge for obtaining enhanced infectivity with herpes simplex virus type 1 DNA. *J. Gen. Virol.* 33:447−458 (1976).
8. Gorman, C. M. and Howard, B. H. Expression of recombinant plasmids in mammalian cells is enhanced by sodium butyrate. *Nucl. Acids Res.* 11:7631−7648 (1983).
9. Luthman, H. and Magnusson, G. High efficiency polyoma DNA transfection of chloroquine treated cells. *Nucl. Acids Res.* 11:1295−1308 (1983).
10. McCutchan, J. H. and Pagano, J. S. Enhancement of the infectivity of simian virus 40 deoxyribonucleic acid with diethyl-aminoethyl-dextran. *J. Natl. Canc. Inst.* 41:351−357 (1968).
11. Sompayrac, L. M. and Dana, K. J. Efficient infection of monkey cells with DNA of simian virus 40. *Proc. Natl. Acad. Sci. USA* 78:7575−7578 (1981).
12. Schaffner, W. Direct transfer of cloned genes from bacteria to mammalian cells. *Proc. Natl. Acad. Sci. USA* 77:2163−2167 (1980).
13. de Saint Vincent, R. B., Delbruck, S., Eckhart, W., Meinkoth, J., Vitto, L., and Wahl, G. The cloning and re-introduction into animal cells of a functional CAD gene, a dominant amplifiable genetic marker. *Cell* 27:267−277 (1981).
14. Felgner, P. L., Gadek, T. R., Holm, M., Roman, R., Chan, H. W., Wenz, M., Northrop, J. P., Ringold, G. M., and Danielsen, M. Lipofection: A highly efficient, lipid-mediated DNA-transfection procedure. *Proc. Natl. Acad. Sci. USA* 84:7413−7417 (1987).

15. Capecchi, M. R. High efficiency transformation by direct microinjection of DNA into cultured mammalian cells. *Cell* 22: 479−488 (1980).
16. Potter, H., Weir, L., and Leder, P. Enhancer-dependent expression of human λ-immunoglobulin genes introduced into mouse pre-B lymphocytes by electroporation. *Proc. Natl. Acad. Sci. USA* 81:7161−7165 (1984).
17. Fromm, M., Taylor, L. P., and Walbot, V. Expression of genes transferred into monocot and dicot plant cell by electroporation. *Proc. Natl. Acad. Sci. USA* 82:5824−5828 (1985).
18. Ecker, J. R. and Davis, R. W. Inhibition of gene expression in plant cells by expression of antisense RNA. *Proc. Natl. Acad. Sci. USA* 83:5372−5376 (1986).
19. Wigler, M., Sweet, R., Sim, G. K., Wold, B., Pellicer, A., Lacy, E., Maniatis, T., Silverstein, S., and Axel, R. Transformation of mammalian cells with genes from procaryotes and eucaryotes. *Cell* 16:777−785 (1979).
20. Thomas, K. R. and Capecchi, M. R. Site-directed mutagenesis by gene targeting in mouse embryo-derived stem cell. *Cell* 51:503−512 (1987).
21. Littlefield, J. W. The use of drug-resistant markers to study the hybridization of mouse fibroblasts. *Exp. Cell Res.* 41: 190−196 (1966).
22. Wigler, M., Silverstein, S., Lee, L. S., Pellicer, A., Cheng, Y., and Axel, R. Transfer of purified herpes virus thymidine kinase gene to cultured mouse cells. *Cell* 11:223−232 (1977).
23. Roberts, J. M. and Axel, R. Gene amplification and gene correction in somatic cells. *Cell* 29:109−119 (1982).
24. Mercola, K. E., Stang, H. D., Browne, J., Salser, W., and Cline, M. J. Insertion of a new gene of viral origin into bone marrow cells of mixe. *Science* 208:1033−1035 (1980).
25. Mulligan, R. C. and Berg, P. Selection for animal cells that express the *Escherichia coli* gene coding for xanthine-guanine phosphoribosyl transferase. *Proc. Natl. Acad. Sci. USA* 78: 2072−2076 (1981).
26. Chapman, A. B., Costello, M. A., Lee, F., and Ringold, G. M. Amplification and hormone-regulated expression of a mouse mammary tumor virus-eco gpt fusion plasmid in mouse 3T6 cells. *Mol. Cell. Biol.* 3:1421−1429 (1983).
27. Urlaub, G. and Chasin, L. A. Isolation of Chinese hamster ovary cell mutants deficient in dihydrofolate reductase activity. *Proc. Natl. Acad. Sci. USA* 77:4216−4220 (1980).
28. Alt, F. W., Kellems, R. E., Bertino, J. R., and Schimke, R. T. Selective multiplication of dihydrofolate reductase genes in methotrexate-resistant variants of cultured murine cells. *J. Biol. Chem.* 253:1357−1370 (1978).

29. Schimke, R. T. (ed.). *Gene Amplification*, Cold Spring Harbor, New York (1982).
30. Schmike, R. T. Gene amplification in cultured cells. *J. Biol. Chem.* 263:5989–5992 (1988).
31. Stark, G. R. and Wahl, G. M. Gene amplification. *Ann. Rev. Biochem.* 53:447–491 (1984).
32. Murray, M. J., Kaufman, R. J., Latt, S. A., and Weinberg, R. A. Construction and use of a dominant, selectable marker: A Harvey sarcoma virus-dihydrofolate reductase chimera. *Mol. Cell. Biol.* 3:32–43 (1983).
33. Kim, S. K. and Wold, B. J. Stable reduction of thymidine kinase activity in cells expressing high levels of anti-sense RNA. *Cell* 42:129–138 (1985).
34. Simonsen, C. C. and Levinson, A. D. Isolation and expression of an altered mouse dihydrofolate reductase cDNA. *Proc. Natl. Acad. Sci. USA* 80:2495–2499 (1983).
35. Kemp, T. D., Swryrd, E. A., Bruist, M., and Stark, G. R. Stable mutants of mammalian cells that overproduce the first three enzymes of pyrimidine nucleotide biosynthesis. *Cell* 9:541–550 (1976).
36. Wahl, G. M., de Saint Vincent, B. R., and DeRose, M. L. Effect of chromosomal position on amplification of transfected genes in animal cells. *Nature* 307:516–520 (1984).
37. Patterson, D. and Carnright, D. V. *Somatic Cell. Genet.* 3:483–495 (1977).
38. de Saint Vincent, B. R. Delbruck, S., Eckhaert, W., Meinkoth, J., Vitto, L., and Wahl, G. The cloning and reintroduction into animal cells of a functional CAD gene, a dominant amplifiable genetic marker. *Cell* 27:267–277 (1981).
39. Ruiz, J. C. and Wahl, G. M. *Escherichia coli* aspartate transcarbamylase: A novel marker for studies of gene amplification and expression in mammalian cells. *Mol. Cell. Biol.* 6:3050–3058 (1986).
40. Kaufman, R. J., Murtha, P., Ingolia, D. E., Yeung, C.-Y., and Kellems, R. E. Selection and amplification of heterologous genes encoding adenosine deaminase in mammalian cells. *Proc. Natl. Acad. Sci. USA* 83:3136–3140 (1986).
41. Yeung, C.-Y., Ingolia, D. E., Bobonis, C., Dunbar, B. S., Riser, M. E., Siciliano, J. J., and Kellems, R. E. *J. Biol. Chem.* 258:8338–8345 (1983).
42. Yeung, G.-Y., Riser, M. E., Kellems, R. E., and Siciliano, M. J. Increased expression of one of two adenosine deaminase alleles in a human choriocarcinoma cell line following selection with adenine nucleosides. *J. Biol. Chem.* 258:8330–8337 (1983).

43. Fernandez-Mejia, C., Debatisse, M., and Buttin, G. Adenosine-resistant Chinese hamster fibroblast variants with hyperactive adenosine-deaminase: An analysis of the protection against exogenous adenosine afforded by increased activity of the deamination pathway. *J. Cell. Physiol.* 20:321–328 (1984).

44. Hoffee, P. A., Hunt, S. W., III, and Chiang, J. (1982). Isolation of deoxycoformycin-resistant cells with increased levels of adenosine deaminase. *Somatic Cell. Genet.* 8:465–477 (1982).

45. Ingolia, D. E., Yeung, C.-Y., Orengo, I. F., Harrison, M. L., Frayne, E. G., Rudolph, F. B., and Kellems, R. E. Purification and characterization of adenosine deaminase from a genetically enriched mouse cell line. *J. Biol. Chem.* 260: 13261–13267 (1985).

46. Riordan, J. R., Deuchars, K., Kartner, N., Alon, N., Trent, J., and Ling, V. Amplification of P-glycoprotein genes in multidrug-resistant mammalian cell lines. *Nature* 316:817–819 (1985).

47. Ames, G. F.-L. The basis of mutidrug resistance in mammalian cells: Homology with bacterial transport. *Cell* 47:323–324 (1986).

48. Gros, P., Croop, J., and Housman, D. Mammalian multidrug resistance gene: Complete cDNA sequence indicates strong homology to bacterial transport proteins. *Cell* 47:371–380 (1986).

49. Chen, G.-J., Chin, J. E., Ueda, K., Clark, D. P., Pastan, I., Gottesman, M. M., and Roninson, I. B. Internal duplication and homology with bacterial transport proteins in the mdrl (P-glycoprotein) gene from multidrug-resistant human cells. *Cell* 47:381–389 (1986).

50. Cros, P., Neriah, Y. B., Croop, J. M., and Housman, D. E. Isolation and expression of a complementary DNA that confers multidrug resistance. *Nature* 323:728–731 (1986).

51. Tabor, C. W. and Tabor, H. Polyamines. *Ann. Rev. Biochem.* 53:749–790 (1984).

52. Chiang, T.-R. and McConlogue, L. Amplification and expression of heterologous ornithine decarboxylase in Chinese hamster cells. *Mol. Cell. Biol.* 8:764–769 (1988).

53. Steglich, C., Grens, A., and Scheffler, I. E. Chinese hamster cells dificient in ornithine decarboxylase activity: Reversion by gene amplification and by azacytidine treatment. *Sum. Cell. Mol. Genet.* 11:11–23 (1985).

54. Andrulis, I., Chen, J., and Ray, P. Isolation of human cDNAs for asparagine synthetase and expression in jensen rat sarcoma cells. *Mol. Cell. Biol.* 7:2435–2443 (1987).

55. Cartier, M., Chang, M., and Stanners, C. Use of the escherichia coli gene for asparagine synthetase as a selective marker in a shuttle vector capable of dominant transfection and amplification in animal cells. *Mol. Cell. Biol.* 7:1623–1628 (1987).

56. Kavathas, P. and Herzenberg, L. A. Stable transformation of mouse L cells for human membrane T-cell differentiation antigens, HLA and β_2-microglobulin: Selection by fluorescence-activated cell sorting. *Proc. Natl. Acad. Sci. USA* 80:524–528 (1983).

57. McClelland, A., Kuhn, L. C., and Ruddle, F. H. The human transferrin receptor gene: Genomic organization, and the complete primary structure of the receptor deduced from a cDNA sequence. *Cell* 39:267–274 (1984).

58. Kavathas, P. and Herzenberg, L. A. Amplification of a gene coding for human T-cell differentiation antigen. *Nature* 306:385–387 (1983).

59. Kaufman, R. J., Bertino, J. R., and Schimke, R. T. Quantitation of ihydrofolate reductase in individual parental and methotrexate-resistant murine cells. *J. Biol. Chem.* 253:5852–5860 (1978).

60. Miller, A. G. and Whitlock, J. P., Jr. Heterogeneity in the rate of benzol[a] pyrene metabolism in single cells: Quantitation using flow cytometry. *Mol. Cell. Biol.* 2:625–632 (1982).

61. Kaufman, R. J., Brown, P. C., and Schimke, R. T. Amplified dihydrofolate reductase genes in unstably methotrexate-resistant cells are associated with double minute chromosomes. *Proc. Natl. Acad. Sci. USA* 76:5669–5673 (1979).

62. Kaufman, R. J. and Schimke, R. T. Amplification and loss of dihydrofolate reductase genes in a Chinese hamster ovary cell line. *Mol. Cell. Biol.* 1:1069–1076 (1981).

63. Kaufman, R. J., Brown, P. C., and Schimke, R. T. Loss and stabilization of amplified dihydrofolate reductase genes in mouse sarcoma S-180 cell lines. *Mol. Cell. Biol.* 1:1084–1093 (1981).

64. Nunberg, J. H., Kaufman, R. J., Schimke, R. T., Urlaub, Urlaub, G., and Chasin, L. A. Amplified dihydrofolate reductase genes are localized to a homogeneously staining region of a single chromosome in a methotrexate-resistant Chinese hamster ovary cell line. *Proc. Natl. Acad. Sci. USA* 75:5553–5556 (1978).

65. Brown, P. C., Beverly, S. M., and Schimke, R. T. Evolution of chromosomal regions containing transfected and amplified dihydrofolate reductase sequences. *Mol. Cell. Biol.* 1:1077–1083 (1981).

66. Kaufman, R. J., Sharp, P. A., and Latt, S. A. Evolution of chromosomal regions containing transfected and amplified dihydrofolate reductase sequences. *Mol. Cell. Biol.* 3:699−711 (1983).

67. Carroll, S., DeRose, M., Gaudray, P., Moore, C., Needham-Vandevanter, D., Van Hoff, D., and Wahl, G. Double minute chromosomes can be produced from precursors derived from a chromosomal deletion. *Mol. Cell. Biol.* 8: 1525−1533 (1988).

68. Carroll, S., Gaudray, P., DeRose, M., Emery, J., Meinkoth, J., Nakkim, E., Subler, M., Von Hoff, D., and Wahl, G. Characterization of an episome produced in hamster cells that amplify a transfected CAD gene at high frequency: Functional evidence for a mammalian replication origin. *Mol. Cell. Biol.* 7:1740−1750 (1987).

69. Kaufman, R. J., Wasley, L. C., Spiliotes, A. T., Gossels, S. D., Latt, S. A., Larsen, G. R., and Kay, R. M. Co-amplification and coexpression of human tissue-type plasminogen activator and murine dihydrofolate reductase sequences in Chinese hamster ovary cells. *Mol. Cell. Biol.* 5:1730−1759 (1985).

70. McClintock, B. Zea mays. *Genetics* 26:234−282 (1941).

71. Cowell, J. K. and Miller, O. J. Occurrence and evolution of homogeneously staining regions may be due to breakage-fusion-bridge cycles following telomere loss. *Chromosoma* 88: 216−221 (1983).

72. Robins, D. M., Axel, R., and Henderson, A. S. Chromosome structure and DNA sequence alterations associated with mutation of transformed genes. *J. Mol. Appl. Genet.* 1:191−203 (1981).

73. Fendrock, B., Destremps, M., Kaufman, R. J., and Latt, S. A. Cytological, flow cytometric, and molecular analysis of the rapid evolution of mammalian chromosomes containing highly amplified DNA sequences. *Histochemistry* 84:121−130 (1986).

74. Andrulis, J. L. and Siminovitch, L. *Gene Amplication* (R. T. Schimke, ed.), Cold Spring Harbor Laboratory, New York, pp. 75−78 (1982).

75. Kaufman, R. J. and Sharp, P. A. Amplification and expression of sequences contransfected with a modular dihydrofolate reductase complementary DNA gene. *J. Mol. Biol.* 159:601−621 (1982).

76. Roberts, J. M., Buck, L. B., and Axel, R. A structure for amplified DNA. *Cell.* 33:53−63 (1983).

77. Haynes, J. and Weissman, C. Constitutive, long-term production of human interferons by hamster cells containing

multiple copies of a cloned interferon gene. *Nucl. Acids Res.*
11:1687–706 (1983).

78. Schahill, S. J., Devos, R., Heyden, J. V., and Fiers, W.
Expression and characterization of the product of a human
immune interferon cDNA gene in Chinese hamster ovary cells.
Proc. Natl. Acad. Sci. USA 80:4654–4658 (1983).

79. McCormack, F., Trahey, M., Innis, M., Dieckmann, B., and
Ringold, G. Inducible expression of amplified human beta
interferon genes in CHO cells. *Mol. Cell. Biol.* 4:166–172
(1984).

80. Kaufman, R. J., Wasley, L. C., Furie, B. C., Furie, B., and
Schoemaker, C. Expression, purification, and characterization
of recombinant γ-carboxylated factor IX synthesized in Chinese
hamster ovary cells. *J. Biol. Chem.* 261:9622–9628 (1986).

81. Kaufman, R. J., Wasley, L. C., and Dorner, A. J. Synthesis,
processing, and secretion of recombinant human factor VIII
expressed in mammalian cells. *J. Biol. Chem.* 263:6352–6362
(1988).

82. Berman, P. W., Dowbenko, D., Simonsen, C. C., and Laskey,
L. A. Detection of antibodies to herpes simplex virus with a
continuous cell line expressing cloned glycoprotein D. *Science*
222:524–527 (1983).

83. Kaetzel, D. M., Browne, J. K., Wondisford, F., Nett, T. M.,
Thomason, A. R., and Nilson, J. H. *Proc. Natl. Acad. Sci.
USA* 82:7280–7283 (1985).

84. McKnight, S. and Tjian, R. Transcriptional selectivity of viral
genes in mammalian cells. *Cell* 46:795–805 (1986).

85. Breathnach, R. and Chambon, P. Organization and expression
of eucaryotic split genes coding for proteins. *Ann. Rev.
Biochem.* 50:349–383 (1981).

86. Benoist, C., O'Hare, K., Breathnach, R., and Chambon, P.
The ovalbumin gene-sequence of putative control regions.
Nucl. Acids Res. 8:127–142 (1980).

87. Serfling, E., Jasin, M., and Schaffner, W. Enhancers and
eukaryotic gene transcription. *Trends Genet.* 1:224–230
(1985).

88. deVilliers, J., Olson, L., Banerji, J., and Schaffner, W.
Analysis of the transcriptional enhancer effect. *Cold Spring
Harbor Symp. Quant. Biol.* 47:911–919 (1983).

89. Laimins, L. A., Khoury, G., Gorman, C., Howard, B., and
Gruss, P. Host-specific activation of transcription by tandem
repeats from simian virus 40 and Moloney murine sarcoma virus.
Proc. Natl. Acad. Sci. USA 79:6453–6457 (1982).

90. Neuhaus, G., Neuhaus-Uri, G., Gruss, P., and Schweiger,
H.-G. Enhancer-controlled expression of the simian virus 40

T-antigen in the green alga *Acetabularia*. *EMBO J.* 3: 2169–2172 (1984).

91. Gorman, C. M., Merlino, G. D., Willingham, M. C., Pastan, I., and Howard, B. H. The Rous sarcoma virus long terminal repeat is a strong promoter when introduced into a variety of eukaryotic cells by DNA-mediated transfection. *Proc. Natl. Acad. Sci. USA* 79:6777–6781 (1982).

92. Boshart, M., Weber, F., Gerhard, J., Dorsch-Hasler, K., Fleckenstein, B., and Schaffner, W. A very strong enhancer is located upstream of an immediate early gene of human cyto-megalovirus. *Cell* 41:521–530 (1985).

93. Banerji, J., Olson, L., and Schaffner, W. A lymphocyte-specific cellular enhancer is located downstream of the joining region in immunoglobulin heavy chain genes. *Cell* 33:729–740 (1983).

94. Gillies, S. D., Morrison, S. L., Oi, V. T., and Tonegawa, S. A tissue-specific transcription enhancer element is located in the major intron of a rearranged immunoglobulin heavy chain gene. *Cell* 38:717–728 (1983).

95. Queen, C. and Baltimore, D. Immunoglobulin gene transcription is activated by downstream sequence elements. *Cell* 33: 741–748 (1983).

96. Ohlsson, H. and Edlund, T. Sequence-specific interactions of nuclear factors with the insulin gene enhancer. *Cell* 45: 35–44 (1986).

97. Lengyel, P. Biochemistry of interferons and their actions. *Ann. Rev. Biochem.* 51:251–282 (1982).

98. Goodbourn, S., Burnstein, H., and Maniatis, T. The human β-interferon gene enhancer is under negative control. *Cell* 45: 601–610 (1986).

99. Enoch, T., Zinn, K., and Maniatis, T. Activation of the human β-interferon gene requires an interferon-inducible factor. *Mol. Cell. Biol.* 6:801–810 (1986).

100. Pelham, H. R. B. Activation of heat-shock genes in eukaryotes. *Trends Genet.* 1:31–35 (1985).

101. Pelham, H. R. B. and Bienz, M. A synthetic heat-shock pro-moter elements confers heat-inducibility on the herpes simplex virus thymidine kinase gene. *EMBO J.* 1:1473–1477 (1982).

102. Topol, J., Ruden, D. M., and Parker, C. S. Sequences required for in vitro transcriptional activation of a *Drosophila* hsp 70 gene. *Cell* 42:527–537 (1985).

103. Wu, C. Activating protein factor binds in vitro to upstream control sequences in heat shock gene chromatin. *Nature* 311: 81–84 (1984).

104. Parker, C. S. and Topol, J. A *Drosophila* RNA polymerase II transcription factor binds to the regulatory site of an hsp 70 gene. *Cell* 37:273–283 (1984).

105. Wurm, F. M., Gwinn, K. A., and Kingston, R. E. Inducible overproduction of the mouse c-myc protein in mammalian cells. *Proc. Natl. Acad. Sci. USA* 83:5414–5418 (1986).

106. Schlesinger, M. J. Heat shock proteins: The search for functions. *J. Cell. Biol.* 103:321–325 (1986).

107. Hunziker, P. E. and Kaegi, J. H. R. Metalloproteins. *Top. Mol. Struct. Biol.* 149–181 (1985).

108. Durnam, D. M. and Palmiter, R. D. Transcriptional regulation of the mouse metallothionein-I gene by heavy metals. *J. Biol. Chem.* 256:5712–5716 (1981).

109. Hager, L. J. and Palmiter, R. D. Transcriptional regulation of mouse liver metallothionein-I gene by glucocorticoids. *Nature* 291:340–342 (1981).

110. Friedman, R. L., Manly, S. P., McMahon, M., Kerr, I. M., and Stark, G. R. Transcriptional and posttranscriptional regulation of interferon-induced gene expression in human cells. *Cell* 38:745–755 (1984).

111. Serfling, E., Lubbe, A., Dorsch-Hasler, K., and Schaffner, W. Metal-dependent SV40 viruses containing inducible enhancers from the upstream region of metallothionein genes. *EMBO J.* 4:3851–3859 (1985).

112. Karin, M., Haslinger, A., Holtgreve, H., Cathala, G., Slater, E., and Baxter, J. D. Activation of a heterologous promoter in response to dexamethasone and cadmium by metallothionein gene 5'-flaning DNA. *Cell* 36:371–379 (1984).

113. Searle, P. F., Davison, B. L., Stuart, G. W., Wilkie, T. M., Norstedt, G., and Palmiter, R. D. Regulation, linkage, and sequence of mouse metallothionein I and II genes. *Mol. Cell. Biol.* 4:1221–1230 (1984).

114. Seguin, C., Felber, B. K., Carter, A. D., and Hamer, D. H. Competition for cellular factors that activate metallothionein gene transcription. *Nature* 312:781–785 (1984).

115. Searle, P. F., Stuart, G. W., and Palmiter, R. D. Building a metal-responsive promoter with synthetic regulatory elements. *Mol. Cell. Biol.* 5:1480–1489 (1985).

116. Pavlakis, G. N. and Hamer, D. H. Regulation of a metallothionein-growth hormone hybrid gene in bovine papilloma virus. *Proc. Natl. Acad. Sci. USA* 80:397–401 (1983).

117. Ringold, G. M. Steroid hormone regulation of gene expression. *Ann. Rev. Pharmacol. Toxicol.* 25:529–566 (1985).

118. Chandler, V. L., Maler, B. A., and Yamamoto, K. R. DNA sequences bound specifically by glucocorticoid receptor in vitro render A heterologous promoter hormone responsive in vivo. *Cell* 33:489–499 (1983).

119. Hynes, N., van Ooyen, A. J. J., Kennedy, N., Herrlich, P., Ponta, H., and Groner, B. Subfragments of the large terminal repeat cause glucocorticoid-responsive expression of mouse mammary tumor virus and of an adjacent gene. *Proc. Natl. Acad. Sci. USA* 80:3637–3641 (1983).
120. Major, J. and Varmus, H. E. A small region of the mouse mammary tumor virus long terminal repeat confers glucorticoid hormone regulation on a linked heterologous gene. *Proc. Natl. Acad. Sci. USA* 80:3637–3641 (1983).
121. Karin, M., Haslinger, A., Holtgreve, H., Richards, R. I., Krauter, P., Westphal, H. M., and Beato, M. Characterization of DNA sequences through which cadmium and glucocorticoid hormones induce human metallothionein-IIA gene. *Nature* 308:513–519 (1984).
122. Payvar, F., DeFranco, D., Firestone, G. L., Edgar, B., Wrange, O., Okret, S., Gustaffson, J. A., and Yamamoto, K. R. Sequence-specific binding of glucocorticoid receptor to MTV DNA at sites within the upstream of the transcribed region. *Cell* 35:381–392 (1983).
123. Govindan, M., Spiess, E., and Majors, J. Purified glucocorticoid receptor-hormone complex from rat liver cytosol bind specifically to cloned mouse mammary tumor virus long terminal repeats in vitro. *Proc. Natl. Acad. Sci. USA* 79:5157–5161 (1982).
124. Scheidereit, C. and Beato, M. Contacts between hormone receptor and DNA double helix within a glucocorticoid regulatory element of mouse mammary tumor virus. *Proc. Natl. Acad. Sci. USA* 81:3029–3033 (1984).
125. Lee, F., Mulligan, R., Berg, P., and Ringold, G. Glucocorticoids regulate expression of dihydrofolate reductase cDNA in mouse mammary tumour virus chimaeric plasmids. *Nature* 294:228–232 (1981).
126. Chapman, A. B., Costello, M. A., Lee, F., and Ringold, G. M. Amplification and hormone-regulated expression of a mouse mammary tumor virus-eco gpt fusion plasmid in mouse 3T6 cells. *Mol. Cell. Biol.* 3:1421–1429 (1983).
127. Ringold, G. M., Lasfargues, E. Y., Bishop, J. M., and Varmus, H. E. Production of mouse mammary tumor virus by cultured cells in the absence and presence of hormones: Assay by molecular hybridization. *Virology* 65:135–147 (1975).
128. Klessig, D. F., Brough, D., and Cleghon, V. Introduction, stable integration, and controlled expression of a chimeric adenovirus gene whose product is toxic to the recipient human cell. *Mol. Cell. Biol.* 4:1354–1362 (1984).
129. Miesfeld, R., Rusconi, S., Godowski, P. J., Maler, B. A., Okret, S., Wilkstrom, A.-C., Gustafsson, J.-A., and

Yamamoto, K. R. Genetic complementation of a glucocorticoid receptor deficiency by expression of cloned receptor cDNA. *Cell* 46:389–399 (1986).

130. Giguere, V., Hollenberg, S. M., Rosenfeld, M. G., and Evans, R. M. Functional domains of the human glucocorticoid receptor. *Cell* 46:645–652 (1986).

131. Hollenberg, S. M., Weinberger, C., Ong, E. S., Cerelli, G., Oro, A., Lebo, R., Thompson, E. B., Rosenfeld, M. G., and Evans, R. M. Primary structure and expression of a functional human glucocorticoid receptor cDNA. *Nature* 318: 635–641 (1985).

132. Walter, P., Green, S., Greene, G., Krust, A., Bornert, J., Jeltsch, J. M., Staub, A., Jensen, E., Scrace, G., Waterfield, M., and Chambon, P. Cloning of the human estrogen receptor cDNA. *Proc. Natl. Acad. Sci. USA* 82: 7889–7893 (1985).

133. Jeltsch, J. M., Krozowski, Z., Quirin-Stricker, C., Gronemeyer, H., Simpson, R. J., Garnier, J. M., Krust, A., Jacob, F., and Chambon, P. Cloning of the chicken progesterone receptor. *Proc. Natl. Acad. Sci. USA* 83:5424–5428 (1986).

134. Webster, N., Jin, J.-R., Green, S., Hollis, M., and Chambon, P. The yeast UAS$_G$ is a transcriptional enhancer in human HeLa cells in the presence of the GAL4 transactivator. *Cell* 52:169–178 (1988).

135. Leach, K., Anderson, K., and Brent, R. DNA-bound Fos proteins activate transcription in yeast. *Cell* 52:179–184 (1988).

136. Kakidani, H. and Ptashne, M. GAL4 activates gene expression in mammalian cells. *Cell* 52:161–167 (1988).

137. Sharp, P. A. Slicing of messenger RNA precursors. *Science* 235:766–771.

138. Buchman, A., Ph.D. thesis, Stanford University (1982).

139. Gruss, B. and Khoury, G. *Nature* 286:634–637 (1980).

140. Miller, J. H. and Calos, M. P. *Current Communications in Molecular Biology*, Cold Spring Harbor Laboratory, New York (1987).

141. Brinster, R., Allen, J., Behringer, R., Gelinas, R., and Palmiter, R. Introns increase transcriptional efficiency in transgenic mice. *Proc. Natl. Acad. Sci. USA* 85:836–840 (1988).

142. Birnstiel, M. L., Busslinger, M., and Strub, K. Transcription termination and 3′ processing: The end is in site! *Cell* 41:349–359 (1985).

143. Frayne, E. G., Leys, E. J., Crouse, G. F., Hook, A. G., and Kellems, R. E. Transcription of the mouse dihydrofolate reductase gene proceeds unabated through seven polyadenyla-

tion sites and terminates near a region of repeated DNA. *Mol. Cell. Biol.* 4:2921–2924 (1984).

144. Falck-Pederson, E., Logan, J., Shenk, T., and Darnell, J. E., Jr. Transcription termination within the E1A gene of adenovirus induced by insertion of the mouse β-major globin terminator element. *Cell* 40:897–905 (1985).

145. Birchmeier, C., Schumperli, D., Sconzo, G., and Birnstiel, M. L. 3' editing of mRNAs: Sequence requirements and involvement of a 60-nucleotide RNA in maturation of histone mRNA precursors. *Proc. Natl. Acad. Sci. USA* 81:1057–1061 (1984).

146. Frayne, E. G. and Kellems, R. E. Structural features of the murine dihydrofolate reductase transcription termination region: Identification of a conserved DNA sequence element. *Nucleic Acids Res.* 14:4113–4125 (1986).

147. Cullen, B. R., Lomedico, P. T., and Ju, G. Transcriptional interference in avian retroviruses — Implications for the promoter insertion model of leukaemogenesis. *Nature* 307:241–245 (1984).

148. Proudfoot, N. J. Transcriptional interference and termination between duplicated α-globin gene constructs suggest a novel mechanism for gene regulation. *Nature* 322:562–565 (1986).

149. Izant, J. G. and Weintraub, H. Constitutive and conditional suppression of exogenous and endogenous genes by antisense RNA. *Science* 229:345–352 (1985).

150. Melton, D. A. Accurate cleavage and polyadenylation of exogenous RNA substrate. *Proc. Natl. Acad. Sci. USA* 82:144–148 (1985).

151. Moore, C. L. and Sharp, P. A. Accurate cleavage and polyadenylation of exogenous RNA substrate. *Cell* 41:845–855 (1985).

152. Kaufman, R. J. and Sharp, P. A. Construction of a modular dihydrofolate reductase cDNA gene: Analysis of signal utilized for efficient expression. *Mol. Cell. Biol.* 2:1304–1319 (1982).

153. Montell, C., Fisher, E. F., Carauthers, M. H., and Berk, A. J. Inhibition of RNA cleavage but not polyadenylation by a point mutation in mRNA 3' consensus sequence AAUAAA. *Nature* 305:600–605 (1983).

154. Higgs, D. R., Goodbourn, S. E. Y., Lamb, J., Clegg, J. B., Weatherall, D. J., and Proudfoot, N. J. Alpha-thalassaemia caused by a polyadenylation signal mutation. *Nature* 306:398–400 (1983).

155. Fitzgerald, M. and Schenk, T. The sequence 5'-AAAUAA-3' forms part of the recognition site for polyadenylation of late SV40 mRNAs. *Cell* 24:251–260 (1981).

156. Wickens, M. and Stephenson, P. Role of the conserved AAUAAA sequence: Four AAUAAA point mutants prevent messenger RNA 3' end formation. *Science* 226:1045−1051 (1984).

157. McDevitt, M. A., Imperiale, M. J., Ali, H., and Nevins, J. R. Requirement of a downstream sequence for generation of a poly(A) addition site. *Cell* 37:993−999 (1984).

158. Sasofsky, M., Connelly, S., Manley, J. L., and Alwine, J. C. Identification of a sequence element on the 3' side of AAUAAA which is necessary for simian virus 40 late mRNA 3'-end processing. *Mol. Cell. Biol.* 5:2713−2719 (1985).

159. Gil, A. and Proudfoot, N. Position-dependent sequence elements downstream of AAUAAA are required for efficient rabbit β-globin mRNA 3' end formation. *Cell* 49:399−406 (1987).

160. Danner, D. and Leder, P. Role of an RNA cleavage/poly(A) addition site in the production of membrane-bound and secreted IgM mRNA. *Proc. Natl. Acad. Sci. USA* 82:8658−8662 (1985).

161. Mason, P. J., Elkington, J. A., Malgorzata, L. M., Jones, M. B., and Williams, J. G. Mutations downstream of the polyadenylation site of a xenopus β-globin mRNA affect the position but not the efficiency of 3' processing. *Cell* 46:263−270 (1986).

162. Zarkower, D. and Wickens, M. A functionally redundant downstream sequence in SV40 late pre-mRNA is required for mRNA3'-end formation and for assembly of a precleavage in vitro. *J. Biol. Chem.* 263:5780−5788 (1988).

163. Pain, V. Initiation of protein synthesis in mammalian cells. *Biochem. J.* 235:625−637 (1986).

164. Kozak, M. Point mutations define a sequence flanking the AUG initiator codon that modulates translation by eukaryotic ribosomes. *Cell* 44:283−292 (1986).

165. Liu, C.-C., Simonsen, C. C., and Levinson, A. D. Initiation of translation at internal AUG codons in mammalian cells. *Nature* 309:82−85 (1984).

166. Kozak, M. Selection of initiation sites by eucaryotic ribosomes: Effect of inserting AUG triplets upstream from the coding sequence for preproinsulin. *Nucl. Acids Res.* 12:3873−3893 (1984).

167. Pelletier, J. and Sonenberg, N. Insertion mutagenesis to increase secondary structure within the 5' noncoding region of a eukaryotic mRNA reduces translational efficiency. *Cell* 40:515−526 (1985).

168. Kozak, M. Influences of mRNA secondary structure on initiation by eukaryotic ribosomes. *Proc. Natl. Acad. Sci. USA* 83:2850−2854 (1986).
169. Pelletier, J. and Sonenberg, N. The involvement of mRNA secondary structure in protein synthesis. *Biochem. Cell Biol.* 65:576−581 (1987).
170. Kaufman, R. J. Identification of the components necessary for adenovirus translational control and their utilization in cDNA expression vectors. *Proc. Natl. Acad. Sci. USA* 82: 689−693 (1985).
171. Kitajewski, J., Schneider, R. J., Safer, B., Munemitsu, S. M., Samuel, C. E., Thimmappaya, B., and Schenk, T. Adenovirus VAI RNA antagonizes the antiviral action of interferon by preventing activation of the interferon-induced eIF-2α kinase. *Cell* 45:195−200 (1986).
172. O'Malley, R. P., Mariano, T. M., Siekierka, J., and Mathews, M. B. A mechanism for the control of protein synthesis by adenovirus VA RNA. *Cell* 44:391−400 (1986).
173. Svensson, C. and Akusjarvi, G. Adenovirus VA RNA_I mediates a translational stimulation which is not restricted to the viral mRNAs. *EMBO J.* 4:957−964 (1985).
174. Kaufman, R. J. and Murtha, P. Translational control mediated by eucaryotic initiation factor-2 is restricted to specific mRNAs in transfected cells. *Mol. Cell. Biol.* 7:1568−1571 (1987).
175. Peabody, D. and Berg, P. Termination-reinitiation occurs in the translation of mammalian cell mRNAs. *Mol. Cell. Biol.* 6: 2695−2703 (1986).
176. Kozak, M. Bifunctional messenger RNAs in eukaryotes. *Cell* 47:481−483 (1986).
177. Kaufman, R. J., Murtha, P., and Davies, M. Translational efficiency of polycistronic mRNAs and their utilization to express heterologous gene in mammalian cells. *EMBO J.* 6: 187−193 (1987).
178. Kelly, R. B. Pathways of protein secretion in eukaryotes. *Science* 230:25−31 (1985).
179. Moore, H. P., Walker, M., Lee, F., and Kelly, R. B. Expressing a human prinsulin cDNA in a mouse ACTH-secreting cell. Intracellular storage, proteolytic processing, and secretion on stimulation. *Cell* 35:531−538 (1983).
180. Burgess, T. L. and Kelly, R. B. Sorting and secretion of adrenocorticotropin in a pituitary tumor cell line after perturbation of the level of a secretary granule-specific protoglycan. *J. Cell. Biol.* 99:2223−2230 (1984).

181. Gething, M. J., Doyle, C., Roth, M. G., and Sambrook, J. *Curr. Topics Membr. Transport* 23:17–41 (1985).

182. Garoff, H. Using recombinant DNA techniques to study protein targeting in the eucaryotic cell. *Ann. Rev. Cell. Biol.* 1:403–445 (1985).

183. Kornfeld, R. and Kornfeld, S. Assembly of asparagine-linked oligosaccharides. *Ann. Rev. Biochem.* 54:631–664 (1985).

184. Farquhar, M. G. Progress in unraveling pathways of Golgi traffic, 11. *Ann. Rev. Cell Biol.* 1:447–488 (1986).

185. Griffiths, G. and Simons, K. The trans golgi network: Sorting at the exit site of the golgi complex. *Science* 234: 438–443 (1986).

186. Sly, W. S. and Fischer, H. D. The phosphomannosyl recognition system for intracellular and intercellular transport of lysosomal enzymes. *J. Cell Biochem.* 18:67–85 (1982).

187. Lasky, L. A., Groopman, J. E., Fennie, C. W., Benz, P. M., Capon, D. J., Dowbenko, D., Nakamura, G. R., Nunes, W. M., Renz, M. E., and Berman, P. W. Neutralization of the AIDS retrovirus by antibodies to a recombinant envelope glycoprotein. *Science* 233:209–212 (1986).

188. Rose, J. K. and Bergman, J. E. Expression from cloned cDNA of cell-surface secreted forms of the glycoprotein of vesicular stomatitis virus in eucaryotic cells. *Cell* 30:753–762 (1982).

189. Gething, M. J. and Sambrook, J. Construction of influenza haemagglutinin genes that code for intracellular and secreted forms of the protein. *Nature* 300:598–603 (1982).

190. Lodish, H. F., Kong, N., Snider, M., and Strous, G. J. Hepatoma secretory proteins migrate from rough endoplasmic reticulum to Golgi at characteristic rates. *Nature* 304:80–83 (1983).

191. Doyle, C., Roth, M. G., Sambrook, J., and Gething, M. J. Mutations in the cytoplasmic domain of the influenza virus hemagglutinin affect different stages of intracellular transport. *J. Cell Biol.* 100:704–714 (1985).

192. Roth, M. G., Gething, M. J., and Sambrook, J. *The Influenza Viruses* (in press).

193. Bole, D. G., Hendershot, L. M., and Kearney, J. F. Post-translational association of immunoglobulin heavy chain binding protein with nascent heavy chains in nonsecreting and secreting hybridomas. *J. Cell Biol.* 102:1558–1566 (1986).

194. Gething, M. J., McCammon, K., and Sambrook, J. Expression of wild-type mutant forms of influenza hemagglutinin:

The role of folding in intracellular transport. *Cell* 46:939−950 (1986).

195. Munro, S. and Pelham, H. R. B. An Hsp70-like protein in the ER: Identity with the 78 kd glucose-regulated protein and immunoglobulin heavy chain binding protein. *Cell* 46: 291−300 (1986).

196. Dorner, A. J., Bole, D. G., and Kaufman, R. J. The relationship of N-linked glycosylation and heavy chain-binding protein association with the secretion of glycoproteins. *J. Cell Biol.* 105:2665−2674 (1987).

197. Pelham, H. Coming in from the cold. *Nature* 332:776−777 (1988).

198. Sadler, J. E. and Davie, E. W. Hemophilia A, hemophilia B, and von Willebrand's disease. In *The Molecular Basis of Blood Disease* (Stamatoyannopoulos, Nienhuis, Leder, and Majerus, eds.), WB Saunders, Philadelphia, pp. 575−630 (1987).

199. Mann, K. G. Membrane-bound enzyme complexes in blood coagulation. In *Progress in Hemostasis and Thrombosis*, Vol. 7 (T. H. Spaet, ed.), Grune and Stratton, Orlando, pp. 1−23 (1984).

200. Toole, J. J., Knopf, J. L., Wozney, J. M., Sultzman, L. A., Buecker, J., Pittman, D. D., Kaufman, R. J., Brown, E., Shoemaker, C., Orr, E. C., Amphlett, G. W., Foster, W. B., Coe, M. L., Knutson, G. J., Fass, D. N., and Hewick, R. M. Molecular cloning of a cDNA encoding human anti-hemophilic factor. *Nature* 312:342−347 (1986).

201. Gitschier, J., Wood, W. I., Goralka, T. M., Wion, K., Chen, E. Y., Eaton, D. L., Vehar, G. A., Capon, D. J., and Lawn, R. M. Characterization of the human factor VIII gene. *Nature* 312:326−330 (1984).

202. Wood, W. I., Capon, D. J., Simonsen, C. C., Eaton, D. L., Gitschier, J., Keyt, B., Seeburg, P. H., Smith, D. H., Hollingshead, P., Wion, K. L., Delwart, E., Tuddenham, E. G. D., Vehar, G. A., and Lawn, R. M. Expression of active human factor VIII from recombinant DNA cloned. *Nature* 312:330−337 (1984).

203. Vehar, G. A., Keyt, B., Eaton, D., Rodrequez, H., O'Brien, D. P. O., Rotblat, F., Oppermann, H., Keck, R., Wood, W. I., Harkins, R. W., Tuddenham, E. G. D., Lawn, R. M., and Capon, D. J. Structure of human factor VIII. *Nature* 312:337−342 (1984).

204. Toole, J. J., Pittman, D. D., Orr, E. C., Murtha, P., Wasley, L. C., and Kaufman, R. J. A large region (=95 kDa)

of human factor VIII is dispensable for in vitro procoagulant activity. *Proc. Natl. Acad. Sci. USA* 83:5939–5942 (1986).

205. Eaton, D. L., Wood, W. I., Eaton, D., Hass, P. E., Hollingshead, P., Wion, K., Mather, J., Lawn, R. M., Vehar, G. A., and Gorman, C. Construction and characterization of an active factor VIII variant lacking the central one-third of the molecule. *Biochemistry* 25:8342–8347 (1986).

206. Kaufman, R. J., Amphlett, G., Schrier, J., Booth, J., Pittman, D., Murtha, P. J., and Giles, A. R. A large region (95 kDa) of factor VIII is dispensable for in vivo procoagulant activity. *Blood* 68:349a (1987).

207. Weiss, H. J. and Hoyer, L. W. Von Willebrand factor: dissociation from antihemophilic factor procoagulant activity. *Science* 182:1149–1157 (1973).

208. Titani, K., Kumar, S., Takio, K., Ericsson, L. H., Wade, R. D., Ashida, K., Walsh, K. A., Chopek, M. W., Sadler, J. E., and Fujikawa, K. Amino acid sequence of human von Willebrand factor. *Biochemistry* 25:3171–3184 (1986).

209. Tuddenham, E. G. D., Lane, R. S., Rotblat, F., Johnson, A. J., Snape, T. J., Middleton, S., and Kernoff, P. B. A. Response to infusions of polyelectrolyte fractionated human factor VIII concentrate in human hemophilia A and von Willebrand's disease. *Br. J. Haematol.* 52:259–267 (1982).

210. Over, J., Sixma, J. J., Bruine, M. H., Trieschnigg, M. C., Vlooswijk, R. A., Beeser-Visser, N. H., and Bouma, B. N. Survival of iodine-125-labeled factor-VIII in normals and patients with classic hemophilia: Observations on heterogeneity of human factor-VIII. *J. Clin. Invest.* 62:223–234 (1978).

211. Douglas, A. S. Antihemophilic globulin assay following plasma infusion in hemophilia. *J. Lab. Clin. Med.* 51:850–859 (1958).

212. Weiss, H. J., Sussman, I. I., and Hoyer, L. W. Stabilization of factor VIII in plasma by the von Willebrand factor. Studies on posttransfusion and dissociated factor VIII and in patients with von Willebrand's disease. *J. Clin. Invest.* 60:390–404 (1977).

213. Brinkhous, K. M., Sandberg, H., Garris, J. B., Mattson, C., Palm, M., Griggs, T., and Read, M. S. Purified human factor VIII procoagulant protein: Comparative hemostatic response after infusions into hemophiliac and von Willebrand disease dogs. *Proc. Natl. Acad. Sci. USA* 82:8752–8756 (1985).

214. Wion, K. L., Kelly, D., Summerfield, J. A., Tuddenham, E. G. D., and Lawn, R. M. Distribution of factor VIII

mRNA and antigen in human liver and other tissues. *Nature* 317:726–729 (1985).

215. Pittman, D. D. and Kaufman, R. J. The protedytic requirements for thrombin activation of anti-hemophilic factor (factor VIII). *Proc. Natl. Acad. Sci. USA* 85:2429–2433 (1988).

216. Wise, R., Pittman, D. D., Handin, R. I., Kaufman, R. J., and Orkin, S. H. The propeptide of von Willebrand factor independently mediates the assembly of von Willebrand multimers. *Cell* 52:229–236 (1988).

217. Pelham, H. R. B. Speculations on the functions of the major heat shock and glucose-regulated proteins. *Cell* 46:959–961 (1986).

218. Trimble, R. B. and Maley, R. Optimizing hydrolysis of N-linked high-mannose oligosaccharides by endo-β-N-acetyl-glucosaminidase H. *Anal. Biochem.* 141:515–522 (1984).

219. Tarentino, A. L., Gomez, C. M., and Plummer, T. H., Jr. Deglycosylation of asparagine-linked glycans by peptide: N-glycosidase F. *Biochem.* 24:4665–4671 (1985).

220. Kornfeld, R. and Kornfeld, S. Assembly of asparagine-linked oligosaccharides. *Ann. Rev. Biochem.* 54:631–664 (1985).

221. Eaton, D. L., Rodriguez, H. R., and Vehar, G. A. Proteolytic processing of human factor VIII. Correlation of specific cleavages by thrombin, factor Xa, and activated protein C with activation and inactivation of factor VIII coagulant activity. *Biochemistry* 25:505–512 (1986).

222. Eaton, D. L., Hass, P. E., Riddle, L., Mather, J., Wiebe, M., Gregory, T., and Vehar, G. A. Characterization of recombinant human factor VIII. *J. Biol. Chem.* 262:3285–3290 (1987).

223. Toole, J. J., Pittman, D., Murtha, P., Wasley, L. C., Wang, J., Amphlett, G., Hewick, R., Foster, W. B., Kamen, R., and Kaufman, R. J. Exploration of structure-function relationships in human factor VIII by site-directed mutagenesis. *Cold Spring Harbor Symp. Quant. Biol.* 51:543–549 (1986).

224. Pittman, D. D., Wasley, L. C., Murray, B. L., Wang, J. H., and Kaufman, R. J. Analysis of structure requirements for factor VIII function using site-directed mutagenesis. *Thrombosis and Hemostasis* 58:344a (1987).

225. Kaufman, R. High level production of proteins in mammalian cells. *Genetic Engineering Principles and Methods*, Vol. 9 (J. Setlow, ed.), Plenum Press, New York, p. 155 (1987).

226. Dorner, A. J., Krane, M. G., and Kaufman, R. J. Reduction of endogenous GRP78 levels improves secretion of a heterologous protein in CHO cells. *Mol. Cell. Biol.* 8:4063–4070 (1988).

3

Large-Scale Animal Cell Culture: A Biological Perspective

MELVIN S. OKA
SmithKline Beecham Pharmaceuticals, King of Prussia, Pennsylvania

RANDALL G. RUPP
Somatogen Corporation, Broomfield, Colorado

I. INTRODUCTION

Animal cell culture has become an important aspect of biotechnology. Molecules are being cloned and expressed in a few cultured animal cell lines for in vitro and in vivo applications. Thus far these molecules are high-value-added proteins and therefore can be made using processes that have not been optimized. In general, the cost of producing these proteins ranges from hundreds to thousands of dollars per gram. However, if products from animal cells are to find continued widespread use, it is important that we improve the economics of production. Until now research and development has concentrated on the type of bioreactors used for culture and the various matrices on or in which cells can grow (1−3). These efforts attempt to ensure that the process is stable and scalable with minimal alterations to existing bioreactor hardware (4,5). Optimization of cell growth and production usually occurs after the reactor type has been selected. This process leads to the selection of cell types that lend themselves to growth in a particular bioreactor

with minimal regard to the selection of an efficient and appropriate
cell type for the secretion of the product. Therefore, suspension
culture has been the scale-up method of choice because monitoring
and control allows near linear increases in volume. In addition,
there is a great deal of experience in these types of systems from
microbial fermentation. There are efficiencies yet to be achieved in
these approaches by increasing cell densities and improving the efficiency of the culture media. However, the greatest increases in
efficiencies of production are to be gained by improving the expression of product per cell per time (specific productivity).

The focus of this chapter is to outline the possible areas in
basic research and development which, when solved, will permit the
development of more efficient, cost-effective processes for cell culture and downstream processing. We have addressed the two major
biological problems encountered today in large-scale cell culture:
the control of cell growth and the optimization of specific cellular
productivity.

II. FACTORS AFFECTING SPECIFIC PRODUCTIVITY

A general review of the cellular mechanisms involved in protein secretion is not within the scope of this chapter. However, the
reader is referred to some recent reviews that cover this topic in
more detail (6–10). Both secretory and membrane-bound proteins
are believed to follow similar pathways, starting with the synthetic
site in the rough endoplasmic recticulum, progressing to the Golgi
apparatus and finally to the cell surface. The proteins following
this pathway may be covalently modified (e.g., glycosylation, oligosaccharide modification, acylation, or partial proteolysis). Interference with some of these modifications can inhibit the movement or
appearance of the molecule at the cell surface. Not all the steps in
the secretory pathway have been completely understood nor have all
aspects of the process been elucidated.

A. Cell Type

In the past, cells selected for the manufacture of heterologous proteins have generally been limited to those cell types that grow well
in suspension (11,12), e.g., Chinese hamster ovary (CHO) cells.
The selection of cell types which are known to be efficient secretors
in vivo has received only limited attention. For example, myeloma
cells have been exploited by a few laboratories in an attempt to develop cell lines with demonstrated secretory properties and effective
growth in suspension culture (13).

Efforts to increase the expression of heterologous proteins within a cell type already expressing low levels of the protein have been attempted on a limited basis. Inserting genes into cells with demonstrated synthetic and secretory machinery may be advantageous. These cells contain the appropriate genetic controls for the production of proteins with the correct configuration and activity.

In our opinion, however, for the efficient, economical production of recombinant proteins from animal cells it is important to develop a limited number of model cells for expression of heterologous genes. These cells would be the recipients of recombinant genes coding for many recombinant products. By selecting only a few model cells for expression, we can channel research efforts to yield a more thorough understanding of the physiology of the cells and their relationship to the culture environment. In addition, this would permit a more objective assessment of the various types of bioreactors. Often one bioreactor is touted as superior to others when in fact the important differences are due to cells with high specific productivities rather than to bioreactor performance. Until a few model cells are used by many laboratories and in many bioreactors, it will remain unclear as to which bioreactor(s) is performing optimally. Some suggested properties of the "ideal" culture cell are as follows:

1. Is easily transfected
2. Has a low requirement for oxygen
3. Produces low quantities of lactic acid
4. Grows in and can be maintained in inexpensive, serum-free defined media
5. Does not produce virus or viral enzymes
6. Has a stable and efficient secretory process
7. Resists shear
8. Has a high specific productivity (e.g., $3.0-5.0$ pg/cell/min)
9. Has an extended, viable plateau phase after the exponential growth phase

Obtaining a cell with all of these characteristics may be difficult. It is important that the cell or cells of choice satisfy most of these criteria. We have not limited the ideal cell to an anchorage-independent cell even though there are clear advantages to this. For example, anchorage-independent cells attain high densities, are relatively easy to subculture, and can be grown in stirred-tank reactors.

Exploitation of other cell types, especially adherent cells, has been limited due to the relative difficulty of culture methods that will support large-scale adherent cell culture. However, with the advent of methodologies that permit attachment-dependent cell lines to attain high densities in long-term culture, the use of adherent lines should be reevaluated. Clearly, more information about cell-substrate interactions are required to optimize adherent cells for large-scale culture. The points we discuss in this chapter refer to either anchorage-dependent or anchorage-independent cells since we feel that both may have an important role in large-scale cell culture.

B. Environmental Factors

The range of environmental factors affecting cellular secretion to be evaluated when new biological manufacturing processes are being developed must be expanded. Shear forces, oxygen supply, and pH control are important components of any culture system. However, considerations of other, more subtle factors which influence secretion must also be considered.

Extracellular matrices (ECM) have been reported to contain glycosaminoglycans, proteoglycans, collagens, and secreted adhesion molecules such as fibronectin and laminin (14-17). The effects of cell substratum on growth regulation and maintenance of normal cellular functions in vitro have been extensively studied (18-24). There exists a close relationship between the substrate on which a cell rests, cellular architecture (25), and expression of differentiated functions (26). The effects of cell substrates on cellular secretion are well known.

A well-documented paradigm of the relationship between cell secretion and the substratum is found in studies that investigated casein secretion in the mammary cell (27-30). Casein, a milk protein, is not secreted at high levels by normal mammary epithelial cells in vitro. However, when the cells are cultured on collagen substrates rather than on polystyrene, a 4- to 20-fold increase in casein production is detected. Further increases are seen with other matrices. The substratum appears to affect only those secretory functions that are already functioning in the cell type.

The specific mechanisms by which ECM influences the secretory mechanisms of the cell remain unclear. Cell surface receptors have been identified that mediate some of the cell-substratum interactions by binding specific components in the extracellular environment (31-33). One class of molecules that is probably involved in cellular adhesion and signaling is the cell surface glycosyltransferases. When glycosyltransferases were first detected on the cell surface, the suggestion was that they could participate in cellular interactions by recognizing and binding to their specific glycoconjugate substrates on adjacent cell surfaces and in the extracellular matrix (34-36). The number and kinds of signals that could be accommodated would be limited only to the large number of potential transferases available to the cell. Normally the transferase would form a stable complex with its complementary substrate present on an adjacent cell or the ECM. However, if a sugar nucleotide were made available, the cells would dissociate and the resulting catalysis could present some regulatory signal to the target cell (37). There is some evidence to support this model (38,39).

Transformed cells can also respond to the substratum. For example, pheochromocytoma PC-12 cells release more dopamine into the medium when cultured on a corneal extracellular matrix than when grown on polystyrene (40). Rat pituitary tumor cells respond to the presence of extracellular matrix by secreting increased levels of prolactin (41). Whatever relationship exists between an untransformed cell and the ECM to some degree exists with transformed cells. In addition, inappropriate substrata may adversely affect secretory properties of the cell. Culture of MOSER colon carcinoma cells on extracellular matrix laid down by other colon lines reduced levels of secreted urokinase by over 50% (42). It appears that the specific component responsible for the reduction in protein synthesis/ secretion in these cells is fibronectin (43). Such studies underscore the need for careful selection of the cell substrata in scale-up.

For long-term cultures, such as those conducted in perfusion-type reactors, the advent of matrix-based reactor technology provides a means for stabilization and control of appropriately chosen secreting cells throughout the production process. The research work in the area of extracellular matrices clearly demonstrates that the maintenance of cellular characteristics is influenced by the physical support on which the cell rests and therefore must be carefully considered when developing the system.

Although there have been some efforts to use the information on cell substrata composition in the formulation of microcarriers, the focus of these efforts has simply been to find a substratum on which the cells can attach and grow with little regard to the expression of differentiated function. Most substrates available in quantities sufficient for large-scale cell culture are microspheres of dextran, collagen-coated dextran, denatured collagen, or inert materials. Only recently have efforts been extended to developing substrates consisting of undenatured, physiological molecules which may be useful in large-scale culture. Development of macroporous beads with variations in the matrix composition represents an opportunity to apply some of the basic research findings in cell-substrate interactions to the optimization of product synthesis and secretion.

Another major component of the culture environment is the nutrient medium in which the cells are cultured. Extensive work in the optimization of defined media for a variety of cell types has been performed (44−47). Most serum-free media have been developed to expedite the discovery of novel proteins and cellular mechanisms.

The advent of large-scale cell culture as a tool to produce novel therapeutics has amplified the need for serum-free media. This

need is not driven solely by cost since many of the serum-free
media are quite expensive with costs-to-produce ranging from $3 to
$35 per liter. A more important aspect of serum-free media develop-
ment is that some media dramatically affect specific productivities.
It has been shown that cells grow in two different serum-free media,
both of which support cell growth equally well, can synthesize
and/or secrete more product in one medium than in the other (48).
Cell growth should not be the only criteria for the selection of culture media. In fact, the impact of serum-free media on downstream
processing is much more important than the cost of the media, at a
certain scale of manufacturing. (As the size of the reactors and
manufacturing capacity increase, the impact of raw material costs
becomes more significant than when manufacturing capacity is low.
At low capacities the overhead costs have a much greater impact on
costs-to-produce than do the raw material costs.) In order to purify secreted products with high yields and purities, it is important
to minimize the number and concentrations of unwanted, extraneous
proteins in the conditioned medium containing the product. Also,
to meet specifications on final product purity it is often *essential*
that no serum be used in the production process.

C. Cell Physiology

While many aspects of cell physiology are well defined, two areas
which might be relevant to future applied research but which are highly speculative include the elimination of recently identified inhibitors
of secretion and the understanding of intracellular protein trafficking.
The identification of the signals and transport proteins might enable
us to manipulate them to significantly increase cellular secretion.
    When cells are stressed in vitro, they often respond by altering
their metabolism (49,50). These alterations can include modifications
to protein synthetic activity. Stress proteins, a family of proteins
that are synthesized in response to dramatic changes in the culture
environment, increase due to changes in temperature, pH, or other
environmental factors. This specific change in protein synthesis
may be a general response to stress that emphasizes the synthesis
of stress proteins at the expense of others, e.g., recombinant molecules. Such redirection of synthetic activity may be undesirable.
    A specific class of stress proteins is the glucose-regulated pro-
teins (GRPs). These are a family of proteins with molecular weights
of approximately 78−80 and 94−100 kDa. They are induced in
animal cells grown in medium with suboptimal concentrations of glucose, inhibitors of glycosylation, and/or a variety of other stress
inducers (51−53). GRP78 shares a consensus sequence with four
members of the heat shock protein (HSP), 70-kDa family. Approximately 60% of the amino acid sequence of the GRP78 protein is

homologous to the HSP70 protein (54). The GRP78 protein is immunologically identical to the immunoglobulin heavy-chain binding protein, BiP (55). The function of BiP is unknown but since it is localized in the lumen of the endoplasmic reticulum (ER) it is inferred to somehow affect protein secretion. BiP is ubiquitously expressed in many cell types (56). Since it is found in high concentrations in cells actively secreting antibodies, it is speculated that it may be normally involved in the stabilization of membrane-bound and secreted proteins. Alternatively, it may have a role in preventing the transport of unassembled heavy chains from the ER to the Golgi. It has been shown that myeloma cells will secrete heavy chains without their accompanying light chains when the association of the heavy chains with BiP is minimal (57,58). Furthermore, it has been demonstrated that GRP78 also binds proteins in the ER to a greater degree when protein folding is blocked or there is abnormal N-linked glycosylation (59-62). The reduction of GRP78 by encoding an GRP78 antisense sequence in CHO cells secreting recombinant tissue plasminogen activator (rtPA) improves the level of rtPA secretion by twofold (63). This increase was seen without any increases in rtPA mRNA over controls. These results suggest that the alteration of the quantity of GRP78 in the cells may significantly affect protein synthetic rates. However, the total suppression of cellular "quality control" mechanisms in an effort to enhance secretion may change the ratio of desired protein product to "defective" proteins. This may make it difficult to purify a well-characterized, therapeutic protein.

Until recently sublethal effects of environmental stress in vitro on cellular metabolism were a specialized area with little impact on scale-up. Along with physiological stress, mechanical stress has been shown to alter the glucose metabolism of erythrocytes (64). Furthermore, shear stress ($15-25$ dynes/cm^2) on cultured human epithelial cells stimulated tPA secretion $2.1-3$ times greater than basal secretion rates (65). Such effects clearly have implications for the design of bioreactors.

Earlier studies indicated that for some secreted proteins, the addition of carbohydrate side chains is essential for normal secretion (66-69). Some investigators hypothesized that the secretion of these proteins could be impaired by improper folding of the protein, which could be partially controlled by the carbohydrate residues (62,67,69-71). Others have shown that a single-site mutation in the protein backbone can also affect secretion (72). For example, they speculate that the substitution of a neutral amino acid for a positively charged amino acid at the critical site may alter the folding of the heavy immunoglobulin chain. Such alterations may affect protein solubility or movement within one of the synthetic compartments. Elucidation of the factors affecting protein glycosylation is therefore important to increasing cellular productivity.

III. CELL PROLIFERATION

The overall control of cell proliferation has received much attention because of the specific need to control inappropriate cell proliferation in cancer. Although significant progress has been made in the understanding of the mechanisms of cellular proliferation, many basic mechanisms are as yet not understood (74).

Clarification of the mechanisms involved in cellular proliferation may permit their manipulation to increase specific productivities. Stimulation of cell proliferation is not only important to the attainment of high cell densities but may be essential when cellular productivity and proliferation are coupled. On the other hand, it would be extremely useful to reduce cell proliferation while maintaining high cell viability and productivity at high cell densities. (Maintenance of the cells in a nonproliferating state may extend the lifetime of the culture, enable the cell to produce larger quantities of material since the synthetic machinery would focus on products of interest, produce the "correct" molecule, and alter cellular metabolism so that nutritional requirements are reduced.)

An understanding of the signal pathway(s) involved in moving a cell from a dividing to a nondividing state and vice versa and the mechanisms which are involved in the transmission of that signal might offer insights into how culture growth may be controlled. Transformed cells appear to require lower concentrations or fewer types of growth factors to survive in vitro than normal cells. The autocrine hypothesis (74,75) proposes that transformed cells secrete growth factors which act in an autostimulatory manner on receptors present on the secreting cell (74−76).

Recent research indicates that it might be possible to genetically modify cells, which normally require exogenous growth factors, so that they synthesize and secrete the needed factors. Such experiments were recently performed using bFGF as the autocrine signal (77). These investigators demonstrated that hamster kidney cells could be released into autonomous growth by the insertion of a bFGF construct into the cell line.

Care must be exercised, however, before inserting a gene for growth factor production into a host cell line. For autocrine factors to function, there must be a membrane receptor for the factor, which in the case of the PDGF receptor is probably under very complex control (78). Results indicate that PDGF receptor expression varies in normal tissues and that fibroblasts and smooth muscle cells cultured in vitro are induced to express PDGF receptors in response to removing the cells from the tissue and establishing them in culture. The number of receptors rises over a 3-day period as the cells are propagated in tissue culture. The mechanism by which the expression of the receptor is upregulated is unknown. Serum factors or the disruption of normal cellular contact with substratum and adjacent cells may be involved in the upregulation.

Elevated levels of EGF receptor expression, following gene amplification events, have been detected in human epithelial and glial tumor cells (79–81), suggesting that overexpression of a normal receptor protein can confer proliferative advantage. The transfection of the normal EGF receptor gene into a nonexpressing host cell line might lead to the same end point (82–84). However, there is some controversy over the ability of normal EGF receptors to initiate the transformed phenotype in some cell lines. Several researchers have noted that various human tumor cells have increased numbers of EGF receptors (85–89), that the number of EGF receptors correlates with tumor growth in vitro (90), and that the increased number of EGF receptors correlates with poor patient survival (91). Direct transfection of normal EGF receptors into cells which are then injected into nude mice have conferred the transformed phenotype on some cell types (82) and not on others (92). Recent data on the number of EGF receptors in human mammary epithelial cell lines suggest that the number of EGF receptors in many tumor lines may not be overexpressed unless the genes are amplified (93).

The EGF receptor is an integral membrane protein of about 170 kDa and has tyrosine kinase activity which is activated upon binding of EGF (94–96). The *erb B* oncogene product bears a great deal of sequence homology to the EGF receptor; however, the external domain is truncated and the cytoplasmic domain lacks either 32 or 71 amino acids (97,98). The v-*erb B* product does retain its tyrosine kinase activity and is constitutively expressed. Presumably, the transfection of the v-*erb B* oncogene into appropriate host cells may alleviate the need for the addition of EGF to the culture environment to maintain cell proliferation. Results of such experiments have shown that such transfections enable hematopoietic cells that normally do not form tumors in nude mice to do so (92). Such cells possess the necessary metabolic pathways which can lead to continuous cell proliferation. Therefore it should be possible to engineer, through recombinant technology, cells that can proliferate in vitro without the addition of exogenous growth factors to the culture medium.

Interactions of growth factors in the physiological cascade are not clearly understood. Occupation of growth factor receptors by their ligands can cause other receptors of the same type to be decreased in number (downregulated), thereby reducing the cell's sensitivity to the presence of more growth factor (86). Other recent work (99) indicates that there may be coordinated downregulation of several receptors under certain conditions. This process, referred to as density-induced downregulation of growth factors, could be an important phenomenon in high-density cell cultures. The coordinated downregulation of several receptors could make the culture insensitive to the presence of these growth factors in the medium. Thus the addition of higher concentrations of growth factors once the cells have reached a certain density might be ineffective in increasing cell proliferation.

Other growth factors have been shown to affect the activities of several species of growth receptors. Constitutive activation of normal EGF receptors by TGF-alpha, a potent agonist for the EGF receptor, results in cells escaping from normal proliferative constraints (100). This is analogous to autocrine stimulation of the PDGF-like oncogene products in v-*sis* transformed cells. On the other hand, TGF-beta inhibits the proliferative response of human bone marrow fibroblast to PDGF (101). The activity of one receptor affecting another receptor function is known as receptor modulation (75). This phenomenon has also been demonstrated in other cell systems (102–104). Careful consideration should be given to this phenomenon when designing a medium that includes growth factors.

Recently, a lens epithelial cell line was shown to be responsive to PDGF (105). In these experiments, the PDGF had to be delivered in pulses to maintain growth and differentiated function in isolated rat neonatal lens. Lens maintained without PDGF or in the continuous presence of PDGF failed to promote growth or function. Such results indicate that in some cell types, the growth promotion effect of PDGF is modulated by the method of delivery. As methods in large-scale cell culture mature, it may be necessary to incorporate more subtle delivery strategies for many components of the medium in order to obtain totally optimal production and control of the cell culture.

As previously discussed, cell matrices can have profound effects on the maintenance of cellular phenotypes. The extracellular matrix can also have profound effects on cell proliferation (26,106, 107). The deprivation of adequate attachment sites for anchorage-dependent cells can prevent the cells from proliferating (108). Ingber and Folkman (109) suggested that the extracellular matrix is a "solid state regulator of soluble growth factor action." They provide evidence that DNA synthesis is dependent on the type of matrix to which the cell is attached. Work by other researchers indicates that corneal and mammary epithelial cells respond differently to stimulation by EGF depending on the culture substrata (110,111). Recently, investigators demonstrated that the ECM is capable of decreasing the sensitivity of cells to PDGF (107). Human skin fibroblasts plated on collagen gels grew more slowly than cells grown on polystyrene even though both were exposed to PDGF. Since they found no evidence of PDGF binding directly to the collagen substrate, they speculate that the ECM might downregulate the number of receptors on the cell surface. Therefore the final function of the matrix or substratum on which the cell rests may involve increasing or decreasing the sensitivity of cells to exogenous growth factors. However, the presence of growth factors also influences the type of extracellular matrix the cells secrete (112). Therefore, a fine understanding of the interrelationship among the cells, growth factors, and substratum must exist to correctly optimize in vitro production systems. The optimization of the cell substratum becomes important in those culture systems that require the use of adherent cell lines. Furthermore, the optimization of the substratum might also lower the requirements of exogenous growth factors needed in the medium.

The examples cited above have been primarily concerned with the growth and creation of autostimulatory cell lines. However, the controlled inhibition of cell growth is also desirable. Culturing cells to high cell densities and then inhibiting continued growth (provided viability and expression of differentiated functions is not adversely affected) could be advantageous. Several approaches to growth inhibition are available at this time. Media manipulations have been used to synchronize cells in vitro. Perhaps the best method of cell synchrony has been described by Tobey et al. (113, 114). A more specific inhibition may be made possible by the removal of the growth factors from the medium when cessation of growth is called for. Unfortunately, many of the growth factors are needed to maintain cell viability as well as proliferation. Perhaps specific inhibitors that affect signal processes but not the overall metabolic state of the cell can be used. A series of low molecular weight protein tyrosine kinase inhibitors, tyrphostins, have been synthesized. They are potent inhibitors of EGF receptor autophosphorylation (115). The inhibitory effects of these compounds is reversed when the compounds are removed from the culture. Such targeted compounds may be useful in the regulation of growth in vitro.

IV. SUMMARY

It is generally recognized that no one cell culture system can be universally applied to all cell types commonly used for biopharmaceutical manufacture. The analogous concept that no single cell type may be useful for the expression of all biopharmaceutical products may also gain credence in the biotechnology community. It may be that like specialized bioreactors, there will come to exist a variety of cell types that will be used for the production of different types of biopharmaceutical products. In addition, it may not be enough in the future just to demonstrate the stability of expression of the amino acid backbone of the protein only; the carbohydrate portion of the molecule may become the subject of real scrutiny. Questions such as how the carbohydrate side chain affects the performance of the molecule in vivo are being asked of more DNA constructs. The next question becomes, how can we control the expression of carbohydrate moieties on the molecule? Such questions are in the future of the biotech manufacturing field.

Aside from those examples mentioned above dealing with the insertion of receptors, other more subtle attempts at modifying cellular metabolism are taking place. It was reported at a recent meeting that the sialyltransferase gene was inserted into a CHO line which did not normally express this enzyme (116). The transfected line was capable of expressing the transferase and, more importantly, the enzyme functioned correctly in sialylating glycoproteins. Other very complex relationships exist between the substratum and the cell that could have very direct consequences on culture maintenance. For example, researchers recently published results indicating that collagenase synthesis and secretion is stimulated in rabbit fibroblasts by autocrine factors (117). They determined that these autocrine proteins had sequence homology to serum amyloid-A and beta-2-microglobulin. It may be that using serum supplements in the medium in those systems that couple fibroblast and collagen substratum may not be prudent, especially for long-term culture.

The traditional selection of a cell type for expressing heterologous proteins has generally been limited to the more "common" cell types such as CHO cells, C127 cells, and myeloma cells. In many cases these cell types were selected because there was a great deal of preexisting literature on the cell type (i.e., "cookbook" methods of transfection for the cell) or the cell was simply being carried in the lab at the time the effort was made to express a biopharmaceutical product. In many cases factors which affect the downstream (in this case, beyond the T-75 flask) side of the manufacturing scale-up were not considered before selecting the cell line as the host for the expression system. Such attitudes are no longer tolerated in the very competitive biopharmaceutical industry. The search for more efficient cell lines and the creation of these lines has begun.

REFERENCES

1. Adamson, S. R. and Schmidli, B. Industrial mammalian cell culture. *Can. J. Chem. Eng.* 64:531–539 (1986).
2. Ratafia, M. Issues in mammalian cell culture production. *Pharmaceut. Technol.* 11:48–56 (1987).

3. Arathoon, W. R. and Birch, J. R. Large-scale cell culture in biotechnology. *Science* 232:1390–1395 (1986).
4. Tyo, M. A. and Spier, R. E. Dense cultures of animal cells at the industrial scale. *Enzyme Microb. Technol.* 9:514–520 (1987).
5. Bleim, R. and Katinger, H. Scale-up engineering in animal cell culture technology. I. *Trends Biotechnol.* 6:190–205 (1988).
6. Kelly, R. B. Pathways of protein secretion in eukaryotes. *Science* 230:25–32 (1985).
7. Singer, S. J., Maher, P. A., and Yaffe, M. P. On the translocation of proteins across membranes. *Proc. Natl. Acad. Sci. USA* 84:1015–1019 (1987).
8. Verner, K. and Schatz, G. Protein translocation across membranes. *Science* 241:1307–1313 (1988).
9. Pfeffer, S. R. and Rothman, J. E. Biosynthetic protein transport and sorting by the endoplasmic reticulum and Golgi. *Ann. Rev. Biochem.* 56:829–852 (1987).
10. Kornfeld, R. and Kornfeld, S. Assembly of asparagine-linked oligosaccharides. *Ann. Rev. Biochem.* 54:631–664 (1985).
11. Sanders, P. G., Hussein, A., Coggins, L., and Wilson, R. Gene amplification: The Chinese hamster glutamine synthetase gene. *Dev. Biol. Standard.* 66:55–63 (1987).
12. Stark, G. R. and Wahl, G. M. Gene amplification. *Ann. Rev. Biochem.* 53:447–492 (1984).
13. Wiedle, U. H. and Buckel, P. Establishment of stable mouse myeloma cells constitutively secreting human tissue-type plasminogen activator. *Gene* 57:131–141 (1987).
14. Yamada, K. M. Cell surface interactions with extracellular materials. *Ann. Rev. Biochem.* 52:761–799 (1983).
15. Martin, G. R. and Timpl, R. Laminin and other basement membrane components. *Ann. Rev. Cell Biol.* 3:57–85 (1987).
16. Iozzo, R. V. Cell surface heparan sulfate proteoglycan and the neoplastic phenotype. *J. Cell. Biochem.* 37:61–78 (1988).
17. Werz, W. and Schachner, M. Adhesion of neural cells to extracellular matrix constituents: Involvement of glycosaminoglycans and the cell adhesion molecules. *Dev. Brain Res.* 43:225–234 (1988).
18. Gospodarowicz, D., Greenburg, G., and Birdwell, C. Determination of cellular shape by the extracellular matrix and its correlation with the control of cellular growth. *Cancer Res.* 38:4155–4171 (1978).
19. Lippman, M. E., Dickson, R. B., Gelmann, E. P., Rosen, N., Knabbe, C., Bates, S., Bronzert, D., Huff, K., and

Kasid, A. Growth regulation of human breast carcinoma occurs
through regulated growth factor secretion. *J. Cell. Biochem.*
35:1–16 (1987).

20. Kleinman, H. K., Graf, J., Iwamoto, Y., Kitten, G. T., Ogle,
R. C., Sasaki, M., Yamada, Y., Martin, G. R., and
Lukenbill-Edds, L. Role of basement membranes in cell dif-
ferentiation. *Ann. NY Acad. Sci.* 513:134–145 (1987).

21. Hay, E. D. (ed.) *Cell Biology of the Extracellular Matrix.*
Plenum Press, New York (1981).

22. Tamada, K. M. and Akiyama, S. K. The interactions of cells
with extracellular matrix components. In *Cell Membranes,*
Vol. 2 (E. Elson, W. Frazier, and L. Glaser, eds.), Plenum
Press, New York, pp. 77–147 (1984).

23. Fujita, M., Spray, D. C., Choi, H., Saez, J., Jefferson, M.,
Hertzberg, E., Rosenberg, L. C., and Reid, L. M. Extra-
cellular matrix regulation of cell–cell communication and tissue-
specific gene expression in primary liver cultures. In *Cellular
Endocrinology: Hormonal Control of Embryonic and Cellular
Differentiation* (G. Serrero and J. Hayashi, eds.), Alan R.
Liss, New York, pp. 333–360 (1986).

24. Martin, G. R. Laminin and other basement membrane components. *Ann. Rev. Cell Biol.* 3:57–85 (1987).

25. Armstrong, P. B. Stabilization of tissue architecture: Involvement of the extracellular matrix. In *Developmental Mechanisms, Normal and Abnormal* (J. Lash and L. Saxen, eds.),
Alan R. Liss, New York, pp. 87–107 (1985).

26. Gospodarowicz, D. J. Extracellular matrices and the control
of cell proliferation and differentiation in vitro. In *New Approaches to the study of benign prostatic hyperplasia* (F. A.
Kimball, A. E. Buhl, and D. B. Carter, eds.), Alan R. Liss,
New York, pp. 103–128 (1984).

27. Yang, J. and Nandi, S. Growth of cultured cells using
collagen as substrate. *Int. Rev. Cytol.* 81:249–286 (1983).

28. Lippman, M. E., Dickson, R. B., Gelman, E. P., Rosen, N.,
Knabbe, C., Bates, S., Bronzert, D., Huff, K., and
Kasid, A. Growth regulation of human breast carcinoma
occurs through regulated growth factor secretion. *J. Cell.
Biochem.* 35:1–16 (1987).

29. Aggeler, J., Park, C. S., and Bissell, J. J. Regulation of
milk protein and basement membrane gene expression: The
influence of the extracellular matrix. *J. Dairy Sci.* 71:
2830–2842 (1988).

30. Bissell, M. J. and Aggeler, J. Dynamic reciprocity: How do
extracellular matrix and hormones direct gene expression. In

Mechanisms of Signal Transduction by Hormones and Growth
Factors (M. Cabot and W. McKeehan, eds.), Alan R. Liss,
New York, pp. 251–262 (1987).
31. von der Mark, K. and Kuhl, U. Laminin and its receptor.
    Biochem. Biophys. Acta 823:147–160 (1985).
32. Malinoff, H. L. and Wicha, M. S. Isolation of a cell-surface
    receptor protein for laminin from murine fibrosarcoma cells.
    J. Cell Biol. 96:1475–1479 (1988).
33. Lotan, R. and Raz, A. Endogenous lectins as mediators of
    tumor cell adhesion. *J. Cell. Biochem.* 37:107–117 (1988).
34. Roseman, S. The synthesis of complex carbohydrates by
    multiglycosyltransferase systems and their potential function in
    intercellular adhesion. *Chem. Physics Lipids* 5:270–297 (1970).
35. Roth, S., McGuire, E. J., and Roseman, S. J. Evidence for
    cell-surface glycosyltransferases: Their potential role in
    cellular recognition. *J. Cell. Biol.* 51:536–547 (1971).
36. Runyan, R. B., Versalovic, J., and Shur, B. D. Functionally
    distinct laminin receptors mediate adhesion and spreading:
    The requirement for surface galactosyltransferase in cell spread-
    ing. *J. Cell. Biol.* 107:1863–1871 (1988).
37. Eckstein, D. J. and Shur, B. D. Cell surface galactosyl-
    transferase interaction with laminin during cell migration on
    basal lamina. In *Altered Glycosylation in Tumor Cells* (C. L.
    Reading, S.-I. Hakomori, and D. M. Marcus, eds.), Alan R.
    Liss, New York, pp. 217–229 (1988).
38. Roth, S. and White, D. Intercellular contact and cell-surface
    galactosyl transferase activity. *Proc. Natl. Acad. Sci. USA*
    69:485–489 (1972).
39. Klohs, W. D., Wilson, J. R., and Weiser, M. M. UDP-galactose
    inhibition of BALB/3T12-3 cell growth. *Exp. Cell Res.* 141:
    365–374 (1982).
40. Betha, C. and Kozak, S. Further characterization of sub-
    straum influence on PC12 cell shape and dopamine processing.
    Mol. Cell. Endocrinol. 43:59–72 (1985).
41. Prysor-Jones, R., Silverlight, J., and Jenkins, J. Differen-
    tial effects of extracellular matrix on secretion of prolactin and
    growth hormone by rat pituitary tumor cells. *Acta Endocrinol.*
    108:156–160 (1985).
42. Boyd, D., Florent, G., Chakrabarty, S., Brattain, D., and
    Brattain, M. G. Alterations of the biological characteristics
    of a colon carcinoma cell line by colon-derived substrata ma-
    terial. *Cancer Res.* 48:2825–2831 (1983).
43. Boyd, D., Florent, G., Childress-Fields, K., and Brattain, M.
    Alteration in the behavior of a colon carcinoma cell line by

extracellular matrix components. *Cancer Lett.* 41:81−90 (1988).

44. Spier, R. Environmental factors: Medium and growth factors. *Animal Cell Biotechnol.* 3:29−53 (1988).

45. Mizrahi, A. and Lazar, A. Media for cultivation of animal cells: An overview. *Cytotechnology* 1:199−200 (1988).

46. Griffiths, J. B. Serum and growth factors in cell culture media: An introductory review. *Dev. Biol. Standard.* 66: 155−160 (1985).

47. Barnes, D., Sirbasku, D., and Sato, G. (eds.). *Cell Culture Methods for Molecular and Cell Biology*, Vols. 1−4, Alan R. Liss, New York (1984).

48. Rupp, R., Tate, E., and Peterson, L. The use of the fluorescent-activated cell sorter to monitor changes in specific productivity. In *Proc. Eur. Soc. Animal Cell Technol.* (in press).

49. Schlesinger, M. Stress response in avian cells. In *Changes in Eucaryotic Gene Expression in Response to Environmental Stress* (B. Atkinson and D. Walden, eds.), Academic Press, Orlando, pp. 183−195 (1985).

50. Hammond, G., Lai, Y., and Market, C. Diverse forms of stress lead to new patterns of gene expression through a common and essential metabolic pathway. In *Proc. Natl. Acad. Sci. USA* 79:3484−3488 (1982).

51. Lee, A. Coordinated regulation of a set of genes by glucose and calcium ionophores in mammalian cells. *Trends Biochem. Sci.* 12:20−23 (1987).

52. Lee, A. The accumulation of three specific proteins related to glucose-regulated proteins in a temperature sensitive hamster mutant cell line K12. *J. Cell. Physiol.* 106:119−125 (1981).

53. Shui, R., Pouyssegur, J., and Pasten, I. Glucose depletion accounts for the induction of two transformation-sensitive membrane proteins in Rous sarcoma virus-transformed chick embryo fibroblasts. In *Proc. Natl. Acad. Sci. USA* 74:3840−3844 (1977).

54. Munro, S. and Pelham, H. An HSP70-like protein in the ER: Identity with the 78kD glucose-regulated protein and immunoglobulin heavy chain binding protein. *Cell* 46:291−300 (1986).

55. Hightower, L. and White, F. Preferential synthesis of rat heat shock and glucose regulated proteins in stressed cardiovascular cells. In *Heat Shock from Bacteria to Man* (M. Schlesinger, M. Ashburner, and A. Tissieres, eds.), Cold Spring Harbor Press, New York, pp. 369−377 (1982).

56. Resendez, E., Attenello, J., Grafsky, A., Chang, C., and Lee, A. Calcium ionophore A23187 induces expression of glucose-regulated genes and their heterologous fusion genes. *Mol. Cell. Biol.* 5:1212–1219 (1985).
57. Hendershot, L., Bole, D., Kohler, G., and Kearney, J. Assembly and secretion of heavy chains that do not associate posttranslationally with immunoglobulin heavy chain-binding protein. *J. Cell Biol.* 104:761–767 (1987).
58. Pollok, B., Anker, R., Elderidge, P., Hendershot, L., and Levitt, D. Molecular basis of the cell-surface expression of immunoglobulin mu chain without light chain in human B lymphocytes. *Proc. Natl. Acad. Sci. USA* 84:9199–9203 (1987).
59. Bole, D., Hendershot, L., and Kearney, J. Postranslational association of immunoglobulin heavy chain binding protein with nascent heavy chains in non-secreting and secreting hybridomas. *J. Cell Biol.* 102:1558–1566 (1986).
60. Sharman, S., Rogers, L., Bandsma, J., Gething, M., and Sambrook, J. SV40 T antigen and the exocytic pathway. *EMBO J.* 4:1479–1489 (1985).
61. Dorner, A., Boyle, D., and Kaufman, R. The relationship of N-linked glycosylation and heavy chain binding protein associated with the secretion of glycoproteins. *J. Cell Biol.* 105:2665–2674 (1987).
62. Gething, M., McCammon, K., and Sambrook, J. Expression of wild-type and mutant forms of influenza hemagglutinin: The role of the folding in intracellular transport. *Cell* 46:939–950 (1986).
63. Dorner, A., Krane, M., and Kaufman, R. Reduction of endogenous GRP78 levels improves secretion of a heterologous protein in CHO cells. *Mol. Cell. Biol.* 8:4063–4070 (1988).
64. Kodieck, M. Enhanced glucose consumption in erythrocytes under mechanical stress. *Cell. Biochem. Function* 4:153–155 (1986).
65. Diamond, S. L., Eskin, S. G., and McIntire, L. V. Fluid flow stimulates plasminogen activator secretion by cultured human endothelial cells. *Science* 243:1483–1485 (1989).
66. Strous, G. and Lodish, H. Intracellular transport of secretory and membrane proteins in hepatoma cells infected by vesicular stomatitis virus. *Cell* 22:709–717 (1980).
67. Gibson, R., Kornfeld, S., and Schlesinger, S. The effect of oligosaccharide chains of different sizes on the maturation and physical properties of the G protein of vesicular stomatitis virus. *J. Biol. Chem.* 256:456–462 (1981).

68. Hickman, S., Kulczycki, A., Jr., Lynch, R., and
    Kornfeld, S. Studies of the mechanisms of tunicamycin inhibi-
    tion of IgA and IgE secretion by plasma cells. *J. Biol. Chem.*
    252:4402–4408 (1977).
69. Fitting, T. and Kabat, D. Evidence for a glycoprotein "sig-
    nal" involved in transport between subcellular organelles.
    J. Biol. Chem. 257:14011–14017 (1982).
70. Leavitt, R., Schlesinger, S., and Kornfeld, S. Impaired
    intracellular migration and altered solubility of nonglycosylated
    glycoproteins of vesicular stomatitis virus and sindbis virus.
    J. Biol. Chem. 252:9018–9023 (1977).
71. Domas, R., Ruusala, A., Machamer, C., Helenius, J.,
    Helenius, A., and Rose, J. Different effects of mutation in
    three domains on folding, quaternary structure and intra-
    cellular transport of vesicular stomatitis virus G protein.
    J. Cell Biol. 107:89–90 (1988).
72. Wu, G., Hozumi, N., and Murialdo, H. Secretion of a
    lambda-2 immunoglobulin chain is prevented by a single amino
    acid substitution in its variable region. *Cell* 33:77–83 (1983).
73. Pardee, A. The yang and yin of cell proliferation: An over-
    view. *J. Cell. Physiol.* (Suppl.) 5:107–110 (1987).
74. Sporn, M. and Todaro, G. Autocrine secretion and malignant
    transformation of cells. *N. Engl. J. Med.* 301:878–880 (1980).
75. Sporn, M. and Roberts, A. Autocrine growth factors and
    cancer. *Nature* 313:745–747 (1985).
76. Hannik, M. and Donaghue, D. J. Autocrine stimulation by the
    v-sis gene product requires a ligand-receptor interaction at
    the cell surface. *J. Cell Biol.* 107:287–298 (1988).
77. Neufeld, G., Mitchell, R., Ponte, P., and Gospodorowicz, D.
    Expression of human basic fibroblast growth factor cDNA in
    baby hamster kidney-derived cells, results in autonomous cell
    growth. *J. Cell Biol.* 106:1385–1394 (1988).
78. Terracio, L., Ronnstrand, L., Tingstrom, A., Rubin, K.,
    Claesson-Welsh, L., Funa, K., and Heldin, C. Induction of
    platelet-derived growth factor receptor expression in smooth
    muscle cells and fibroblast upon tissue culturing. *J. Cell
    Biol.* 107:1947–1957 (1988).
79. King, C., Kraus, M., Williams, L., Merlino, G., Pastan, I.,
    and Aaronson, S. Human tumor cell lines with EGF receptor
    gene amplification in the absence of aberrant sized mRNA's.
    Nucleic Acid Res. 13:8477–8486 (1985).
80. Libermann, T., Nusbaum, M., Razon, N., Kris, R., Lax, I.,
    Soreq, H., Whittle, N., Waterfield, M., Ullrich, A., and
    Schlessinger, J. Amplification enhanced expression and

possible rearrangement of EGF receptor gene in primary human brain tumors of glial origin. *Nature* 313:144–147 (1985).
81. Werner, M., Humphrey, P., Bigner, D., and Bigner, S. Growth effects of epidermal growth factor (EGF) and monoclonal antibody against the EGF receptor on four glioma cell lines. *Acta Neuropathologica* 77:196–201 (1988).
82. Velu, T., Bequinot, L., Vass, W., Willingham, M., Merlino, G., Pastan, I., and Lowy, D. Epidermal growth factor-dependent transformation by a human EGF receptor proto-oncogene. *Science* 238:1408–1410 (1988).
83. Collins, M., Downward, J., Miyajima, A., Maruyama, K., Arai, K., and Mulligan, R. Transfer of functional EGF receptors to an IL-3 dependent cell line. *J. Cell. Physiol.* 137: 293–298 (1988).
84. DiFiore, P., Pierce, J., Fleming, T., Mazan, R., Ulrich, A., King, C., Schlessinger, J., and Aaronson, S. Overexpression of the human EGF receptor confers an EGF-dependent transformed phenotype of NIH 3T3 cells. *Cell* 51:1063–1070 (1987).
85. King, C., DiFiore, P., Pierce, J., Segatto, O., Kraus, M., and Aaronson, S. Oncogenic potential of the erbB-2 gene: Frequent overexpression in human mammary adenocarcinomas and induction of transformation in vitro. In *Growth Regulation of Cancer* (M. E. Lippman, ed.), Alan R. Liss, New York, pp. 189–199 (1988).
86. Hirai, M., Gamou, S., Minoshima, S., and Shimizu, N. Two independent mechanisms for escaping epidermal growth factor-mediated growth inhibition in epidermal growth factor receptor-hyperproducing human tumor cells. *J. Cell Biol.* 107:791–799 (1985).
87. Merlino, G., Xu, Y., Ishii, S., Clark, A., Semba, K., Toyoshima, K., Tamamoto, T., and Pastan, I. Amplification and enhanced expression of the epidermal growth factor receptor gene in A431 human carcinoma cells. *Science* 224: 417–419 (1984).
88. Lin, C., Chen, W., Kruiger, W., Stolarsky, L., Weber, W., Evans, R., Verman, I., Gill, G., and Rosenfeld, M. Expression cloning of human EGF receptor complementary DNA: Gene amplification and three related messenger RNA products in A431 cells. *Science* 224:843–848 (1984).
89. Yamamoto, T., Kamat, N., Kawano, H., Shizu, S., Kuroki, T., Toyoshima, K., Rikimaru, K., Nomura, N., Ishizaki, R., Pastan, I., Gamou, S., and Shimizu, N. High incidence of amplification of epidermal growth factor receptor gene in

human squamous carcinoma cell lines. *Cancer Res.* 46:414−416 (1986).

90.   Gill, G., Santon, J., and Bertics, P.   Regulatory features of the epidermal growth factor receptor. *J. Cell. Physiol.* (suppl.) 5:31−35 (1987).

91.   Hendler, F., Shum-Siu, A., Nanu, L., and Ozanne, B.   Over-expression of EGF receptors in squamous tumors is associated with poor survival in growth factors and their receptors: Genetic control and rational application. *J. Cell. Biochem.* (suppl.) 12A:105 (1988).

92.   Pierce, J., Ruggiero, M., Fleming, T., DiFiore, P., Greenberger, J., Varticovski, L., Schlessinger, J., Rovera, G., and Aaronson, S.   Signal transduction through the EGF receptor transfected in IL-3 dependent hematopoietic cells. *Science* 239:628−631 (1988).

93.   Zajchowski, D., Band, V., Pauzie, N., Tager, A., Stampfer, M., and Sager, R.   Expression of growth factors and oncogenes in normal and tumor-derived human mammary epithelial cells. *Cancer Res.* 48:7041−7047 (1988).

94.   Bertics, P., Hubler, L., Chen, W., Carpenter, C., Rosenfeld, M., and Gill, G.   Epidermal growth factor receptor: Structure and regulation by self-phosphorylation. In *Growth Regulation of Cancer* (M. E. Lippman, ed.), Alan R. Liss, New York, pp. 157−167 (1988).

95.   Basu, M., Biswas, R., and Das, M.   42,000-Molecular weight EGF receptor has protein kinase activity. *Nature* 311:477−480 (1984).

96.   Carpenter, G.   Receptors for epidermal growth factor and other polypeptide mitogens. *Ann. Rev. Biochem.* 56:881−914 (1987).

97.   Downward, J., Yarden, Y., Mayes, E., Scrace, G., Totty, N., Stockwell, P., Ullrich, A., Schlessinger, J., and Waterfield, M.   Close similarity of epidermal growth factor receptor and v-erb-B oncogene protein sequences. *Nature* 307: 521−527 (1984).

98.   Ullrich, A., Coussens, L., Hayflick, J., Dull, T., Gray, A., Tam, A., Lee, J., Yarden, Y., Libermann, T., Schlessinger, J., Downward, J., Mayes, E., Whittle, N., Waterfield, M., and Seeburg, P.   Human epidermal growth factor receptor cDNA sequence and aberrant expression of the amplified gene in A431 epidermoid carcinoma cells. *Nature* 309:418−425 (1984).

99.   Rizzino, A., Kazakoff, R., Ruff, E., Kuszynski, C., and Nebelsick, J.   Regulatory effects of cell density on the

binding of transforming growth factor beta, epidermal growth factor, platelet-derived growth factor, and fibroblast growth factor. *Cancer Res.* 48:4266–4271 (1988).

100. Rosenthal, A., Linquist, P., Bringman, T., Goeddel, D., and Derynck, R. Expression in rat fibroblast of a human transforming growth factor-alpha cDNA results in transformation. *Cell* 46:4701–4705 (1986).

101. Bryckaert, M., Lindroth, M., Lonn, A., Tobelem, G., and Wasteson, A. Transforming growth factor (TGF beta) decreases the proliferation of human bone marrow fibroblasts by inhibiting the platelet-derived growth factor (PDGF) binding. *Exp. Cell Res.* 179:311–321 (1988).

102. Bowen-Pope, D., Dicorleto, P., and Ross, R. Interactions between the receptors for platelet-derived growth factor and epidermal growth factor. *J. Cell Biol.* 96:679–683 (1983).

103. Zackary, F. and Rosengurt, E. Modulation of the epidermal-growth-factor receptor by mitogenic ligands: Effects of bombasin and role of protein kinase-C. *Cancer Surv.* 4: 729–765 (1985).

104. Davis, R. J. and Czech, M. P. Tumor promoting phorbol diesters cause the phosphorylation of epidermal growth factor receptors in normal fibroblasts at threonine-654. *Proc. Natl. Acad. Sci. USA* 81:4080–4084 (1985).

105. Brewitt, B. and Clark, J. Growth and transparency in the lens, an epithelial tissue, stimulated by pulses of PDGF. *Science* 242:777–779 (1988).

106. Terranova, V. P. and Wikesjo, U. M. E. Extracellular matrices and polypeptide growth factors as mediators of functions of cells of the periodontium: A review. *J. Periodontol.* 58:371–380 (1987).

107. Rhudy, R. W. and McPherson, J. M. Influence of the extracellular matrix on the proliferative response of human skin fibroblasts to serum and purified platelet-derived growth factor. *J. Cell. Physiol.* 137:185–191 (1988).

108. Folkman, J. and Moscona, A. Role of cell shape in growth control. *Nature* 273:345–349 (1978).

109. Ingber, D. E. and Folkman, J. Regulation of endothelial growth factor action: Solid state control by extracellular matrix. In *Mechanisms of Signal Transduction by Hormones and Growth Factors* (M. C. Cabot and W. L. McKeehan, eds.), Alan R. Liss, New York, pp. 273–282 (1987).

110. Gospodarowicz, D., Greenburg, G., and Birdwell, C. Determination of cellular shape by the extracellular matrix and its correlation with the control of cellular growth. *Cancer Res.* 38:4155–4171 (1978).

111. Salomon, D., Kiotta, L., and Kidwell, W. Differential re-
     sponse to growth factor by rat mammary epithelium plated on
     different collagen substrata in serum-free medium. *Proc.
     Natl. Acad. Sci. USA* 78:382–386 (1981).
112. Kidwell, W. R., Mohananm, S., Sanfilippo, B., and Solomon,
     D. S. Tissue organization and cancer: Role of autocrine
     growth factors in extracellular matrix biosynthesis. In
     Growth Regulation of Cancer (M. E. Lippman, ed.), Alan R.
     Liss, New York, pp. 145–150 (1988).
113. Tobey, R., Valdez, J., and Crissman, H. Synchronization of
     human diploid fibroblasts at multiple stages of the cell cycle.
     Exp. Cell Res. 179:400–416 (1988).
114. Tobey, R. and Ley, K. Isoleucine-mediated regulation of
     genome replication in various mammalian cell lines. *Cancer
     Res.* 31:46–51 (1971).
115. Yaish, P., Gazit, A., Gilon, C., and Levitzki, A. Blocking
     of EFG-dependent cell proliferation by EGF receptor kinase
     inhibitors. *Science* 242:933–935 (1988).
116. Paulson, J., Colley, K., Lee, E., and Roth, J. Characteriza-
     tion of the expression of beta-galactoside alpha-2,6-sialyl-
     transferase in Chinese hamster ovary and COS-1 cells. Pro-
     grams and Abstracts for the 17th Annual Meeting of the So-
     ciety for Complex Carbohydrates, San Antonio, Texas,
     3–5 November 1988.
117. Brinckerhoff, C., Mitchell, T., Karmilowicz, M., Kluve-
     Beckerman, B., and Benson, M. Autocrine induction of
     collagenase by serum amyloid A-like and beta-2-microglobulin-
     like proteins. *Science* 243:655–657 (1989).

4

Mammalian Cell Physiology

JAMES N. THOMAS
Immunex Corporation, Seattle, Washington

I. INTRODUCTION

Mammalian cells are gaining popularity as expression systems for complex proteins of human pharmaceutical interest in the biotechnology industry. This is largely due to their ability to make proteins essentially identical to those made in vivo, in contrast to prokaryotic systems that do not glycosylate or perform other complex posttranslational modifications. A disadvantage of the mammalian system is the relatively high cost of production. For some products a benefit may be realized in reduced purification costs, but large-scale cell culture can still be an expensive proposition. Factors that contribute to its cost are relatively low product yields per volume of medium, complex medium design, and the need for ultrapure biochemicals and water.

How can the cost of this technology be reduced? Many avenues are currently being explored, from basic research into expression vectors, to improved, more efficient manufacturing processes. A fruitful area of work in the future will be to gain a better understanding of the physiology of cells. This information can then be applied to process optimization and/or the development of improved cell lines for use as expression vehicles. A broad definition of cell physiology might be all the vital functions, both physical and chemical, that interact to collectively form a living cell. A cell, in reality, is a "biological machine" with all the necessary machinery (enzymes, transporters, pumps, etc.) to perform the desired work. In our case the work is to make large quantities of complex, glycosylated proteins. To perform this work, cells must have sufficient energy for all their biosynthetic and maintenance needs, they must be able to transport and metabolize substrates for macromolecular synthesis, and they must possess the type and quantity of metabolic (enzymatic) machinery needed to accomplish the task.

What are the rate-limiting chemical or physical processes in cells that govern the amount of genetically engineered protein made? Placing powerful expression vectors into mammalian cells and amplifying gene copies is all well and good, but do the cells possess the necessary internal machinery to process the message and make large quantities of the desired protein? Can they transform enough of their energy substrates into ATP to fuel the synthesis of large quantities of genetically engineered proteins? Are flux-regulating steps in metabolic pathways limiting the synthesis of macromolecules like mRNA and DNA? These are but a few of the many questions related to the physiology of mammalian cells that have only just begun to be asked. Many critical steps lie between transcription of the engineered gene and its secretion as the finished product. Identification of the rate-limiting steps, and their subsequent elimination through genetic engineering, could pay great dividends by increasing the amount of product synthesized per cell.

Can we increase cell yield per volume of medium in mammalian cell culture? Mammalian cells reach only a small fraction of the dry cell mass that yeast and bacterial fermentations achieve in conventional batch fermentations. Cell mass yields per volume of medium for bacteria and yeast are at least 10–50 times those now achievable by mammalian cells. In addition, bacteria and yeast can accomplish these yields several times faster. What regulates cell growth rate and cell densities in mammalian cell culture? Can we alter the growth rate by altering flux-limiting steps in key metabolic pathways? Answering these and other relevant questions could increase cell yields per volume of medium, providing another way of increasing product yield per volume of medium without altering the specific production of the protein per cell.

The subject of cell physiology is obviously a rather broad topic and would take several volumes to cover adequately. Therefore this chapter will touch only on aspects of cell physiology that include the transport and metabolism of key nutrients, specifically amino acids and carbohydrates. Also included in this chapter is a discussion of the role of oxygen in cell culture. The metabolism of these nutrients is central to the overall health and function of mammalian cells, and they represent, on a molar basis, the most rapidly used nutrients in the cell culture environment. Some understanding of how cells utilize these nutrients should help in developing strategies for their optimization in cell culture systems. In addition, much of the discussion will be directed toward events or factors that alter the transport and/or metabolism of the key nutrients mentioned. This information will hopefully stimulate some thought as to how the study of cell physiology, and specifically the study of cell metabolism, might lead to improvements in the production of genetically engineered proteins.

II. CARBOHYDRATES

Carbohydrates perform a variety of functions in mammalian cells, from precursors of other small molecules and macromolecules to oxidizable energy substrates. This section will briefly describe which carbohydrates are most readily utilized by cultured cells and what factors or conditions influence their transport and metabolism. In general, the pathways of carbohydrate metabolism will not be discussed in detail since the reader can find this information in general biochemistry texts.

A. Utilization

The early work of Eagle et al. (1) demonstrated that both normal and transformed cells were able to grow in medium containing a carbohydrate other than glucose. The cell lines used in their experiments included normal cells derived from adult human conjunctiva and liver; three normal lines derived from human embryonic intestine, heart, and connective tissue; and four transformed lines which included HeLa, KB, H.Ep. 1, and mouse L cells. They found that two disaccharides (trehalose and turanose), three hexoses (D-fructose, D-galactose, and D-mannose), and two phosphate esters (D-glucose-1-phosphate and D-glucose-6-phosphate) promoted growth similar to glucose. The presence of serum was probably beneficial in utilization of the disaccharides, and the presence of 1 mM pyruvate was necessary for proliferation with some carbohydrates. An early review on carbohydrate and energy metabolism by Paul (2) describes growth of several mammalian cells on similar substrates. In addition, some cells were also able to use xylose, ribose, sorbitol, cellobiose, and melibiose as energy substrates. An extensive study of 93 carbohydrates was conducted by Burns et al. (3) using five different cell lines [CHO-K1, a CHO-K1 mutant lacking glucose-6-

phosphate dehydrogenase and hypoxanthine-guanine phosphoribosyl-transferase, mouse LM (TK), B-16 (mouse pigmented melanoma), and (HTC hepatoma cells resistant to 8-azaguanine)]. Glucose oxidase and catalase were added to the culture medium to eliminate glucose contributed by enzymatic hydrolysis of polymers and glucose present in reagents as a contaminate. A total of 15 carbohydrates of the 93 were found to support proliferation. These were similar to those previously found (1,2), with the exception of maltotriose, dextrin, glycogen, amylose, and D-galacturonate. In general, the di-, tri-, and polysaccharides found to support growth contained glucose monomers (3). The addition of pyruvate (1 mM) to the culture medium was necessary to support cell growth in the presence of some carbohydrates. Of special interest is the utilization of polysaccharides such as dextrin, glycogen, and amylose as energy substrates. Cells apparently were able to transport and intracellularly cleave these polysaccharides to liberate glucose. Cell proliferation with these polymers, in the presence of 1 mM pyruvate, was often better than in the presence of glucose alone.

Scannell and Morgan (4) determined that the growth-promoting effects of starch and maltose in CHO-K1 cultures was due to the activity of serum enzymes that hydrolyzed the polymers into glucose monomers. They found that growth on maltose and starch was greatly diminished when serum was heat-treated, although maltose- and starch-utilizing variants of CHO-K1 cells have been isolated which evidently contain amylase and maltase activity (5).

A review by Morgan and Faik (6) describes a list of carbohydrates that support the growth of mammalian cells in culture (Table 1). For a more extensive review of growth-promoting carbohydrates and their metabolism, see also Morgan and Faik (7).

Table 1 Substrates That Support the Growth of Cultured Animal Cells

Hexoses	D-fructose, D-galactose, D-glucose, D-mannose
Pentoses	D-ribose, D-talose, D-xylose
Disaccharides	Cellobiose, melibiose, turanose, maltose, trehalose
Trisaccharides	Maltotriose
Polysaccharides	Amylose, dextrin, glycogen
Alcohols	Sorbitol
Sugar phosphates	Glucose-1-phosphate, glucose-6-phosphate

Source: Adapted from Ref. 6.

B. Transport

The glucose transport system has been the most extensively studied transport system in animal cells. Since relatively little information is available describing other carbohydrate transport systems, the following discussion will be limited to the transport of glucose.

Transport of glucose across the plasma membrane of animal cells is mediated through a protein transporter that functions as part of a facilitated diffusion or an active transport process (8–19). Human erythrocytes and rat adipocytes are known to transport glucose by a facilitated diffusion system (8,9), whereas epithelia have been shown to transport glucose across their plasma membranes by an active transport mechanism (12). Other cells such as L6 rat myoblasts and chick embryo fibroblasts may contain both facilitated diffusion and active transport systems (13,14). In studies with L6 myoblasts and chick embryo fibroblasts, two glucose analogues, 2-deoxyfluoro-D-glucose (2-DOG) and 3-O-methyl-D-glucose (3-OMG), were used to distinguish between high- and low-affinity transporters. The 2-DOG analog is specific for the high-affinity or active transport system and 3-OMG is specifically carried by a low-affinity or facilitated diffusion transporter (13,14). In these studies energy uncouplers inactivated 2-DOG transport but had no effect on 3-OMG transport.

Western blot analysis of the glucose transporter for 3T3-L1 adipocytes indicates that the protein is a molecule with an average molecular weight of 55,000 (15). Two proteins have also been identified in chicken embryo fibroblasts with approximate molecular weights of 41,000 and 82,000 that are believed to be the monomer and dimer forms of the chicken embryo cell transporter (16). Recent work elucidating the amino acid sequence of the glucose transport proteins from human hepatoma (19) and rat brain (18) suggest a heterogeneously glycosylated protein of approximately 55,000 molecular weight that possesses a common sequence homology of 96.7%. Maiden et al. (17) compared the arabinose-H^+ and the xylose-H^+ membrane transport of *E. coli* to the human hepatoma and rat brain glucose transporters and found an amino acid sequence homology of about 40%. They predict from the hydropathic profiles of the molecules that 12 transmembrane regions are present. The similarities of transporters from such divergent sources may underline their critical function in cells.

A number of factors, including changes in the physical culture environment, can influence the transport of glucose into mammalian cells. Toyoda et al. (20) observed a threefold increase in the rate of 3-O-methyl-D-glucose uptake in adipocytes when the pH of the incubation buffer was raised from 7 to 8. They also observed an approximately 40% increase in glucose transport in hypertonic buffer (presence of 300 mM sorbitol). Both of these observations were in the absence of insulin; in the presence of insulin neither condition increased transport over insulin-containing controls. Miller et al. (21) found that

utilization of glucose by chick embryo fibroblasts is strongly influenced by the presence of bicarbonate (25mM). A higher concentration of glucose (16.7mM compared to 2.7mM) in the presence of
insulin was also shown to enhance the transport of glucose. Enhancement (18–65% of basal) was due to an increase in V_{max} without
a significant change in K_m of transport. Glutamine and serum have
also been shown to increase the transport of glucose in NIL hamster
fibroblasts (22).

Insulin is known to increase the transport of glucose in a variety of tissues (23,24). Some believe that the action of insulin may
be mediated through changes in intra- and extracellular calcium
concentrations (25,26). Another current theory for the stimulation
of glucose uptake by insulin in fat and muscle cells involves its enhancement of the translocation of glucose transporters from intracellular membranes to the plasma membrane (15,27–32). When cells
undergo subcellular fractionation and are then assayed for the
presence of the transporter, it can be found in both the plasma
membrane and the light microsome fractions. If these cells are subjected to insulin stimulation, then a progressive increase in the
amount of transporter is seen in the plasma membrane fraction, with
a corresponding decrease of transporter in the light microsome fraction. Vesicles, similar in size and density to endosomes, have been
purified from cell homogenates that are enriched in the glucose
transport protein (29). These vesicles are believed to be involved
in the translocation process of the transporter from an intracellular
pool to the cell membrane.

The addition of transforming growth factor β and/or epidermal
growth factor to cultures of 3T3 cells, NRK-49F (rat kidney cells),
or human fibroblasts has also been shown to stimulate the uptake of
glucose as measured by 2-deoxyglucose transport (33,34). Apparently, phorbol esters will also increase the transport of glucose
in rat adipocytes (35) and in chicken embryo fibroblasts (36). An
increased number of glucose transport proteins were found in the
plasma membrane fraction of these cells and this correlated well with
an elevated V_{max} and a constant K_m.

Gene expression of the facilitated glucose transporter protein
has been recently studied in BALB/c3T3, NIH3T3, and rat-2 cells
(37). The addition of 15% calf serum to confluent cultures induced
a five- to 10-fold increase in glucose transporter mRNA. This result is presumably due to the growth factors and hormones contained
in serum. An accumulation of glucose transporter mRNA was also
demonstrated in BALB/c3T3 cells in response to fibroblast growth
factor, platelet-derived growth factor (PDGF), epidermal growth
factor (EGF), and 12-O-tetradecanoylphorbol-13-acetate (TPA).
Other investigators found that PDGF increased both the rate of
glucose transport and the level of membrane-associated glucose

transporter by 1.7-fold in BALB/c3T3 cells (38). This result was accompanied by a five-fold increase in the accumulation of glucose transporter mRNA.

Horner et al. (39) discovered a correlation between dexamethasone concentration and the amount of glucose transporter protein in the plasma membrane fraction of human diploid fibroblasts. They found that cells treated with 100 nM dexamethasone for 4 hr had 40% less plasma membrane-associated glucose transporters than controls. A drop of 48% in the amount of glucose transported was also noted, and these changes could be reversed by treating cells with 200 nM insulin for 30 min. The authors concluded that dexamethasone caused a translocation of glucose transport proteins from the plasma membrane to an internal location and that insulin reversed this translocation process.

Transport of glucose can also be stimulated by viral transformation with DNA or RNA tumor viruses. Hatanaka et al. (40) demonstrated an enhanced D-glucose uptake in normal mouse embryo cells after infection with three strains of murine sarcoma virus (MSV). Increased transport correlated well with morphological changes. Mannose and galactose transport are also increased in sarcoma virus-infected cells (41). Lang and Weber (42), using a temperature-sensitive Rous sarcoma virus, demonstrated an increased uptake of both 2-deoxyglucose and 3-O-methylglucose with chicken embryo fibroblasts. When cells were shifted to permissive conditions for transformation, an enhanced rate of transport of both glucose analogs occurred within 3 hr. Hexose transport decreased when normal cell cultures reached confluency, but virally transformed cells transported hexose at higher rates in growing and confluent cultures (43,44). In general, transport of 2-deoxy-D-glucose and/or 3-O-methyl-D-glucose have been shown to increase in a number of cell lines transformed by either RNA or DNA virus (40−47), and increased transport activity seems to be dependent on an increase in V_{max} with no change in K_m.

Recently, Shawver et al. (48) studied the turnover of glucose transport proteins in cultured fibroblasts transformed with the *src* oncogene. They found that transformation by *src* did not increase the synthesis of these proteins but instead reduced their turnover rate. This led to an accumulation of glucose transporter proteins in the transformed fibroblasts.

C. Metabolism

Glycolytic activity (the production of lactate from glucose) of cells in culture will depend on a number of conditions including the concentration of dissolved oxygen in the culture environment (49−52), the type of carbohydrate (53,54), exposure of cells to growth factors or mitogens (55−58), and whether the culture is primary, established, or transformed (59−61).

Warburg (52) noted long ago that both normal and tumor cells convert glucose to lactic acid, a process that he called glycolysis. He found that normal cells metabolize through glycolysis in the absence of oxygen but that lactate production almost disappears in the presence of oxygen. Warburg (52) also noted that tumor cells produce lactic acid in the presence or absence of oxygen, but the formation of lactic acid is generally lower in the presence of adequate oxygen concentrations (49−51) (as discussed later in Section V).

Lactate formation by some types of cells can be decreased by switching from glucose as a primary source of carbohydrate to other monosacharrides, such as fructose or galactose (53,54,62−65). The reason for this reduced lactate formation is presumably the slower transport and metabolism of fructose and galactose by cultured cells. This would reduce the amount of carbon entering the glycolytic pathway, reduce cellular pyruvate pools, and effectively reduce the amount of lactic acid formed. Since glycolysis is diminished in these cultures, they must adjust their metabolism to favor an aerobic production of ATP, presumably by using another oxidizable energy substrate such as glutamine. Reitzer et al. (66) found that HeLa cells were able to grow equally well on fructose, galactose, or glucose. They calculated that in the most extreme case, cells cultured on 2 mM fructose, the glycolytic activity of HeLa cells was reduced by approximately 900-fold. Their conclusion was that under these conditions almost all the fructose carbon passed through the pentose phosphate pathway for use as a biosynthetic precursor. In another paper, Reitzer et al. (67) suggested that the major function of the pentose phosphate pathway in HeLa cells was the generation of ribose-5-phosphate for nucleic acid synthesis. An earlier paper by Zielke et al. (68) showed the feasibility of growing normal human fibroblasts in the absence of glucose (for one or two cell divisions) when the medium was supplemented with hypoxanthine, thymidine, and uridine. Glutamine was the energy substrate used by HeLa cells (66,67) and by normal human fibroblasts (68,69) when their glycolytic rate was slowed. A discussion of the metabolism of glutamine and other amino acids will be covered in the next section of this chapter.

Cells that are able to use "alternate" carbohydrates such as galactose or fructose generally produce less lactate, consume more glutamine and oxygen, and produce more ammonia. A more stable pH brought about by these changes in metabolism has been exploited through the development of bicarbonate-free media that can be used to culture cells outside of a CO_2-controlled environment (53,62,70). A modified L-15 medium was used to culture Vero cells in microcarrier perfusion cultures to densities approaching 3.0×10^7 cells/ ml without pH control (65).

Bissell et al. (59) studied the metabolism of glucose in primary cultures of adult hepatocytes, and in two permanent cell lines [Buffalo rat liver (BRL) and transplantable rat hepatoma (HTC) cells derived from normal rat liver and rat hepatoma, respectively]. They found that primary cultures consumed glucose 40 times more slowly than the permanent cell lines. In addition, the primary cultures converted 3 - 6 times more glucose to CO_2 than permanent cultures. These results suggest that the primary cultures were able to obtain much of their energy requirements through the oxidation of glucose. As time passed the primary cultures also underwent adaptation, gradually changing their metabolic pattern to resemble those of the permanent cell lines.

Exposure of cells to growth factors, such as EGF and TGFβ have been shown to stimulate glycolysis (55,56). Matrisian et al. (71) found that when quiescent rat fibroblasts were stimulated to grow in the presence of EGF or serum, mRNAs for lactate dehydrogenase and other glycolytic enzymes were elevated. Brand et al. (57) observed an increase in glucose uptake in rat thymocytes exposed to the mitogen concanavalin A. A large percentage of the glucose was converted to lactate, but the mitogen also caused an increase in glucose oxidation. In another study, Brand (58) found that cultured rat thymocytes, stimulated to proliferate by concanavalin A and interleukin 2, increased their glucose metabolism 53-fold. The stimulated (proliferating) thymocytes converted 90% of metabolized glucose to lactate, whereas resting thymocytes converted only 53%.

Racker et al. (56) transfected rat-1 cells with *ras* and *myc* oncogenes. They found a high rate of glycolysis associated with cells transfected with a *ras* oncogene, while *myc*-transfected cells showed little increase in glycolysis. When the rat-1 cells transfected with the *myc* oncogene were subsequently exposed to TGFβ, they also showed a higher rate of glycolysis.

The growth rate or biochemical activity of some cells is correlated with their rate of glycolysis (72 - 74). Elicited (inflammatory) macrophages show a dramatic increase in glycolytic activity that may be related to their increased phagocytic or secretory activities (74). Freshly isolated colonocytes from rat colon, normally rapidly growing cells in vivo, were shown to convert 83% of their utilized glucose to lactate (72). Resting and stimulated lymphocytes also exhibit high glycolytic activity, converting very little of the consumed glucose to CO_2 (75,76). It was speculated that higher rates of glycolysis better enable cells to respond quickly to stimuli that call for an increased growth rate or an increased biosynthetic capacity.

Table 2   Comparison of Specific Activities of Glycolytic and Gluconeogenic Enzymes in Two Rapidly Growing Tumors to Normal Liver and Kidney Tissue

Enzymes	Hepatoma 3683 (% of normal liver)	MK-3 kidney tumor (% of normal kidney)
Glycolysis		
Hexokinase	1430	632
6-Phosphofructokinase	277	505
Pyruvate kinase	499	782
Gluconeogenesis		
Glucose-6-phosphate	<1	10
Fructose-biphosphate	<1	18
PEP carboxykinase	<1	—
Pyruvate carboxylase	<1	—

Source: Adapted from Ref. 59.

     The early work of Weber (77), Aisenberg and Morris (78), and Lin et al. (79) suggests a correlation between growth rate and the rate of glycolysis in hepatoma. They compared the growth rate of several transplantable hepatocellular carcinomas to glycolytic activity and found that intermediate and fast-growing hepatomas exhibited higher rates of glycolysis than normal hepatocytes or slow-growing, more differentiated hepatomas. This prompted Weber and others (61, 80–85) to examine the activity of enzymes involved in glycolysis, gluconeogenesis, the pentose phosphate pathway, purine and pyrimidine synthesis, and DNA synthesis. Comparisons of glycolytic and gluconeogenic enzyme activities in normal liver and in liver tumors provided strong evidence that upon transformation the activities of key glycolytic enzymes were greatly increased and those of key gluconeogenic enzymes were decreased. This relationship also held true for the other major gluconeogenic tissue in the body, the renal cortex. A comparison of these enzymatic activities in liver and kidney tumors is given in Table 2.
     The activities of key enzymes in the pentose phosphate pathway were also found to be elevated in tumor as compared to normal liver tissue but were generally not linked to growth rate. However, the activity of phosphoribosylpyrophosphate (PRPP) synthetase was

Table 3   Comparison of Specific Activities of Pentose Phosphate
Enzymes from Transformed and Normal Tissue of Liver and Kidney
Origin

Enzymes	Hepatoma 3683F (% of normal liver)	MK-3 kidney tumor (% of normal kidney)
Pentose phosphate synthesis		
Direct oxidation		
Glucose-6-phosphate DH[a]	903	1103
6-Phosphogluconate DH[b]	203	867
Indirect oxidation		
Transaldolase	255	454
Transketolase	277	198
Pentose phosphate utilization		
PRPP synthetase	278	259

[a]Glucose-6-phosphate DH stands for glucose-6-phosphate dehydrogenase.

[b]6-Phosphogluconate DH stands for 6-phosphogluconate dehydrogenase.
Source: Adapted from Ref. 85.

found to be positively correlated with the tumor growth rate.
PRPP can be used in both salvage and de novo pathways of purine
and pyrimidine biosynthesis, functioning as an allosteric activator
of the rate-limiting enzyme of de novo pyrimidine biosynthesis.
Activation of these enzymes would therefore greatly favor DNA
replication and growth. Activities of enzymes involved in pentose
phosphate synthesis and utilization are compared in transformed
and normal tissues from both liver and kidney in Table 3. Weber
(84,85) concluded that an imbalance in the carbohydrate metabolism
of cancer cells is the result of reprogramming of gene expression.
He suggests that the expression of key enzymes will alter the
phenotype of cancer cells, giving them a growth advantage over
normal tissue.

Clearly, there are many factors that can alter the rate of carbohydrate uptake and glycolysis in mammalian cells. The significance of an altered glycolytic rate has been debated for many years and will not be resolved in this chapter. The strong correlation between increased carbohydrate consumption, elevated glycolytic activity, and increased growth rate in mammalian cells is nevertheless intriguing, and difficult to explain as merely coincidental.

III. AMINO ACIDS

Amino acids play a central role in the metabolism of animal cells in culture. They perform many diverse functions, from building materials for protein synthesis to acting as oxidizable energy substrates. Many amino acids serve as precursors for important small molecules within cells. Examples of these include polyamine synthesis from arginine, glutathione synthesis from glycine, choline synthesis from serine, and purine synthesis from glycine, glutamine, and aspartate (86). This chapter will briefly discuss amino acids most often utilized from cell culture media, their systems of transport into cells, and some factors and conditions that alter their transport and metabolism.

A. Utilization

The early work of Eagle (87) demonstrated a need for 12 amino acids to support the proliferation of strain L mouse fibroblasts in medium containing $0.25-2\%$ dialyzed horse serum. Cells would die within $1-3$ days in the absence of any one of the 12 amino acids. Glutamine was later added to this list (88), which also includes arginine, cyst(e)ine, histidine, isoleucine, leucine, lysine, methionine, phenylalanine, threonine, tryptophan, tyrosine, and valine. These amino acids, with a few exceptions, form the backbone of many serum-containing as well as serum-free media in use today.

The rates at which these and other amino acids are utilized or produced can vary dramatically between cell lines, as illustrated in Table 4. The relative concentration of amino acids and serum in the culture medium and other conditions of the culture environment will also influence the rates of utilization or production of specific amino acids.

In general, the utilization of glutamine from the culture medium is at least $5-10$ times faster than that of other amino acids $(94-99)$. Amino acids other than glutamine that are almost always utilized are leucine, isoleucine, lysine, threonine, valine, tyrosine, methionine, histidine, tryptophan, arginine, and phenylalanine. Amino acids that will be produced or consumed, depending on the cell line and to a certain degree the culture conditions, are alanine, serine, glycine, proline, aspartic acid, and glutamic acid.

B. Transport

Christensen (100) originally defined systems A, ASC, L, and Ly^+ for the transport of amino acids into Ehrlich ascites tumor cells. Systems A, ASC, and Ly^+ are all sodium ion-dependent and system L is sodium ion-independent. This nomenclature is still used with the exception of the Ly^+ system which is now referred to as system y^+. Neutral amino acids are believed to be transported primarily in systems A, ASC, and L in human diploid fibroblasts (101). The same three systems are operative in Chinese hamster ovary (CHO) cells (102,103), and most of the neutral amino acids are transported to a varying degree by all three transporters. The transport of glutamine and alanine in rat colonocytes appears to be through a common carrier system, similar to system A found in other cells (104). The author suggests that in these cells intracellular alanine might exchange for extracellular glutamine. Glutamine is primarily taken up by system ASC in CHO cells (102) and in rat mesenteric lymphocytes (105), but in rat liver cells transport of glutamine is exclusively through system N (106,107). Amino acid transport system N is sodium-dependent and apparently specific for amino acid amides like glutamine and asparagine and also for L-histidine.

The aromatic amino acids tyrosine, phenylalanine, and tryptophan are transported in rat liver cells by two sodium-independent systems, the T and L transporters (108,109). The T-transport system was originally defined by Rosenberg et al. (110) to be a sodium-independent system specific for aromatic amino acids in human red blood cells.

The transport of cationic amino acids in human skin fibroblasts has been shown to be through the y^+ transporter (111). When cells were incubated in physiological concentrations of arginine and homoarginine, distribution ratios (the ratio of intracellular concentration to extracellular concentration) of over 20 were achieved. System y^+ also transports lysine in cultured human fibroblasts (112) and is probably identical to the Ly^+ system which functions in other cell types (111).

The other class of transporters found in mammalian cells are those for the transport of anionic amino acids. Bannai and Kitamura (113) found that human diploid fibroblasts (IMR-90) transported L-glutamate and L-cystine via a sodium ion-independent system. This system is probably similar or identical to the x_c^- transporter found in normal human fibroblasts (114). Another anionic system

Table 4   Disappearance or Production of Amino Acids in Different Cell Lines ($\times 10^{-9}$ μM/cell/hr)

Cell type	Jenson sarcoma[a]	MOPC-31 C/R mouse plasmacytoma[b]	L-M mouse fibroblast[c]	Normal human embryo intestine[d]
Glutamine	−59.4	−26.20	−57.60	−62.0
Leucine	−7.15	−3.7	−14.9	−14.9
Isoleucine	−6.68	−15.10	−3.6	−10.4
Lysine	−7.65	−8.70	−1.5	−9.8
Threcnine	−2.65	nc	−1.8	−17.5
Valine	−3.93	−1.6	−3.0	−4.3
Tyrosine	−1.53	−7.0	−1.1	−0.3
Phenylalanine	−2.45	−1.3	−1.1	−0.3
Asparagine	−14.22	nc	nc	nm
Arginine	−2.73	nc	−1.7	−28.3
Methionine	−2.6	−1.9	−1.0	−5.0
Histidine	−2.4	nc	−0.5	−6.7
Alanine	+3.25	nc	+0.4	+9.5
Serine	−7.63	+2.6	+0.3	nc
Glycine	−0.4	+1.9	+6.9	+1.6
Tryptophan	nm	nm	nm	−0.3
Cysti(e)ne	−1.68	nm	−0.8	+0.1
Proline	−4.5	+1.5	+2.9	+1.1
Aspartic acid	+6.6	+0.8	+2.0	nc
Glutamic acid	+14.1	+3.1	+5.0	+4.4

[a]Calculated from Ref. 89.
[b]Calculated from Ref. 90.
[c]Calculated from Ref. 91.
[d]Calculated from Ref. 92.
[e]Calculated from Ref. 93.
Note: nc = no change, nm = not measured.

Normal human pituitary[d]	HeLa Gey- cervical carcinoma[d]	KB, Eagle- squamous cell carcinoma of the lip[d]	H. Ep. No. 2 Fjelde human epitheloma of larynx[d]	BHK-21 Clone 13 (suspension)[e]
-25.6	-51.2	-58.0	-71.7	-6.9
-19.4	-11.7	-10.2	-13.6	-2.3
-14.2	-9.9	-9.0	-12.1	-2.3
-10.5	-9.0	-10.2	-9.8	-1.8
-3.4	-1.3	-3.1	-3.5	-1.2
-19.3	-8.8	-0.5	-9.7	-2.3
-5.0	-5.8	-2.4	-4.6	-1.5
-5.0	-5.1	-2.4	-1.5	-0.3
nm	nm	nm	nm	+2.6
-23.5	-20.3	-17.1	-4.0	-1.9
-13.4	-0.3	-0.3	-3.1	-1.1
-5.3	-5.6	-5.6	-2.9	nm
+11.1	+1.4	+15.8	+9.0	+5.8
-2.6	+0.4	nc	nc	-3.4
-10.5	+5.0	+1.5	+0.3	+2.9
-1.3	-2.0	-3.1	-0.5	-0.30
-1.3	+0.5	-5.4	-3.8	-0.30
-5.4	+1.5	+2.6	+1.4	+0.3
nc	+3.1	+1.1	nc	-1.71
+3.2	+16.4	+7.8	+15.2	-5.4

Table 5   Major Transport Systems Found in Mammalian Cells

Designation	Na^+-dependent	Preferred amino acid class
A	Yes	Neutral amino acids[a]
ASC	Yes	Neutral amino acids[a]
L	No	Neutral amino acids[a]
y^+	Yes	Cationic amino acids[b]
T	No	Aromatic amino acids[c]
N	Yes	Amino acid amides and histidine[d]
X^-_{AG}	Yes	Anionic amino acids[e]
X^-_c	No	Anionic amino acids[e]

[a]From Refs. 100−102 and 105.
[b]From Refs. 111 and 112.
[c]From Refs. 108−110.
[d]From Refs. 106 and 107.
[e]From Refs. 113−115.

has been found in rat hepatocytes and human skin fibroblasts which is sodium ion-dependent (114,115). This second anionic amino acid transporter has been provisionally named the X^-_{AG} system. Table 5 lists the major transporters that have been covered in this text. This is not an exhaustive list but major classes of transporters described in the literature are represented.

The transport of amino acids into mammalian cells can be regulated by nutritional, hormonal, or other environmental factors, or by changes within cells like transformation. The intracellular or extracellular concentration of amino acids probably has the most profound influence on the efficiency or capacity of transport in animal cells. Transport regulation by amino acid concentration can be divided into two categories. The first is termed *trans* effects on transport which can be further subdivided into *trans* inhibition and *trans* stimulation (112). These mechanisms of inhibition or stimulation can be defined as processes by which a substrate on one side of a membrane inhibits or stimulates the passage of another identical or analogous substrate present on the other side of a membrane. The transport of neutral amino acids by system A can be regulated by *trans* inhibition whereas systems L, ASC, and y^+ can be regulated by *trans* stimulation (101,112,116,117).

Amino acid concentration can also affect the transport of amino acids by a mechanism termed *adaptive regulation* (112,116). The system that seems most sensitive to this type of regulation is system A, although systems N, L, and x_c^- may also be regulated in this way, but to a lesser degree (112,114,116−118). The terms *repression* and *derepression* are used in the literature to describe how adaptive regulation works. Normally, when amino acids to be transported are present in excess, the system is repressed and vice versa. The repression or derepression of system A transport in cultured human fibroblasts involves active mRNA transcription and protein synthesis (119). Starvation-induced derepression may increase the number of functioning transporters in the plasma membranes of these cells, resulting in an increased V_{max} with no change in K_m. Supplementation of human fibroblasts with system A amino acid substrates results in a repression of transport, possibly through the synthesis of a repressor protein that acts directly on the transporter (119). Fafournoux et al. (120) examine the repression of system A transport activity in rat hepatocytes induced by insulin and glucagon. They found that amino acid substrates for system A were effective, as was a nonmetabolizable substrate (2-aminoisobutyrate), in repressing the induction provided by glucagon. Apparently, the metabolism of system A amino acids is not needed for the repression mechanism. When actinomycin was used to inhibit mRNA synthesis, system A amino acids failed to repress induction by glucagon, also suggesting that a repressor protein is responsible for a decrease in amino acid transport through system A. Moffett and Englesberg (121,122) proposed a model for the regulation of system A transport activity from their work with CHO-K1 mutants. Their model suggests that an apo repressor-inactivator is synthesized by a regulatory gene and can be converted into the active repressor-inactivator in the presence of system A amino acids or their structural analogs. In the model, this repressor-inactivator can also act at the level of gene transcription for the system A transporter itself, resulting in fewer transporter proteins when the transporter substrates are present. Gazzola et al. (123) examined the effect of serum on regulation of system A transport activity in fetal human fibroblasts and found evidence for posttranslational control of transport. An increase in V_{max} was apparent in the presence of serum that was not dependent on protein synthesis. The authors speculate that serum is essential for a vectorial posttranslational event which leads to insertion of presynthesized transport proteins into the cell membrane. This mechanism would be analogous to the transport of glucose, as previously discussed in this chapter.

Other changes in the culture environment can also affect the rate of amino acid transport. An increase in medium osmolarity has been shown to increase the transport of certain amino acids in chick embryo fibroblasts (124–126). In these studies system A transport activity seems to be primarily affected, although the resulting enlarged intracellular amino acid pool may enhance the transport of amino acids through system L (*trans* stimulation). Transport of amino acids can also be affected by the pH and temperature of the culture medium (117).

It has been known for several years that hormones play a critical role in the transport of amino acids into cells. This literature is voluminous and covers the spectrum of tissues and hormones that are currently known. Some generalizations can be made regarding the hormone and target tissue. Hormones tend to increase the transport of amino acids in their specific target tissue in response to an increase in the metabolic needs of that tissue, such as an increase in macromolecular synthesis or an increase in energy production (116,117). The system that is consistently affected by addition of hormones to the culture environment is primarily system A, although other transporters, such as systems L and x_c^-, are known to be regulated by hormones in certain cell types (112,127). An example of this type of hormone-transport interaction is evident in cultured human fibroblasts. Longo et al. (127) studied the effect of insulin on six transporters identified in these cells. They found that of the six, two transport systems (A and x_c^-) are strongly stimulated by insulin in a dose-dependent manner and four transport systems (ASC, X_{AG}^-, y^+, and L) are essentially unaffected by the hormone. The increase in transport was a result of an increase in V_{max} without a significant change in K_m. Leoni et al. (128) studied the effect of glucagon and insulin on system A activity in rat hepatocytes freshly isolated at different stages of pre- and postnatal development. They found no response to the hormones before the 18th day of fetal life and in the perinatal period. Cells isolated from 18- to 20-day-old fetuses and from adult animals showed an increased in system A activity in response to the hormones. The enhancement in 18- to 20-day fetal tissue was through an increase in V_{max} of transport with no change in K_m. An interesting point of this study is that cells isolated from fetal liver prior to 18 days of development showed the highest rate of amino acid transport through system A. Transport seemed to be turned on and unregulated in response to the high metabolic need and rapid growth of the more undifferentiated fetal liver.

Growth factors have also been shown to affect the transport of amino acids. Epidermal growth factor, fibroblast growth factor, platelet-derived growth factor, nerve growth factor, and multiplication-stimulating activity have all been shown to increase the

transport of amino acids into mammalian cells (116). The system
that seems to be enhanced in the presence of these factors is system A, and its stimulation is via an increase in V_{max} with little or
no change in K_m of transport.

There is some evidence that cell density can influence the rate
of transport of amino acids in animal cells. Borghetti et al. (129)
studied the transport of amino acids through systems A, ASC, L,
and Ly^+ (y^+) in Balb/c3T3, SV3T3, and SV3T3 revertant cell lines.
Transport activity of the sodium ion-dependent systems A and ASC
were markedly decreased with increasing cell density in all three
cell lines. Systems L and Ly^+ (y^+), both sodium ion-independent,
were not affected by cell density in any of the lines tested. The
transformed SV3T3 line and the SV3T3 revertant lines had higher
rates of transport through system A and ASC at all cell densities
than the normal Balb/c3T3 line. Boerner and Saier (130) investigated the influence of cell density on amino acid transport through
systems A, ASC, L, and N in Madin-Darby canine kidney (MDCK)
cells grown in a defined medium. They found that when cells were
grown to elevate densities a decrease in the activities of systems A
and ASC were evident with little if any change in systems L and N.
System A activity is also inhibited in high-density cultures of chick-
en embryo fibroblasts (CEF) compared to normal gorwing or Rous
sarcoma virus-transformed CEF cultures (16).

Oncogenic transformation of cells is known to increase the transport of certain amino acids into animal cells, although enhancement
in the transport of amino acids is not as dramatic as that for glucose (131). No enhancement of transport was evident in Rous sarcoma virus-transformed CEF cells compared to nontransformed cells
when they were examined immediately after removal from normal
culture medium (16). When normal and transformed cells were deprived of amino acids for 2 hr, an elevation of two- to threefold in
system A transport was observed in the Rous transformed cells
compared to their normal counterparts. No rate changes were observed in the other amino acid transport systems. Borghetti et al.
(129) observed an increase in system A and ASC transport in SV-40
transformed Balb/c 3T3 cells as already mentioned. This increase
in transport over normal controls was also maintained in revertant
SV3T3 cells that had resumed normal density-dependent inhibition
of growth. The activity of system A has been shown to increase
by three- to sevenfold in a MDCK cell line that has undergone chemical transformation (130). Enhancement of transport activity through
system A was due to an increase in V_{max} rather than an increase
in K_m of transport. Elevated amino acid transport through system A as a result of transformation does not occur in all cells.
Gazzola et al. (132) compared the transport of neutral amino acids
in normal human fibroblasts with cells obtained from two fibroblastoid

tumors at different stages of differentiation. They found little difference between the neoplastic and normal cells in basal rates of
amino acid transport.

C. Metabolism

The tricarboxylic acid (TCA) cycle functions as a major route for
the synthesis and oxidation of most amino acids and is the primary
route through which amino acid carbon flows in the synthesis of
other small molecules. Figure 1 depicts the probable points of entry

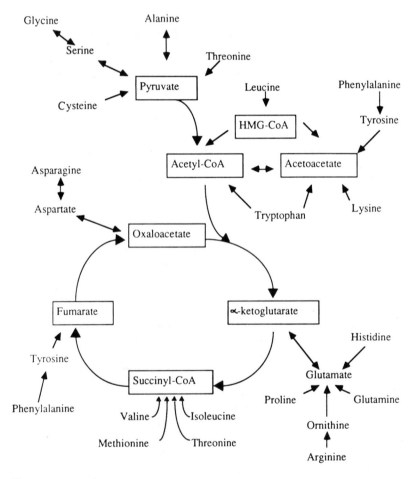

Figure 1   Points at which amino acids enter or exit the TCA cycle
in mammalian cells.

of several amino acids into the TCA cycle. Before entering the TCA cycle, the α-amino groups are usually removed by transferring to α-ketoglutarate to form glutamate and the α-keto acid of the amino acid (133). This is called an aminotransferase or transamination reaction. Examples of two common transaminase reactions occurring in mammalian cells are aspartate transaminase and alanine transaminase as illustrated below.

$$\text{L-alanine} + \alpha\text{-ketoglutarate} \longleftrightarrow \text{pyruvate} + \text{L-glutamate}$$

$$\text{L-aspartate} + \alpha\text{-ketoglutarate} \longleftrightarrow \text{oxaloacetate} + \text{L-glutamate}$$

Lehninger (133) points out that the use of the α-ketoglutarate/L-glutamate couple in many transamination reactions is an example of the convergence of catabolic pathways that result in the accumulation of nitrogen in one amino acid (glutamate), an event that facilitates the excretion of excess nitrogen in animals. Since transaminase reactions are freely reversible, the concentration of substrates can affect the reaction direction. In vitro, cells tend to consume more glutamine than their in vivo counterparts (61). The amount of glutamate formed from glutamine by cells cultured in vitro in a glutamine-rich environment would cause the above transamination reactions to favor L-alanine and L-aspartate formation. The branched-chain amino acids are also converted to their keto analogs by an amino transferase with α-ketoglutarate as amino acceptor (134). Animal tissues contain at least three isoenzymes of the branched-chain amino acid amino transferases, although the strength of their activities can vary widely from one tissue to another (135). Isolated cells in tissue culture may therefore vary widely in their ability to use branched-chain amino acids as oxidative substrates or as precursors for the synthesis of other molecules.

The metabolism of glutamine by cells in culture has been vigorously studied due to its importance as a respiratory fuel and as a biosynthetic precursor. Windmueller and Spaeth (136) found that as much as 30% of the plasma glutamine was removed from blood after just a single pass through the small intestine of man and many animals. Of the amount removed, 50% was converted to CO_2. In perfused jejunal segments glutamine and glucose were utilized in equimolar amounts, but the amount of glutamine oxidized was fivefold greater than glucose (137,138). Ardawi and Newsholme (70) studied the rate of O_2 consumption in isolated colonocytes in the presence of glutamine, n-butyrate, glucose, or ketone bodies. They observed an increase in the consumption of O_2 with each substrate except glucose. The uptake of glucose and glutamine were similar, but lactic acid accounted for 83% of the glucose consumed. The major end products of glutamine metabolism besides CO_2 in isolated rat colonocytes and in perfused rat small intestine were glutamate, aspartate, alanine, and ammonia (70,139).

Tildon and Roeder (140) compared the oxidation of glutamine in dissociation brain cells and in brain cell homogenates from 2-, 10-, 15-, 25-, and 90-day-old rats. No difference was found in glutamine oxidation in the homogenates, but a twofold increase was observed in whole cells from 2-day-old rats when compared to older rats. They concluded that glutamine may play an important role in the provision of energy in brain and that membrane transport might be involved in the regulation of glutamine oxidation.

The oxidation of glutamine, glucose, ketone bodies, and fatty acids were compared in human diploid fibroblast (141). The rate of glutamine oxidation in these cells was 98 nmol/hr/mg cell protein compared to 2 nmol/hr/mg cell protein or less for acetoacetate, D-3-hydroxybutyrate, octanoic acid, and palmitic acid. Human diploid fibroblasts have also been shown to undergo from one to two cell divisions in the absence of glucose when their medium was supplemented with 10% dialyzed fetal calf serum and combinations of hypoxanthine, thymidine, uridine, and inosine (66,142). Under these conditions, glutamine utilization from the medium was twice the normal rate observed in the presence of 5.5 mM glucose. Cultured chick pigment epithelial cells can undergo about two cell doublings in the absence of glucose by increasing their rate of glutamine utilization from two- to sixfold (143). Zielke et al. (144) suggested that the major end products of glutamine metabolism in many mammalian cells are lactic acid, CO_2, and glutamate. As much as 13% of the glutamine in the culture medium of human diploid fibroblasts is metabolized to lactatic acid (145).

Switching the carbohydrate source from glucose to galactose or fructose forced HeLa cells to derive 98% of their energy from glutamine oxidation instead of the normal 50% in the presence of high glucose (64). These authors found that less than 2% of glutamine was incorporated directly into protein, 18−25% was found in other macromolecules, 13% was found as lactate, and 35% was oxidized to CO_2.

Brand et al. (55) studied the metabolism of glutamine and glucose in freshly prepared resting and concanavalin A-stimulated rat thymocytes. They found that the major end products of glutamine metabolism in resting and stimulated thymocytes were glutamate, aspartate, CO_2, and NH_3. Stimulation of thymocytes with concanavalin A caused an increase in all metabolic end products except glutamate, suggesting an increased oxidation of glutamine in stimulated cultures. In another study, proliferating cultures of rat thymocytes stimulated by concanavalin A and Lymphocult T (interleukin 2) were shown to increase their utilization of glutamine by eightfold over controls (56). The major end products were again glutamate, aspartate, CO_2, and ammonia. Addition of 4 mM glucose or 2 mM malate strongly decreased the utilization of glutamine in proliferating cultures.

Mouse macrophages also utilize glutamine at an elevated rate, but they appear to oxidize less than 10% of the total consumed (71). The major end products of glutamine metabolism in these cells are glutamate, aspartate, and lactate. Lymphocytes can alter their metabolism of glutamine in response to an immune stimulus (146,147). They experience a burst of metabolic activity that coincides with a rapid increase in cell division and production of mediators of the immune system including antibodies, lymphotoxin, chemotactic factors, and mitogenic agents. Ardawi and Newsholme (148) isolated lymphocytes from rat mesenteric lymph nodes and found that after stimulation with the mitogen concanavalin A, glutamine utilization was stimulated by 51%. The major end products of glutamine metabolism in lymphocytes are ammonia, glutamate, and aspartate.

Pardridge and Casanello-Ertl (149) incubated 10-day-old muscle cells in the absence of glutamine and found a progressive depletion of intracellular aspartate and glutamate. Concomitant with the depletion of these amono acids was a 15-fold increase in the intercellular lactate/pyruvate ratio. These authors suggest that depletion of glutamate and aspartate, resulting from the absence of glutamine, affected the normal functioning of the malate-aspartate shuttle used in the transport of reducing equivalents from the cytosol to the mitochondria.

The first enzymatic step in the utilization of glutamine is the removal of the amide group of glutamine to form glutamic acid. Since this reaction involves a large drop in free energy, it is probably irreversible and could be the rate-limiting step for glutamine utilization in most mammalian cells (150,151). There is some evidence to suggest that glutaminase activity may be related to growth rate in both normal and neoplastic cells (152,153). Knox et al. (153) noted a correlation of the growth rate of rat neoplasms with glutaminase activity; in general, faster growth rates correspond to higher glutaminase activity in these tissues. In cultures of human fibroblasts the activity of glutaminase increased threefold and remained elevated for 2 days after cells were subcultured into fresh medium (152). Once the fibroblasts reached confluency, a decrease in glutaminase activity was observed independent of glutamine concentration.

Figure 2 illustrates some of the primary routes of glutamine metabolism in animal cells. No attempt has been made to assign subcellular compartments for the various reactions, although glutaminase activity seems to be associated with the mitochondria in many cells (151,154,155). Glutamine carbon enters the TCA cycle as α-ketoglutarate, formed either through transamination reactions as discussed

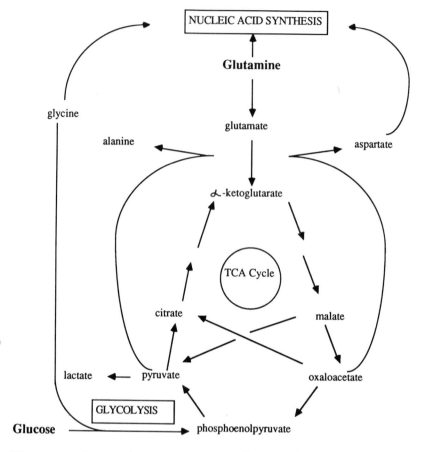

Figure 2 Schematic representation of some of the interrelationships of glucose and glutamine metabolism in mammalian cells.

earlier or through an oxidative deamination step catalyzed by glutamate dehydrogenase. Aminotransferases or transamination reactions
are believed to be the major route for the second step in glutamine
utilization in lymphocytes (156), intestinal cells (138,139), and
tumor cells (157). As seen in Fig. 2, the major products from
transamination reactions involving glutamate (which originates from
glutamine) are alanine and aspartate. As previously discussed, the
formation of these products will be dependent on the relative abundance of alanine transaminase or aspartate transaminase present in
a cell line or tissue.

Lactate is another metabolic end product from glutamine metabolism by some cells (71,145), which indicates that glutamine can contribute to the intracellular pool of pyruvate. The pathway used to convert TCA cycle intermediates derived from glutamine to pyruvate is still not well understood and probably varies between cell types. Two of the best candidates include conversion of malate to pyruvate by malic enzyme and conversion of oxaloacetate to phosphoenolpyruvate by phosphoenolpyruvate carboxykinase (PEPCK) (61,151,158).

The reasons for elevated glutamine utilization and metabolism in mammalian cell culture are not completely understood. As this section has discussed, most animal cells in culture utilize significant quantities of glutamine, especially when compared to other amino acids. Glutamine definitely contributes to the generation of energy in many cultured cells but undoubtably serves in other cellular capacities as well. A discussion of the role that glutamine may play in rapidly growing cells, along with glucose, will be discussed in the next section.

IV. METABOLIC CHANGES THAT ACCOMPANY INCREASED PROLIFERATIVE AND BIOSYNTHETIC CAPACITIES

A. Altered Glucose Metabolism

In general, cells that are rapidly growing or have the capacity to grow rapidly also tend to rapidly remove glucose from the culture medium (55,56,58,70,73,74,76,146,151). Much of the glucose in these cells is converted to lactic acid in the presence of sufficient oxygen, which is generally referred to as aerobic glycolysis. A notable example of elevated aerobic glycolysis was demonstrated by Brand (56) and by Brand et al. (55) in resting and mitogenically stimulated thymocytes. They found that thymocytes stimulated with concanavalin A and/or interleukin 2 increased their metabolism of glucose 53-fold. The stimulated (proliferating) thymocytes converted 90% of the metabolized glucose to lactate whereas resting thymocytes converted only 53%. This is an example of a normal cell population altering its metabolism and growth rate in response to mitogenic stimulation.

Many mitogenic stimuli or events previously discussed in this chapter are known to increase the transport of glucose into cells. Some of these include hormones like insulin (23,24), growth factors like TGFβ and EGF (33,34), phorbol esters (35,36), and viral transformation (38,42). Glycolytic activity is also known to increase in some cells when exposed to mitogens such as EGF or TGFβ (53,54) or concanavalin A and interleukin 2 (55,56). Reiss et al. (159)

demonstrated that purified EGF receptors phosphorylate tyrosine residues in several glycolytic enzymes, including the key regulatory enzymes glyceraldehyde-3-phosphate dehydrogenase and phosphofructokinase. Transformation of fibroblasts by Rous sarcoma virus (encoding a tyrosine-specific protein kinase) has also been shown to phosphorylate glycolytic enzymes including enolase, lactate dehydrogenase, and phosphoglycerate mutase (160,161). The significance of these tyrosine phosphorylations is not completely clear, although they might be a mitogenic signal for activating key enzymes of the glycolytic pathway. The work of Weber (58) and others (59,75,82−85) strongly suggests that a correlation exists between elevated activities of key glycolytic and pentose phosphate enzymes and growth rates of tumors derived from several different tissues.

Several lines of evidence suggest that cell growth rate and changes in glucose uptake and metabolism are related. Fast-growing cells tend to take up glucose at an accelerated rate and in turn convert much of this glucose to lactate. Mitogen-stimulated cells and transformed cells will also mimic this metabolism. Not only do mitogens and transformation influence the rate of glucose uptake, but they may also directly affect the activity of key glycolytic enzymes through changes in phosphorylation states or other mechanisms. The still unanswered question is, "Are these metabolic changes simply a consequence of rapid growth or do they confer some growth advantage?"

Other evidence suggests that rapid glycolysis is not essential for growth. Reduction of the glucose concentration in cell culture medium will reduce the rate at which some cells convert glucose to lactate (1,74). Altering the carbohydrate source from glucose to fructose or galactose will also reduce the glycolytic activity of some cells and may or may not affect proliferative rate (51,60,64,68). The ability of cells to grow in a medium containing a carbohydrate other than glucose is cell type-dependent and may be related to the ability of a cell population to transport, phosphorylate, and metabolize a given carbohydrate. MDCK cells grow at a slightly slower rate on fructose than on glucose (51), whereas Vero cells grow equally well on fructose, galactose, or glucose (60). Interestingly, Vero cells produce increasing amounts of lactate as the concentration of fructose is increased; this is similar to glucose, although production of lactate from fructose on a molar basis is less significant. HeLa cells exhibit a doubling time of 28 hr on 10 mM glucose and 37 hr on 10 mM galactose or 10 mM fructose (64). We have experienced little growth and poor viability in CHO-K1 lines cultured in medium containing only fructose or galactose as a carbohydrate source (162). When these CHO lines are cultured in medium containing increasing concentrations of pyruvate, some growth and improved viability are observed. We might conclude from these data that growth rate is not strictly dependent on glycolytic activity, but a reduced amount of substrate passing through the glycolytic pathway may slow or prevent growth. The ability of cells to channel glycolytic intermediates into critical biosynthetic and/or energy-yielding pathways may influence the growth rate of a cell population, especially under restricted glucose or "alternate" carbohydrate conditions.

B. Altered Glutamine Metabolism

Like glucose, the metabolism of glutamine is also accelerated in rapidly growing cells (55, 56, 64, 70, 71, 73, 136 − 139, 142, 146 − 148). This pattern of elevated glutamine utilization in fast-growing cells has been termed "glutaminolysis" by McKeehan (151, 163). The uptake and metabolism of glutamine increases eightfold in mitogenically stimulated thymocytes (56). Lymphocytes increase their metabolism of glutamine in response to an immune stimulus prior to undergoing a rapid increase in cell division and protein synthesis (146, 147). The major end products of glutamine metabolism are generally CO_2, ammonia, lactate, glutamate, aspartate, and alanine. The relative concentrations of these metabolites will depend on the metabolic needs of a given cell population and probably on the cell enzymology.

Glutamine functions as an oxidizable source of energy for many cells cultured in vitro, and the metabolic end products of glutamine metabolism are similar for a variety of cells. As discussed earlier, Reitzer et al. (64) found that 98% of the energy requirement of HeLa cells could be derived from glutamine when galactose or fructose was substituted for glucose in the culture medium. They showed that 35% of the glutamine was metabolized to CO_2 under these conditions. Human diploid fibroblasts will grow for up to two cell doublings in the absence of carbohydrate when their medium is supplemented with hypoxanthine, thymidine, and either uridine or inositol (66, 142). Cultured chick pigment epithelial cells will also grow in the absence of glucose for up to two cell doublings (143). In both cases, the uptake of glutamine was increased from two- to eightfold above control levels, suggesting that glutamine could partially compensate for the lack of carbohydrate, probably by meeting the energy requirements of these cells. Addition of 5.5 mM glucose inhibited the rate of L[1-^{14}C]- and L[5-^{14}C]glutamine oxidation by 80% in cultures of normal human diploid fibroblast (67). The reciprocal relationship also occurred in that D[6-^{14}C] glucose oxidation was inhibited by 90% when 2 mM glutamine was added to the incubation medium. These data suggest that glutamine can play a major role as an energy substrate in cultured cells and that the rate of its utilization and oxidation may depend on the type and concentration of carbohydrate in the culture medium.

Besides its roles in protein synthesis and as an oxidizable energy substrate, glutamine can play other important roles in cellular
metabolism, e.g., its role in purine metabolism. Raivio and
Seegmiller (164) suggest that the availability of glutamine in the
culture medium of normal and hypoxanthine phosphoribosyltrans-
ferase-deficient fibroblasts may be rate limiting for de novo purine
synthesis in vitro. The first "committed step" in purine synthesis
is the reaction of 5-phosphoribosyl-1-pyrophosphate (PRPP) with
glutamine to form phosphoribosylamine, an example of a glutamine-
dependent amination (86). A second glutamine-dependent amination
and an aspartate-dependent amination also occur for each molecule
of inosinic acid formed. Aspartate then donates another amino group
to inosinic acid to form adenosine monophosphate (AMP), which can
then be converted to adenosine diphosphate (ADP) and eventually to
adenosine triphosphate (ATP) (86). Glycine also contributes significantly to purine synthesis by contributing two carbon atoms and
one nitrogen atom to each purine molecule formed. Aspartate is a
major byproduct of glutamine metabolism in some cells (70,71) and
glycine can be formed from 3-phosphoglycerate, a glycolytic intermediate (Fig. 2). The contributions of glutamine, aspartate, and
glycine to the purine molecule adenine are shown in Fig. 3.

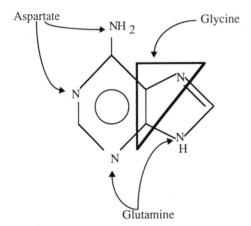

Figure 3   Contributions of aspartate, glycine, and glutamine to
the synthesis of the purine molecule adenine. (Adapted from
Ref. 86.)

C. Control of Branched Pathways

Many theories and explanations have been put forward in an attempt to explain the high rates of glycolysis in cells with high proliferative potential (74,75,151,165 − 170), but a consensus of thought in this area still does not exist. A recent hypothesis by Newsholme et al. (75) explains the high rates of glutamine and glucose utilization by fast-growing cells as necessary for sensitive control of branched pathways leading to the synthesis of macromolecules. In their system they describe three fluxes (J, J_a, and J_b) as illustrated by Fig. 4. The sum of J_a plus J_b is J, which is the maximum flux through the branched pathway and is determined by the maximum rate of enzyme 1 (E1) or concentration of substrate A. E2 and E3 are enzymes catalyzing flux through J_a and J_b and no direct feedback occurs from these pathways to modulate E1. If X, a positive modulator of E2, increases flux through J_a, then there will be a decreased flux through J_b because less substrate B will be available. A reduced concentration of substrate B will eventually decrease flux through J_a because under these circumstances, flux through J_a and J_b are directly related to the concentration of precursor substrate B. Newsholme et al. (75) refer to this effect as "opposition" to effector X, thereby reducing a cell's sensitivity to the positive modulator (X). A cell might decrease this opposition by greatly increasing the flux through J_b relative to J_a. If only 4 − 5% of substrate B, compared to say 40 − 50%, is deflected through J_a, then it seems less probable that a reduced concentration of substrate B will significantly "oppose" flux through J_a. Flux through glycolysis or glutaminolysis might be substituted for J_b and flux through biosynthetic pathways used for macromolecular synthesis could be J_a. A

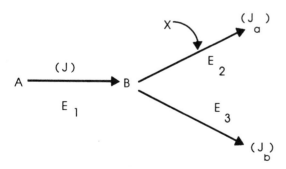

Figure 4 Schematic representation of flux through branched metabolic pathways. The control of flux by X is described in the text. (From Ref. 73.)

positive modulator (X) could be a growth factor, transformation, or some other mitogenic event. The positive modulator might affect the flux through a biosynthetic pathway by directly modifying the activity of a rate-limiting enzyme (i.e., phosphorylating or de-phosphorylating), or by affecting its concentration through trans-criptional or translational control. This hypothesis does not explain all the available data, like decreased glycolysis in low glucose-con-taining medium and lowered glycolytic activity in the presence of alternate carbohydrates (1,51,60,64,68,74); but it does suggest a logical approach to the study of cell metabolism and how flux through metabolic pathways might influence cell growth rates.

A quantitative analysis of metabolism coupled with strong ex-perimental data seems a logical approach for studying the modulation of cell metabolism. Crabtree and Newsholme (171) recently outlined this type of approach in great detail. The complexity of this type of analysis is formidable, but the payoff might be a better under-standing of growth and biosynthetic regulation, and their modulation by everything from hormones to oncogenes.

The metabolism of glucose and glutamine by animal cells in cul-ture is strongly influenced by molecular oxygen. The next section will discuss the role oxygen plays in animal cell culture and its in-fluence on cell metabolism.

V. OXYGEN

The primary function of oxygen in a cell culture environment is to serve as the terminal acceptor of electrons in the electron transport chain (ETC). This enables cells to efficiently convert chemical en-ergy derived from dehydrogenation reactions into high-energy phos-phate bonds in the form of ATP. An insufficient supply of oxygen causes cells to switch from aerobic metabolism (respiration) to anaerobic metabolism (glycolysis), a much less efficient method of energy generation. One role of glycolysis is therefore to buffer the effects of inadequate oxygen concentrations (172). Respiration-deficient mutants, which consume little or no oxygen, have been characterized that require high glucose concentrations for growth (173–176). Growth rates of wild-type and respiration-deficient mutants were similar, suggesting that energy derived from glycoly-sis was sufficient, although inefficiently obtained. A total de-pendence on glycolysis for the provision of energy is the exception rather than the rule. Most cells, even those originating from tumor tissue, derive a significant amount of their energy from oxidative phosphorylation (177). In order to better understand the role of oxygen in cell metabolism, especially in the generation of ATP, a general overview of oxidative phosphorylation and its regulation will be discussed.

A. Oxidative Phosphorylation

The generation of ATP in animal cells is accomplished through the mitochondrial ETC located in the inner mitochondrial membrane. It consists of a multienzyme complex which accepts reducing equivalents from NADH and $FADH_2$ and transfers them through several oxidation-reduction reactions to molecular oxygen, with the subsequent formation of water and ATP (179). The transfer of reducing equivalents occurs down an electrochemical potential gradient which liberates energy through exergonic redox reactions. This liberated energy is used for the synthesis of ATP (179). Steady-state systems which balance the production and utilization of energy in the form of ATP exist in all living organisms. A precise dynamic balance is maintained between reactions that produce ATP within cells and those that use it. As a result, the concentration of ATP within cells should remain constant under nonlimiting conditions. The rate of ATP synthesis in man can vary from 0.4 g ATP/min/kg body weight at rest to 9.0 g ATP/min/kg body weight during strenuous exercise (179). This wide variation in rate of synthesis illustrates the capacity of cells in vivo to adjust to substantial utilization of ATP for work.

The regulation of mitochondrial oxidative phosphorylation can probably best be explained by the near-equilibrium hypothesis first postulated by Wilson et al. (180). This hypothesis states that the energy generated through the transfer of two reducing equivalents from intramitochondrial NADH to cytochrome *c* is equal, within experimental error, to that required for the synthesis of 2 mol ATP (179–181). Therefore, oxidative phosphorylation from NADH to cytochrome *c* is near equilibrium, which results in almost 100% efficiency in the transduction of electrochemical energy for ATP synthesis at the first two phosphorylation sites. The final phosphorylation site occurs at cytochrome *c* oxidase, a complex membrane-bound, multisubunit enzyme that reduces molecular oxygen to form water (182). Only half the negative free enthalpy change of this reaction is utilized for ATP synthesis; the other 50% negative free energy is lost as heat. This reaction is therefore highly irreversible and is responsible for regulating the rate of mitochrondrial phosphorylation (181).

An alternate hypothesis for control of oxidative phosphorylation is the "translocase hypothesis" as put forth by Klingenberg (183, 184). The exchange of adenine nucleotides across the mitochrondrial inner membrane to the cytosol takes place by a stoichiometric exchange with cytosolic ADP and is catalyzed by the enzyme adenine nucleotide translocase. The hypothesis suggests that the enzyme adenine nucleotide translocase is rate determining for the overall respiratory process. Forman and Wilson (185) found a substantial excess of adenine nucleotide translocase activity in rat heart and rat liver which suggests, along with other data (178), that this enzyme is not rate limiting for respiration.

While reduction of cytochrome c and its subsequent reduction of molecular oxygen to water are apparently the rate-limiting steps of respiration, other conditions also influence this rate. These can be summarized by four factors that interact to govern mitochondrial respiration:

1. The concentration of the respiratory chain proteins
2. The concentration of molecular oxygen
3. The concentration of intramitochondrial substrate levels and metabolic controls that determine intramitochondrial $[NAD^+]/[NADH]$
4. The rate of utilization of cellular ATP which determines cytosolic $[ATP]/[ADP][P_i]$

Heart mitochondria exhibit high respiratory activity and also contain more redox carrier per mg of protein than other tissues (178). This is indirect evidence that the concentration of respiratory chain proteins influences respiratory activity. Dudley et al. (186) increased the mitochondrial content of rat skeletal muscle by exercise training or decreased it by hypothyroidism. They found that as the mitochondrial content increased, oxidative capacity increased and respiratory control became more sensitive. A smaller change in a cytosolic modulator like ADP was necessary to elicit respiratory control, and accumulation of lactate was lower in tissues with higher mitochondrial content.

Since molecular oxygen is the ultimate acceptor of electrons in oxidative phosphorylation, and the transfer of electrons from cytochrome c to molecular oxygen is probably the rate-limiting step, the intracellular concentration of oxygen will greatly influence respiratory activity (178, 181, 182, 187 – 189).

It also stands to reason that the availability of intramitochondrial NADH would have a profound influence on respiratory activity within cells. The electrons that reduce cytochrome c, and that subsequently reduce molecular oxygen to form water, must originate from substrates involved in dehydrogenation reactions. As discussed in the previous sections, glucose and glutamine serve as the substrates most readily involved in energy-generating pathways in cultured mammalian cells.

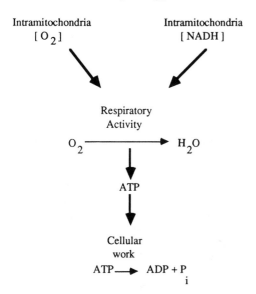

Figure 5   Major intracellular influences on cellular respiratory activity.

The last major factor that influences respiratory activity is cytosolic [ATP], [ADP], and [P_i]. Pools of these reactants in the cytosol of cells remain at fairly constant levels under normal conditions. An increase in the level of work performed by cells increases the utilization of ATP and therefore its synthesis from ADP and inorganic phosphate. This work might be an increase in cellular growth rate or an increase in the synthesis of engineered proteins. Figure 5 provides a very schematic representation of the components that influence cellular energy metabolism.

The concentration of respiratory proteins in most cells are at fixed concentrations that are determined by the genetic profiles of cells. Therefore, the three components that probably have the most influence on respiratory activity are intramitochondria [NADH], intramitochondria [O_2], and the rate of ATP hydrolysis for cellular work. If our desire is to have cells make and secrete large quantities of highly complex proteins, which would increase the demand on the cytosolic ATP pool, then we must be sure that intramitochondria [NADH] and [O_2] are nonlimiting.

B. "Optimum" Dissolved Oxygen Concentration

The ability of most cells to maintain high cytosolic $[ATP]/[ADP][P_i]$
will be dependent in part on intramitochondria molecular oxygen
concentration. Mitochondria are efficient reducers of molecular
oxygen, but if oxygen concentrations fall to very low levels, the
rate of ATP synthesis will be decreased with a concomitant lowering of $[ATP]/[ADP][P_i]$ (190). This is probably one of the homeostatic mechanisms that activates glycolysis since ADP and P_i are
known activators of glycolysis and ATP is a known inhibitor (172).
Cells containing multiple gene copies of constitutively produced proteins should have greater energy needs than their nontransfected
counterparts. Inefficient production of energy under low-oxygen
conditions might limit the ability of cells to maximally produce a
genetically engineered protein, although cell growth rates may remain normal. We have seen some evidence of this in CHO cultures
grown under different dissolved oxygen concentrations (191). Cultures grown under low dissolved oxygen conditions converted almost
100% of the glucose supplied to lactate, whereas CHO cultures grown
under higher "optimum" concentrations converted less than 40% of
consumed glucose to lactate. Growth rates and cell viabilities were
similar, but cultures grown under lower dissolved oxygen levels
produced 40% less of the engineered protein.

Elevated oxygen concentrations in cell culture also stimulate glycolysis. Rueckert and Mueller (47) found that HeLa cells cultured
under 95% oxygen (approximately 720 mm Hg) consumed more glucose, which they metabolized almost totally to lactate, compared to
cultures grown under 20% (150 mm Hg) oxygen. Valin et al. (48)
also observed a shift to a more anaerobic metabolism when WI-38
cells were grown at elevated (291 mm Hg) oxygen levels. We have
observed similar effects in cultures of CHO cells (191).

The mechanisms involved in the above metabolic shift are probably a result of cellular lipid peroxidation. Enzymes involved in
oxidative phosphorylation are associated with the mitochondrial membrane and are probably vulnerable to damage by lipid peroxidation.
The primary mechanism of protection in vivo against oxygen toxicity
and lipid peroxidation is the maintenance of low tissue pO_2 (192).
This should also be the case for cell culture, but even low pO_2 will
not protect cells that have been depleted of antioxidants (192–194).
The optimal oxygen concentration in the culture environment will
therefore be a function of the cells' sensitivity to peroxidative damage as well as the concentration of oxygen needed to drive oxidative
phosphorylation.

Much of the published work involving the definition of optimum
oxygen concentrations in cell culture has been obtained using un-
mixed monolayer cultures, so it is possible that microgradients of
dissolved oxygen above the monolayer were present in those

experiments. Werrlein and Glinos (195) identified the presence of microgradients in monolayer cultures of WRL-10A cells, a subline of the mouse L-cell fibroblast, using microoxygen cathodes. They observed oscillations of microenvironmental pO_2 in the medium overlaying attached cultures; the amplitude of these oscillations increased with the depth of the medium overlaying the cultures. When rotenone was introduced to uncouple respiration the oscillations immediately ceased, providing evidence that respiratory activity was responsible for oxygen depletion and oxygen replacement was due to diffusion.

Some work has been devoted to determining the oxygen uptake rate (OUR) values of various cells in culture, and these generally range from 0.05 to 0.5 mmol $O_2/10^9$ cells/hr (62,196). Very little work has been devoted to examining the influence of culture conditions in OUR. Fleischaker (197) found that FS-4 cells (normal human fibroblasts) cultured in galactose-containing medium consumed more oxygen than those grown on glucose. They also examined the influence of varying dissolved oxygen concentrations on the OUR of FS-4 cells (196). Their conclusions were that with FS-4 cells the OUR is independent of dissolved oxygen concentration in the region between 25 and 75% of air saturation. Whether this trend would hold true for epithelial cells that tend to consume more oxygen (198) or for cells transfected with a constitutively produced, highly amplified protein is currently unknown.

What constitutes an adequate dissolved oxygen concentration? Ultimately it will be that concentration of extracellular oxygen that provides a nonlimiting intramitochondria molecular oxygen concentration. Oxygen must diffuse through a polar intracellular cytosolic environment before it reaches the mitochondria, the site of oxidative phosphorylation. This may not be trivial since the solubility of oxygen in water is low (Table 6). Recent evidence indicates that intracellular oxygen gradients may exist in cardiac myocytes and hepatocytes under some conditions (200−204). These gradients may not be physiologically significant at elevated pO_2 values, but at low pO_2 concentrations intracellular gradients may be substantial. Oxygen and ATP concentrations have been studied in cells by measuring endogenous enzyme activities with different subcellular localizations. By using enzymes as in situ probes it was possible to predict intracellular oxygen and ATP gradients (204). Mathematical modeling and electron microscopy were also used to indicate that mitochondrial clustering and distribution within the cytosol are major factors in determining the magnitude and location of concentration gradients. Jones (204) speculates that mitochondria may be localized in areas of high energy demand and that microheterogeneity of metabolite concentrations can occur in cells which might determine the rates of high-flux processes. A genetically engineered,

Table 6   Solubility of Oxygen in
Perfluorochemicals and Other Solvents at
25°C and 1 atm

Solvent	Solubility of oxygen (mM)[a]
Water	1.3
Ethanol	9.2
Acetone	10.5
Carbon tetrachloride	11.4
Benzene	8.4
F-benzene	19.5
F-butyltetrahydrofuran	21.9
F-decalin	16.5
Bis(F-butyl)-ethene[b]	19.7

[a]Numbers represent means of values where
applicable.

[b]Measured at 37°C.

Source: Adapted from Ref. 199.

constitutively produced protein could be an example of a high-flux
process that would place heavy demands on cytosolic ATP concentrations.

The delivery of oxygen in vivo to cardiac tissue and possibly
other tissue is probably controlled by mitochondrial oxidative phosphorylation (205, 206). When the oxygen tension in cardiac myocytes is inadequate to meet energy demands, cytosolic [ATP]/
[ADP][P_i] decreases, which subsequently decreases vascular resistance and increases coronary blood flow. This effectively increases the transport of oxygen from the lungs to the tissues.
Oxidative phosphorylation in vivo can therefore be called a tissue
oxygen sensor.

The delivery of oxygen in conventional bioreactors is far from
this level of sophistication. We tend to empirically test for the
optimum concentration, searching for the pO_2 that does not promote
peroxidative damage and at the same time is not limiting oxidative
phosphorylation. Finding the truly "optimum" dissolved oxygen

concentration may be a difficult task without taking advantage of more sensitive oxygen sensors, like oxidative phosphorylation, as feedback systems for oxygen delivery. The solubility of oxygen in a polar environment may also be a limitation and may require the use of oxygen carriers, similar to myoglobin or hemoglobin in vivo. Perfluorocarbons are beginning to be used in a variety of ways in biotechnology (207) and may find some applications in large-scale cell culture as oxygen carriers. The solubilities of oxygen in water, perfluorochemicals, and other solvents at 25°C and 1 atm are given in Table 6.

VI. CONCLUSIONS

To fully utilize mammalian cells as expression systems for recombinant proteins a better understanding of the control of metabolic fluxes through biosynthetic and energy producing pathways is needed. Future work in this area might lead to methods of increasing cell proliferation rates, increasing maximum cell yields per volume of medium, or increasing the synthetic rates of genetically engineered proteins. Advances in any one of these areas would lead to increased volumetric yields of genetically engineered proteins, something that will be needed to improve the long-term cost effectiveness of mammalian cell culture.

This chapter has reviewed some aspects of cell physiology pertaining to the metabolism of carbohydrates, amino acids, and oxygen by animal cells in culture. Carbohydrates and amino acids were discussed with a special emphasis on factors that alter their transport and metabolism, and oxygen was discussed from the perspective of its function in animal cells. A better fundamental understanding of the metabolism of these nutrients will hopefully provide the researcher with a logical starting point for the development of efficient cell culture processes.

REFERENCES

1. Eagle, H., Barban, S., Levy, M., and Schuze, H. O. The utilization of carbohydrates by human cell cultures. *J. Biol. Chem.* 233(3):551−558 (1958).
2. Paul, J. Carbohydrate and energy metabolism. In *Cells and Tissue Culture* (E. N. Willmer, ed.), Academic Press, New York, pp. 239−268 (1965).
3. Burns, R. L., Rosenberger, P. G., and Klebe, R. J. Carbohydrate preferences of mammalian cells. *J. Cell. Phys.* 88: 307−316 (1976).

4. Scannell, J. and Morgan, M. J.   The regulation of carbohy-
    drate metabolism in animal cells: Isolation of starch- and
    maltose-utilizing variants. *Biosci. Rep.* 2:99−106 (1982).
5. Scannell, J. and Morgan, M. J.   The regulation of carbohy-
    drate metabolism in animal cells: Growth on starch and mal-
    tose. *Biochem. Soc. Trans.* 8:633−634 (1980).
6. Morgan, M. J. and Faik, P.   Carbohydrate metabolism in cul-
    tured animal cells. *Bios. Rep.* 1:669−686 (1981).
7. Morgan, M. J. and Faik, P.   The utilization of carbohydrates
    by animal cells. In *Carbohydrate Metabolism in Cultured Cells*
    (M. J. Morgan, ed.), Plenum Press, New York, pp. 29−75
    (1986).
8. Baldwin, S. A. and Lienhard, G. E.   Glucose transport across
    plasma membranes: Facilitated diffusion systems. *Trends
    Biochem. Sci.* 6:205−208 (1981).
9. Czech, M. P.   Regulation of the D-glucose transport system in
    isolated fat cells. *Mol. Cell. Biochem.* 11:51−63 (1976).
10. Graff, J. C., Wohlhueter, R. M., and Plagemann, P. G. W.
    Deoxyglucose and 3-O-methylglucose transport in untreated
    and ATP-depleted Novikoff rat hepatoma cells. Analysis by a
    rapid kinetic technique, relationship to phosphorylation and the
    effects of inhibitors. *J. Cell. Phys.* 96:171−188 (1978).
11. Klip, A., Logan, W. J., and Li, G.   Hexose transport in L6
    muscle cells: Kinetic properties and the number of [^{3}H] cyto-
    chalasin B binding sites. *Biochim. Biophys. Acta* 687:265−280
    (1982).
12. Moran, A., Turner, R. J., and Handler, J. S.   Regulation of
    sodium-coupled glucose transport by glucose in a cultured
    epithelium. *J. Biol. Chem.* 258:15087−15090 (1983).
13. D'Amore, T. and Lo, T. C. Y.   Hexose transport in L6 rat
    myoblasts. 1. Rate-limiting step, kinetic properties, and
    evidence for two systems. *J. Cell. Phys.* 127:95−105 (1986).
14. Christopher, C. W., Kohlbacher, M. S., and Amos, H.   Trans-
    port of sugars in chick-embryo fibroblasts: Evidence for a
    low-affinity and high-affinity system for glucose transport.
    Biochem. J. 158:439−450 (1976).
15. Biber, J. W. and Lienhard, G. E.   Isolation of vesicles con-
    taining insulin-responsive, intracellular glucose transporters
    from 3T3-L1 adipocytes. *J. Biol. Chem.* 261:16180−16184
    (1986).
16. Weber, M. J., Evans, P. K., Johnson, M. A., McNair, T. F.,
    Nakamura, K. D., and Salter, D. W.   Transport of potassium,
    amino acids, and glucose in cells transformed by Rous sar-
    coma virus. *Fed. Proc.* 43:107−112 (1984).

17. Maiden, M. C. J., Davis, E. O., Baldwin, S. A., Moore, D. C. M., and Henderson, P. J. F. Mammalian and bacterial sugar transport proteins are homologous. *Nature* 325:641–643 (1987).
18. Birnbaum, M. J., Haspel, H. C., and Rosen, O. M. Cloning and characterization of a cDNA encoding the rat brain glucose-transporter protein. *Proc. Natl. Acad. Sci. USA* 83: 5784–5788 (1986).
19. Mueckler, M., Caruso, C., Baldwin, S. A., Panico, M., Blench, I., Morris, H. R., Allard, W. J., Lienhard, G. E., and Lodish, H. F. Sequence and structure of a human glucose transporter. *Science* 229:941–945 (1985).
20. Toyoda, N., Robinson, F. W., Smith, M. M., Flanagen, J. E., and Tetsuro, K. Apparent translocation of glucose transport activity in rat epididymal adipocytes by insulin-like effects of high pH or hyperosmolarity. *J. Biol. Chem.* 261(5):2117–2122 (1986).
21. Miller, M. H., Amos, H., and Sens, D. A. Regulation of glucose utilization in chick embryo fibroblasts by bicarbonate ion. *J. Cell. Phys.* 107:295–302 (1981).
22. Kalckar, H. M., Ullrey, D. B., and Laursen, R. A. Effects of combined glutamine and serum deprivation on glucose control of hexose transport in mammalian fibroblast cultures. *Proc. Natl. Acad. Sci. USA* 77:5958–5961 (1980).
23. Morgan, H. E. and Whitfield, C. F. In *Current Topics of Membranes and Transports*, Vol. 4 (F. Bronner and A. Kleinzeller, eds.), Academic Press, New York, pp. 255–303 (1974).
24. Clausen, T. In *Current Topics of Membranes and Transport*, Vol. 6 (R. Bronner and A. Kleinzeller, eds.), Academic Press New York, pp. 169–226 (1975).
25. Schudt, C., Gaertner, U., and Pette, D. Insulin action on glucose transport and calcium fluxes in developing muscle cells in Vitro. *Eur. J. Biochem.* 68:103–111 (1976).
26. Draznin, B., Sussman, K., Kao, M., Lewis, D., and Sherman, N. The existence of an optimal range of cytosolic free calcium for insulin-stimulated glucose transport in rat adipocytes. *J. Biol. Chem.* 262:14385–14388 (1987).
27. Cushman, S. W. and Wardzala, L. J. Potential mechanism of insulin action on glucose transport in the isolated rat adipose cell. *J. Biol. Chem.* 255:4758–4762 (1980).
28. Kono, T., Suzuki, K., Dansey, L. E., Robinson, F. W., and Blevins, T. L. Energy-dependent and protein synthesis-independent recycling of the insulin-sensitive glucose transport mechanism in fat cells. *J. Biol. Chem.* 256:6400–6407 (1981).

29. James, D. E., Lederman, L., and Pilch, P. F. Purification of insulin-dependent exocytic vesicles containing the glucose transporter. *J. Biol. Chem.* 262:11817−11824 (1987).

30. Kono, T., Robinson, F. W., Blevins, T. L., and Ezake, O. Evidence that translocation of the glucose transport activity is the major mechanism of insulin action on glucose transport in fat cells. *J. Biol. Chem.* 257:10942−10947 (1982).

31. Oka, Y. and Czech, M. P. Photoaffinity labeling of insulin-sensitive hexose transporters in intact rat adipocytes. *J. Biol. Chem.* 259:8125−8133 (1984).

32. Ezaki, O., Kasuga, M., Akanuma, Y., Takata, K., Hirano, H., Fujita-Yamaguchi, Y., and Kasahara, M. Recycling of the glucose transporter, the insulin receptor and insulin in rat adipocytes. *J. Biol. Chem.* 261:3295−3305 (1986).

33. Barnes, D. and Colowick, S. P. Stimulation of sugar uptake in cultured fibroblasts by epidermal growth factor (EGF) and EGF-binding arginine esterase. *J. Cell. Physiol.* 89:633−640 (1976).

34. Inman, W. H. and Colowick, S. P. Stimulation of glucose uptake by transforming growth factor β: Evidence for the requirement of epidermal growth factor-receptor activation. *Proc. Natl. Acad. Sci. USA* 82:1346−1349 (1985).

35. Martz, A., Mookerjee, B. K., and Jung, C. Y. Insulin and phorbol esters affect the maximum velocity rather than the half-saturation constant of 3-O-methylglucose transport in rat adipocytes. *J. Biol. Chem.* 261:13606−13609 (1986).

36. Yamada, K., Tillotson, L. G., and Isselbacher, K. J. Regulation of hexose transport in chicken embryo fibroblasts: Stimulation by the phorbol ester TPA leads to increase numbers of functioning transporters. *J. Cell. Phys.* 127:211−215 (1986).

37. Hiraki, Y., Rosen, O. M., and Birnbaum, M. J. Growth factors rapidly induce expression of the glucose transporter gene. *J. Biol. Chem.* 263:13655−13662 (1988).

38. Rollins, B. J., Morrison, E. D., Usher, P., and Flier, J. S. Platelet-derived growth factor regulates glucose transporter expression. *J. Biol. Chem.* 263:16523−16526 (1988).

39. Horner, H. C., Munck, A., and Lienhard, G. E. Dexamethasone causes translocation of glucose transporters from the plasma membrane to an intracellular site in human fibroblasts. *J. Biol. Chem.* 262:17696−17702 (1987).

40. Hatanaka, M., Huebner, R. J., and Gilden, R. V. Alterations in the characteristics of sugar uptake by mouse cells transformed by murine sarcoma viruses. *J. Natl. Cancer Inst.* 43:1091−1096 (1969).

41. Hatanaka, M. Transport of sugars in tumor cell membranes. *Biochim. Biophys. Acta* 355:77–104 (1974).

42. Lang, D. R. and Weber, M. J. Increased membrane transport of 2-deoxyglucose and 3-O-methylglucose is an early event in the transformation of chick embryo fibroblasts by Rous sarcoma virus. *J. Cell. Phys.* 94:315–320 (1978).

43. Bose, S. K. and Zlotnick, B. J. Growth and density-dependent inhibition of deoxyglucose transport in Balb 3T3 cells and its absence in cells transformed by murine sarcoma virus. *Proc. Natl. Acad. Sci. USA* 70:2374–2378 (1973).

44. Hatanaka, M. Saturable and nonsaturable process of sugar uptake: Effect of oncogenic transformation in transport and uptake of nutrients. *J. Cell. Phys.* 89:745–749 (1976).

45. Isselbacher, K. J. Increased uptake of amino acids and 2-deoxy-D-glucose by virus-transformed cells in culture. *Proc. Natl. Acad. Sci. USA* 69:585–589 (1972).

46. Kletzien, R. F. and Perdue, J. F. Sugar transport in chick embryo fibroblasts. II. Alterations in transport following transformation by a temperature sensitive mutant of the Rous Sarcoma virus. *J. Biol. Chem.* 249:3375–3382 (1974).

47. Ventuta, S. and Rubin, H. Sugar transport in normal and Rous sarcoma virus-transformed chick embryo fibroblasts. *Proc. Natl. Acad. Sci. USA* 70:653–657 (1973).

48. Shawver, L. K., Olson, S. A., White, M. K., and Weber, M. J. Degradation and biosynthesis of the glucose transporter protein in chicken embryo fibroblasts transformed by the *src* oncogene. *Mol. Cell. Biol.* 7(6):2112–2118 (1987).

49. Rueckert, R. R. and Mueller, G. C. Effect of oxygen tension on HeLa cell growth. *Cancer Res.* 20:944–949 (1960).

50. Balin, A. K., Goodman, B. P., Rasmussen, H., and Cristofalo, V. J. The effect of oxygen tension on the growth and metabolism of WI-38 cells. *J. Cell. Physiol.* 89:235–250 (1976).

51. Kilburn, D. G., Lilly, M. D., Self, D. A., and Webb, F. C. The effect of dissolved oxygen partial pressure on the growth and carbohydrate metabolism of mouse LS cells. *J. Cell Sci.* 4:25–37 (1969).

52. Warburg, O. *The Metabolism of Tumours.* Constable, London (1930).

53. Imamura, T., Crespi, C. L., Thilly, W. G., and Brunengraber, H. Fructose as a carbohydrate source yields stable pH and redox parameters in microcarrier cell culture. *Anal. Biochem.* 124:353–358 (1982).

54. Cristofalo, V. J. and Kritchevsky, D. Growth and glycolysis in the human diploid cell strain WI-38. *Proc. Soc. Exp. Biol. Med.* 118:1109–1112 (1965).

55. Boerner, P., Resnick, R. J., and Racker, E. Stimulation of glycolysis and amino acid uptake in NRK-49F cells by transforming growth factor β and epidermal growth factor. *Proc. Natl. Acad. Sci. USA* 82:1350–1353 (1985).

56. Racker, E., Resnick, R. J., and Feldman, R. Glycolysis and methylaminoisobutyrate uptake in rat-1 cells transfected with *ras* or *myc* oncogenes. *Proc. Natl. Acad. Sci. USA* 82: 3535–3538 (1985).

57. Brand, K., Williams, J. F., and Weidemann, M. J. Glucose and glutamine metabolism in rat thymocytes. *Biochem. J.* 221: 471–475 (1984).

58. Brand, K. Glutamine and glucose metabolism during thymocyte proliferation. *Biochem. J.* 228:353–361 (1985).

59. Bissell, D. M., Levine, G. A., and Bissell, M. J. Glucose metabolism by adult hepatocytes in primary culture and by cell lines from rat liver. *Am. J. Phys.* 234:C122–C130 (1978).

60. Weber, G. Behavior of liver enzymes during hepatocarcinogenesis. *Adv. Cancer Res.* 6:403–494 (1961).

61. Weber, G. Differential carbohydrate metabolism in tumor and host. In *Molecular Interrelations of Nutrition and Cancer* (M. S. Arnott, J. van Eys, and Y.-M. Wang, eds.), Raven Press, New York, pp. 191–208 (1982).

62. Barngrover, D., Thomas, J., and Thilly, W. G. High density mammalian cell growth in Leibovitz bicarbonate-free media formula: Effects of fructose and galactose in culture biochemistry. *J. Cell Sci.* 78:173–189 (1985).

63. Thilly, W. G., Barngrover, D., and Thomas, J. N. Microcarriers and the problem of high density cell culture. In *From Gene to Protein: Translation into Biotechnology.* Academic Press, New York, pp. 75–103 (1982).

64. Thomas, J. Nutrients, oxygen and pH. In *Mammalian Cell Technology* (W. G. Thilly, ed.), Butterworths, Boston, pp. 109–130 (1986).

65. Nahapetian, A. T., Thomas, J. N., and Thilly, W. G. Optimization of environment for high density Vero cell culture: Effect of dissolved oxygen and nutrient supply on cell growth and changes in metabolities. *J. Cell Sci.* 81:65–103 (1986).

66. Reitzer, L. J., Wice, B. M., and Kennell, D. Evidence that glutamine, not sugar, is the major energy source for cultured HeLa cells. *J. Biol. Chem.* 254:2669–2676 (1979).

67. Reitzer, L. J., Wice, B. M., and Kennell, D. The pentose cycle: Control and essential function in HeLa cell nucleic acid synthesis. *J. Biol. Chem.* 255:5616–5626 (1980).

68. Zielke, H. R., Ozand, P. T., Tildon, J. T., Sevdalian, D. A., and Cornblath, M. Growth of human diploid fibroblasts in the

absence of glucose utilization. *Proc. Natl. Acad. Sci. USA*
73:4110–4114 (1976).

69. Zielke, H. R., Ozand, P. T., Tildon, J. T., Sevdalian,
    D. A., and Cornblath, M. Reciprocal regulation of glucose
    and glutamine utilization by cultured human diploid fibroblasts.
    J. Cell. Phys. 95:41–48 (1976).

70. Leibovitz, A. The growth and maintenance of tissue-cell cul-
    tures in free gas exchange with the atmosphere. *Am. J. Hyg.*
    78:173–180 (1963).

71. Matrisian, L. M., Rautman, G., Magun, B. E., and Breath-
    nanch, R. Epidermal growth factor or serum stimulation of rat
    fibroblasts induces an elevation in mRNA levels for lactate de-
    hydrogenase and other glycolytic enzymes. *Nucl. Acids Res.*
    13:711–726 (1985).

72. Ardawi, M. S. M. and Newsholme, E. A. Fuel utilization in
    colonocytes of the rat. *Biochem. J.* 231:713–719 (1985).

73. Newsholme, P., Gordon, S., and Newsholme, E. A. Rates of
    utilization and fates of glucose, glutamine, pyruvate, fatty
    acids and ketone bodies by mouse macrophages. *Biochem. J.*
    242:631–636 (1987).

74. Newsholme, P., Curi, R., Gordon, S., and Newsholme, E. A.,
    Metabolism of glucose, glutamine, long-chain fatty acids and
    ketone bodies by murine macrophages. *Biochem. J.* 239:121–
    125 (1986).

75. Newsholme, E. A., Crabtree, B., and Ardawi, M. S. M. The
    role of high rates of glycolysis and glutamine utilization in
    rapidly dividing cells. *Bios. Rep.* 5:393–400 (1985).

76. Hume, D. A., Radik, J. L., Ferber, E., and Weidemann,
    M. J. Aerobic glycolysis and lymphocyte transformation.
    Biochem. J. 174:703–709 (1978).

77. Weber, G., Morris, H. P., Love, W. C., and Ashmore, J.
    Comparative biochemistry of hepatomas. II. Isotope studies
    of carbohydrate metabolism in Morris hepatoma 5123. *Cancer
    Res.* 21:1406–1411 (1961).

78. Aisenberg, A. C. and Morris, H. P. Energy pathways of
    hepatoma No. 5123. *Nature* 191:1314–1315 (1961).

79. Lin, Y. C., Elwood, J. C., Rosado, A., Morris, H. P., and
    Weinhouse, S. Glucose metabolism in a low-glycolysing
    tumour, the Morris hepatoma 5123. *Nature* 195:153–155 (1962).

80. Elsford, H. L., Freese, M., Passamani, E., and Morris, H. P.
    Ribonucleotide reductase and cell proliferation. I. Variations
    of ribonucleotide reductase activity with tumor growth rate in a
    series of rat hepatomas. *J. Biol. Chem.* 245:5228–5233 (1970).

81. Harkrader, R. J., Jackson, R. C., Ross, D. A., and
    Weber, G. Increase in liver and kidney deoxycytidine kinase

activity linked to neoplastic transformation. *Biochem. Biophys. Res. Commun.* 96:1633–1639 (1980).

82. Jackson, R. C., Lui, M. S., Boritzki, T. J., Morris, H. P., and Weber, G. Purine and pyrimidine nucleotide patterns of normal, differentiating and regenerating liver and of hepatomas in rats. *Cancer Res.* 40:1286–1291 (1980).

83. Kizaki, H., Williams, J. C., Morris, H. P., and Weber, G. Purine and pyrimidine nucleotide patterns of normal, differentiating and regenerating liver and of hepatomas in rats. *Cancer Res.* 40:1286–1291 (1980).

84. Weber, G., Olah, E., Denton, J. E., Lui, M. S., Takeda, E., Tzeng, D. Y., and Ban, J. Dynamics of modulation of biochemical programs in cancer cells. *Adv. Enzym. Reg.* 19:87–102 (1981).

85. Weber, G., Olah, E., Lui, M. S., Kizaki, H., Tzeng, D. Y., and Takeda, E. Biochemical commitment to replication in cancer cells. *Adv. Enzym. Reg.* 18:3–26 (1980).

86. Metzler, D. E. The metabolism of nitrogen-containing compounds. In *Biochemistry, the Chemical Reactions of Living Cells.* Academic Press, New York, pp. 805–889 (1977).

87. Eagle, H. Nutrition needs of mammalian cells in tissue culture. *Science* 122:501–504 (1955).

88. Eagle, H. Amino acid metabolism in mammalian cell cultures. *Science* 130:432–437 (1959).

89. Kruse, P. F., Jr., Miedema, E., and Carter, H. C. Amino acid utilizations and protein synthesis at various proliferation rates, population densities, and protein contents of perfused animal cell and tissue cultures. *Biochemistry* 6:949–955 (1967).

90. Roberts, R. S., Hsu, H. W., Lin, K. D., and Yang, T. J. Amino acid metabolism of myeloma cells in culture. *J. Cell Sci.* 21:609–615 (1976).

91. Stoner, G. D. and Merchant, D. J. Amino acid utilization by L–M strain mouse cells in a chemically defined medium. *In Vitro* 7:330–343 (1972).

92. McCarthy, K. Selective utilization of amino acids by mammalian cell cultures. *Exp. Cell Res.* 27:230–240 (1962).

93. Arathoon, W. R. and Telling, R. C. Uptake of amino acids and glucose by BHK 21 clone 13 suspension cells during cell growth. *Dev. Biol. Standard* 50:145–154 (1982).

94. Mohberg, J. and Johnson, M. J. Amino acid utilization by 929-L fibroblasts in chemically defined media. *J. Natl. Cancer Inst.* 31:611–625 (1963).

95. Griffiths, J. B. and Pirt, S. J. The uptake of amino acids by mouse cells (strain LS) during growth in batch culture and

chemostat culture: The influence of cell growth rate. *Proc. Roy. Soc. B.* 168:421−438 (1967).

96. Blaker, G. J., Birch, J. R., and Pirt, S. J. The glucose insulin and glutamine requirements of suspension cultures of HeLa cells in a defined culture medium. *J. Cell Sci.* 9: 529−537 (1971).

97. Lambert, K. and Pirt, S. J. The quantitative requirements of human diploid cells (strain MRC-5) for amino acids, vitamins and serum. *J. Cell Sci.* 17:397−411 (1975).

98. Butler, M. and Thilly, W. G. MDCK microcarrier cultures: Seeding effects and amino acid utilization. *In Vitro* 18: 213−219 (1982).

99. Butler, M., Imamura, T., Thomas, J., and Thilly, W. G. High yields from microcarrier cultures by medium perfusion. *J. Cell Sci.* 61:351−363 (1983).

100. Christensen, H. N. Some special kinetic problems of transport. *Adv. Enzymol.* 32:1−20 (1969).

101. Gazzola, G. C., Dall'Asta, V., and Guidotti, G. G. The transport of neutral amino acids in cultured human fibroblasts. *J. Biol. Chem.* 255:929−936 (1980).

102. Shotwell, M. A., Jayme, D. W., Kilberg, M. S., and Oxender, D. L. Neutral amino acid transport systems in Chinese hamster ovary cells. *J. Biol. Chem.* 256:5422−5427 (1981).

103. Shotwell, M. A., Lobaton, C. D., Collarini, E. J., Moreno, A., Giles, R. E., and Oxender, D. L. Genetic studies of leucine transport in mammalian cells. *Fed. Proc.* 43:2269−2272 (1984).

104. Ardawi, M. S. M. The transport of glutamine and alanine into rat colonocytes. *Biochem. J.* 238:131−135 (1986).

105. Ardawi, M. S. M. and Newsholme, E. A. The transport of glutamine into rat mesenteric lymphocytes. *Biochim. Biophys. Acta* 856:413−420 (1986).

106. Kilberg, M. S., Handlogten, M. E., and Christensen, H. N. Characteristics of an amino acid transport system in rat liver for glutamine, asparagine, histidine, and closely related analogs. *J. Biol. Chem.* 255:4011−4019 (1980).

107. Haussinger, D., Soboll, S., Mailer, A. J., Gerok, W., Tager, J. M., and Sies, H. Role of plasma membrane transport in hepatic glutamine metabolism. *Eur. J. Biochem.* 152:597−603 (1985).

108. Salter, M., Knowles, R. G., and Pogson, C. I. Transport of the aromatic amino acids into isolated rat liver cells. *Biochem. J.* 233:499−506 (1986).

109. Handlogten, M. E., Weissbach, L., and Kilberg, M. S. Heterogeneity of Na⁺-independent 2-aminobicyclo-(2,2,1)-heptane-2-carboxylic acid and L-leucine transport in isolated rat hepatocytes in primary culture. *Biochem. Biophys. Res. Commun.* 104:307−313 (1982).

110. Rosenberg, R., Young, J. D., and Ellroy, J. C. L-tryptophan transport in human red blood cells. *Biochim. Biophys. Acta* 598:375−384 (1980).

111. White, M. F., Gozzola, G. C., and Christensen, H. N. Cationic amino acid transport into cultured animal cells. *J. Biol. Chem.* 257:4443−4449 (1982).

112. Guidotti, G. G. The transport of amino acids and its regulation in animal cells. In *Membranes in Tumour Growth* (T. Galeotti, A. Cittadini, G. Neri, and S. Papa, eds.), Elsevier, Amsterdam, pp. 381−386 (1982).

113. Bannai, S. and Kitamura, E. Transport interaction of L-cystine and L-glutamate in human diploid fibroblasts in culture. *J. Biol. Chem.* 255:2372−2376 (1980).

114. Dall'Asta, V., Gazzola, G. C., Franchi-Gazzola, R., Bussolati, O., Longo, N., and Guidotti, G. G. Pathways of L-glutamic acid transport in cultured human fibroblasts. *J. Biol. Chem.* 258:6371−6379 (1983).

115. Gazzola, G. C., Dall'Asta, V., Bussolati, O., Makowske, M., and Christensen, H. N. A stereoselective anomaly in dicarboxylic amino acid transport. *J. Biol. Chem.* 256:6054−6059 (1981).

116. Shotwell, M. A., Kilberg, M. S., and Oxender, D. L. The regulation of neutral amino acid transport in mammalian cells. *Biochim. Biophys. Acta* 737:267−284 (1983).

117. Guidotti, G. G., Borghetti, A. F., and Gazzola, G. C. The regulation of amino acid transport in animal cells. *Biochim. Biophys. Acta* 515:329−366 (1978).

118. Bracy, D. S., Handlogten, M. E., Barber, E. F., Han, H.-P., and Kilberg, M. S. cis-Inhibition, trans-inhibition, and repression of hepatic amino acid transport mediated by system A. *J. Biol. Chem.* 261:1514−1520 (1986).

119. Gazzola, G. C., Dall'Asta, V., and Guidotti, G. G. Adaptive regulation of amino acid transport in cultured human fibroblasts. *J. Biol. Chem.* 256:3191−3198 (1981).

120. Fafournoux, P., Remesy, C., and Demigne, C. Control by amino acids of the activity of system A-mediated amino acid transport in isolated rat hepatocytes. *Biochem. J.* 231:315−320 (1985).

121. Moffett, J. and Englesberg, E. Regulation of the A system of amino acid transport in Chinese hamster ovary cells,

CHO-K1: The difference in specificity between the apo-repressor inactivator (apo-ri) and the transporter and the characterization of the proposed apo-ri. *J. Cell. Phys.* 126: 421–429 (1984).

122. Moffett, J. and Englesberg, E. Recessive constitutive mutant Chinese hamster ovary cells (CHO-K1) with an altered A system for amino acid transport and the mechanism of gene regulation of the A system. *Mol. Cell. Biol.* 4:799–808 (1986).

123. Gazzola, G. C., Dall'Asta, V., Franch-Gazzola, R., Bussolati, O., Longo, N., and Guidotti, G. G. Post-translational control by carrier availability of amino acid transport in fetal human fibroblasts. *Biochem. Biophys. Res. Commun.* 120:172–178 (1984).

124. Tramacere, M., Petronini, P. G., and Borghetti, A. F. Effect of hyperosmolarity on the activity of amino acid transport system L in avian fibroblasts. *J. Cell. Phys.* 121:81–86 (1984).

125. Tramacere, M., Petronini, P. G., Severini, A., and Borghetti, A. F. Osmoregulation of amino acid transport activity in cultured fibroblasts. *Exp. Cell Res.* 151:70–79 (1984).

126. Petronini, P. G., Tramacere, M., Kay, J. E., and Borghetti, A. F. Adaptive response of cultured fibroblasts to hyperosmolarity. *Exp. Cell Res.* 165:180–190 (1986).

127. Longo, N., Franch-Gazzola, R., Bussolati, O., Dall'Asta, V., Foa, P. P., Guidotti, G. G., and Gazzola, G. C. Effect of insulin on the activity of amino acid transport systems in cultured human fibroblasts. *Biochim. Biophys. Acta* 844:216–223 (1985).

128. Leoni, S., Spagnuolo, S., Dini, L., and Devirgiliis, L. C. Regulation of amino acid transport in isolated rat hepatocytes during development. *J. Cell. Phys.* 130:103–110 (1987).

129. Borghetti, A. F., Piedimonte, G., Tramacere, M., Severini, A., Ghiringhelli, P., and Guidotti, G. G. Cell density and amino acid transport in 3T3, SV3T3, and SV3T3 revertant cells. *J. Cell. Physiol.* 105:39–49 (1980).

130. Boerner, P. and Saier, M. H., Jr. Growth regulation and amino acid transport in epithelial cells: Influence of culture conditions and transformation on A, ASC, and L transport activities. *J. Cell. Physiol.* 113:240–246 (1982).

131. Parnes, J. R. and Isselbacher, K. J. Transport alterations in Virus-transformed cells. *Prog. Exp. Tumor Res.* 22: 79–122 (1978).

132. Gazzola, G. C., Dall'Asta, V., Franchi-Gazzola, R., Bussolati, O., Longo, N., and Guidotti, G. G. Amino acid transport in normal and neoplastic cultured human fibroblasts.

In *Cell Membranes and Cancer* (T. Galeotti, A. Cittadini, G. Neri, S. Papa, and L. A. Smets, eds.), Elsevier, Amsterdam, pp. 169–174 (1985).

133. Lehninger, A. L. Oxidation-reduction enzymes and electron transport. In *Biochemistry, the Molecular Basis of Cell Structure and Function.* Worth, New York, pp. 447–506 (1977).

134. Krebs, H. A. and Lund, P. Aspects of the regulation of the metabolism of branched-chain amino acids. *Adv. Enzym. Reg.* 15:375–394 (1977).

135. Ichihard, I., Noda, C., and Ogawa, K. Control of leucine metabolism with special reference to branched-chain amino acid transaminase isozymes. *Adv. Enzym. Reg.* 11:155–166 (1973).

136. Windmueller, H. G. and Spaeth, A. E. Uptake and metabolism of plasma glutamine by the small intestine. *J. Biol. Chem.* 249:5070–5079 (1974).

137. Windmueller, H. G. and Spaeth, A. E. Identification of ketone bodies and glutamine as the major respiratory fuels *in vivo* for postabsorptive rat small intestine. *J. Biol. Chem.* 253:69–76 (1978).

138. Windmueller, H. G. Glutamine utilization by the small intestine. *Prog. Enzymol.* 53:201–237 (1982).

139. Hanson, P. J. and Parsons, D. S. Metabolism and transport of glutamine and glucose in vascularly perfused small intestine of the rat. *Biochem. J.* 166:509–519 (1977).

140. Tildon, J. T. and Roeder, L. M. Glutamine oxidation by dissociated cells and homogenates of rat brain: Kinetics and inhibitor studies. *J. Neurochem.* 42:1069–1076 (1984).

141. Sumbilla, C. M., Zielke, C. L., Reed, W. D., Ozand, P. T., and Zielke, H. R. Comparison of the oxidation of glutamine, glucose, ketone bodies and fatty acids by human diploid fibroblasts. *Biochim. Biophys. Acta* 675:301–304 (1981).

142. Zielke, H. R., Zielke, C. L., and Ozand, P. T. Glutamine: A major energy source for cultured mammalian cells. *Fed. Proc.* 43:21–125 (1984).

143. Barbehenn, E. K., Masterson, E., Koh, S.-W., Passonneau, J. V., and Chader, G. J. An examination of the efficiency of glucose and glutamine as energy sources for cultured chick pigment epithelial cells. *J. Cell. Phys.* 118:262–266 (1984).

144. Zielke, H. R., Sumbilla, C. M., Zielke, C. L., Tildon, J. T., and Ozand, P. T. Glutamine metabolism by cultured mammalian cells. In *Glutamine Metabolism in Mammalian Tissues* (D. Haussinger and H. Sies, eds.), Springer-Verlag, Berlin, pp. 247–254 (1984).

145. Zielke, H. R., Sumbilla, C. M., Sevdalian, D. A., Hawkins, R. L., and Ozand, P. T. Lactate: A major product of glutamine metabolism by human diploid fibroblasts. *J. Cell. Physiol.* 104:433−441 (1980).

146. Newsholme, E. A., Crabtree, B., and Ardawi, M. S. M. Glutamine metabolism in lymphocytes: Its biochemical, physiological and clinical importance. *Quart. J. Exp. Phys.* 70: 473−489 (1985).

147. Ardawi, M. S. M. and Newsholme, E. A. Glutamine metabolism in lymphoid tissues. In *Glutamine Metabolism in Mammalian Tissues* (D. Haussinger and H. Sies, eds.), Springer-Verlag, Berlin, pp. 235−246 (1984).

148. Ardawi, M. S. M. and Newsholme, E. A. Glutamine metabolism in lymphocytes of the rat. *Biochem. J.* 212:835−842 (1983).

149. Pardridge, W. M. and Casanello-Ertl, D. Effects of glutamine deprivation on glucose and amino acid metabolism in tissue culture. *Am. J. Physiol.* 236:E234−E238 (1979).

150. Meister, A. Glutamic acid and glutamine. In *Biochemistry of the Amino Acids*. Academic Press, New York, Vol. 2, pp. 617−635 (1965).

151. McKeehan, W. L. Glutaminolysis in animal cells. In *Carbohydrate Metabolism in Cultured Cells* (M. J. Morgan, ed.), Plenum Press, New York, pp. 111−150 (1984).

152. Sevdalian, D. A., Ozand, P. T., and Zielke, H. R. Increase in glutaminase activity during the growth cycle of cultured human diploid fibroblasts. *Enzyme* 25:142−144 (1980).

153. Knox, W. E., Horowitz, M. L., and Friedell, G. H. The proportionality of glutaminase content to growth rate and morphology of rat neoplasms. *Cancer Res.* 29:669−680 (1969).

154. Curi, R., Newsholme, P., and Newsholme, E. A. Intracellular distribution of some enzymes of the glutamine utilisation pathway in rat lymphocytes. *Biochem. Biophys. Res. Commun.* 138:318−322 (1986).

155. Krebs, H. A. Glutamine metabolism in the animal body. In *Glutamine: Metabolism, Enzymology, and Regulation* (J. Mora and R. Palacios, eds.), Academic Press, New York, pp. 319−329 (1980).

156. Ardawi, M. S. M. and Newsholme, E. A. Maximum activities of some enzymes of glycolysis, the tricarboxylic acid cycle and ketone-body and glutamine utilization pathways in lymphocytes of the rat. *Biochem. J.* 208:743−748 (1982).

157. Moreadith, R. W. and Lehninger, A. L. The pathways of glutamate and glutamine oxidation by tumor cell mitochondria.

Role of mitochondrial NAD(P)$^+$-dependent malic enzyme. *J. Biol. Chem.* 259:6215−6221 (1984).

158. Sumbilla, C. M., Ozand, P. T., and Zielke, H. R. Activities of enzymes required for the conversion of 4-carbon TCA cycle compounds to 3-carbon glycolytic compounds in human diploid fibroblasts. *Enzyme* 26:201−205 (1981).
159. Reiss, N., Kanety, H., and Schlessinger, J. Five enzymes of the glycolytic pathway serve as substrates for purified epidermal-growth-factor-receptor kinase. *Biochem. J.* 239: 691−697 (1986).
160. Cooper, J. A., Esch, F. S., Taylor, S. S., and Hunter, T. Phosphorylation sites in enolase and lactate dehydrogenase utilized by tyrosine protein kinases *in vivo* and *in vitro*. *J. Biol. Chem.* 259:7835−7841 (1984).
161. Cooper, J. A., Reiss, N. A., Schwartz, R. J., and Hunter, T. Three glycolytic enzymes are phosphorylated at tyrosine in cells transformed by Rous sarcoma virus. *Nature* 302:218−223 (1983).
162. Rice, G. and Thomas, J. T. (unpublished observations).
163. McKeehan, W. L. Glycolysis, glutaminolysis and cell proliferation. *Cell Biol. Int. Rep.* 6:635−647 (1982).
164. Raivio, K. O. and Seegmiller, J. E. Role of glutamine in purine synthesis and in guanine nucleotide formation in normal fibroblasts and in fibroblasts deficient in hypoxanthine phosphoribosyltransferase activity. *Biochim. Biophys. Acta* 299: 283−292 (1973).
165. Wenner, C. E. Regulation of energy metabolism in normal and tumor tissues. In *Cancer*, Vol. 3 (F. F. Becker, ed.). Plenum Press, New York, pp. 389−401 (1975).
166. Sols, A. The Pasteur effect in the allosteric era. In *Reflections on Biochemistry* (A. Kornberg, B. L. Horecker, L. Cornudella, and J. Oro, eds.). Pergamon Press, Elmsford, NY, pp. 199−206 (1976).
167. Wang, T., Marquardt, C., and Folker, J. Aerobic glycolysis during lymphocyte proliferation. *Nature* 261:701 (1976).
168. Racker, E. Why do tumor cells have a high aerobic glycolysis? *J. Cell. Physiol.* 89:697−700 (1976).
169. Lazo, P. A. and Sols, A. Energetics of tumour cells: Enzyme basis of aerobic glycolysis. *Biochem. Soc. Trans.* 8: 579 (1980).
170. Gregg, C. T. In *Growth, Nutrition and Metabolism of Cells in Culture*, Vol. 1 (G. H. Rothblat and U. J. Cristofalo, eds.), Academic Press, New York, pp. 88−136 (1972).
171. Crabtree, B. and Newsholme, E. A. A quantitative approach to metabolic control. *Curr. Topics Cell. Reg.* 25:21−76 (1985).

172. Sussman, I., Erecinska, M., and Wilson, D. F. Regulation of cellular energy metabolism, The Crabtree effect. *Biochim. Biophys. Acta* 591:209−223 (1980).
173. Donnelly, M. and Scheffler, I. E. Energy metabolism in respiration-deficient and wild type Chinese hamster fibroblasts in culture. *J. Cell. Physiol.* 89:39−52 (1976).
174. Breen, G. A. M. and Scheffler, I. E. Respiration-deficient Chinese hamster cell mutants: Biochemical characterization. *Som. Cell Gen.* 5:441−451 (1979).
175. Soderberg, K., Mascarello, J. T., Breen, G. A. M., and Scheffler, I. E. Respiration-deficient Chinese hamster cell mutants: Genetic characterization. *Som. Cell Gen.* 5(2): 225−240 (1979).
176. Scheffler, I. E. Biochemical genetics of respiration-deficient mutants of animal cells. In *Carbohydrate Metabolism in Cultured Cells* (M. J. Morgan, ed.), Plenum Press, New York, pp. 77−109 (1986).
177. Pedersen, P. L. Tumor mitochondria and the bioenergetics of cancer cells. *Prog. Exp. Tumor Res.* 22:190−274 (1978).
178. Erecinska, M. and Wilson, D. F. Regulation of cellular energy metabolism. *J. Memb. Biol.* 70:1−14 (1982).
179. Erecinska, M. and Wilson, D. F. The mitochondrion and its functions. In *Inhibitors of Mitochondrial Functions. International Encyclopedia of Pharmacology and Therapeutics.* Pergamon Press, Elmsford, NY, Section 107, pp. 1−17 (1981).
180. Wilson, D. F., Stubbs, M., Veech, R. L., Erecinska, M., and Krebs, H. A. Equilibrium relations between the oxidation-reduction reactions and the adenosine triphosphate synthesis in suspensions of isolated liver cells. *Biochem. J.* 140:57−64 (1974).
181. Wilson, D. F. The mitochondrial respiratory chain. In *Bioelectrochemistry 1, Biological Redox Reactions* (G. Milazzo and M. Blank, eds.), Plenum Press, New York, pp. 249−282 (1983).
182. Azzi, A. Mitochondria: The utilization of oxygen for cell life. *Experientia* 40(9):901−906 (1984).
183. Klingenberg, M. Metabolite transport in mitochondria: An example for intracellular membrane function. In *Essays in Biochemistry,* Vol. 6 (P. N. Campbell and F. Dickens, eds.), Academic Press, New York, pp. 119−159 (1970).
184. Klingenberg, M. The ADP-ATP translocation in mitochondria, a membrane potential controlled transport. *J. Membr. Biol.* 56:97−105 (1980).
185. Forman, N. G. and Wilson, D. F. Dependence of mitochondrial oxidative phosphorylation on activity of the adenine nucleotide translocase. *J. Biol. Chem.* 258:8649−8655 (1983).

186. Dudley, G. A., Tullson, P. C., and Terjung, R. L. Influence of mitochondrial content on the sensitivity of respiratory
control. *J. Biol. Chem.* 262:9109-9114 (1987).

187. Wilson, D. F. Regulation of in vivo mitochondrial oxidative
phosphorylation. In *Membranes and Transport*, Vol. 1 (A. N.
Martonosi, ed.). Plenum Press, New York, pp. 349-355
(1982).

188. Wilson, D. F., Owen, C. S., and Holian, A. Control of mitochondrial respiration: A quantitative evaluation of the roles
of cytochrome C and oxygen. *Arch. Biochem. Biophys.* 182:
749-762 (1977).

189. Sugano, T., Nozomu, O., and Chance, B. Mitochondrial functions under hypoxic conditions. The steady states of cytochrome c reduction and of energy metabolism. *Biochim.
Biophys. Acta* 347:340-358 (1974).

190. Wilson, D. F., Erecinska, M., Drown, C., and Silver, I. A.
The oxygen dependence of cellular energy metabolism. *Arch.
Biochem. Biophys.* 195:485-493 (1979).

191. Malaska, T. and Thomas, J. T., unpublished observations.

192. Hornsby, P. J. and Crivello, J. F. The role of lipid peroxidation and biological antioxidants in the function of the adrenal
cortex. 1. A background review. *Mol. Cell. Endocrinol.*
30:1-20 (1983).

193. Hornsby, P. J. and Gill, G. N. Regulation of glutamine and
pyruvate oxidation in cultured adrenocortical cells by cortisol,
antioxidants, and oxygen: Effects on cell proliferation.
J. Cell. Physiol. 109:111-120 (1981).

194. Chance, B., Boveria, A., Nakase, Y., and Sies, H. Hydro-
peroxide metabolism: An overview. In *Functions of Glutathione in Liver and Kidney* (H. Sies and A. Wendal, eds.),
Springer-Verlag, New York, pp. 95-106 (1978).

195. Werrlein, R. J. and Glinos, A. D. Oxygen microenvironment
and respiratory oscillations in cultured mammalian cells.
Nature 251:317-319 (1974).

196. Fleischaker, R. J. and Sinskey, A. J. *Eur. J. Appl.
Microbiol. Biotech.* 12:193-197 (1981).

197. Fleischaker, R. j., Ph.D. thesis, Massachusetts Institute of
Technology, Cambridge (1982).

198. Taylor, W. G. and Camalier, R. F. Modulation of epithelial
cell proliferation in culture by dissolved oxygen. *J. Cell.
Physiol.* 111:21-27 (1982).

199. Riess, J. G. and LeBlanc, M. Solubility and transport phenomena in perfluorochemicals relevant to blood substitution
and other biomedical applications. *Pure Appl. Chem.* 54:
2383-2406 (1982).

200. Jones, D. P. and Kennedy, F. G.   Intracellular oxygen supply during hypoxia. *Am. J. Physiol.* 243:C247−C253 (1982).
201. Jones, D. P. and Kennedy, F. G.   Intracellular O_2 gradients in cardiac myocytes. Lack of a role for myoglobin in facilitation of intracellular O_2 diffusion. *Biochem. Biophys. Res. Commun.* 105(2):419−424 (1982).
202. Jones, D. P.   Effect of mitochondrial clustering on O_2 supply in hepatocytes. *Am. J. Physiol.* 247:C83−C89 (1984).
203. Kennedy, F. G. and Jones, D. P.   Oxygen dependence of mitochondrial function in isolated rat cardiac myocytes. *Am. J. Physiol.* 250:C374−C383 (1986).
204. Jones, D. P.   Intracellular diffusion gradients of O_2 and ATP. *Am. J. Physiol.* 250:C663−C675 (1986).
205. Wilson, D. F. and Erecinska, M.   Effect of oxygen concentration on cellular metabolism. *Chest* 88(4):229s−232s (1985).
206. Wilson, D. F., Erecinska, M., and Silver, I. A.   Metabolic effects of lowering oxygen tension in vivo. *Adv. Exp. Med. Biol.* 159:293−301 (1983).
207. Mattiasson, B. and Adlercreutz, P.   Perfluorochemicals in biotechnology. *Trends Biotechnol.* 5:250−254 (1987).

5

Cell Banking

MICHAEL E. WIEBE and LAURIE H. MAY
Genentech, Inc., South San Francisco, California

I. INTRODUCTION

The cryopreservation of cells began in the 1950s and early 1960s with
the discovery that slowly freezing cells in the presence of a cryo-
protectant (glycerol, dimethyl sulfoxide) would allow cells to be preserved indefinitely at low temperatures, and that they could later be
thawed with full restoration of cell viability and function (1–11).
This has had a considerable impact in the banking of rare human
blood cells for transfusion (12–14); in the storage of sperm for use
in artificial insemination (15,16); in the availability of diploid strains
and continuous cell lines for use in basic research, principally in
the fields of virology, cellular and molecular biology, biochemistry,
and genetics; and for use in the manufacture of vaccines and biopharmaceuticals.

    Central repositories of animal cells maintain very large collec-
tions of reference cells which are generally available to the scien-
tific community. For example, the American Type Culture Collection
(ATCC) now has in its repository some 2800 characterized cell lines
and hybridomas derived from approximately 75 different species (17).
The ATCC collection is the national repository of cell lines in the
United States and also functions as the WHO International Reference
Center for Cell Cultures. Cell repositories supported by the U.S.
Department of Health and Human Services include the National

Institute of General Medical Sciences (NIGMS) Human Genetic Mutant
Cell Repository, containing over 4400 characterized cell cultures,
and the National Institute on Aging (NIA) Cell Repository with approximately 1050 characterized cell cultures (18,19). These two col-
lections are housed at the Coriell Institute for Medical Research in
Camden, New Jersey. The European Collection of Animal Cell Cultures (ECACC) currently maintains over 600 cell lines. ECACC was
established in 1984 as a joint venture between the British Public
Health Laboratory Service (PHLS) and the Department of Trade and
Industry, and is located at the PHLS Center for Applied Microbiology
and Research, Porton Down, Salisbury, Wiltshire, U.K. (20).

Manufacturers that use diploid strains or continuous cell lines
to produce vaccines or biopharmaceuticals usually generate, characterize, and maintain their own cell banks. This chapter will focus
on the use of cultured cells in biopharmaceutical manufacturing, the
central role of cell banks in process standardization, the special
strategies and conditions which must be employed in the production
and maintenance of these cell banks, and a brief description of the
characterization of cell banks.

II. USE OF CULTURED CELLS IN BIOPHARMACEUTICAL
MANUFACTURING

Historically, human biopharmaceuticals have been extracted and
purified from human and animal fluids and tissues. For example,
these include albumin, clotting factors VIII and IX, gamma globulins,
and hepatitis B vaccine from human serum or plasma; urokinase and
several peptide hormones from human urine; growth hormone from
pituitaries of human cadavers; and insulin, lipase, amylase, trypsin,
chymotrypsin, and glucagon from porcine or bovine pancreas. Vaccines have been manufactured by the culturing of virus in a variety
of animal organs or tissues in vivo or primary cells in vitro. These
include embryonated hen or duck eggs, and primary cells derived
from a variety of animal species including primates.

The use of serially passaged cultured cells to produce biologicals
has its origins in the early 1960s when Hayflick and Moorhead (21)
described the establishment of human diploid cell strains derived
from several normal fetal tissues and suggested that they might be
useful for the preparation of human virus vaccines. One of these
cell strains, WI-38, was carefully characterized by karyological
studies and by a thorough search for viral contaminants, and was
found to be normal and free of exogenous infectious agents. Despite
this, there was great reluctance to use WI-38 cells during the 1960s

largely due to fears that unknown "cancer-causing" viruses or other
"cancer-causing" components would be transmitted to vaccine recipients. The controversy over the use of human diploid cells raged
on for 10 years before the first diploid cell product, poliomyelitis
vaccine, was approved for use in the United States. The history of
the acceptance of human diploid cell strains has been documented in
a number of recent reviews (22–24). Human diploid cells (WI-38 or
MRC-5) have been used to produce licensed vaccines against poliovirus, adenovirus types 4 and 7, rubella virus, rubeola virus, and
rabies. Recipients of these vaccines number in the tens of millions
and there are no reports of untoward effects that are due to the
cell substrate.

The use of cultured cells to produce biologicals has now expanded to include continuous heteroploid cell lines. Technical developments responsible for this include the following: (1) the finding
that some continuous cell lines can produce potentially useful biopharmaceuticals, e.g., Namalwa cells produce lymphoblastoid interferon; (2) myeloma cells can be fused to antibody-producing B cells
to generate hybridomas which produce highly specific monoclonal
antibodies for indefinite periods of time; (3) the development of a
variety of procedures that are economical for the mass culture of
cells (reviewed in Chapters 8–13) including microcarriers, hollow-
fiber systems, adaptation of traditional deep-tank fermenter technology, continuous perfusion systems, fluidized bed reactors, and many
others; and (4) the use of recombinant DNA technology to generate
genetically engineered mammalian cells to produce large quantities of
a protein normally foreign to that cell.

The primary advantage of using continuous cell lines for manufacture of biopharmaceuticals, compared to diploid cell strains, is
their indefinite life span. Human diploid cells, such as WI-38 or
MRC-5, begin to reach senescence after 40–60 population doublings
(17,24,25). Since these cells double approximately every 24 hr,
cells derived from the stocks currently available (at population
doubling levels of 10–15) could be in continuous culture for no more
than 1–4 months. The development of genetically engineered mammalian cells requires considerably longer periods of continuous culture for processes which include transfection, cloning, gene amplification, adaptation to production culture conditions, and preparation
of cells for banking. Additionally, the ability to culture cells in a
production "seed train" for extended periods of time after thawing
an ampule has obvious advantages in extending the life of a cryopreserved cell bank.

There is now a consensus that continuous cell lines are acceptable for the production of biopharmaceuticals, with evaluation on a

Table 1 Conferences on the Use of Continuous Cell Lines to Produce Vaccines and Biopharmaceuticals

Conference	Location	Year	Ref.
Cell Substrates: Their Use in the Production of Vaccines and Other Biologicals	Lake Placid, New York	1978	26
Production and Exploitation of Existing and New Animal Cell Substrates	Gardone Riviera, Italy	1984	27
NIH Workshop on Abnormal Cells, New Products, and Risk	Bethesda, Maryland	1984	28
ESACT Meeting on Advances in Animal Cell Technology	Baden, Austria	1985	29
Biotechnology of Mammalian Cells: Development for the Production of Biologicals	Yokohama, Japan	1986	30
WHO Study Group on Biologicals	Geneva, Switzerland	1986	31
Banbury Conference on Therapeutic Peptides and Proteins: Assessing the New Technologies	Cold Spring Harbor, New York	1987	32
Modern Approaches to Animal Cell Technology	Tiberias, Israel	1987	33
Continuous Cell Lines as Substrates for Biologicals	Arlington, Virginia	1988	34
ESACT Meeting on Advances in Animal Cell Biology and Technology for Bioprocesses	Knokke, Belgium	1988	35

Table 2   Licensed Biologicals Produced in Continuous Cell Lines

Biological	Cell line	Countries where licensed
Lymphoblastoid interferon	Namalwa	U.K., Canada, Japan
Rabies vaccine	Vero	France, Thailand
Polio vaccine	Vero	France, Belgium
OKT3	Murine hybridoma	U.S.A., West Germany, France, Italy, Switzerland
Recombinant tPA	CHO	U.S.A., Canada, Philippines, New Zealand, France, Austria, South Korea, Brazil, Luxembourg, West Germany, Belgium, Mexico, South Africa, Singapore, Cyprus, Spain, Netherlands, Taiwan, Australia, Colombia, Hong Kong, Denmark, Bahrain, Switzerland, Thailand, Italy, U.K.

case-by-case basis. The primary concern about the use of continuous cell lines has been that they are abnormal cells, many of them being derived from tumors, wherein the possibility exists that cancer-causing DNA, proteins, or viruses could be transmitted to recipients of products made in these cells. A number of conferences have been held to consider these issues over the last 12 years (Table 1).

National control authorities and other organizations have generated guides or "points to consider" to assist manufacturers that are developing processes that employ continuous cell lines (36–40). Several human biologicals produced in continuous cell lines are now licensed by national control authorities (Table 2).

III. CENTRAL ROLE OF CELL BANKS IN PROCESS STANDARDIZATION

One of the most important advantages of using serially cultured cells (diploid cell strains or continuous cell lines) to produce vaccines or biopharmaceuticals is the ability to have a characterized common starting source of product for each production lot, i.e., the cryopreserved master cell bank (MCB). Exhaustive characterization of

the MCB allows the manufacturer to fully define the product source
with regard to (1) freedom from adventitious and endogenous agents,
(2) range of susceptibility to adventitious agents, and (3) potential
molecular contaminants (DNA, protein). It is not practical nor
economical for primary cells, or human or animal fluids or tissues,
to be evaluated as thoroughly prior to product extraction. Furthermore, the establishment of a MCB and thus the availability of a
single common product source makes possible more definitive studies
of manufacturing consistency. For example, one can establish that
cells can be cultured for a given length of time after thaw from the
MCB without changes occurring in process parameters and product
quality. In short, cryopreservation of the cell substrate as a
thoroughly characterized MCB provides the manufacturer with the
opportunity to define and standardize the production process to a
level not otherwise possible.

IV. STRATEGIES AND PROCEDURES

Numerous activities precede the banking of a genetically engineered
cell line. A crucial step is the choice of a suitable cell line into
which genetic material can be transfected, causing the cells to produce the protein desired. This process can be streamlined if several candidate parental cell lines have themselves been banked and
characterized. The advantages of using a characterized parental
cell line are many, including (1) the knowledge that the cell line
can be successfully transfected with foreign genetic material, (2)
previous demonstration of the lack of adventitious agents, (3) some
advance information regarding the presence or absence of endogenous retroviruses or retrovirus-like particles, and (4) data supporting the identity of the cell line. Following transfection of the cells
there are cloning and gene amplification steps, often followed by
additional cloning before or after adaptation to culture and production conditions.

As soon as possible after a cell line is demonstrated to have potential utility as a production line, it is prudent to establish a small
preliminary cell bank consisting of 20 – 50 ampules so that early testing and characterization can begin. Generally there are many preliminary banks made for any one product, as a result of ongoing
research which is directed toward finding the best plasmid, the best
substrate, and the most productive cell line. An ampule from the
preliminary bank may be thawed and cells cultured for initial characterization, including sterility and mycoplasma testing as well as
ultrastructural examination by electron microscopy.

Once a cell line is chosen for the manufacture of a biological and
its preliminary cell bank has been shown to be free of microbial and

mycoplasmal contamination, cell banking can proceed. The concept of a two-tiered cell bank originated in the early 1960s (41) and is generally accepted as the most practical approach for the cryopreservation of product-secreting mammalian cells. In accordance with this, a MCB is made first, usually from cultures expanded from an ampule of the preliminary cell bank. A manufacturer's working cell bank (MWCB) is then made by expanding one or more ampules from the MCB. Usually no more than 3 weeks of culture is required to obtain enough cells for a MWCB. It is the MWCB which is used to provide cells for manufacturing product. The utility of the two-tiered cell-banking system may be appreciated by the following example: If a MCB of 200 ampules is prepared, at least 20 ampules will likely be used for cell bank characterization, leaving 180 ampules for manufacturing. Suppose that 20 ampules per year are used for manufacturing. If ampules from the MCB are used directly for the manufacturing process, only 9 years of production would be possible. Although the life of the MCB could be extended by making it larger from the beginning (e.g., a MCB of 500 ampules should have a 24-year life span in the above example), it is prudent to give the MCB an "indefinite" life span. This can be accomplished by the two-tiered cell bank system. In the above example, the 180 ampules remaining following MCB characterization would only be employed to create MWCBs as needed. If each of these MWCBs were derived from a single ampule from the MCB and consisted of only 200 ampules, a total of 36,000 ampules could ultimately be available for manufacturing. If 20 ampules per year were required for production, cryopreserved cells would not be exhausted for 1800 years, a period of time which would undoubtedly exceed the useful life of the product.

Cells are prepared for banking by expanding cultures in progressively larger vessels until a pool of cells can be obtained which is sufficient to prepare several hundred ampules for the MCB or for the MWCB, each ampule containing 1×10^7 to 5×10^7 cells. The medium used for growth of the cell cultures is chosen based on preliminary development work on the cell line to be banked. Media are prepared in the cell culture facility following Good Manufacturing Practices (GMPs). Cells are most often frozen in serum-containing growth medium to which dimethyl sulfoxide (DMSO) is added (10% v/v), but we have found that some hybridoma cultures are more successfully recovered from a freeze medium containing 90% fetal bovine serum and 10% glycerol.

The ampules used for all cell banks are made of borosilicate glass and are designed to permit storage in the liquid or vapor phase of liquid nitrogen. Ampules are visually inspected and are labeled using an offset printing machine and special, permanent ink. The label generally consists of the cell number (a unique log-in number assigned sequentially), the cell name, and a lot number. Cells

from a single pool are dispensed into preprinted, sterile ampules
which are then flame-sealed and tested for integrity by submersion
into a solution of methylene blue in alcohol. Intact ampules are
frozen to $-80°C$ in a controlled-rate freezing apparatus, then transferred to permanent storage in the vapor phase of liquid nitrogen.
Following completion of a cell bank, an ampule is thawed to conﬁrm
the ability of the frozen cells to recover.

The stability of MCBs cryopreserved and maintained as described
above has been found to be excellent. No significant loss of viability,
as estimated by the trypan blue exclusion method, has been detected in our MCBs of genetically engineered cells. Cells from one
of our earliest MCBs have shown no loss in viability when tested
during the course of 5.5 years of storage. This is not surprising
since some of the oldest cell lines available were cryopreserved and
stored under similar conditions at the American Type Culture Collection and remain viable after storage for more than 20 years (17).
The WI-38 MCB reveals viable cells after 27 years of storage (42,43).

Every cell line undergoing cultivation for the preparation of a
MCB or MWCB is grown in an individual, self-contained cell culture
laboratory in which no other cell lines are being carried. This
greatly reduces the risk of cell line cross-contamination. Careful
consideration is made regarding the laboratory design, especially
with respect to the air-handling system. Important features include
limited access, cleanable surfaces, biological safety cabinets for open
transfer, avoidance of dust traps, HEPA-filtered supply air, and
positive air pressurization. Personnel must be trained in proper
laboratory procedures, including (1) Good Manufacturing Practices;
(2) use of gowns, gloves, masks, head and shoe covers; (3) use of
mechanical pipetting aids; (4) good cell culture practice; and (5)
aseptic technique.

Computerized records of all frozen cell lines are maintained in a
database. The information in this database includes the derivation
of the cells, the product being made, a laboratory notebook reference, and the number and location of ampules. Hard copies of
the database are available in the laboratory as well as in the curator's office for easy reference. Periodically, the ampules remaining
in a MCB or MWCB are counted by physical inspection to validate
the accuracy of the ongoing record keeping. Other records include
log books to document removal of ampules from a bank, for which
prior authorization is required; laboratory usage records posted
outside each cell culture room to document the history of the room
with respect to the cell lines grown there; liquid nitrogen level
records for tracking weekly manual level checks on all nitrogen
storage units; laboratory notebooks for recording daily cell culture
work; and standard operating procedure (SOP) manuals containing
the appropriate SOPs and test procedures used in the laboratory.

Numerous measures are taken to ensure the physical safety of the cell banks once they have been made. These include (1) the use of automatic filling liquid nitrogen tanks, (2) alarm systems which sound both locally and at a remote site staffed 24 hr a day to alert responsible individuals of equipment failure, (3) manual level checks at intervals designed to eliminate the risk of total loss of liquid nitrogen and to provide a backup to the automatic alarm system, and (4) distribution of the ampules in at least two storage tanks within the facility as well as in remote storage sites both inside and outside the United States.

Finally, the stability of the cryopreserved cells of the MCBs is monitored periodically to demonstrate control over storage conditions. Cells are evaluated for viability by the trypan blue exclusion method and for stability of growth characteristics.

V. CHARACTERIZATION OF CELL BANKS

Exhaustive characterization of the MCB allows standardization of the starting source for all subsequent production. Although cell characterization is reviewed in detail in Chapter 18, it will be discussed here briefly as it relates to cells in and derived from MCBs and MWCBs. National control authorities and other health organizations have published guidelines or "points to consider" which outline recommendations for such characterization (36 – 40). Although these documents do not always agree in every detail, the principles are the same and can be summarized as follows:

1. The origin and general characteristics of the cells should be known. This includes the origin (species, sex, organ, tissue) and passage history of the cell line; a description of any deliberate changes made to the cell line, such as genetic engineering, cloning, and the application of selective pressure; morphology; karyology; confirmation of species identity; growth characteristics; and tumorigenicity.
2. Cells derived from the MCB should be shown to be free of adventitious agents including bacteria, fungi, mycoplasma, and exogenous viruses.
3. Expression of endogenous retroviruses should be determined. The presence of these virus-like particles should be sought under conditions known to cause their induction, and by a variety of detection systems including assays for specific enzymatic activity (reverse transcriptase), biological activity, isotopic labeling of physical particles, the use of genetic and immunological probes, if available, and electron microscopy. If specific probes are available, they can be used to validate removal of the endogenous retrovirus-like particles

during product purification. Otherwise, one must rely on
spiking experiments with high-titered, biologically active,
model retroviruses to validate inactivation during the purification process (see Chapter 19).

4. A maximum limit for continuous cell culture during production should be defined and cells should be characterized at
and beyond the maximum limit as well as at intermediate
levels. Characterization at the various times in culture
should at least include cell growth rate, morphology, level
of endogenous retrovirus expression, specific productivity,
and product quality.

5. Cell culture-derived impurities of concern (i.e., DNA, proteins, endogenous virus particles) should be identified and
assays for these contaminants constructed and qualified.
These assays are required to validate removal of the impurities during purification and for quality control.

Although the various guidelines are not completely consistent as
to whether the MCB or the MWCB should be exhaustively characterized, we have chosen to conduct exhaustive characterization of the
MCB and to conduct limited characterization of each MWCB which
includes tests for sterility, mycoplasma, and identity. We reasoned
that if there were no changes in cell characteristics (growth rate,
morphology, level of endogenous retrovirus expression, specific
productivity, and product quality) during extended passage experiments (see 4 above), the several weeks of cell culture required to
produce a MWCB (which would be well within the maximum cell culture time limit) would also not result in changes to these characteristics. On the other hand, tests for adventitious agents should be
conducted to assure that the cells were not contaminated during cell
passage and preparation of the MWCB. Additionally, tests for
identity should be run to demonstrate that the MWCB was not contaminated with cells from another line during preparation.

VI. SUMMARY

The use of cultured cells (diploid strains or continuous cell lines)
to produce biopharmaceuticals provides a level of standardization to
the manufacturing process that cannot be attained by the more traditional methods of biological extraction from animal or human fluids,
tissues, or primary cells. The key to this advantage is the ability
to cryopreserve the production cell line as a master cell bank. This
bank serves as the common starting source of a given product for
the lifetime of the manufacture of that product. Since the MCB is
the common and only starting source, it can be exhaustively

characterized with regard to contamination by adventitious and endogenous agents. Assays can be developed for cellular components that are potential contaminants of the product. The removal of these components during product purification can be validated and their removal confirmed by rigorous quality control. In conclusion, characterized cell banks are central to the standardization of biopharmaceutical manufacturing processes and give rise to the production of high-quality products not attainable by the traditional methods of extraction of product from sources which are continually changing.

REFERENCES

1. Scherer, W. F. and Hoogasian, A. C. Preservation at sub-zero temperatures of mouse fibroblasts (strain L) and human epithelial cells (strain HeLa). *Proc. Soc. Exp. Biol. Med.* 87: 480–487 (1954).
2. Swim, H. E., Haff, R. F., and Parker, R. F. Some practical aspects of storing mammalian cells in the dry-ice chest. *Cancer Res.* 18:711–717 (1958).
3. Stuberg, C. S., Soule, H., and Berman, L. Preservation of human epithelial-like and fibroblast-like cell strains at low temperatures. *Proc. Soc. Exp. Biol. Med.* 98:428–431 (1958).
4. Stulberg, C. S., Rightsel, W. A., Page, R. H., and Berman, L. Virologic use of monkey kidney cells preserved by freezing. *Proc. Soc. Exp. Biol. Med.* 101:415–418 (1959).
5. Lovelock, J. E. and Bishop, M. W. H. Prevention of freezing damage to living cells by dimethyl sulphoxide. *Nature* 183: 1394–1395 (1959).
6. Craven, C. The survival of stocks of HeLa cells maintained at −70°C. *Exp. Cell Res.* 19:164–174 (1960).
7. Ferguson, J. Long term storage of tissue culture cells. *Aust. J. Exp. Biol.* 38:389–394 (1960).
8. Takano, K., Yamada, M., and Hirokawa, Y. Long-term frozen storage of mammalian cell lines. *Jap. J. Med. Sci. Biol.* 14: 27–37 (1961).
9. Evans, V. J., Montes de Oca, H., Bryant, J. C., Schilling, E. L., and Shannon, J. E. Recovery from liquid-nitrogen temperature of established cell lines frozen in chemically defined medium. *J. Natl. Cancer Inst.* 29:749–756 (1962).
10. Porterfield, J. S. and Ashwood-Smith, M. J. Preservation of cells in tissue culture by glycerol and dimethyl sulphoxide. *Nature* 193:548–550 (1962).
11. Nagington, J. and Greaves, R. I. N. Preservation of tissue culture cells with liquid nitrogen. *Nature* 194:993–994 (1962).

12. Moss, G. S. Preservation of red cells by freezing. In *Surgery Annual*, Vol. 2 (P. Cooper and L. M. Nyhus, eds.), Appleton-Century-Crofts, New York, pp. 35−50 (1970).

13. Luyet, B. and Rapatz, G. A review of basic researches on the cryopreservation of red blood cells. *Cryobiology* 6:425−482 (1970).

14. Sloviter, H. A. First transfusions of red blood cells previously frozen and thawed. *Vox Sang.* 48:254−256 (1985).

15. Sherman, J. K. Synopsis of the use of frozen human semen since 1964: State of the art of human semen banking. *Fertil. Steril.* 24:397−412 (1973).

16. Alfredsson, J. H., Gudmundsson, S. P., and Snaedal, G. Artificial insemination by donor with frozen semen. *Obst. Gyn. Surv.* 38:305−313 (1983).

17. Hay, R. J., Macy, M. L., Chen, T. R., McClintock, P., and Reid, Y. *American Type Culture Collection Catalogue of Cell Lines and Hybridomas*, 6th ed., American Type Culture Collection, Rockville, Maryland (1988).

18. *1986/1987 Catalog of Cell Lines. NIGMS Human Genetic Mutant Cell Repository*, U.S. Department of Health and Human Services, Bethesda, Maryland (1986).

19. *National Institute of Aging 1986 Catalog of Cell Lines*, U.S. Department of Health and Human Services, Bethesda, Maryland (1986).

20. Doyle, A., *European Collection of Animal Cell Cultures Catalogue*, 3rd ed., ECACC, Salisbury, U.K. (1988).

21. Hayflick, L. and Moorhead, P. S. The serial cultivation of human diploid cell strains. *Exp. Cell Res.* 25:585−621 (1961).

22. Hayflick, L., Plotkin, S. A., and Stevenson, R. E. History of the acceptance of human diploid cell strains as substrates for human virus vaccine manufacture. *Dev. Biol. Standard* 68:9−17 (1987).

23. Hopps, H. E., Cell substrate issues: A historical perspective. In *Abnormal Cells, New Products and Risk*, 6th ed. (H. E. Hopps and J. C. Petricciani, eds.), Tissue Culture Association, Gaithersburg, Maryland, pp. 13−17 (1985).

24. Hayflick, L. History of cell substrates used for human biologicals. *Dev. Biol. Standard* 70:11−26 (1989).

25. Jacobs, J. P., Garrett, A. J., and Merton, R. Characteristics of a serially propagated human diploid cell designated MRC-9. *J. Biol. Standard* 7:113−122 (1979).

26. Petricciani, J. C., Hopps, H. E., and Chapple, P. J. *Cell Substrates: Their Use in the Production of Vaccines and Other Biologicals*, Plenum Press, New York (1979).

27. Spier, R., Griffiths, B., and Hennessen, W. *Dev. Biol. Standard.*, Vol. 60, S. Karger, Basel (1985).
28. Hopps, H. E. and Petricciani, J. C. *In Vitro*, Monograph No. 6, Tissue Culture Association, Gaithersburg, Maryland (1985).
29. Spier, R. and Hennessen, W. *Dev. Biol. Standard.*, Vol. 66, S. Karger, Basel (1987).
30. Umeda, M. *Biotechnology of Mammalian Cells: Development for the Production of Biologicals*, Springer-Verlag, New York (1986).
31. Petricciani, J. C. and Hennessen, W. *Dev. Biol. Standard.*, Vol. 68, S. Karger, Basel (1987).
32. Marshak, D. R. and Liu, T. Y. *Therapeutic Peptides and Proteins: Assessing the New Technologies*, Cold Spring Harbor Laboratory, New York (1988).
33. Spier, R. E. and Griffiths, J. B. *Modern Approaches to Animal Cell Technology*, Butterworths, London (1987).
34. Hayflick, L. and Hennessen, W. *Dev. Biol. Standard.*, Vol. 70, S. Karger, Basel (1989).
35. Spier, R. E., Griffiths, J. B., Stephenne, J., and Crooy, P. J. *Advances in Animal Cell Biology and Technology for Bioprocesses*, Butterworths, London (1989).
36. *Points to Consider in the Characterization of Cell Lines Used to Produce Biologicals (November 1987)*, Office of Biologics Research and Review, U.S. Food and Drug Administration, Bethesda, Maryland (1987).
37. *Points to Consider in the Manufacture and Testing Of Monoclonal Antibody Products for Human Use (1987)*, Office of Biologics Research and Review, Center for Drugs and Biologics, U.S. Food and Drug Administration, Bethesda, Maryland (1987).
38. *Points to Consider in The Production and Testing of New Drugs and Biologicals Produced by Recombinant DNA Technology (1985)*, Office of Biologics Research and Review, U.S. Food and Drug Administration, Bethesda, Maryland (1985).
39. *Notes to Applicants for Marketing Authorizations on the Production and Quality Control of Medicinal Products Derived by Recombinant DNA Technology*, Ad Hoc Working Party on Biotechnology/Pharmacy, Committee for Proprietary Medicinal Products, Commission of the European Communities, Brussels (1987).
40. *Acceptability of Cell Substrates for Production of Biologicals. Report of a WHO Study Group*, Technical Report Series 747, World Health Organization, Geneva (1987).

41. Minimum requirements for the selection and use of human diploid cell strains in the production of virus vaccines. In *Proceedings of the Symposium on the Characterization and Uses of Human Diploid Cell Strains,* Permanent Section of Microbial Standardization, Opatija, Yugoslavia, Biostandards, Case Postale 229, CH-1211, Geneva 4, Switzerland, pp. 709–733 (1963).
42. Hayflick, L. The limited in vitro lifetime of human diploid cell strains. *Exp. Cell Res.* 37:614–636 (1965).
43. Hayflick, L. Personal communication (1989).

6

Expression of Cloned Proteins in Mammalian Cells: Regulation of Cell-Associated Parameters

JENNIE P. MATHER and MARY TSAO
Genentech, Inc., South San Francisco, California

I. INTRODUCTION

The last decade has seen an explosive growth in both our knowledge of molecular biology and the commercial exploitation of that knowledge. A great deal of effort has gone into the cloning and expression of genes coding for proteins that were previously available in limited quantities (such as h-GH) or not at all (i.e., an in vitro mutated gene). Initially, attempts were made to produce these proteins in bacterial or yeast expression systems. However, experience has shown that many of these proteins can only be expressed in the desired form in animal cell culture. Factors that may influence the decision to use cell culture are the desire for appropriate glycosylation of the protein, ease of purification of secreted products, and the need for protein processing with complex folding and disulfide bond formation.

The selection of the mammalian cell to be used as an expression system for a cloned gene will depend on the characteristics of the protein to be expressed, the expression systems available, regulatory and safety issues, and the type of cell culture process desired. This decision is best made with the input of cell biologists,

molecular biologists, and biochemists since much of the nature of the
subsequent work in process design will depend on the initial choice.
Once the gene of choice is expressed in a mammalian cell line, the
process of optimizing production, as well as that of developing and
scaling up a true production system, involves a number of issues in
the realm of cell biology. This chapter will deal with the optimization of these cell culture parameters, which can lead to a significant
increase in product yields. These increases could be thought of as
arising from one of three different categories of mechanisms which
we call (1) genetic/molecular, (2) biochemical, or (3) cellular.

Examples of the types of gains in product yields falling in category 1 would be increases in specific productivity of the cells
through an increase in the number of gene copies per cell (gene
amplification), an increase in the amount of mRNA per cell (induction or increased expression), or an increased translational efficiency. There are a growing number of expression systems in which
the gene of choice can be amplified by placing selective pressure on
a linked gene, the best known of these being the dihydrofolate reductase system. The topic of optimization of the genetic/molecular
aspects of product secretion deserves a separate treatment of its own
and is beyond the scope of this chapter.

Biochemical improvements in yield may be obtained by increasing
product stability, increasing the amount of the product expressed in
the desired form, or removing undesired substances which may make
the product difficult to recover or unstable during the duration of
culture. In the third category we would include increases in yields
obtained by increasing maximum cell density, improving the functional viability of the cells, or extending the time period that viable
cells remain in production. These improvements may be cell type-
specific or apply to the production of a specific product in any mammalian cell culture expression system.

The optimization of the parameters of the culture system refers
to the optimization of the environment for a specific cell type.
These parameters are generally similar for a given cell expressing
any recombinant product. There are a wide range of cell culture
systems available in different configurations. Each system from
tissue culture plates to large-scale fermenters has its own strengths
and weaknesses. Several of these systems may be used in the
course of developing a production system. The choice of culture
system will depend on the properties of the cells, the development
time required, the potential for scalability, and the capitalization
required to alter or expand existing facilities or build new ones. If
a specific culture configuration is required (e.g., high-density suspension culture), the cell line to be used can sometimes be selected

to enhance the desired properties. This may, however, take considerable time and be accomplished only with a concomitant reduction in other desirable properties such as specific productivity. It is highly desirable to have several different cell types expressing a given product in order to fully evaluate the pros and cons of different production systems for a specific product.

For convenience we will divide the discussion on cellular parameters into the optimization of the physicochemical, nutritional, and hormonal environment of the cell. All of these factors interrelate and changes in one can alter the optimum for another. It is therefore wise to reassess all of the parameters when any major alteration is made to the culture system. Finally, the optimization of these parameters is cell type-specific. Frequently, optimal conditions may vary significantly for different cell lines or even clones of the same cell type expressing different proteins.

II. OPTIMIZATION OF PHYSICOCHEMICAL FACTORS

Mammalian cells in vivo are in a carefully balanced homeostatic environment. In many respects this precise control is for the benefit of the organism as a whole rather than any particular cell type. Nevertheless, most cell types have stringent requirements as to the range of physical parameters (e.g., pH or temperature) which are for growth and performance. For best results, therefore, culture systems have to be developed in which the physical environment matches the cells' requirements, and in which the changes induced in the environment by cell growth and metabolism can be corrected or compensated to maintain homeostasis.

Important physicochemical parameters which are controlled to a greater or lesser extent in different cell culture systems are temperature, pH, pO_2, pCO_2, redox potential, and osmolarity (1,2). For suspension cultures shear, mixing time, and agitation rates are also of concern. The method and level of control of these parameters may be quite different in different types of culture systems or at different scales. As with the other parameters discussed below, these may be interactive and interdependent. It is most convenient to optimize these parameters in a large number of small-culture vessels in which the environment can be automatically monitored, recorded, and controlled to predetermined set points. It is especially important in these studies to be aware of interdependent variables such as the increased osmolarity caused by the addition of base to control pH. To examine the effect of only one variable at a time, it is necessary to use a chemostat-type culture under steady-state conditions in which all parameters but the one being studied can be kept constant (3,4). This technique can be applied to the investigation of nutritional and hormonal requirements of mammalian cells but it is technically more demanding than using multiple small-scale cultures.

Temperature is obviously a factor of overriding importance, having an effect on all aspects of metabolism and protein synthesis and degradation. Most cells are cultured at 37°C, which is human body core temperature. However, cells derived from human peripheral tissues and cells derived from other species may normally function at temperatures that are significantly different from 37°C (5). In addition, secreted protein products may be subject to temperature-dependent degradation during a production run. In this case, lowering the temperature may lead to lower protein synthetic rates but better product characteristics. The optimal temperature must be determined experimentally, preferably after adapting the cells to be studied to each different temperature over a period of several weeks (Fig. 1).

A special case in which temperature control is of paramount importance would be mammalian cells where heat shock promoters are used to express the recombinant product (6). In such a system the optimal temperatures (and timing) of the growth phase, the heat shift to initiate transcription, and the production (translation) phase need to be determined separately.

All mammalian cells produce lactic and carbonic acids as metabolic byproducts which decrease the pH of the cell culture medium. The rate of production of these compounds and the optimal and acceptable pH range will vary from one type of cell line to another and depend on the composition of the medium used. In batch cultures ranging in scale from plates to large fermenters, pH is controlled by adding buffers to the medium. The optimal pH range is usually $7.0 - 7.5$ (7). The most common buffering system is a sodium bicarbonate/CO_2 system, frequently used in combination with a phosphate buffering system. The bicarbonate system is especially useful in large-scale fermentation systems where pH can be maintained automatically by adding more base in response to the drop in pH produced by cell metabolites. When serum-supplemented medium is used, the serum itself provides significant buffering capacity. When culturing cells in serum-free medium additional buffering capacity can be provided by the addition of organic buffers, although this may be prohibitively expensive on a large scale.

In the traditional CO_2 incubator-plate culture system it is extremely difficult to control pO_2 and redox potential, so there is much less information available concerning these parameters. Most cells in the body are exposed to a pO_2 significantly less than that of air and studies of cells grown in suspension in fermenters confirm

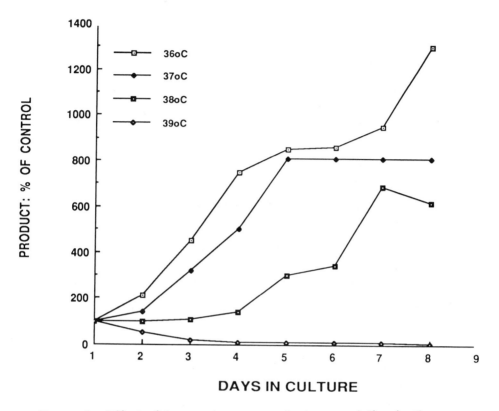

Figure 1 Effect of temperature on product accumulation in the medium. Cells were grown for two passages at the indicated temperature before the start of the production run. The temperatures used were 36, 37, 38, and 39°C, as indicated. Values are expressed as percentage of the day 1 titer at 37°C.

that the optimum pO_2 is generally lower than that of air. This should still be determined experimentally, however, for cells grown in fermenters or other systems where the pO_2 can be controlled. In addition, some cells, especially those grown in serum-free medium, have been shown to require antioxidants or reducing compounds for optimal function (8–10). Vitamins E and C can function as antioxidants as well as act as vitamins, and can be screened in addition to other nonnutritional antioxidants. Reducing compounds compatible with cell culture are β-mercaptoethanol, dithiothreitol, and cysteine (an amino acid frequently included in media formulations).

Animal cells are grown attached to a substrate or in suspension
depending on the cell type, and some can be grown in either mode.
Mammalian cells are fragile in suspension, at least compared to
microbes. However, with few exceptions, this fragility is not well
studied or understood (11). This greater fragility, however, makes
the minimization of shear forces and other mechanical damage of
considerable importance for some types of cells including hybridomas.
Suspension cultures, whether of free cells or cells attached to microcarriers, must be stirred to provide mixing of nutrients, oxygen,
and metabolic byproducts, and to keep the cells in a homogeneous
suspension so that control of pH and other parameters, can be maintained. This mixing can be provided while minimizing cell damage
by optimizing vessel design (12) and agitation rates.

All existing established cell lines have been derived under a
specific set of physicochemical conditions, which has generally been
static culture at 37°C in an air/CO_2 atmosphere. These conditions
may be used as a starting point for the optimization of the nutritional and hormonal environment as discussed below. The physical
parameters should then be optimized in the final culture configuration to be used for production. As mentioned above, this can be
especially important in scaling up a suspension culture system for
a relatively fragile mammalian cell.

III. OPTIMIZATION OF NUTRITIONAL FACTORS

The nutritional requirements of cells are usually provided in the
standard powdered media formulations. However, the performance
of some of the simpler media, such as Eagle's minimal essential medium (MEM), may depend on the addition of complex biological extracts such as serum to provide required nutrients and growth
factors. In addition, some nutrients are unstable on storage or removed during filtration, and thus may not be present in adequate
concentrations at the time of use even though they are included in
the medium formulation. Nutrients can be divided into several categories: amino acids, amino acid derivatives, fatty acids, complex
lipids, complex carbohydrates, sugars, vitamins and coenzymes,
nucleic acid derivatives, salts, and trace elements (13). A nutrient
may be essential for, enhance, or have no effect on a specific cell's
growth and survival (Fig. 2). Not only the absolute requirements
but the relative concentrations of nutrients must be optimized for
each individual cell type. These requirements can be so stringent
that the nutrient and hormonal environment can be adjusted so as
to allow the growth of only one desired cell type in a mixture of
cells (13–15). The optimal nutrient mixture and concentrations
will depend on the cell type to be used, the culture configuration
(e.g., low or high density and suspension vs. attached culture),

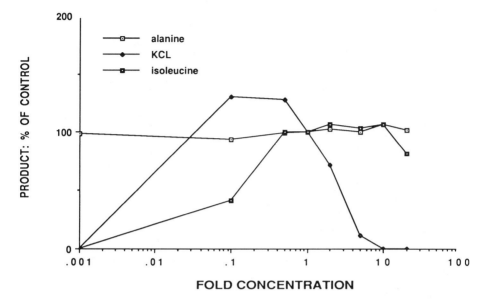

Figure 2   Product secretion as a function of nutrient concentration. Three examples are given from different classes of nutrients to illustrate different types of responses, isoleucine, KC1, and alanine. Values are normalized to percentage of control, where 100% is the value obtained at the initial (one time) concentration of the nutrient.

and the medium supplements to be used.   Amino acid concentrations which are optimal for plating density may be entirely inadequate for a high-density perfusion culture.   Alternatively, cells at high density may actually secrete significant amounts of amino acids thus lowering or eliminating the requirement for that amino acid.   Finally, serum is not only a source of hormones and growth factors but is a complex mixture of nutrients including vitamins, complex carbohydrates, fatty acids and complex lipids, and sugars; and can provide part or all of the cell requirements for these nutrients.

Two approaches can be used to optimize the nutrient environment for a given cell line.   The first is to screen a series of commercially available nutrient formulas or mixtures of such formulas, or to supplement existing media with specific components (e.g., a high-glucose medium).   While this approach is rapid and will frequently lead to an improvement in cell performance, the specificity of nutrient mixtures for given cell types and the interaction of the

nutrients with each other make it unlikely that this approach will
lead to the optimal medium for a specific application.

The second approach will lead to a true optimization of the nutritional environment for a specific cell. All screening should be done
under the conditions to be optimized to the extent that this is technically and financially feasible. Thus optimization should be carried
out using the cell density, temperature, osmolarity, medium supplements, and time of culture of choice. If optimum protein titers are
desired, then the titer should be determined under the various experimental conditions. Other parameters such as cell density and
specific productivity are of scientific interest but could be mislead-
ing if used as the sole source of primary data. Once these initial
conditions and the parameter to be measured have been determined,
the approach is to screen each nutrient independently to optimize
its concentration and then rescreen each, holding the others at the
previously derived optimal concentration. Since there are many components to some of the more complex media, this is obviously a more
time-consuming approach. In addition, this optimization is best done
in the presence of no or the minimal amount of extraneously added
undefined components which might provide nutrients in a nonrepro-
ducible fashion. Thus the nutrient and hormonal optimization discussed below will frequently go hand-in-hand with an initial hormonal optimization required to decrease or eliminate the need for
serum in order to optimize the nutrient component of the medium.
This in turn might decrease the requirement for a specific hormonal
factor or enhance cell productivity with the previously determined
factors, so that the hormonal supplements required should be reas-
sessed after nutrient optimization.

IV. OPTIMIZATION OF "HORMONAL FACTORS"

Most cell types will not grow and/or secrete proteins optimally in
medium consisting only of nutrients, even when these are optimized
as discussed above. Such cells may require the addition of hormones, growth factors, transport proteins, and attachment factors
(for anchorage-dependent cells). Traditionally these have been
provided as part of complex biological mixtures such as serum or
organ extracts. However, recent work has shown that such complex mixtures can be replaced by more highly purified fractions or
mixtures of defined factors (16,17). The optimization of this "hormonal" environment may be undertaken for a variety of reasons,
such as the reduction or elimination of the need for undefined
growth factors, removal of inhibitory factors, or provision of critical hormones at desirable levels. It may decrease the total protein in the cell culture fluid, thus improving the ease of purification of the protein of interest, and increase reproducibility by

having a completely defined system. Finally, the process of defining the optimal cell culture environment for a known cell type and
comparing the optimal growth environment for different cell types
leads to a greater understanding of the regulation of cell growth
and function.

As discussed above, the nutrient portion of the medium becomes
more critical as one removes more of the undefined protein portion
of the medium. Serum-free defined cell culture in general requires
the use of one of the more complete nutrient mixtures. In addition,
the medium should be made using ultrapure water to avoid the introduction of toxic contaminants (18) and should not be stored in
liquid form for more than 2–3 weeks. Toxic contaminants, whether
introduced via impure water or other medium components or through
the breakdown of required medium components during prolonged or
improper storage, are a greater concern in defined medium due to
the extremely low protein content of the medium and the absence of
protective components of serum such as metal-binding proteins.

If the cell line to be used will survive in an unsupplemented
serum-free nutrient mixture, then screening of the hormonal factors
may begin at once. Most cell types require insulin to survive
serum-free and this hormone should be tested initially (17). Transport proteins such as transferrin (19) (the plasma iron transport
protein), ceruloplasmin (20), and high-density lipoprotein (a lipid
carrier) will frequently increase cell growth in vitro. These may,
in some cases, be replaced by the nutrient that they transport if
it is added in the correct form and in adequate quantities at the
time of addition of the cells (13,19). In addition, cells frequently
require one or more hormones from each of the following groups:
steroids, prostaglandins, growth factors, pituitary hormones, and
peptide hormones. The set of optimal hormones (and the optimal
concentration for each) will vary for each cell type (21,22), although they are likely to be similar for cells from different species
derived from the same cell type.

For example, Table 1 shows the cell number and tissue plasminogen activator production from the human Bowe's melanoma
cell line in either serum-free medium supplemented with various hormones or in the presence of serum. The hormones selected in this
case were those shown to stimulate the growth of a mouse melanoma
line (23). As discussed above, some of the factors increase cell
number, and some increase tPA production of cell. Cells grown in
serum-supplemented medium reached a higher cell density and thus
produced higher titers but had a lower specific productivity than
the serum-free case. Overall, the hormone requirements for this
line are strikingly similar to those which were growth promoting for
the mouse melanoma line (23). For comparison, Fig. 3 shows the
increase in titer of a product produced from a recombinant gene

Table 1   Cell Number and tPA Production by Bowe's Melanoma
After 5 Days in Culture[a]

	Cell number (×105)	tPA	
Condition		ng/105 cells	ng/ml cells
Serum-free	0.69	<30	lts
+ ins	0.86	<30	lts
+ ins + TF	1.08	<30	lts
+ ins + TF + NGF	1.06	54	57
+ ins + TF + NGF + Prog	1.09	55	59
7.5% FBS	6.00	30	178

[a]lts, less than standard; ins, insulin; TF, human transferrin;
NGF, nerve growth factor; Prog, progesterone; FBS, fetal bovine
serum.

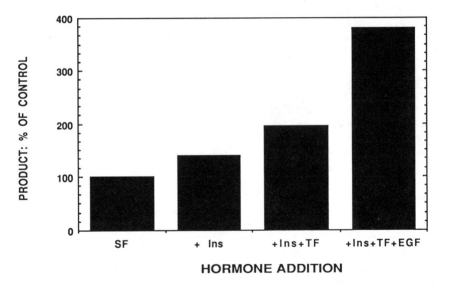

Figure 3   Product secretion as a function of hormonal supplementation of the medium. The hormones insulin (ins), transferrin (TF),
and epidermal growth factor (EGF) are added to the medium sequentially. Each addition further increases the titer over that obtained in the serum-free medium.

which can be obtained with hormone supplementation of serum-free medium.

If the cell line used will not attach in serum-free medium, then attachment factors such as fibronectin, laminin, and serum spreading factor should be screened (24). If the cells still do not survive, other undefined low-protein mixtures may be tested. These include pituitary or other organ extracts, medium conditioned by other cell types at high density, and cell feeder layers. All of the above are generally too expensive or too labor-intensive to be considered for a large-scale production process. Recently, however, collagen-coated microcarrier beads (Cytodex 3) have become available commercially. As such systems come into wider use it will be easier to evaluate the advantages of commercial production of pharmaceuticals using cell culture systems which benefit from or require growth on such matrices or in the presence of undefined mixtures other than serum. Meanwhile, these small-scale studies can provide a valuable insight into the mechanisms regulating cell growth and product secretion, and help define the major problems which may then be overcome in other ways.

One such approach is to use undefined serum fractions such as serum albumin or fetuin, or reduced concentrations of serum in combination with some of the purified hormones described above. Another is to use one of the commercially available "serum extenders" or "serum substitutes," which are designed to reduce total serum protein added to the medium. If this approach is taken, as many as possible of the available mixtures should be screened for each cell type to be used. As with the screening of the commercial nutrient mixtures, the result will reflect the best cell performance with the mixtures tested, not necessarily the maximal performance possible with the cell.

V. CHOOSING THE CELL CULTURE SYSTEM

Plate systems, in which the cells can be grown attached to a surface are in many respects the most convenient. The most extensively used are tissue culture dishes, but the same principles apply from the microtiter plates at the small scale to roller bottles and multisurface propagators, having surfaces of up to 0.5 m^2/unit, for larger volumes. These systems are convenient and, although labor-intensive, require a minimum of engineering expertise and sophisticated equipment, compared to perfusion or fermentation-type systems. They are, however, relatively inefficient and are useful only for supplying limited quantities of material or for research studies. With recombinant cells, for example, these systems may be the most convenient for producing proteins from transient expression

systems. Tissue culture plates are also most often used for the
screening studies required for medium optimization since a very
large number of conditions must be tested.

Solid supporting matrices have been used extensively at larger
scales for the production of greater volumes of material from attachment-dependent cells (25–28). In these systems the support matrix,
which may be of steel, glass, organic polymer, or ceramic material,
is contained and sterilized within the culture chamber before medium
and cells are introduced. After the cells have attached, medium is
circulated through the matrix and an external loop, which often contains a surge vessel, where sensors and control systems can be
used to regulate the conditions of culture. These systems do not
necessarily support homogeneous culture conditions and zones often
develop within them where there is little cell growth or, alternatively,
excessive growth leading to necrosis. In addition, it is often difficult to monitor conditions within the chambers directly and external
indicators have to be developed and relied on.

Suspensions of microcarrier beads with attached anchorage-
dependent cells (29), or of cells grown within or trapped in suspended bead matrices (30,31), offer many of the advantages of suspension culture. Included among these are culture homogeneity
(allowing conditions to be monitored directly) and large-scale potential. Both of these systems have their own special considerations (discussed in Chapters 4 and 10) in the optimization of culture nutrient
and hormonal parameters.

The most extensively used systems for production of material
from animal cells are suspension cultures with anchorage-independent
cells. These systems have been scaled to volumes of many thousands of liters (32). Such cultures are probably the most easy to
study on a large scale, especially for purposes of monitoring specific
consumption or production rates and for studies involving continuous
culture. Because of these advantages suspension culture is the
method of choice for most applications. However, it is not the best
solution in all cases. Some cells can grow and survive but do not
produce recombinant proteins at as high a level in suspension as
when they are attached. Other cells cannot be adapted to grow in
suspension or are so delicate that the conditions of a large-scale
suspension culture cause unacceptable levels of cell damage.

VI. SCALE-UP AND PURIFICATION

One of the main purposes of industrial scale cell culture is to support research on and the manufacture of valuable proteins of pharmaceutical interest. Frequently these need to be produced in quantities requiring very large-scale cultures (32). Among these are
enzymes (e.g., factor VIII and tissue plasminogen activator),

hormones and growth factors (e.g., human growth hormone and erythropoietin), proteins for use as vaccines (e.g., HIV and hepatitis proteins), and monoclonal antibodies for in vivo imaging, in vitro diagnostics, or potential cancer therapy. Most if not all of the products in this realm are parenterals and their use will be regulated accordingly by various government agencies around the world. For a discussion of regulatory concerns related to the production of recombinant proteins in mammalian cells, see Chapter 19.

It is also important that the cell culture medium generated be compatible with the purification process. Frequently, changes in the cell culture environment can lead to significant savings of time and effort in downstream processing. This can be accomplished not only by increasing titers but providing a significant increased specific activity and stability of the desired product in the starting material, thus eliminating some purification steps and increasing yield. In an optimized, serum-free, low-protein medium, the product may comprise 10−75% of the protein in the cell culture fluid.

VII. HOST CELL SUBSTRATES AND RELATED CONCERNS

The type of animal cell chosen for expression of natural or recombinant proteins depends on the purpose of the work. If the aim is to express a molecule for biochemical studies alone, the only real constraints are whether the product is appropriate in terms of folding, glycosylation, final purity, etc. For these studies, any convenient cell line and vector expression system can be used. If the reason for making the product is to conduct clinical trials or to market it as a therapeutic, there are many more criteria to be considered (33,34). To date published work examining continuous cell lines, whether for natural or recombinant products, has not brought to light any areas of serious concern.

No matter which continuous cell line is used for expression of a cloned gene, if the product is to be used in human beings there will be regulations concerning the limits of residual impurities permitted. The impurities can be considered in several categories: cellular protein, endogenous virus particles, cellular DNA, extraneous protein contaminants (e.g., bovine serum proteins or material leached off purification apparatus), antibiotics, and endotoxins.

Some of these categories have a bearing on the choice of cells and systems to be used. For instance, proteins expressed from cells containing stably integrated genes are likely to have fewer cellular and DNA contaminants than those expressed as part of a process which destroys the host cell, e.g., as happens with some viral vectors and inducible systems which result in death and lysis of most of the cells. The use of defined serum-free media

formulations eliminates serum proteins as possible contaminants of
the final product. The use of some antibiotics is not permitted by
regulatory authorities and the need for antibiotics can be eliminated
with appropriate steps taken to ensure sterility. Endotoxin levels
can be maintained at acceptable levels by screening all medium components to ensure low initial endotoxin levels in the medium and by
careful handling and storage of the harvested cell culture fluid,
preferably maintaining sterility throughout as much of the purification process as possible.

VIII. CONCLUSIONS

The cloning of a gene and construction of an expression vector are
only the first steps in the production of pharmaceutical products in
mammalian cell culture systems. Recent experience has shown that
optimization of the cell culture environment plays a major role in increasing yields even for proteins expressed under the control of
viral promoters. Thus optimizing the cell culture environment can
lead to as much as a 10- to 100-fold increase in product yields as
well as allowing control of several aspects of the quality of the
product. These studies on the interrelationship of the cell host,
the recombinant protein, the expression system, and the culture
environment are in their infancy. However, given the importance of
these factors in developing economic production systems for cell culture products, it is expected that there will be significant progress
in this field in the coming years.

ACKNOWLEDGMENTS

We thank our colleagues at Genentech, Inc., with whom we worked
on developing processes for recombinant protein expression for the
past 6 years. The support of the operations and assay groups was
crucial in obtaining the information which facilitated the writing of
this chapter. Working with others in the Cell Culture R&D and
Molecular Biology departments has been both a pleasure and a continuing source of scientific stimulation.

REFERENCES

1. Arathoon, W. R. and Telling, R. C. Uptake of amino acids and
    glucose by BHK-21 clone 13 suspension cells during cell growth.
    Dev. Biol. Standard. 50:145−154 (1981).

2. Weymouth, C. Osmolality of mammalian cells. *In Vitro* 6:109 109 – 127 (1970).
3. Pirt, S. J. *Principles of Microbe and Cell Cultivation.* Blackwell, Oxford (1975).
4. Birch, J. R. et al. *Proceedings of the NATO Advanced Studies Institute.* Plenum Press, New York (1985).
5. Freshney, R. I. *Culture of Animal Cells: A manual of Basic Technique.* Alan R. Liss, New York, pp. 65 – 66 (1983).
6. Wurm, F. W., Gwinn, K. A., and Kingston, R. E. Inducible overproduction of the mouse c-myc protein in mammalian cells. *Proc. Natl. Acad. Sci. USA* 83:5414 – 5418 (1986).
7. McKeehan, W. L. The role of nutrients in control of normal and malignant cell growth. In *Molecular Interrelationship of Nutrition and Cancer* (M. S. Arnott, J. van Eys, and Y.-M. Wang, eds.), Raven Press, New York, pp. 249 – 263 (1982).
8. Knight, D. R., Hunt, T. K., Scheuenstuhl, H., Halliday, B. J., Werb, Z., and Banda, M. J. Oxygen tension regulates the expression of angiogenesis factor by macrophages. *Science* 221:1283 – 1285 (1983).
9. Hornsby, P. J. and Gill, G. N. Regulation of glutamine and pyruvate oxidation in cultured adrenocortical cells by cortisol, antioxidants, and oxygen: Effects on cell proliferation. *J. Cell. Physiol.* 109:111 – 120 (1981).
10. Mather, J. P., Saez, J. M., Dray, F., and Haour, F. Vitamin E prolongs survival and function of procine Leydig cells in culture. *Acta Endocrinol.* 102:470 – 475 (1983).
11. Tramper, J., Joustra, D., and Vlack, J. M. Bioreactor design for growth of shear-sensitive insect cells. In *Plant and Animal Cells: Process possibilities* (C. Webb and F. Mavituna, eds.). Ellis Horwood, Chichester, England, pp. 125 – 136 (1987).
12. Tolbert, W. R., Lewis, C., Jr., White, P. J., and Feder, J. Perfusion culture systems for production of mammalian cell biomolecules. In *Large Scale Mammalian Cell Culture* (J. Feder and W. R. Talbot, eds.). Academic Press, New York, pp. 97 – 123 (1985).
13. Ham, R. G. and McKeehan, W. L. Media and growth requirements. In *Methods in Enzymology*, Vol. 58 (W. B. Jakoby and I. R. Pastan, eds.). Academic Press, New York, pp. 44 – 93 (1979).
14. McKeehan, W. L., McKeehan, K. A., Hammond, S. L., and Ham, R. G. Improved medium for clonal growth of human dip]loid fibroblasts at low concentrations of serum protein. *In Vitro* 13:399 – 416 (1977).

15. Tsao, M. T., Walthall, B. J., and Ham, R. G. Multiplication of normal human keratinocytes in a defined medium. *J. Cell. Physiol.* 110:219−229 (1982).

16. Murakami, H., Yamane, I., Barnes, D. W., Mather, J. P., Hayashi, I., and Sato, G. H. *Growth and Differentiation of Cells in Defined Environment.* Springer-Verlag, Tokyo (1985).

17. Sato, G. H., Pardee, A. B., and Sirbasku, D. A. *Growth of Cells in Hormonally Defined Media.* Cold Spring Harbor Press, New York (1982).

18. Mather, J. P., Kaczarowski, F., Gabler, R., and Wilkins, F. Effects of water purity and addition of common water contaminants on the growth of cells in serum-free media. *Biotechnology* 4:57−63 (1986).

19. Pwrez-Infante, V. and Mather, J. P. The role of transferrin in the growth of testicular cell lines in serum-free medium. *Ex. Cell Res.* 142:325−332 (1982).

20. Mather, J. P. Ceruloplasmin, a copper-transport protein, can act as a growth promoter for some cell lines in serum-free medium. *In Vitro* 18:990−996 (1982).

21. Barnes, D. and Sato, G. Serum-free culture: A unifying approach. *Cell* 22:649−655 (1980).

22. Mather, J. P. *Mammalian Cell Culture: The Use of Hormone Supplemented Media.* Plenum Press, New York (1984).

23. Mather, J. P. and Sato, G. J. The growth of mouse melanoma cells in hormone-supplemented, serum-free medium. *Ex. Cell Res.* 120:191−200 (1979).

24. Barnes, D. Attachment factors in cell culture. In *Mammalian Cell Culture: The Use of Hormone Supplemented Media* (J. P. Mather, ed.). Plenum Press, New York, pp. 195−238 (1984).

25. Robinson, J. H., Butlin, P. M., and Imrie, R. C. Growth and characteristics of human diploid fibroblasts in packed beds of glass beads. *Dev. Biol. Standard.* 46:173−181 (1980).

26. Merk, W. A. Large scale production of human fibroblast interferon in cell fermenters. *Dev. Biol. Standard.* 50:137−140 (1982).

27. Brown, P. C., Costello, M. A. C., Oakley, R., and Lewis, J. L. Applications of the mass culturing technique (MCT) in the large scale growth of mammalian cells. In *Large Scale Mammalian Cell Culture* (F. Feder and W. R. Talbot, eds.). Academic Press, New York, pp. 59−72 (1985).

28. Lydersen, B. K., Putnam, J., Bognar, E., Patterson, M., Pugh, G. C., and Noll, L. A. The use of ceramic matrix in a large scale culture system. In *Large Scale Mammalian Cell Culture* (F. Feder and W. R. Talbot, eds.). Academic Press, New York, pp. 39−58 (1985).

29. Reuveny, S. Microcarriers for culturing mammalian cells and their applications. In *Advances in Biotechnological Processes*, Vol. 2 (A. Mizrahi and A. L. van Wezel, eds.). Alan R. Liss, New York, pp. 1–32 (1983).

30. Jarvis, A. and Gardina, T. Production of biologicals from microencapsulated living cells. *Biotechniques* 1:22–27 (1983).

31. Karkare, S. B., Phillips, P. G., Burke, D. H., and Dean, R. C., Jr. Continuous production of monoclonal antibodies by chemostatic and immobilized hybridoma culture. In *Large Scale Mammalian Cell Culture* (F. Feder and W. R. Talbot, eds.). Academic Press, New York, pp. 127–150 (1985).

32. Arathoon, W. R. and Birch, J. R. Large scale cell culture in biotechnology. *Science* 232:1390–1395 (1986).

33. Petricciani, J. C., Hopps, H. E., and Chapple, P. J. (eds.). *Cell Substrates: Their Use in the Production of Vaccines and Other Biologicals*. Plenum Press, New York (1979).

34. Osborn, J. E. Biological risk of viral agents endogenous to cell substrates. In *Abnormal Cells, New Products and Risk* (H. E. Hopps and J. P. Petricciani, eds.). TCA, Gaithersburg, pp. 174–175 (1985).

35. Finster, N. B. and Fantes, K. H. The purity and safety of interferon prepared for clinical use: The case for lymphoblastoid interferon. *Interferon* (I. Gresser, ed.). Academic Press, New York, p. 65 (1980).

36. Lubiniecki, A. S. *Safety Considerations for Cell Culture Derived Biologicals in Large Scale Cell Culture Technology* (B. K. Leydersen, ed.). C. Hauser, Munich (in press).

7

Assay Requirements for Cell Culture Process Development

MARY B. SLIWKOWSKI and EDWARD T. COX
Genentech, Inc., South San Francisco, California

I. INTRODUCTION

Initially, cell culture process development efforts were focused solely on maximizing the amount, or titer, of desired product. This
emphasis has changed as the use of large-scale cell culture for production of commercial proteins has increased. Maximizing titer will
always be a driving force, but there is an increasing recognition
of the importance of product quality. This chapter provides an
overview of the assays most useful for achieving maximum production of high-quality products.

The products of interest in commercial applications of mammalian
cell culture are usually proteins. Thus, product quality refers to
protein structure and function. Mammalian cells are desirable systems for the production of pharmaceutical proteins largely because
of their ability to correctly synthesize and secrete glycoproteins
and to generate disulfide bridges leading to properly folded, biologically active products. Therefore, this chapter concentrates on
assays which directly assess relevant aspects of protein structure
and function, or which monitor processes known to affect these
properties. The emphasis here is not on exhaustive protein characterization, but rather on quick screening of product quality

during the development of the production process. For the sake of brevity, we have assumed that the products of interest are secreted proteins expressed at $\geqslant 1$ mg/liter. With slight modifications, most of the assays described in this chapter would also apply to membrane proteins. The methods required for solubilization of such proteins prior to further characterization are discussed in several excellent reviews (1,2).

II. CELL HEALTH

A. Cell Lysis—Lactate Dehydrogenase Release

Membrane damage and subsequent cell lysis cause the release of intracellular proteases and glycosidases which can degrade secreted products. Lysis also releases other intracellular macromolecules into the culture fluid, complicating subsequent purification of the product. Microscopic examination of cells with dyes or fluorescent labels is a common method of quantitating cell viability (3). In addition, we find it useful to monitor the release of intracellular marker proteins.

Lactate dehydrogenase (LDH) is one such cytosolic enzyme which is readily and accurately measured spectrophotometrically by following the formation of NADH at 340 nm.

$$\text{Lactate} + \text{NAD} \xleftarrow{\quad \text{LDH} \quad} \text{pyruvate} + \text{NADH}$$

A representative assay for cell lysis using LDH activity as the marker enzyme is shown in Fig. 1. To determine LDH activity per cell, saponin is employed to release LDH from the cell pellet (4). Provided that the value for LDH activity per cell remains constant throughout the production period, comparison of activity in the supernatant and the cell pellet can be used to quantitate lysis.

Since bovine serum contains LDH activity, proper controls are essential. We observed that stability of LDH activity in the culture supernatant from different cell lines varies widely (Fig. 2). In cases where LDH activity is unstable, other enzyme markers may be more useful. We also found that uptake of the exclusion stain, trypan blue, precedes release of LDH from lysed Chinese hamster ovary (CHO) cells. This is probably a reflection of the size difference between LDH and trypan blue (M_r = 140,000 and 961, respectively) and demonstrates that the two methods can yield different information.

Once a suitable assay of cell lysis is developed, it can be used for evaluation of culture conditions, including medium composition,

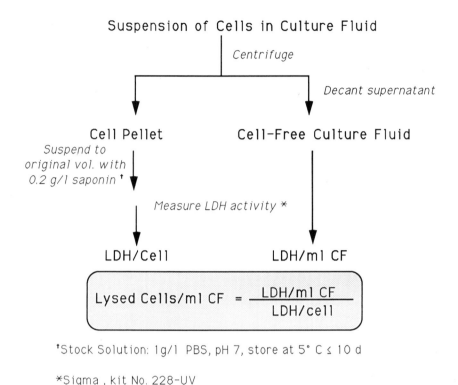

Figure 1 Release of lactate dehydrogenase (LDH) as a measure of cell lysis.

reactor geometry, and time of harvest. The effect of product accumulation on cell health may also be investigated.

B. Two-Dimensional Gels on Cell Pellets

Recent evidence with attached cells suggests that production related stresses, such as shear forces or nutrient depletion, may adversely affect cell metabolism, protein synthesis, or protein secretion (6). Two-dimensional (2D) electrophoresis is a powerful technique for monitoring these aspects of cell health. High resolution is achieved in 2D gels by combining the separation power of two different electrophoretic methods. These techniques are discussed further in

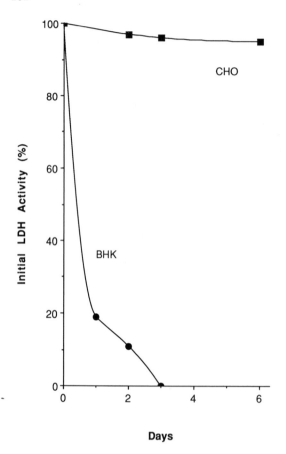

Figure 2   Stability of LDH activity at 37°C in culture supernatants of baby hamster kidney (BHK) and CHO cells. (From Ref. 5, copyright 1986 by the AAAS.)

Section V.   Silver-stained 2D gels of lysed cell pellets show up to ~800 individual protein spots with a detection sensitivity of ~1 ng per spot. These gels are rather labor-intensive but complete gel* and analytical services are commercially available (7). Figure 3 shows a 2D gel of a CHO cell lysate.

*Note added in proof: Gel services no longer available.

IEF

acidic   basic   SDS-
                                                                    PAGE

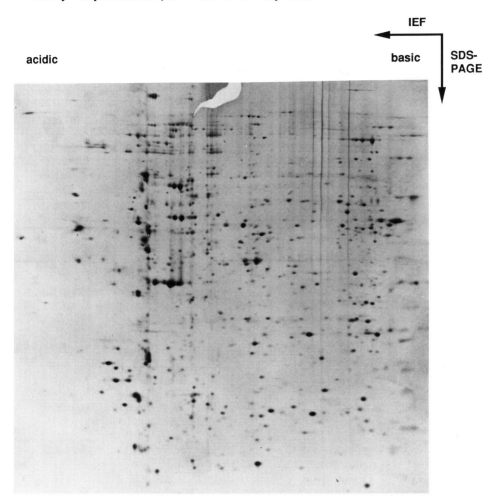

Figure 3   Silver-stained 2D gel of CHO cell lysate.   1st dimension:
IEF in 8 M urea, pH 3–10; 2nd dimension: SDS-PAGE in 10% T.
Gel was run by Protein Databases, Inc., Huntington Station, New
York, and silver-stained (55).

III. NUTRIENTS AND METABOLIC
    INTERMEDIATES

Cell culture medium contains several types of small molecules which
serve as nutrients for the growth and maintenance of cells. Medium
optimization is an important phase of process development (see Chapter 6) and several assays are of great benefit in this effort. Amino
acid analysis provides a method for monitoring depletion of these
nutrients. This technique also yields information regarding the
utilization of glutamine, one of the two major carbon energy sources
used by cells (see Chapter 4). Glucose is the other major carbon
energy source. Additional information about the metabolism of cells
under a given set of growth conditions can be obtained from monitoring glucose utilization and lactate production. Additionally, glucose starvation has been shown to alter glycosylation patterns (8);
therefore, cultures should be monitored for glucose depletion.

A. Amino Acids

Several methods of amino acid analysis are currently in use and the
choice of method depends on the application. (For a more complete
discussion, see Ref. 9.) Conventional analysis, involving ion exchange chromatography and postcolumn derivatization of amino
groups with ninhydrin, was first developed in 1958 (10). Since
then many advances have been made and dedicated amino acid ana-
lyzers are commercially available. The ion exchange–ninhydrin
method is very reliable, accurate, and reproducible and is thus well
suited to quality control laboratories. For many other applications,
including those described here, faster and more sensitive systems
have been developed in the last decade.

The newer techniques involve precolumn derivatization and separation by reversed phase high-performance liquid chromatography
(RP-HPLC). These methods do not require a dedicated instrument;
the same HPLC system can be used to perform many of the assays
discussed in this chapter. UV detection following derivatization with
phenylisothiocyanate (PITC) is an excellent method for determining
amino acid composition of purified proteins and peptides (11,12).
However, for complex mixtures such as culture fluids, fluorescent
detection after o-phthaldialdehyde (OPA) derivatization is the method of choice.

The OPA method combines high sensitivity (<1 pmol) with speed
(<30 min). It is fully automatable and free from interference by
the many other compounds found in culture fluids. The original
OPA method had several disadvantages which have since been resolved (13). Analysis of secondary amino acids (e.g., proline)
can be obtained by including a second derivatization step, with

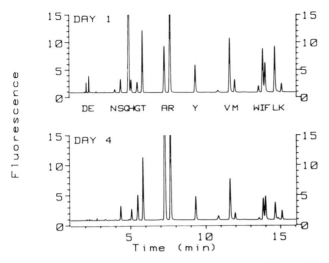

Amino Acid	% Change	Amino Acid	% Change
Aspartic acid (D)	−42	Arginine (R)	+8
Glutamic acid (E)	−74	Tyrosine (Y)	−21
Asparagine (N)	−50	Valine (V)	−28
Serine (S)	−5	Methionine (M)	−39
Glutamine (Q)	−99	Tryptophan (W)	−43
Histidine (H)	−22	Isoleucine (I)	−51
Glycine (G)	+81	Phenylalanine (F)	−22
Threonine (T)	−5	Leucine (L)	−62
Alanine (A)	+290	Lysine (K)	−11

Figure 4   Amino acid analysis of culture supernatant samples by *o*-phthaldialdehyde method (14).

9-fluorenylmethylchloroformate (FMOC), in the automated sequence and using dual-wavelength detection (14). If analysis of cysteine is required, manual alkylation with iodoacetic acid can be performed prior to automated derivatization with OPA (15). Figure 4 illustrates the changes in the supernatant amino acid profile observed over time in culture.

B. Glucose and Lactate

Glucose and lactate can be measured in culture supernatants by
enzymatic assays which are available as commercial kits (Sigma). A
common method for glucose assay involves the coupled enzyme system shown here (16):

$$\text{Glucose} + H_2O + O_2 \xrightarrow{\text{glucose oxidase}} \text{gluconic acid} + H_2O_2$$

$$H_2O_2 + o\text{-dianisidine} \xrightarrow{\text{peroxidase}} \text{oxidized } o\text{-dianisidine}$$

The amount of oxidized o-dianisidine produced is measured
spectrophotometrically and is proportional to the amount of glucose
present in the original sample. Similar assays use lactate dehydrogenase to convert lactate and NAD to pyruvate and NADH (17).
The NADH, monitored at 340 nm, is proportional to the lactate concentration. Recently, ion-moderated partitioning high performance
liquid chromatography has been used to measure glucose and lactate
in cell culture media (18). This method appears to be as sensitive
and reproducible as the enzymatic assays.

IV. PROTEIN QUANTIFICATION

Since maximization of product titer is a major objective of process
development, quantitation of both the specific product and total secreted protein is essential. Two different types of product assays,
immunological and functional, are needed for the complete analysis.
In all the assays discussed in this section, culture fluid should be
clarified by centrifugation or filtration before analysis.

A. Immunological Assays

The advantage of immunologically based assays is their ability to
specifically measure the amount of product in a complex mixture
such as culture fluid. Implementation of these assays requires the
generation of antibodies, either mono- or polyclonal. These antibodies can then be used in conjunction with competitive or noncompetitive assays employing radioactive or enzymatic detection
methods.

1. Radioimmunoassay

Immunoassays that depend on radioactivity for the measurement of a
particular protein are referred to as radioimmunoassays or simply
RIAs (19,20). Classically, this term applies to assays in which a

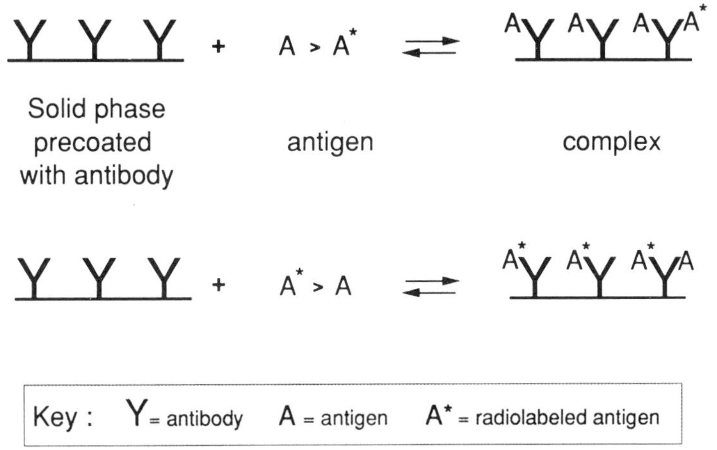

Figure 5 Competing reactions underlying radioimmunoassay.

fixed amount of radiolabeled antigen competes with unlabeled antigen for a limiting amount of specific antibody (Fig. 5). Once equilibrium is obtained, excess free radioactive antigen is removed, and the percentage of radioactive antigen bound to antibody is measured. By comparison with a standard curve generated using known antigen concentrations, the amount of antigen in an unknown sample can be determined. Radioactivity is plotted vs. the log of the antigen concentration. Sensitivity in the picogram range may be achieved in this type of assay provided that the antigen is labeled with a radioactive tracer, such as ^{125}I, to a high specific activity. Antibodies with high specificity and affinity are also required.

Various methods of separation of free and bound antigen are employed, the most common of which involve solid phase methods using antibody precoated on beads, plates, tubes, etc. If purified antigen is very scarce, the antibody can be labeled instead of the antigen in a procedure called immunoradiometric assay (IRMA). It is also possible to use a direct binding radioimmunoassay to measure antigen concentration but this is less common and requires careful standardization. RIA is described in greater detail, including specific procedures, in Ref. 21.

2. Enzyme-Linked Immunosorbent Assays

To avoid the problems of handling and disposal of radioactive materials and the instability of ^{125}I-labeled proteins, enzyme linked immunosorbent assays (ELISA) have been developed (22). These

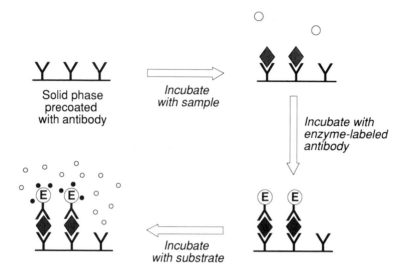

Figure 6   Principles of noncompetitive ELISA.

assays use enzymes rather than radiolabels for detection but are otherwise analogous to RIA. The most common ELISA is a noncompetitive, "sandwich" assay using 96-well plates precoated with antibody (Fig. 6). The plates are incubated with samples containing a limiting amount of antigen, to permit formation of an antigen-antibody complex. A second, enzyme-labeled antibody is then used to tag the complex and the enzymatic activity is determined using a colorimetric reaction. The amount of antigen bound to the wells is determined from a standard curve of absorbance vs. the log of antigen concentration. Suitable care must be taken to thoroughly wash the wells between incubation steps and the samples are often diluted with proteins such as albumin to prevent nonspecific adsorption. Horseradish peroxidase, alkaline phosphatase, and urease are the most commonly used enzymes.

B. Functional Assays

The immunoassays described above are based on the recognition of certain structural determinants, the presence of which may be independent of functional activity. Moreover, these assays require production of specific antibodies, a time-consuming process. Assays measuring product activity are also important in cell culture process development and provide information complementary to that from immunoassays.

Assessment of activity is relatively straightforward for products which are enzymes. For example, the proteolytic activity of tissue plasminogen activator (t-PA) can be measured using the synthetic chromogenic substrate, H-d-isoleucyl-L-prolyl-L-arginine-p-nitro-analine (S-2288, Kabi Diagnostic AB). Plasminogen activator hydro-lyzes this substrate, releasing p-nitroanaline, which can be meas-ured spectrophotometrically at 410 nm (23). As always, a thorough understanding of the reaction is important and proper controls must be included, particularly when assaying crude mixtures such as cul-ture fluid. In this example, the one- and two-chain forms of t-PA show different reaction rates and the relative contributions of each must be considered (24). Assays for many other enzymes have been described (25) and these are often readily adaptable to process de-velopment needs. A 96-well plate format facilitates the handling of large numbers of samples.

For nonenzymatic products, biological activity must be assessed in a different manner. These assays are often cell-based and there-fore more complicated to perform than enzymatic activities. Measure-ment of the antiviral effects of interferon requires development of such an assay. While many different methods have been used, in-hibition by interferon of the cytopathic effect of a virus forms the basis for a relatively simple assay (26). A monolayer of anchorage-dependent mammalian cells, sensitive to a particular strain of virus, is grown in culture plates. The cells are exposed to serially diluted samples containing interferon and then challenged with virus. Cell viability is measured to determine the dilution at which 50% protec-tion of the monolayer occurs. The final value for the sample titer is determined by comparison to a reference standard preparation of interferon. Important variables in this assay include selection of the cell line, since sensitivity to interferon varies, and selection of virus, which should be titrated in the chosen cell line. Cell-based activity assays are often amenable to automation, at least for reagent delivery and viability measurement.

C. Total Protein

Assays which measure total protein are also needed in process de-velopment. The ratio of specific product to total protein provides important information regarding subsequent purification of the prod-uct and may be useful in determining time of harvest. Additionally, such assays can be used to monitor recovery of medium protein com-ponents following preculture manipulations such as filtration.

Several commonly used protein assays are compared in Table 1. The Lowry assay has been a standard method of protein determination

Table 1   Comparison of Assays for Total Protein

	Lowry	Bradford	BCA	UV
Sensitivity (µg)	1	1	1	10
Complexity	Complex	Simple	Simple	Simple
Time (min)	40	5	30	5
Sample destruction	Yes	Yes	Yes	No
Interferences	Many	Few	Least	Few
Automation	Difficult	Easy	Easy	Difficult

for over 30 years (27). The mechanism of this reaction is complex, involving both the biuret reaction of cupric ions with the protein backbone and the reduction of phosphomolybdic and phosphotungstic acids by tyrosine residues. Several reagents must be prepared and accurate results require careful timing of additions and mixing steps. Many common substances interfere with the Lowry assay.

Several newer assays offer significant advantages. The Bradford assay utilizes the color shift observed as Coomassie brilliant blue G-250 binds to hydrophobic regions of proteins (28). This assay is rapid and simple, using only one reagent, is sensitive to 1 µg of protein, and is free from many of the interferences which limit the Lowry's usefulness. More recently, another assay has been described using bicinchoninic acid (BCA), which has the further advantage of involving less interference by detergents (29). Adaptations of both of these assays for 96-well plates have been described (30,31). Protein assays are compared in much greater detail in a recent review (32).

Ultraviolet (UV) absorption can also be used to measure protein concentration (33). The absorbance of a purified protein at 280 nm is a reflection of its relative content of the aromatic amino acids, tyrosine and tryptophan and, to a lesser extent, phenylalanine. Absorbance measurements are nondestructive and rapid, and therefore useful for monitoring the concentration of purified proteins during subsequent manipulations. For use with complex mixtures such as culture fluid, it is necessary to correct for the contributions of nucleic acids and for light-scattering effects of large or insoluble macromolecules (34).

The assays outlined above demonstrate the diversity of methods available to quantitate protein concentration. Appropriate controls must always be included to assess the accuracy, precision, and reproducibility of any assay. Since all of these methods show significant variability from protein to protein, such measurements on protein mixtures must be interpreted with caution. The assay value of a mixture varies with the distribution of individual protein components in the sample and with the protein used to standardize the assay. Interference by other sample components can be monitored by "spiking" culture fluid with known amounts of protein.

V. PROTEIN STRUCTURE/PRODUCT QUALITY

Mammalian cells are used for production of pharmaceutical proteins because of their ability to correctly synthesize and process complex structural features. These features include protein chain length, glycosylation, and intra- and interchain disulfide bonds. The ability of cells to carry out these processes correctly is affected by the conditions under which the cells are grown. Thus, it is important to monitor the integrity of these structural features during process development. This section provides an overview of methods useful for this purpose.

Before undertaking structural characterization of cell culture-derived products, it is important to remember that naturally produced proteins are often heterogeneous. Natural glycoproteins exhibit extensive carbohydrate heterogeneity and secretory proteins are often found in various stages of proteolytic processing. Thus, when proteins are produced by cells in culture, it is reasonable to expect similar amounts of structural variability. At the cell culture process development level, it is often most useful to characterize the full spectrum of molecules produced in culture. This information can then serve as a useful reference for recovery process development efforts.

A. Sample Preparation

In all cases, cells and cell debris should be removed by centrifugation or filtration before further manipulation.

1. Concentration

The most versatile characterization assays are those that can be performed directly on culture fluid samples. Often these samples simply require concentration of total protein prior to analysis. The ratio of specific product to total protein, as well as the product titer, determines which samples can be assayed directly. Several methods useful for concentration of culture fluid are discussed below. With all of these methods, it is important to monitor the recovery of specific product. Losses can be tolerated as long as subpopulations of protein structure are not inadvertently selected.

Volatile organic solvents, such as acetone or acetonitrile, can be used to concentrate proteins by precipitation (35). Two to five volumes of cold solvent are added to an appropriate aliquot of culture fluid. Ten micrograms of total protein is the smallest workable amount for this procedure. After mixing, precipitation is allowed to proceed at −20°C for ⩾1 hr. Precipitated protein is then collected by centrifugation and any remaining solvent removed by evaporation. Depending on their initial concentration, some amino acid components of media may also precipitate.

Ultrafiltration is also useful for the concentration of proteins from dilute solutions. Several devices (Amicon, Millipore) are available for handling small volumes, most designed for use with a centrifuge, and some with an automatic volume stop. We use these routinely for purified samples which often do not contain enough total protein to form a visible pellet in the precipitation methods described above.

2. Purification

In some cases, the ratio of specific product to total protein is too low to allow direct analysis using whole culture fluid. In other cases, purified protein is required for more detailed characterization. In these instances, a simple and rapid purification procedure which is fully or partially automatable is required. Affinity chromatography is usually the method of choice, since it often allows a single-step purification directly from culture fluid.

The general principles of affinity chromatography are illustrated in Fig. 7. A column containing a specific affinity ligand covalently attached to an insoluble support matrix is exposed to culture fluid. The product binds to the ligand, contaminating proteins are washed away, and the product is eluted from the immobilized ligand by disrupting the affinity interaction.

Many different types of ligands can be used for affinity chromatography (36), including substrate or binding site analogs, cofactors, inhibitors, proteins, and even cells. For example, the enzyme tissue-plasminogen activator can be purified with several different affinity ligands, including lysine (37), benzamidine (38), and antibodies (39). The last method is an example of immunoaffinity chromatography (40), a variation of the technique in which specific antibodies, either poly- or monoclonal, are used as ligands.

Affinity ligands must exhibit high specificity to preferentially remove product from the complex mixture of contaminating proteins found in culture fluid. The binding affinity for the ligand should be high enough to prevent loss of product from the column during washing steps, but not so high as to complicate subsequent elution of purified protein. In addition, the ligand should be able to withstand many purification cycles without significant loss of affinity.

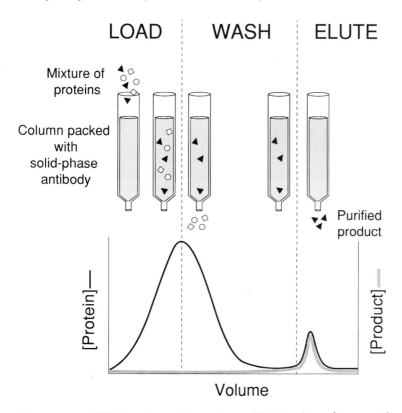

Figure 7 Affinity chromatography: illustration of separation principles and representative elution profile.

This problem is more serious with large ligands, such as proteins, than with small ones.

The support matrix should exhibit good mechanical stability to provide for adequate flow rates. It should be devoid of nonspecific interactions with contaminating proteins of the culture fluid. The support matrix also must provide the chemical groups necessary for covalent attachment of the ligand. The coupling procedure should not alter the binding affinity of the ligand. Often an inert spacer arm is required between the ligand and the support to prevent steric hindrance during the interaction with the protein. The affinity matrix should be stable enough to prevent leaching of the ligand during repeated elution cycles. For a complete survey of immobilization methods and support matrices, see Ref. 41.

B. Polyacrylamide Gel Electrophoresis (PAGE)

Electrophoretic procedures conducted using polyacrylamide gels (42, 43) have become standard methods of protein analysis, since they provide information regarding a wide variety of structural elements. Using different methods of detection, electrophoresis can be performed either directly on culture fluid or on purified proteins. Electrophoretic procedures are rather labor-intensive, however, and interpretation of gel patterns can be somewhat subjective since quantitation is often difficult.

Electrophoretic equipment is available in a number of different sizes and configurations: large gels, minigels, and Phast system (Pharmacia); tube gels vs. slab gels; horizontal vs. vertical systems. Each type has advantages under certain circumstances. Large gels are recommended for high-resolution applications, although the smaller gels are easier to handle and faster to run. Individual tube gels provide flexibility during system development because several different types of gels may be evaluated simultaneously. Once an analytical method is developed, slab gels provide a more uniform background for comparison of samples. Vertical gels can accommodate larger sample volumes and are commonly used for SDS gels, but most isoelectric focusing gels are run in horizontal units. Precast slab gels are now available for many systems and these greatly simplify the work.

1. Detection

Immunoblots. As with the immunoassays described in Section IV.A, immunoblots are useful for detection of specific products in mixtures of proteins such as culture fluid. First, proteins are transferred from gels to a matrix such as nitrocellulose or polyvinylidene difluoride (PVDF). The mechanical strength of PVDF allows direct microsequencing of protein bands cut from these membranes (44). This provides a very powerful technique for the characterization of protein species produced in cell culture and often requires only partial purification prior to electrophoresis.

Transfer can be achieved by electrophoretic methods (45) or by simple diffusion (46). Electrophoretic transfer equipment is of two main types: wet or semidry. Semidry units use smaller buffer volumes, require shorter transfer times, and often maintain better resolution. Prestained or naturally colored protein standards can be used to easily monitor transfer. However, the transfer efficiency should be checked by staining the gel after blotting, especially during methods development. It is also helpful to use two sheets of membrane during initial transfers, since some proteins have been observed to pass through the first sheet under standard conditions.

Once transfer is achieved, development of the blot can be accomplished in several ways. Total protein can be detected with India ink (47), Coomassie blue (48), or gold (49) stains, depending on the sensitivity required. The specific product can be detected by immunomethods. The blot is incubated first with a nonspecific protein such as gelatin (50), to block unused protein binding sites. Then a product-specific antibody is allowed to bind to the protein bands on the blot. The antigen-antibody complex is detected in one of two ways.

In the original "Western" blot procedure (51), ^{125}I-labeled protein A is used, followed by autoradiography. In ELISA, an enzyme-labeled, species-specific second antibody is used and the enzyme is detected in a colorimetric reaction. The most commonly used enzyme labels are horseradish peroxidase (HRP) and alkaline phosphatase. A sensitive substrate for HRP is 4-chloronaphthol. Prolonged exposure of developed blots to light should be avoided since most ELISA patterns will fade.

A critical review of the various methods of transfer and detection is available (52). Figure 8 (discussed further in the next section) illustrates the power of immunodetection for analysis of unpurified culture supernatants.

Protein Stains. For analysis of purified proteins or for detecting total protein following electrophoretic separation, several procedures are available for direct staining of gels. Silver stains provide the highest level of sensitivity, ∿1 ng. Merril (53) recently reviewed the many different silver-staining procedures which have been developed. Some protocols produce different colored bands for many proteins (54). This is useful for identifying uniquely colored proteins in complex patterns, such as seen with 2D gels, but the intensity variations make quantitation more difficult. We find the Morrissey (55) method to give the best combination of speed, sensitivity, and reliability. Response to silver staining varies among individual proteins, and different applications may require different protocols.

Staining with Coomassie blue offers the greatest uniformity and linearity of response, although it is the least sensitive, ∿1 μg. When quantitative results are required and sufficient quantities of protein available, gels can be stained with Coomassie blue (56) and scanned by densitometry.

Stained gels are easily preserved by soaking in 5% glycerol for ∿30 min and then air-drying between sheets of cellophane membrane. Large binder clips are used to hold the gels in acrylic frames while drying.

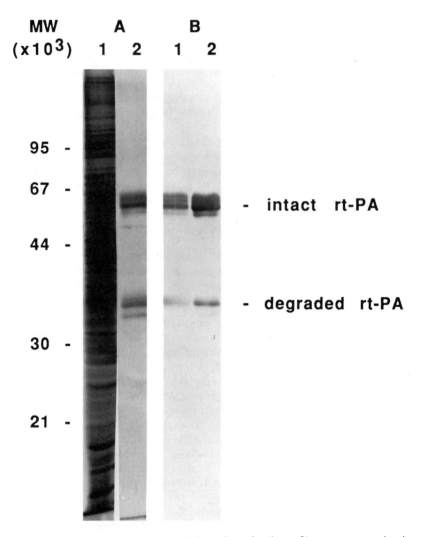

Figure 8 SDS-PAGE of rt-PA. Sample 1, culture supernatant from CHO cells producing rt-PA; sample 2, purified rt-PA, containing intact and degraded forms. Samples were reduced with βME and separated in duplicate 10% T, slab gels according to (57). A constant amount of rt-PA (0.5 μg) was loaded in each lane. (A) Total protein was detected by staining with silver (55). (B) Gel was transferred to nitrocellulose (45) and t-PA was visualized by immunoreaction using goat anti-rt-PA followed by peroxidase-conjugated swine anti-goat IgG. The chromogen 4-chloronaphthol was used for color development.

2. Sodium Dodecyl Sulfate (SDS) Electrophoresis

Electrophoresis in the presence of the ionic detergent SDS is probably the most common method of protein analysis. The procedure
used most often (57) is based on the classical work of Ornstein (58)
and Davis (59). This method utilizes a discontinuous buffer system
with stacking and running gels of slightly different composition (pH
and buffer concentration) to achieve "stacking" of proteins in tight
bands. These gels give sharper bands with better resolution and
allow the use of larger sample volumes than earlier continuous gel
systems (56).

SDS gels provide information on the size of the protein chain.
SDS binds to the peptide backbone in a constant wt/wt ratio and
negates the affects of amino acid charge on the electrophoretic mobility of the proteins. This separation is based solely on size.
Molecular weights can be determined by comparison with the migration of standard proteins of known weights. Glycoproteins often
migrate anomalously, probably because SDS does not bind to the
carbohydrate portions of these molecules (35).

SDS gels can be used to monitor proteolysis, which is necessary
for correct processing of many proteins, but can also be an unwel-
come side effect of cell lysis during culture. Samples run with and
without reducing agents, such as β-mercaptoethanol (βME) or dithiothreitol (DTT), also provide information on disulfide-bonded subunits and aggregates. Glycoproteins often appear as broad smears
rather than sharp bands due to their more extensive heterogeneity.

The gels in Fig. 8 illustrate the usefulness of SDS-PAGE for
analysis of cell culture products. A culture supernatant may contain proteins of many different molecular weights, as revealed by
silver staining after electrophoresis (lane 1A). The state of the
product made under any given conditions can be assessed without
prior purification by using immunoblotting (lane 1B). Under the
conditions used in this example, rt-PA is produced in two forms,
intact protein ($M_r \sim 66,000$) and a proteolytically degraded form
($M_r \sim 33,000$).

3. Isoelectric Focusing

Isoelectric focusing (IEF) provides complementary information to
SDS-PAGE. Separation in IEF is based on charge rather than size.
Thus, IEF provides information on structural features, such as
sialic acid heterogeneity and phosphorylation, as well as on the
integrity of the amino acid side chains of proteins. Some side
chains are susceptible to oxidation, deamidation, or mixed disulfide
formation, which may alter the total charge of the protein and thus
its migration in IEF.

For insoluble proteins, IEF can be conducted in the presence of
denaturants, such as urea. Care must be taken to prevent

carbamylation artifacts, and alterations in pH measurements should be taken into account in estimating isoelectric point (pI) values in gels containing urea. As with SDS, IEF can be performed in the presence or absence of reducing agents. Proteins can be maintained stably in the reduced state by subsequent alkylation of the free sulfhydryl groups (60). Glycoproteins often exhibit complex focusing patterns containing many bands, due to sialic acid heterogeneity.

A thorough treatment of IEF is provided by Righetti (61).

4. Two-Dimensional Gels

Two-dimensional gels, mentioned in Section II.B, combine the resolving and analytical power of two electrophoretic methods. Samples are first analyzed in one type of gel, which is then applied to a second gel for separation by a different method. The most common form involves IEF in the first dimension and SDS-PAGE in the second dimension (62). Although systems have been developed for analyzing multiple samples (63,64), these gels are still rather labor-intensive to be useful in routine screening of process development samples. They are helpful in assigning identities to the individual bands of complex focusing patterns.

Systems which combine liquid chromatography (see Section V.C) in the first dimension with electrophoresis in the second dimension are currently under development* (65). These methods look very promising for reducing the labor involved in traditional 2D analysis.

5. Capillary Electrophoresis

Another promising innovation in electrophoretic technique is capillary electrophoresis (66). This method combines high resolution with short run times to circumvent many of the disadvantages of the classical electrophoretic methods. Capillary electrophoresis is really related to both chromatography and electrophoresis. It involves electrophoretic separations of very small sample volumes, nanoliters, inside capillary tubes. Detection of separated components is conducted in real time using flow-through chromatographic detectors. Liquids, electrophoretic gels, and chromatographic packings can all be used as separation media. The availability of automated capillary electrophoresis instruments may result in the widespread use of these techniques.

C. Liquid Chromatography

In contrast to the electrophoretic methods described above, chromatographic procedures are readily automatable with HPLC systems and the results are easily quantified. Many different chromatographic

---

*Note added in proof: Development of these methods has been discontinued.

techniques are available for the characterization of purified proteins (67). If the product has unique physical properties which allow easy separation, or if expression is high and there are low levels of contaminating proteins, these procedures are directly applicable to culture fluid. The latter situation often occurs with hybridomas.

1. Gel Filtration

Proteins may be separated by size using differential sieving through matrices of limiting porosity. This technique is variously known as gel filtration, molecular sieve chromatography, and size exclusion chromatography. It is most useful with purified products and allows quantitation of the relative amounts of proteolysis which have occurred during culture. Gel filtration is often conducted in the presence of denaturing agents, such as SDS, guanidine hydrochloride, or urea. These agents are used to maintain solubility and to suppress artifactual interactions of proteins with themselves and with column materials during molecular weight determinations.

   Figure 9 illustrates the usefulness of gel filtration for monitoring proteolysis. The rt-PA produced under the two growth

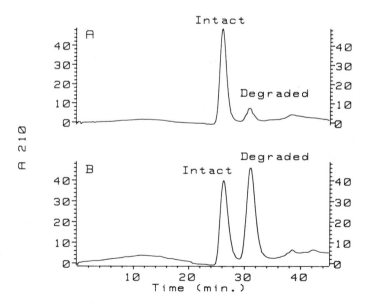

Figure 9   Comparison of gel filtration profiles of rt-PA produced by CHO cells under different culture conditions. Purified samples were reduced with DTT and analyzed on a TSK-3000 SW column with SDS in the elution buffer. (A) 88% intact rt-PA; (B) 45% intact rt-PA.

conditions used in this example exhibited very different amounts of
proteolytic degradation.

Low-pressure gel filtration is thoroughly discussed by Fischer
(68) and the considerations for HPLC methods were recently reviewed (69).

2. Reversed Phase

Separations of proteins with reversed phase columns are based mainly on hydrophobicity, with the most hydrophobic proteins eluting
last (70). Since the order of elution is not as predictable as with
gel filtration, electrophoretic analysis of collected fractions is often
useful for initial identification of the various species separated.
RP-HPLC has very high resolving power and therefore is the most
likely chromatographic method to be useful for the analysis of un-
fractionated culture fluid.

RP-HPLC of intact purified proteins and of the mixture of peptides obtained following digestion with a protease, such as trypsin,
can provide information on many aspects of protein structure. Fine
structural details, such as deamidation or oxidation of side chains
of amino acid residues, may be discerned with reversed phase
analysis. This is especially true with small, nonglycoprotein molecules, such as human growth hormone, which tend to be relatively
homogeneous due to the lack of carbohydrate (71).

D. Carbohydrate Content

The presence of carbohydrate moieties increases solubility of glycoproteins and provides some protection from proteolysis (72). Clearance of glycoproteins from the bloodstream is affected by the nature
of the sugar residues exposed on the surface (73). Thus, it can
be valuable to monitor the monosaccharide content of glycoproteins
during process development. All of the methods presently available
for carbohydrate analysis can only be performed with purified
protein.

Several standard colorimetric procedures are useful for quantitative analysis of carbohydrate content. Sialic acid can be measured
by the thiobarbituric acid method (74) and neutral sugars by the
phenol-sulfuric acid method (75). These colorimetric assays are not
very sensitive (30 nmol/ml) and large amounts of purified glycoprotein are needed, especially when the carbohydrate content is
low. A more sensitive assay (0.3 nmol/ml) for sialic acid involves
derivatization with a fluorescent reagent followed by RP-HPLC (76).

In the development of production processes for some glycoproteins, more extensive carbohydrate analysis may be useful. Many of
the classical methods (77) are insensitive and labor-intensive. In
contrast, a recently developed chromatographic technique for analysis

of mono- and oligosaccharides is both fast and sensitive. Anion exchange chromatography with pulsed amperometric detection appears especially useful for analysis of oligosaccharide chains derived from digests of purified glycoproteins (78). This procedure can provide information regarding the number, type, and composition of the carbohydrate chains.

VI. SUMMARY

Because the structural parameters which influence product "quality" will vary from protein to protein, the screening methods used during development of each production process will also differ. Thus, not all of the methods described above will be needed for each product. One of the challenges of cell culture process development is choosing the correct structural features to monitor.

The next level of process development may involve on-line control of many of the parameters discussed above. This would allow product quality-driven manipulations of culture conditions. The major factor limiting this advance may not be the techniques for real-time analysis, but rather a more thorough understanding of the structural features which describe a high-quality pharmaceutical protein.

ACKNOWLEDGMENTS

We thank Jim Stramondo for guidance during development of the LDH assay and especially for his early recognition of the importance of product quality; Tom Ihrig and Larry Forman for critical evaluation of the manuscript and for helpful discussions; Rob Arathoon for critical reading and general guidance throughout this work; Will Mulhern for amino acid analysis.

REFERENCES

1. Helenius, A. and Simons, K. Solubilization of membranes by detergents. *Biochim. Biophys. Acta* 415:29−79 (1975).
2. Tanford, C. and Reynolds, J. A. Characterization of membrane proteins in detergent solutions. *Biochim. Biophys. Acta* 457: 133−170 (1976).
3. Singh, N. P. and Stephens, R. E. A novel technique for viable cell determinations. *Stain Technol.* 61:315−318 (1986).
4. McEwen, B. F. and Arion, W. J. Permeabilization of rat hepatocytes with *staphylococcus aureus* alpha-toxin. *J. Cell Biol.* 100:1922−1929 (1985).

5. Arathoon, W. R. and Birch, J. R. Large-scale cell culture in biotechnology. *Science* 232:1390–1395 (1986).

6. Goochee, C. F., Passini, C., Lall, R., Morrison, D. R., and Kalmez, E. E. Hydrodynamic stress and intracellular protein synthesis, 194th American Chemical Society National Meeting, New Orleans, August 30–September 4, 1987. *Abstr. Pap. Am. Chem. Soc.* 194:MBTD 52 (1987).

7. Blose, S. H. and Hamburger, S. A. Computer-analyzed high resolution two-dimensional gel electrophoresis: A new window for protein research. *BioTechniques* 3:232–236 (1985).

8. Davidson, S. K. and Hunt, L. A. Sindbis virus glycoproteins are abnormally glycosylated in Chinese hamster ovary cells deprived of glucose. *J. Gen. Virol.* 66:1457–1468 (1985).

9. Ogden, G. and Foldi, P. Amino acid analysis: An overview of current methods. *LC-GC* 5:28–40 (1987).

10. Spackman, D. H., Stein, W. H., and Moore, S. Automatic recording apparatus for use in the chromatography of amino acids. *Anal. Chem.* 30:1190–1206 (1958).

11. Bidlingmeyer, B. A., Cohen, S. A., and Tarvin, T. L. Rapid analysis of amino acids using pre-column derivatization. *J. Chromatogr.* 336:93–104 (1984).

12. Heinrikson, R. L. and Meredith, S. C. Amino acid analysis by reverse-phase high-performance liquid chromatography: Pre-column derivatization with phenylisothiocyanate. *Anal. Biochem.* 136:65–74 (1984).

13. Jones, B. N. and Gilligan, J. P. Amino acid analysis by o-phthaldialdehyde precolumn derivatization and reversed-phase HPLC. *Am. Biotech. Lab.* (Dec):46–51 (1983).

14. Schuster, R. Determination of amino acids in biological, pharmaceutical, plant and food samples by automated precolumn derivatization and high-performance liquid chromatography. *J. Chromatogr.* 431:271–284 (1988).

15. Turnell, D. C. and Cooper, J. D. H. Rapid assay for amino acids in serum or urine by pre-column derivatization and reversed-phase liquid chromatography. *Clin. Chem.* 28:527–531 (1982).

16. Bergmeyer, H. U. and Bernt, E. Determination of glucose with glucose oxidase and peroxidase. In *Methods of Enzymatic Analysis*, Vol. 3 (H. U. Bergmeter, ed.). Academic Press, New York, pp. 1205–1212 (1974).

17. Gutmann, I. and Wahlefeld, A. W. L-(+)-Lactate determination with lactate dehydrogenase and NAD. In *Methods of Enzymatic Analysis*, Vol. 3 (H. U. Bergmeyer, ed.). Academic Press, New York, pp. 1464–1468 (1974).

18. Tedesco, J. L. Analysis of glucose and lactic acid in cell culture media by ion moderated partitioning high performance liquid chomatography. *BioTechniques* 5:46−51 (1987).
19. Yalow, R. S. Radioimmunoassay: A probe for the fine structure of biologic systems. *Science* 200:1236−1245 (1978).
20. Van Vunakis, H. Radioimmunoassays: An overview. *Meth. Enzymol.* 70:201−209 (1980).
21. Johnstone, A. and Thorpe, R. *Immunochemistry in Practice*. Blackwell Scientific, Oxford, pp. 233−252 (1982).
22. Engvall, E. Enzyme immunoassay ELISA and EMIT. *Meth. Enzymol.* 70:419−439 (1980).
23. Smith, R. E. Contributions of histochemistry to the development of the proteolytic enzyme detection system in diagnostic medicine. *J. Histochem. Cytochem.* 31:199−209 (1983).
24. Tate, K. M., Higgins, D. L., Holmes, W. E., Winkler, M. E., Heyneker, H. L., and Vehar, G. A. Functional role of proteolytic cleavage at arginine-275 of human tissue plasminogen activator as assessed by site-directed mutagenesis. *Biochemistry* 26:338−343 (1987).
25. Colowick, S. P. and Kaplan, N. O. (eds.). *Methods in Enzymology*, Vols. 1−163. Academic Press, New York (1955−1988).
26. Armstrong, J. A. Cytopathic effect inhibition assay for interferon: Microculture plate assay. *Meth. Enzymol.* 78:381−387 (1980).
27. Lowry, O. H., Rosebrough, N. J., Farr, A. L., and Randall, R. J. Protein measurement with the Folin phenol reagent. *J. Biol. Chem.* 193:265−275 (1951).
28. Bradford, M. M. A rapid and sensitive method for the quantitation of microgram quantities of protein utilizing the principle of protein-dye binding. *Anal. Biochem.* 72:248−254 (1976).
29. Smith, P. K., Krohn, R. I., Hermanson, G. T., Mallia, A. K., Gartner, F. H., Provenzano, M. D., Fujimoto, E. K., Goeke, N. M., Olson, B. J., and Klenk, D. C. Measurement of protein using bicinchoninic acid. *Anal. Biochem.* 150:76−85 (1985).
30. Simpson, I. A. and Sonne, O. A simple, rapid, and sensitive method for measuring protein concentration in subcellular membrane fractions prepared by sucrose density ultracentrifugation. *Anal. Biochem.* 119:424−427 (1982).
31. Hinson, D. L. and Webber, R. J. Miniaturization of the BCA protein assay. *BioTechniques* 6:14−18 (1988).
32. Davis, E. M. Protein assays: A review of common techniques. *Am. Biotech. Lab.* 6:28−37 (1988).

33. Layne, E. Spectrophotometric and turbidimetric methods for measuring proteins. *Meth. Enzymol.* 3:447–454 (1957).

34. Leach, S. J. Effect of light scattering on ultraviolet difference spectra. *J. Am. Chem. Soc.* 82:4790–4792 (1960).

35. Hames, B. D. An introduction to polyacrylamide gel electrophoresis. In *Gel Electrophoresis of Proteins: A Practical Approach* (B. D. Hames and D. Rickwood, eds.). IRL Press, Oxford, pp. 1–91 (1981).

36. *Affinity Chromatography: Principles and Methods.* Pharmacia Fine Chemicals Co., Piscataway, New Jersey (1983).

37. Radcliffe, R. and Heinze, T. Isolation of plasminogen activator from human plasma by chromatography on lysine-Sepharose. *Arch. Biochem. Biophys.* 189:185–194 (1978).

38. Fuchs, H. E., Berger, H., Jr., and Pizzo, S. V. Catabolism of human tissue plasminogen activator in mice. *Blood* 65: 539–544 (1985).

39. Upshall, A., Kumar, A. A., Bailey, M. C., Parker, M. D., Favreau, M. A., Lewison, K. P., Joseph, M. L., Maraganore, J. M., and McKnight, G. L. Secretion of active human tissue plasminogen activator from the filamentous fungus *Aspergillus nidulans*. *Biotechnology* 5:1301–1304 (1987).

40. Phillips, T. M. High performance immunoaffinity chromatography. *LC* 3:962–972 (1985).

41. Scouten, W. H. A survey of enzyme coupling techniques. *Meth. Enzymol.* 135:30–65 (1987).

42. Chrambach, A. and Rodbard, D. Polyacrylamide gel electrophoresis. *Science* 172:440–451 (1971).

43. Hames, B. D. and Rickwood, D. (eds.). *Gel Electrophoresis of Proteins: A Practical Approach.* IRL Press, Oxford (1981).

44. LeGendre, N. and Matsudaira, P. Direct protein microsequencing from Immobilon[TM]-P transfer membrane. *BioTechniques* 6:154–159 (1988).

45. Towbin H., Staehelin, T., and Gordon, J. Electrophoretic transfer of proteins from polyacrylamide gels to nitrocellulose sheets: Procedure and some applications. *Biochemistry* 76: 4350–4354 (1979).

46. Reinhart, M. P. and Malamud, D. Protein transfer from isoelectric focusing gels: The native blot. *Anal. Biochem.* 123: 229–235 (1982).

47. Hancock, K. and Tsang, V. C. W. India ink staining of proteins on nitrocellulose paper. *Anal. Biochem.* 133:157–162 (1983).

48. Matsudaria, P. Sequence from picomole quantities of proteins electroblotted onto polyvinylidene difluoride membranes. *J. Biol. Chem.* 262:10035–10038 (1987).

49. Moeremans, M., Daneels, G., and De Mey, J. Sensitive colloidal metal (gold or silver) staining of protein blots on nitrocellulose membranes. *Anal. Biochem.* 145:315−321 (1985).

50. Saravis, C. A. Improved blocking of nonspecific antibody binding sites on nitrocellulose membranes. *Electrophoresis* 5:54−55 (1984).

51. Burnette, W. N. "Western blotting": Electrophoretic transfer of proteins from sodium dodecyl sulfate-polyacrylamide gels to unmodified nitrocellulose and radiographic detection with antibody and radioiodinated protein A. *Anal. Biochem.* 112:195−203 (1981).

52. Bers, G. and Garfin, D. Protein and nucleic acid blotting and immunobiochemical detection. *BioTechniques* 3:276−288 (1985).

53. Merril, C. R. Silver-stain detection of proteins separated by polyacrylamide gel electrophoresis. In *New Directions in Electrophoretic Methods* (J. W. Jorgenson and M. Phillips, eds.). ACS Symposium Series 335, American Chemical Society, Washington, D.C., pp. 74−90 (1987).

54. Sammons, D. W., Adams, L. D., and Nishizawa, E. E. Ultrasensitive silver-based color staining of polypeptides in polyacrylamide gels. *Electrophoresis* 2:135−141 (1981).

55. Morrissey, J. H. Silver stain for proteins in polyacrylamide gels: A modified procedure with enhanced uniform sensitivity. *Anal. Biochem.* 117:307−310 (1981).

56. Weber, K., Pringle, J. R., and Osborn, M. Measurement of molecular weights by electrophoresis on SDS-acrylamide gel. *Meth. Enzymol.* 26:3−27 (1972).

57. Laemmli, U. K. Cleavage of structural proteins during the assembly of the head of bacteriophage T4. *Nature* 227:680−685 (1970).

58. Ornstein, L. Disc electrophoresis. I. Background and theory. *Ann. NY Acad. Sci.* 121:321−349 (1964).

59. Davis B. J. Disc electrophoresis. II. Method and application to human serum proteins. *Ann. NY Acad. Sci.* 121:404−427 (1964).

60. Means, G. E. and Feeney, R. E. *Chemical Modification of Proteins.* Holden-Day, San Francisco, pp. 218−219 (1971).

61. Righetti, P. G. *Isoelectric Focusing: Theory, Methodology and Applications* (T. S. Work and R. H. Burdon, eds.). Elsevier, Amsterdam (1983).

62. O'Farrell, P. H. High resolution two-dimensional electrophoresis of proteins. *J. Biol. Chem.* 250:4007−4021 (1975).

63. Anderson, N. G. and Anderson, N. L. Two-dimensional analysis of serum and tissue proteins: Multiple isoelectric focusing. *Anal. Biochem.* 85:331−340 (1978).

64. Anderson, N. L. and Anderson, N. G. Two-dimensional analysis of serum and tissue proteins: Multiple gradient-slab gel electrophoresis. *Anal. Biochem.* 85:341–354 (1978).
65. Nugent, K. D., Slattery, T. K., Lundgard, R. P., Wehr, C. T., Seiter, C. H., and Burton, W. G. Automated two-dimensional protein analysis with an optimized RP-HPLC first dimension. *Biochromatography* 3:168–173 (1988).
66. Compton, S. W. and Brownlee, R. G. Capillary electrophoresis. *BioTechniques* 6:432–439 (1988).
67. Hancock, W. S. (ed.). *CRC Handbook of HPLC for the Separation of Amino Acids, Peptides, and Proteins*, Vol. II. CRC Press, Boca Raton, pp. 303–419 (1984).
68. Fischer, L. *Gel Filtration Chromatography* (T. S. Work and R. H. Burdon, eds.). Elsevier, Amsterdam (1980).
69. Kato, Y. Successful size-exclusion separations of proteins. *LC* 1:540–544 (1983).
70. Hancock, W. S. and Harding, D. R. K. Review of separation conditions. In *CRC Handbook of HPLC for the Separation of Amino Acids, Peptides, and Proteins*, Vol. II (W. S. Hancock, ed.). CRC Press, Boca Raton, pp. 303–312 (1984).
71. Hancock, W. S., Canova-Davis, E., Battersby, J., and Chloupek, R. The use and limitations of reversed phase HPLC for the analysis of recombinant proteins and their tryptic digests. In *Biotechnologically Derived Medical Agents* (J. L. Gueriguian, V. Fattorusso, and D. Poggiolini, eds.). Raven Press, New York, pp. 31–49 (1988).
72. Olden, K., Bernard, B. A., White, S. L., and Parent, J. B. Function of the carbohydrate moieties of glycoproteins. *J. Cell. Biochem.* 18:313–335 (1982).
73. Ashwell, G. and Harford, J. Carbohydrate-specific receptors of the liver. *Ann. Rev. Biochem.* 51:531–554 (1982).
74. Spiro, R. G. Analysis of sugars found in glycoproteins. *Meth. Enzymol.* 8:1–26 (1966).
75. Ashwell, G. New colorimetric methods of sugar analysis. *Meth. Enzymol.* 8:85–95 (1966).
76. Hara, S., Yamaguchi, M., Takemori, Y., Nakamura, M., and Ohkura. Y. Highly sensitive determination of N-acetyl- and N-glycolylneuraminic acids in human serum and urine and rat serum by reversed-phase liquid chromatography with fluorescence detection. *J. Chromatogr.* 377:111–119 (1986).
77. Chaplin, M. F. and Kennedy, J. F. *Carbohydrate Analysis: A Practical Approach.* IRL Press, Oxford (1986).
78. Chen, L.-M., Yet, M.-G., and Shao, M.-C. New methods for rapid separation and detection of oligosaccharides from glycoproteins. *FASEB J.* 2:2819–2824 (1988).

8

Nonperfused Attachment Systems for Cell Cultivation

ARTHUR Y. ELLIOTT
Merck Pharmaceutical Manufacturing Division of Merck & Co., Inc., West Point, Pennsylvania

I. INTRODUCTION

Techniques for the in vitro cultivation of animal cells have advanced dramatically since 1907 when Harrison (1) cultivated the neuroblast of the frog in clotted lymph and observed the growth of the fibrillae from the central body. This work is usually credited as the first successful "tissue culture." In 1912, Carrel (2) published work describing a strain of connective tissue cells from the chick that he kept in a state of active multiplication for many passages. For the next 50 years, many cultures were started from primary tissue explants and kept alive as initial cultures or passaged cells growing on only one surface of either glass or plastic tissue culture vessels.

It was not until the 1950s that enzymes (trypsin, collagenase, hyaluronidase, and pronase) were used to disaggregate tissue masses and thus liberate single cells for planting in culture. This method of enzymatic disaggregation is also employed to liberate monolayer cell culture from the surface of culture vessels so that the cells can either be transferred to additional culture vessels (subculturing) or utilized for experimentation. Similar tissue culture procedures were put to use in the production of both polio (3,4) and measles (5) vaccines whereby primary animal cells were dispersed with trypsin, planted in culture bottles, inoculated with virus, and the culture fluids harvested after a suitable incubation period and held as bulk vaccine. The demand for vaccines far exceeded the supply and thus alternative processes were sought to increase the capacity of vaccine-manufacturing processes that employed cultured cells as the substrate tissue for virus production.

This chapter deals with two specific techniques for the mass cultivation of anchorage-dependent cells. Although other systems are available, these two procedures, roller bottle culture and unit process vessel, have been used with great success for cultivation of both primary and serial diploid cell growth. Specifically, growth of these cells in mass culture has enabled the pharmaceutical industry to have an almost unlimited capacity to supply the need for vaccines such as mumps, measles, and rubella. In addition, the roller bottle culture system has also been shown to be an ideal system for mass cultivation of human cell lines derived from urinary tract tumors.

II. ROLLER BOTTLE CULTURES

The use of roller bottles for the purpose of large-scale culture of mammalian cells has been of benefit to many areas of research. It was possible to grow large numbers of cells for studies in immunology, molecular biology, and virus production. The pharmaceutical industry has taken advantage of the positive features of large-scale cell production in roller bottles to produce virus vaccines. Rubella vaccine [both HPV 77 (6) and RA27/3 strains] is routinely produced in roller bottle cultures.

A. Vaccine Production

In 1969, Merck Sharp & Dohme received a license for the manufacture of the HPV-77 ("Meruvax") strain of live attenuated rubella virus vaccine. This vaccine was manufactured in Pekin duck embryo cells obtained from an isolated flock that was free from all known diseases that infect domestic duck flocks. In 1978, the RA27/3 strain of live attenuated rubella virus vaccine replaced the HPV-77 strain and was licensed for use in the United States. The RA27/3 rubella strain, developed by scientists at the Wistar Institute, was propagated in human diploid fibroblasts (WI-38). Although duck cells were used to grow HPV-77 virus and WI-38 cells used to manufacture RA27/3 virus, both vaccines were produced in roller bottle culture of monolayer cells inoculated with the respective virus. The manufacturing procedures for the two vaccines are very similar and are summarized below.

Tissue culture roller bottles (850 cm^2 polystyrene, Corning Glass Works, Corning, NY) are seeded with an appropriate number of cells to produce a monolayer in approximately 4 days. Cells are planted in bottles in approximately 100 ml of tissue culture growth medium (supplemented with 10% irradiated fetal bovine serum) and incubated at 37°C. When a satisfactory cell monolayer has been obtained, the growth medium is discarded. Virus is inoculated onto the cells, fresh medium added, and the bottles placed back on the roller racks and incubated (Fig. 1). Harvest fluids are removed from inoculated bottles and replaced with fresh medium at 2- to 3-day intervals through approximately 10−12 harvest cycles. By using this production procedure it is possible to manufacture millions of doses from a single production lot consisting of several hundred roller bottles (6).

B. Tumor Cell Culture

In addition to normal mammalian cells, human tumors have also been cultured with great success using roller bottles (7). By growing human tumor cells in roller bottle cultures it was possible to obtain enough cells to do detailed studies such as hamster inoculation for tumor induction analysis, chromosome analysis including Q- and G-banding studies, isoenzyme studies, and immunology studies (8). Glass roller bottles (No. 7000, Bellco disposable bottles; Bellco Glass Co., Vineland, NJ) were found to be superior to plastic bottles for the growth of human epithelial cell tumors of the urinary tract. The reason for the superior cell growth in glass bottles is not completely understood but is thought to be due to the difference in surface charge on the two types of vessels. Before the cells were planted, the surface of each bottle was conditioned by adding 100 ml of outgrowth medium (RPM 1640 Grand Island Biological Company, Grand Island, NY, containing 10% heat inactivated fetal bovine serum) and rotating the bottle at 37°C for 30 min. The medium was then removed from the bottle and replaced with 100 ml of fresh outgrowth medium containing 1×10^7 tumor cells. The bottles were placed in a Bellco roller bottle incubator at 37°C and rotated at 0.5 rpm. Initially, the human epithelial tumor cells grew better on plastic flasks than on disposable glass roller bottles, but after several passages of the cells on plastic flasks, tumor cells seeded in glass roller bottles at a high cell density (1×10^7 cells per bottle or greater) would grow to form a heavy, multilayered cell sheet. Good cell attachment and outgrowth were not obtained if the surface of the glass bottles was not exposed to outgrowth medium before the cells were seeded. The cells grew out more slowly on

Figure 1   Roller bottles on a rack which can be programmed to turn at desired rpm rate.   Rack is easily moved from incubator to laboratory for medium change or harvest of infected fluids.

glass and required 10−14 days to form a monolayer. When confluent, the cells continued to grow in multilayers, and after 3−4 weeks
in culture, 1×10^9 cells could be recovered from a single roller
bottle. This procedure has allowed the user to produce larger numbers of tumor cells in early passage and to store them for future
studies (6).

III. MULTIDISK PROPAGATORS

With the large worldwide demand for vaccines, it became necessary
to develop techniques and hardware for mass cultivation of anchorage-dependent cell culture systems. These mass culture vessels
are required to produce a larger number of cells per system volume
than previous systems and be adaptable to vaccine production activities.

Figure 2   Jacketed stainless steel unit process vessel. Outlet
ports are used for removal of infected harvest fluids and replacement of tissue culture medium.

Figure 3   Titanium disks that are inserted into unit process vessel and for anchorage support for primary cell growth during
vaccine-manufacturing process.

In 1969, Molin and Heden (9) published their studies on large-
scale cultivation of human diploid cells on titanium disks in a special apparatus. The vessel was modified by McAleer and coworkers in 1975 and used for the production of mumps, measles, and
rubella vaccines (10). They described vaccine production from
virus grown by a new unit process method employing titanium
disks in stainless steel tanks for support of cell growth (Figs. 2
and 3). They found that the yields of virus and the clinical findings
in tests of the vaccine were comparable to those obtained when
conventional bottle cell culture methods were employed (11).

Since 1975, both measles ("Attenuvax") and mumps ("Mumpsvax") vaccines have been produced in unit process mass cultivation vessels (6). Both vaccines are produced by similar processes utilizing primary chicken embryo cells from a special isolation flock as the cell substrate for virus growth. Embryos are minced, trypsinized, washed, counted, and diluted in growth medium (199 medium containing 2% irradiated fetal bovine serum) for planting. Cell suspensions are placed in a 10-liter stainless steel unit process vessel containing 100 titanium plates and incubated until a monolayer of cell is achieved on the plates (usually within 48 hr) (Fig. 2). Medium is removed from the vessel and cells are inoculated with the appropriate virus and medium added. The cultures are incubated and multiple harvests are taken over the next 3 weeks of incubation. At each harvest, the medium is removed and replaced with fresh medium. Using this procedure, it is possible to manufacture millions of doses of vaccine in a single vessel. It is estimated that since 1975, approximately 100 million doses of mumps and measles vaccine have been manufactured using this unit process procedure.

IV. NUNC MULTITRAY UNIT

In 1979, Skoda and coworkers (12) described the development and utilization of a tissue culture system that was composed of a basic unit tray of polystyrene with overflow vents in two corners (NUNC single-tray unit). By putting two or more trays together, one or more chambers are formed (NUNC multitray unit). The overflow vents have two functions: first, they provide connection between the chambers; and, second, they prevent the flow of medium into the lower chambers when the unit is placed in a horizontal position. Use of a single-tray unit presents no advantage over existing plastic flasks, but the multitray unit is composed of 11 trays, forming 10 chambers.

The multitray unit is connected via peristaltic pump with three reservoir bottles. The first bottle is filled with a saline solution for rinsing the unit, the second bottle filled with growth medium containing cells, and the third bottle serves as a reservoir for effluent collection.

This unit possesses several advantages over flasks or roller bottles including the labor-saving feature and the fact that the units are supplied assembled, sterile, ready for use. Absence of microbial contamination is assured as they are sterilized by gamma irradiation. In addition, the multitray unit is made of disposable plastic and thus eliminates the problem of washing and sterilization before use.

V. ROTARY DISK SYSTEM

Recently, Harold Lee described the design of a small-scale rotary
disk cell system for anchorage-dependent cells (13). Using his procedure trypsin-dispersed chicken embryo fibroblasts are grown on a
round glass disk of 60 mm in diameter in a petri dish. Growth of
cells on both sides of the disk is accomplished by turning the disk
over in the petri dish and immersing it in the desired cell suspension. When cell attachment is complete, usually within 1 hr, the
disk is removed from the petri dish and placed in a cradle in a culture chamber containing culture medium. The cradle of the chamber
holds 22 culture disks, but the operation of multiple units hooked up
in parallel or serial connection is possible. Medium is pumped
through the unit from a fresh medium source to an effluent reservoir. A motor is used to rotate the cradle holding the disks and a
port is provided for addition of a specific gas mixture.

The inventor notes that the flexibility and versatility can be extended to various bioprocesses, such as insoluble enzymes on agarose, polyacrylamide, or other inert substrata. The system can
also be used in $0-9$ gravity, as in a space craft.

VI. SUMMARY

The pressing need for large-scale culture methods has prompted efforts to develop vessels which will accommodate the growth of large
numbers of anchorage-dependent cells. The large-scale "cell factories" have been used to grow viruses for vaccine production,
nucleic acid studies, and various cancer research projects. Two
particular types of culture vessels used for large-scale production
of anchorage-dependent cells were discussed and examples of their
use in vaccine production given. Development of these large-scale
culture systems has enabled the pharmaceutical companies to (1)
meet the ever-increasing demands worldwide for vaccine, (2) employ
a production process that produces a cost-efficient vaccine product
in cell culture, and (3) produce large volumes of bulk vaccine at a
single campaign, thus allowing multiple usage of a single production
unit over the course of a year. The type of large-scale culture
vessel used depends on the purpose of the culture and the type of
cells that are to be grown in the vessel. The unit process vessel,
with its multidisks, provides superior surface area, but because of
the stainless steel housing, the cells cannot be monitored micro-
scopically. Additionally, the unit process vessel can be equipped
with a jacket for precise temperature regulation and thus eliminate
the need for incubators. The roller bottles, on the other hand,
allow microscopic monitoring of the growing cells but are limited in

surface area available for growth, and the volume of harvest fluid obtainable from a bottle is very limited.

It should be noted that handling of roller bottles is labor-intensive and requires many more manipulations and hence more personnel than employing unit process-type vessels for vaccine-manufacturing operations.

The use of mass cultivation techniques in cancer research has also been of immense benefit. The production of large numbers of neoplastic cells has provided researchers with the raw materials to perform molecular biology experiments such as DNA sequencing, which required milligram quantities of cells before sufficient quantities of DNA could be isolated to allow sequencing studies. Additionally, the technique of mass cultivation also provided sufficient quantities of cells for inoculation into either nude mice or immunosuppressed hamsters in an attempt to determine the tumor induction capacity of the cultured cells. Additional characterization studies which require large numbers of cell, enzyme, and isoenzyme profiles can also be performed on cells grown in roller bottle cultures.

There are numerous factors which limit the productivity of mass cultivation systems and the scientist should be aware of their existence. As discussed previously, initial cell attachment is very important and vital to eventual growth and monolayer formation. The selection of the proper growth medium is of paramount importance and factors such as nutrient depletion, waste product accumulation, and loss of pH optimum as a determinant for medium change can all reduce productivity to a greater or lesser degree depending on the selection of the proper growth medium. Selection of the proper growth or maintenance medium is also critical for maintaining the maximum cell viability. Maintaining cell viability is necessary for retaining high productivity and, in the case of vaccine manufacture, virus production in mass culture systems.

REFERENCES

1. Harrison, R. G. Observations on the living developing nerve fiber. *Proc. Soc. Exp. Biol. Med.* 4:140−143 (1907).
2. Carrel, A. On the permanent life of tissues outside the organisms. *J. Exp. Med.* 15:516−528 (1912).
3. Salk, J. E., Drech, U., Younger, J. S., Bennett, B. L., Lewis, L. J., and Bazeley, P. L. Formaldehyde treatment and safety testing of experimental poliomyelitis vaccines. *Am. J. Publ. Health* 44:563−570 (1954).
4. Sabin, A. B. Oral poliovirus vaccine, recent results and recommendations for optimum use. *Roy. Soc. Health* 2:51−58 (1962).

5. Enders, J. F., Katz, S. L., Milovanovic, M. V., and
    Holloway, A. Studies on an attenuated measles-virus vaccine.
    I. Development and preparation of the vaccine: Techniques
    for assay of effects of vaccination. *N. Engl. J. Med.* 263:
    153–154 (1960).
6. Elliott, A. Y. Manufacture and testing of measles, mumps and
    rubella vaccine. *Proceedings 19th Immunization Conference on
    Immunization,* Boston, MA, May 21–24, 1984, pp. 79–83
    (1984).
7. Elliott, A. Y., Bronson, D. L., Stein, N., and Fraley, E. E.
    In vitro cultivation of epithelial cells derived from tumors of
    the human urinary tract. *Cancer Res.* 36:365–369 (1976).
8. Elliott, A. Y., Bronson, D. L., Cervenka, J., Stein, N., and
    Fraley, E. E. Properties of cell lines established from transi-
    tional cell cancers of the human urinary tract. *Cancer Res.*
    37:1279–1289 (1977).
9. Molin, O. and Heden, C. G. Large-scale cultivation of human
    diploid cells on titanium disks in a special apparatus. *Progress
    in Immunological Standardization,* Vol. 4, Karger, New York,
    pp. 106–110 (1969).
10. McAleer, W. J., Buynak, E. B., Weibel, R. E., Villarejos,
    V. M., Scattergood, E. M., Wasmuth, E. H., McLean, A. A.,
    and Hilleman, M. R. Measles, mumps and rubella virus vac-
    cines prepared from virus produced by unit process. *J. Biol.
    Standard.* 3:381–384 (1975).
11. Scattergood, E. M., Schwartz, J. B., Villarejos, V. M.,
    McAleer, W. J., and Hilleman, M. R. Optimization techniques:
    Studies in cell culture, drug development and industrial
    pharmacy. *Drug Dev. Industr. Pharm.* 9:745–766 (1983).
12. Skoda, R., Hormann, A., Spath, O., and Johansson, A.
    Communicating vessel systems for mass cell culture of anchorage-
    dependent cells. *Dev. Biol. Standard.* 42:121–126 (1979).
13. Lee, H. H. A rotary-disk system to cultivate anchorage-
    dependent cells. *Am. Lab.* (May/June):12–19 (1986).

9

Perfusion Systems for Cell Cultivation

BRYAN GRIFFITHS
Public Health Laboratory Service, Centre for Applied Microbiology and Research, Salisbury, England

I. INTRODUCTION

Perfusion refers to continuous flow at a steady rate, through or over a population of cells, of a physiological nutrient solution. It implies the retention of the cells within the culture unit as opposed to continuous-flow culture which washes the cells out with the withdrawn medium (e.g., chemostat). It is not a new or novel technique —it has been used since 1912 (1) to keep small pieces of tissue viable for extended microscopic observation (for review, see 2). The technique was initiated because it was recognized that cells in vivo are continuously supplied with blood, lymph, or other body fluids to keep them in a constant physiological environment. Without perfusion cells in culture go through alternative phases of being fed and starved, thus limiting full expression of growth and metabolic potentials. The current use of perfusion is in response to the challenge of growing cells at high densities (10^7 to 5×10^8/ml). To increase a cell density beyond $2-4 \times 10^6$/ml (or 2×10^5/cm^2) the medium has to be constantly replaced with a fresh supply in order to make up nutritional deficiencies and to remove toxic products. Perfusion thus allows far better control of the culture environment (pH, oxygen and other nutrient levels, etc.) and is a means of significantly increasing the surface area within a culture for cell attachment. As in the body, when exercise increases the demand for oxygen and the production of toxic metabolites (e.g., lactate) and the heart has to pump harder to increase the blood flow rate, so in culture when the cell density and/or growth rate increases the efficiency of perfusion has to be increased. The difficulties of achieving this increased efficiency and developments toward solving this problem are one of the subjects of this chapter.

II. PERFUSION PARAMETERS

A. Objectives

The prime objective of using perfusion techniques is to achieve a controlled and homogeneous environment within the bioreactor. This is achieved by supplying nutrients and removing toxic metabolites, and in effect mimics the in vivo capillary and lymphatic system which supports over 10^9 cells/cm^3. This allows two major culture developments to be carried out: a reduction in the volume of gas headspace and therefore better utilization of the culture volume, and an increase in unit cell density (per ml or per cm^2) by $1-2$ orders of magnitude (summarized in Fig. 1).

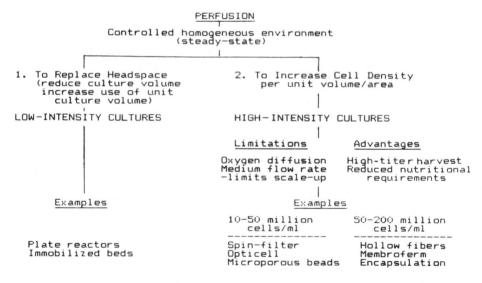

Figure 1   Summary of the benefits of perfusing animal cell cultures.

Low process intensity refers to cell densities of less than
5×10^6/ml, i.e., the standard or classical culture density. The
main use of perfusion in this type of culture is for substrate cells
with a view to increasing the surface area per unit of culture volume. Flasks, roller bottles, etc., usually have a low medium-to-
volume ratio (1:10) to allow sufficient gas exchange to the headspace to delay oxygen exhaustion and low pH inhibition. This obviously means that 90% of the volume is not being directly utilized
for cell growth. Perfusion is used to replace the necessity for this
large gas volume and allows a greatly increased substrate surface
area for cell attachment per unit volume. The medium perfusion
rate has to be sufficient to keep the cells optimally supplied with
oxygen and other nutrients, and to prevent localized acidity, but
not so fast as to shear off cells, particularly mitotic cells that are
always more loosely attached then interphase cells. This can be
easily achieved until significant unit scale-up is attempted or until
the process intensity is increased. Culture units that have been
developed for unit growth of substrate cells at conventional (i.e.,
equivalent to stationary flasks) are based on two concepts — parallel arrangement of flat disks, plates, or membranes, and the use
of a solid matrix material (often glass spheres) packed into a fixed
bed (Section IV). High process intensity refers to the technique
of significantly increasing the cell density per unit volume to values
in the range of $5-20 \times 10^7$ cells/ml. This has the advantage of
producing a higher titer product (which often facilitates downstream
processing), a smaller culture system, and a saving in serum and
other expensive medium additives. If it can be coupled to volumetric
scale-up, then this goes a long way toward closing the productivity
gap with prokaryotic systems and producing an economically satisfactory yield. To support cells at high density needs extremely
efficient perfusion techniques to prevent the development of nonhomogeneity. This means the use of highly sophisticated procedures and apparatus and is thus usually confined to a relatively
small scale (e.g., hollow fiber systems). The alternative approach
is to limit cell densities to about 5×10^7/ml by using more simplistic
techniques which are often capable of extensive volumetric scale-up
(e.g., spin filters) (Section V).

B. Influence of Cell Type

Cells are grown in one of two modes — free suspension or attached
to a substrate — and completely different culture criteria have to be
applied for each. Substrate cell scale-up demands a huge increase
in surface area for cell attachment per unit volume and is limited by

the configuration of the matrix so that adequate, but nondamaging, medium flow patterns can be established. Although it is considered far more difficult to scale up (volumetrically) substrate than suspension cells, they do have the advantage of being immobilized; thus unit density scale-up is theoretically easier. For the growth of suspension cells a mixing system usually based on stirring, but also airlift, is used in order to keep the cells homogeneously suspended with an even distribution of nutrients, gases, etc., throughout the bulk fluid. Volumetric scale-up has been very easy to achieve — there are successful industrial processes at the 8000-liter (stirred) and 2000-liter (airlift) scale. Unit density scale-up is more difficult because if perfusion is used the freely suspended cells have to be separated from the exit medium flow path. This has been achieved with filtration techniques (which support cell densities up to 5×10^7/ml) or by using entrapment and immurement techniques (which permit densities over 10^8 cells/ml) to immobilize the cells (e.g., hollow fibers, membroferm) (see Section V).

C. Culture Kinetics

Perfusion systems can be open, i.e., a flow of fresh medium into the culture balanced by an outward flow to the harvest vessel, or closed recirculating systems (Fig. 2). The culture kinetics are biphasic with a growing phase followed by a steady state, where the cell density remains more or less constant (growth is balanced by cell death). Product expression and nutrient levels remain constant (Fig. 2a). The actual value of these parameters is modulated by the perfusion rate. If perfusion is open, then the medium is not fully utilized. In a recirculating system the medium can be replenished with oxygen and the pH corrected, thus obtaining a far more efficient use of the nutrients, e.g., 1 ml Eagles MEM yields $3-5 \times 10^5$ cells in open systems and $1-2 \times 10^6$ in a recirculating system (3). Media can only support a finite number of cells, so that in a closed system fresh media has to be added and the used medium removed at intervals; thus the kinetic profile shows decreasing nutrient concentrations (Fig. 2b). The usual working practice is to harvest a proportion of the culture at regular intervals (e.g., 50% of the culture every 2 days). This allows repeated product batches to be obtained and allows significant growth of the cells, which is especially important for products that are growth-dependent.

The main difference between this system and the previously used, or alternative, practice of changing the medium (either entirely or proportionally) at intervals is that the environmental conditions and nutrient concentrations remain more or less constant. A media change routine means alternative feeding and fasting and reduces the productivity of the cells.

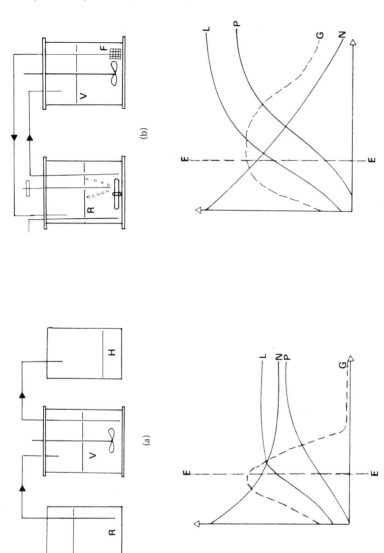

Figure 2 Culture kinetics using open (a) and closed (recirculating) (b) perfusion techniques. L, cells; G, growth rate; N, nutrients; P, product; V, culture vessel; R, reservoir; H, harvested medium; F, filter; E, initiation of perfusion.

D. Oxygen

The first factor to become growth limiting during scale-up is usually
oxygen. This occurs despite a relatively low utilization rate by
cells ($2-6$ $\mu g/10^6$ cells/hr) (4) and is due to the low solubility of
oxygen in media (7.6 mg/liter) and a low rate of diffusion ($2-4 \times$
10^{-5} cm^2/sec) (5,6). It has been calculated that the oxygen supply has to be within $10-30$ cell diameters (e.g., $100-300$ μm) (7,8),
otherwise the concentration falls below a critical level of 1 mg/liter
accompanied by drastic changes in cell metabolism and behavior (9).
Thus in a packed bed of cells, or in a substrate matrix, the diffusion path of oxygen must not exceed 300 μm, i.e., the distance
between oxygen supply sources must not exceed 600 μm. This is
obviously a generalization since in fact oxygen utilization rates vary
between different cell lines/types ($1.3-10.9$ $\mu g/10^6$ cells/hr) (4),
and also between growing and stationary cells.

In low process intensity systems for suspension cells a low aspect (height/diameter) ratio of $1-1.5:1$ may allow surface diffusion
to supply adequate oxygen. Density scale-up has been achieved by
supplying oxygen through sparging, either into the culture or
into a cell-free compartment (10,11), by bubble-free membrane
aeration (12,13), by external loops through an oxygenator, by
hollow fibers (10), or by media perfusion (15). For substrate cells,
once the large gas headspace is lost, then oxygen supply usually
depends on that provided in the perfused medium (hollow fibers are
also used in some instances).

E. Nutrients and Toxic Metabolites

Oxygen is considered the most critical nutrient in scale-up but other
nutrients (e.g., the highly utilized ones such as glucose and glutamine) may also become limiting. Although they are added at high
concentration, this concentration cannot be increased without causing
toxicity in many cases (usually due to metabolic products such as
ammonia and lactate). Thus the perfusion rate has to take account
of removal of unwanted products as well as supplying essential nutrients. This is the value of a recirculating system in that the
waste products can be neutralized in the reservoir conditioning vessel, but more critical monitoring of the growth-promoting ability of
the medium is needed.

F. Culture Configuration

It will be obvious from the previous sections that all the factors
discussed will greatly influence the geometric configuration of the
bioreactor. Dense matrices provide a large surface area but a

Table 1   Yields of Vero Cells in Fixed Beds of Different
Diameter Glass Spheres

Sphere diameter (mm)	Cell yield		
	$cm^2 \times 10^5$	$ml \times 10^6$	$kg \cdot sphere \times 10^8$
2	0.36	2.7	6.7
3	0.78	1.6	4.0
4	1.25	2.0	5.0
5	2.50	3.0	8.0
6	2.20	2.4	6.0
7	2.40	2.0	5.0
8	2.30	1.7	4.3

compromise has to be reached in order for the channels through the
matrix to be large enough to allow a fast nonshearing medium flow
rate through the bioreactor.   This is definitely a limiting factor to
scale-up as demonstrated by the hollow fiber reactors and means
sacrificing some of the process intensity potential of the system.
An example of the effect of reaching a compromise between surface
area and adequate channel size is given in Table 1.   Immobilized
beds of glass spheres have traditionally used 3-mm-diameter spheres
(16,17).   However, because nonhomogeneity patterns were detected
in scale-up (see Section IV.B) a comparison of sphere diameters was
carried out with the result that despite a significant reduction in
surface, 5-mm spheres yielded a greater cell density than 3-mm
spheres (Table 1).   The most feasible explanation is that the greater
channel dimensions allowed better optimization of the conditions in
the bioreactor with the result that higher yields were obtained,
which become more apparent with scale-up.

G. Cell Physiology at High Cell Densities

Little enough is known about cell physiology at conventional cell
densities and at high densities there will be other factors to consider.   On the plus side is the fact that with increasing cell density
more medium components become nonessential.   There are hundreds

of metabolites produced by the cell, or supplied in the serum, which have to be above a critical threshold concentration before the cell can retain them in the intracellular pool against a concentration gradient (10). Obviously, the higher the cell density the higher the extracellular level of these metabolites. This phenomenon has a great cost-saving effect when applied to serum components and usually allows serum-free media to be used above densities of about 5×10^7 cells/ml. On the negative side end-product inhibition and increased accumulation of waste products may limit cell performance. This may be responsible for the often seen phenomenon of diminishing returns, i.e., the cell yield/product expression level falls off per unit of medium as the cell density increases. Another factor which may influence cells at high density is the increased contact of cells and the effect this has on surface receptors and the cell regulatory processes linked to them (19).

The message in this section is that a lot more information is needed on cell physiology at high densities before the process can be optimized and that what data are available from low-density cultures may not be totally relevant.

III. HISTORICAL PERSPECTIVES

The use of perfusion to keep small fragments of tissue viable in microscope chambers developed in complexity between 1912 and 1970 (2), culminating in Rose's dual-rotary circumfusion system (2). The pioneer system for the current biotechnological requirement of supporting high cell densities was the cytogenerator (21). This was a U-shaped tube with side arms of sintered glass surrounded by the medium reservoir (Fig. 3a). Mixing was by a combination of perfusion and a tidal rocking action brought about by pulsing air alternately into the top of the two side arms. The system supported over 2×10^7 cells/ml, an unprecedented density for 1957, and was capable of running continuously for 600–1000 hr before filter blockage forced a termination. The volume was small (300 ml plus a reservoir of 300 ml) and the U-shaped configuration did not allow easy scale-up; otherwise more would have been made of this advanced technique. Gori (22) developed the dialysis fermenter as a simpler and more easily scaled up system, and because dialysis membranes did not encourage cell growth on their cell surface, unlike other materials such as sintered glass. Gori used the system as a means of maintaining maximum growth rates over a long period of time (in a chemostat mode) rather than for achieving high cell density (Fig. 3b). For those interested in evaluating this concept there is a modern version available (23). This membrane fermenter has a Cuprophan dialysis membrane forming an inner chamber within a

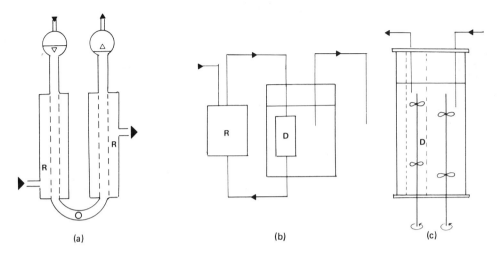

Figure 3 Perfusion culture bioreactors. (a) Graff cytogenerator (21). (b) Dialysis fermenter (22). (c) Membrane fermenter (23). R, reservoir; D, dialysis membrane, ◄, medium perfusion direction.

reactor tank (Fig. 3c). The tank has two independent stirrer shafts, one inside the membrane and the other in the main tank volume. However, for perfusion of suspension cells the most popular techniques are currently based on static systems using membranes (24,25), ultrafiltration fibers (26–30), gel encapsulation (31–33), spin filters (34,35), or various entrapment techniques (36–39) (see Section V).

The cytogenerator also encouraged the development of perfusion systems for substrate or anchorage-dependent cells. The first system was a glass helix perfusion chamber (40) of 40-ml capacity. Perfusion rates of 2–400 vol/day were used and this technique has been developed in many modern laboratories using glass tubing and spheres (16,17,41) at scales up to at least 100 liters (42). It is not only a currently used technology but one that is being rapidly developed (see Section IV.B). However, in the 1960s large-scale growth of substrate cells could only be achieved by using multiple-batch roller bottle culture—a system still in use with many pharmaceutical companies because of its simplicity and reliability. In order to increase the productivity of roller bottles, Kruse et al. (43) developed a swivel cap which allowed medium feed and withdrawal lines to be inserted into the roller bottle and perfusion to be carried out (Fig. 4a). This was an extension of their successful

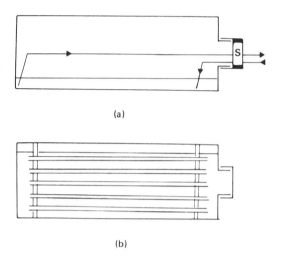

(a)

(b)

Figure 4   (a) Roller bottle with swivel cap (S) for continuous perfusion of medium (43).   (b) Roller bottle with glass tubing, e.g.,
Corbeil (46,47) and Gyrogen (48) systems.

perfusion experiments in flasks which allowed up to 17 cell layers
(45×10^6 cells/cm^2) to be grown using perfusion rates in the order
of 10 vol/day (44).   This development is commercially available (45)
and has been adapted by others to increase the surface area.   Two
such modifications are the Corbeil (46,47) and Gyrogen (48) systems,
which pack the volume of the cylindrical "roller bottle" with glass
tubing (Fig. 4b).   The Corbeil operates in a roller machine which
alternately rotates 360° in each direction, thus eliminating the need
for swivel caps.   The Gyrogen system has been developed into a sophisticated, process-controlled apparatus with 34-m^2 surface area
(49).   These systems have not been used to any great extent.   One
reason is that the only substrate cells needed in great quantities in
the 1960s and 1970s were human diploid cells.   This cell line was
licensed for the manufacture of biologicals because it was considered
"normal" and safe (i.e., nontumorigenic) and a characteristic it had
to retain was monolayer growth.   Multilayering was considered an
indication of abnormality (i.e., malignancy) and was not allowed,
even when this was brought about by perfusion techniques.   Now
that more liberal regulations are coming into force (50) the scale of
production needed does not encourage the use of any multiple
process such as roller bottles.

The low product expression levels and growth rates of cells means that for any industrial process to be economically viable it has to be capable of substantial scaling-up. This can be achieved volumetrically with suspension cells, as processes are in use up to 8000 liters (51), or by increasing the process intensity (i.e., cells and product per unit volume). The aim is to increase the process intensity in volumetric scale-up systems, and this is the challenge which can only be solved with the sophisticated use of perfusion techniques, as described in Section V.

IV. LOW PROCESS INTENSITY SYSTEMS

Low process intensity refers to cell concentrations under 5×10^6/ml or 2×10^5/cm^2 and the use of perfusion, as previously discussed, is primarily to increase the surface area for substrate cells within a unit volume. Perhaps the first large-scale system to achieve this was the IL-410 reactor developed by Jensen (52). A gas-permeable membrane (FEP-Teflon) in the form of a flat tube was wound in a spiral, with suitable spacers. Tubing lengths up to 10 m were used giving a surface area of 25,000 cm^2. The cells attached to the internal surfaces of the membrane which had a high permeability to oxygen and carbon dioxide but would not allow passage of other medium constituents. The unit was kept in a gas-controlled incubator which allowed an equilibrium across the membrane to be maintained. Media is pumped through the lumen of the membrane tube. Multilayering up to seven layers of cells was observed. More conveniently used systems are plate reactors, fixed bed, and microcarrier cultures.

A. Plate Reactors (Fig. 5)

These systems were originally based on a roller culture bottle which was filled with a series of parallel vertical plates a few millimeters apart. The unit was half filled with medium and rotated. The system developed (for reviews, see 49, 53) from a small roller bottle to a large unit based on titanium (54) (e.g., New Brunswick multidisk tissue cell propagator; Biotech cell cultivator) or glass plates (e.g., Connaught multisurface cell propagator). However, inherent problems of getting an even cell distribution on the plates, or confluent cell sheets sliding off, limited the use of this type of culture unit. The development of horizontally packed plates overcame this problem and systems were scaled up to over 200-liter fermenters (55). Perfusion was used to recirculate the media within the stationary vessel and to oxygenate the medium. However, one set of problems was

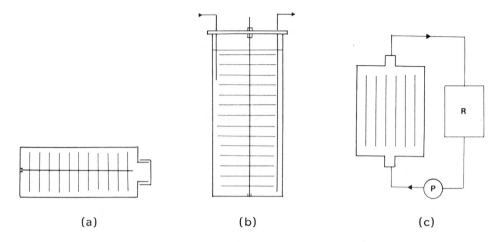

(a)   (b)   (c)

Figure 5   Plate reactors. (a) Roller bottle with vertical plates (49, 53,54). (b) Stacked plate fermenter (55). (c) Vertical plate system (56). R, reservoir; p, pump.

exchanged for another and, in the case of horizontal plates, it was found that to drain the medium from between the plates the unit had to be tilted. This necessitated severe engineering modifications, and costs, to a 200-liter reactor. Also only one side of the plates could be used for cell growth; thus the stacked plate reactor concept seemed inadequate for scale-up, except for a configuration used in plate heat exchangers. However, a vertical plate system using glass plates separated by Teflon gaskets has been successfully used for the production of vaccines (polio, measles) from human diploid cells in vessels up to 10 m^2 (56). The media is pumped vertically up through the parallel plates at a linear velocity of 1.5 – 3 cm/min. The cell density was limited to $2 \times 10^5/cm^2$ for regulatory reasons.

1.   Plate Heat Exchanger

Plate heat exchangers are commonplace in many process plants and are designed to maximize the surface area between two perfusing liquids in order to rapidly heat or cool the process fluid. This seemed an ideal arrangement for cell growth (8,10,57,58) in that a large surface area could have cells and growth medium on one side of the plate, and circulating water at 37°C on the other (Fig. 6). In addition, although the plates are stacked vertically, the individual plates are ribbed or corrugated in order to increase the surface

Plate pack, (a) preheat section

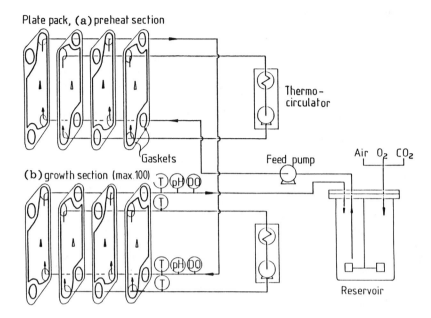

Figure 6   Plate heat exchanger culture system (APV) showing flow path of medium (b) and temperature-controlled water (a) through the individual plates in the culture rig. T, temperature sensor; pH, DO, pH, and dissolved oxygen electrodes.

area still further. This means that most of the surfaces are inclined at a 45° angle rather than being vertical, and thus increases the chances of homogeneous attachment and decreases the risks of sloughing off complete cell sheets.

The apparatus consists of a frame in which the independent stainless steel (grade 316) plates are supported on rails and clamped between a head and a follower. The plates are sealed around the outer edge with a silicon gasket and can be arranged so that either the process fluid or the warming fluid enters the chamber. Oxygen and pH probes can be situated in the medium circulation system by the entrance and exit points to the rig to monitor the culture environment and to control the flow rate. Medium is perfused from a reservoir vessel where pH and oxygen equilibration is carried out. The culture is monitored by the oxygen differential and by glucose utilization. Plates are available from 850 cm^2 to 3.25 m^2 and a rig can hold up to 100 plates so the scale-up potential is enormous.

Also, whatever the size between 1 and 3000 m^2, the linear velocity
of growth medium can be kept constant and nonturbulent.

In practice the plate heat exchanger gives mixed results. Confluent and reasonably homogeneous layers of cells can be obtained
with an average density of $8-12 \times 10^4$ cells/cm^2, i.e., densities compatible with many other substrates and culture units. However,
some cell lines, particularly epithelial, do not give even cell coverage
and have definite concentration gradients. The system is mainly
suited to cells that attach rapidly (1–3 hr), and also to mobile cells
(e.g., fibroblasts) that can make up for initial unevenness in distribution due to settling out in the individual chambers during the inoculation period and before perfusion is started (after 15–18 hr).

In conclusion, the plate heat exchanger is a relatively cheap apparatus, has a high-quality reutilizable substrate, and is simple to
use with good medium flow characteristics. It also allows rapid temperature changes to be made, which is advantageous for some processes (e.g., change from growth to virus production, cell harvesting
by trypsinization). The system is very efficient for harvesting cells,
giving 90–99% recovery. It is at best a low process intensity culture and only suitable for certain cell lines without further modifications to the geometry of the plates.

B. Fixed Beds

The concept of using packed beds for animal cells has been borrowed
from both the chemical industry and wastewater treatment plants.
The aim is to have a large surface area with a low resistance to fluid
flow and homogeneous distribution of fluids throughout the bed. A
vast range of packing materials have been used including steel
springs, Raschig glass rings, glass helices, sponges, and glass
spheres (for review, see 59). The glass sphere has been the most
consistently used material with systems published for BHK cells and
FMD virus up to 100 liters (42) and for human diploid cells (17,60,
61). The advantages of glass spheres are that they provide a consistent and calculatable packing geometry, a high degree of stability
(i.e., the bed is not subject to movement which would act as a mill),
and are available in Pyrex glass a preferred and cheap reutilizable
substrate for most cells. A disadvantage is that spheres have the
lowest possible surface area to volume ratio for solids assayed in a
packed bed.

The culture system is conceptionally simple (Fig. 7) consisting
of a suitable container for the packed bed and medium circulation
from a reservoir, in which full environmental control and monitoring
is carried out, driven by pump or air-lift principles. Traditionally,
3-mm spheres have been used as providing a good compromise between maximizing surface area and keeping an open matrix. However,

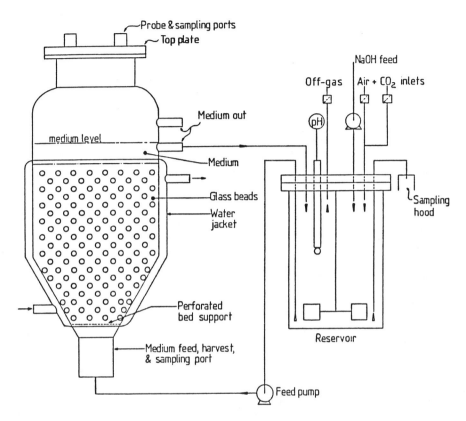

Figure 7 Fixed-bed culture system based on glass spheres with medium circulated through the bed from a reservoir in which the medium environment is monitored and controlled.

recent data (41) have shown better yields to be obtained from 5-mm spheres despite the significant decrease in surface area (Table 1). This work was initiated because it was found that in beds of 20 × 8 cm considerable gradients in cell density were observed (Fig. 8). This was obviously a consequence of uneven inoculation because the bed acted as a depth filter. The problem could be overcome with the use of a fibronectin coating (an expensive option) or a specially designed perforated inoculation tube running vertically through the center of the bed. Having established a high degree of homogeneity ways were sought of increasing the productivity of the system. Reversing the media flow every 24 hr brought about a 100% increase, but the change to 5-mm-diameter spheres brought about the most

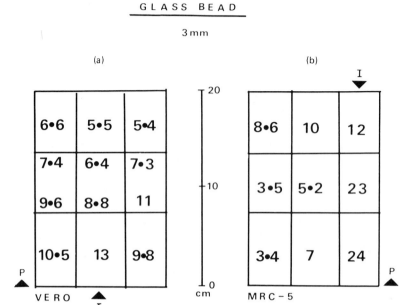

Figure 8   Cell counts in fixed glass sphere beds showing the concentration gradient through the culture after 168-hr culture.
(a) Vero cells, average cell density $8.4 \times 10^4/cm^2$, range $5.4-13.0 \times$
$10^4/cm^2$.   (b) MRC-5 cells, average cell density $6.9 \times 10^4/cm^2$,
range $3.4-23.8 \times 10^4/cm^2$.   P = perfusion direction at $2-5$ linear
cm/min, I = point of inoculation.

dramatic improvement.   The only interpretation for this is that the
increase in channel diameter so created is critical; packed 5-mm
spheres give a channel size of 1-mm diameter and it can be inferred that this size is required to give optimum flow characteristics
through the bed.   In fact, it allows faster flow rates to be maintained without washing out cells and therefore allows better supply
of nutrients and removal of waste products, as well as keeping the
channels open.   Scale-up has given linear results up to 50 liters,
but according to calculations (59), there is no reason why this system should not be equally effective at 2- to 3000-liter volume.
Another advantage of the larger sphere size is that trypsinization

of the cells is facilitated by the more open structure and less chance of clumped cells blocking the bed (as frequently happens with 3-mm beads). The system is more advantageously used as a continuous rather than a batch culture, as demonstrated by Brown et al. (62). These authors, using a reactor of 3-mm glass spheres with a maximum aspect ratio of 1:2, are able to keep cultures continuously for periods from 4 months to over 1 year with densities up to $10^6/cm^2$ (this includes aggregates in the capillary space). This is with reactors up to 60-liter volume, but 200-liter reactors are assumed to be feasible.

To summarize, glass bead reactors represent a simple design concept, are cheap to manufacture with a high-quality reutilizable substrate (Pyrex glass), and are capable of running for long periods of time with minimal supervision. They answer the requirement for simplicity — there is very little likelihood of process breakdown. Productivity is reasonable, with an expectation of $2-5 \times 10^5$ cells/ cm^2. There is also the capability of altering the medium volume-to-surface area/cell number ratio which has advantages in a batch system for obtaining a more concentrated product.

V. HIGH PROCESS INTENSITY SYSTEMS

High process intensity refers to significant increases in cell densities above the standard $5 \times 10^6/ml$ or $2 \times 10^5/cm^2$. The main developments have been for suspension cells using entrapment or immurement techniques often resulting in yields above $2 \times 10^8/ml$ (26–33,38). Some of these developments are now being modified for attached cells. As discussed in Section IV, the problems are to increase surface area and to have an open matrix for efficient media flow. The systems discussed below use the concepts of either protecting the cell from fast-shearing flow rates or designing the culture with shorter flow paths. Some of these technologies are discussed in greater detail in separate chapters of this book.

A. Microcarrier Techniques

The concepts and use of microcarriers for the growth of substrate cells have been well reviewed in (63–65) and in Chapter 11. In this section only the density scale-up of this system is described, since aspects of the approaches used are broadly applicable. Microcarrier culture is the most successful scale-up technology available for substrate cells and is the only system currently considered as an alternative to roller bottles for industrial manufacturing processes. Scale-up has been achieved to 1200 liters for vaccine manufacture and 4000 liters for interferon. The success of this technique is

based on the facts that (1) there are a wide range of carriers available, one of which will always prove suitable for any cell type and
process requirement; and (2) the microcarrier, with cells attached,
can be considered analogous to a suspension cell and thus can be
grown in stirred fermenters with all the advantages of process control, homogeneity, and scale-up that these vessels have.

Volumetric scale-up has proved possible but these systems are
limited to conventional cell densities (under 5×10^6/ml). One advantage of a microcarrier (200 µm) over a suspension cell (10−20
µm) is that its larger size makes separation from the medium easier
by filtration. This has allowed a successful use of the spin filter
technique to scale up the density of microcarrier cultures to $2-5 \times 10^7$/ml.

1. Spin Filter Technology

A spin filter is a filter device (usually of stainless steel, but may be
porcelain or sintered glass) attached to the stirrer shaft of the
bioreactor (Fig. 9). They are effective against blockage because they
have a large surface area (therefore a low flow rate at any one
point of the filter) and as they rotate the boundary effect at the

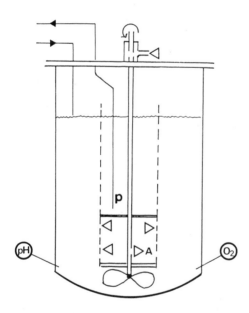

Figure 9   Spin filter reactor. Combined perfusion (p) and aeration (A) spin filter of 65-µm mesh stainless steel used for high-
density microcarrier culture (8,10).

Table 2   Comparison of Oxygenation Methods in a 39-liter Spin
Filter Bioreactor

Aeration system	Stirrer speed (RPM)	Gas	Flow rate (ml/min)	OTR (mg/liter/hr)
Surface	40	Air	300	0.5
Sparging, direct	40	Air	300	4.6
Sparging, spin filter	40	Air	300	3.0
Perfusion (1 vol/hr)	40	Air	300	12.6
Spin filter sparging + per- fusion (1 vol/hr)	80	O_2	1000	92.0

surface of the filter prevents cells and microcarriers from attaching
to the filter surface. The development of spin filters for suspension culture from the original publication by Himmelfarb in 1969 (65)
was recently reviewed (8). For microcarrier culture a filter mesh
of 60–100 μm can be used which allows fast perfusion rates (at
least up to 2 vol/hr) without blockage. This fast perfusion rate
is required to supply oxygen to the culture in a simple yet nondamaging manner. To support cells at 5×10^7/ml a combination
of oxygenation methods must be used which includes surface aeration (aspect ratio 1.5:1), perfusion at 2 vol/hr, and sparging into
the spin filter, which is cell-free (Table 2) (8,10,58). Alternative
methods that are available include the bubble-free membrane systems (12,13).

   The addition of a spin filter to a conventional bioreactor is a
cheap and simplistic means of increasing the process intensity by
10-fold. The system illustrated in Fig. 9 has been proven at the
100-liter scale and there is no apparent reason why it should not
be suitable in much larger reactors. The technique is far more
versatile and useful than the previously used gravitational methods
of separating microcarriers from the medium (67,68), which could
only support slow perfusion rates (1–2 vol/day). An alternative
method with similar potential is the use of hollow fibers removed
from their cartridge and suspended in the bulk culture fluid (69,
70).

2. Microporous Beads

A means of increasing the process intensity of the microcarrier system is to substitute the solid spheres, which have surface growth
only, with porous beads. These beads have a honeycomb matrix
that allows the cells to enter the bead and make use of a greatly
increased surface area for cell attachment. In addition, the cells
are protected within the sphere from mechanical damage (by bead
collision) and other physical agents such as shear caused by stirring, mixing, or perfusion. This gives the potential to not only
scale up the process intensity per bead but to increase the density
of the beads within the culture.

Two systems are currently in use: the Verax CF-IMMO bead
(38,71) and the gelatin bead (Percell) (72). The Verax bead is
made of collagen with a diameter of approximately 500 μm. The
interconnecting pores and channels within the bead are 20–40 μm
diameter and constitute 85% of the bead volume. The cell density
within the microsphere is in the range of $1-4 \times 10^8$ cells/sphere
ml for both suspension and substrate cells. They are used in a
fluidized bed and kept suspended in a high velocity (70 cm/min),
upward flowing culture medium. The bed concentration is almost
4 million microspheres per liter which is 25% by volume solids,
and the estimated total microsphere surface is 300 m^2/liter. By
contrast, solid microcarriers at a concentration of 15 g/liter oc-
cupy about 27.5% of the volume and possess a surface area of 8 m^2;
therefore, the porous microsphere has a 37-fold higher surface area.

In conclusion, the increase in process intensity with microporous
beads means that the volumetric scale-up capacity of microcarrier
can be augmented 30- to 40-fold with the use of a matrix carrier.
This makes the microporous system a very powerful technology with
perhaps the potential to be the dominant culture technology of the
future, especially as it is equally suitable for suspension and attached cells.

B. Fixed (Immobilized) Beds

Immobilized beds of glass spheres have already been described (Section IV.B). The only way to scale up the process intensity of the
system is to take an analogous strategy to that described for microcarriers —and use macroporous beads. The use of large-matrix
spheres (2–6 mm) increases the productivity 4- to 20-fold depending on the diameter of sphere and dimensions of the internal pores
and channels (73). Currently, this methodology has only been investigated using anchorage-dependent cells but there is every possibility that the macroporous spheres will trap suspension cells and

thus be suitable for both cell types. This expectation is based on alternative systems using polyurethane sponge cubes in a fixed bed (74) for the growth of hybridoma cells.

C. Ceramic Matrix Immobilization (Opticell)

The Opticell system uses a ceramic matrix with either a smooth or a porous surface (36,75,76). The smooth surface is used when cells need to be harvested; otherwise the porous surface (50-μm pores) is preferable for both suspension and attached cells as it entraps the former and provides a larger surface area for the latter. The ceramic matrices are cylindrical with 1-mm^2 channels running their length and are available in units of 0.425-, 4.25-, and 12-m^2 sizes (smooth finish), or 1- and 11-m^2 (porous). The ceramic matrix is housed in a plastic cylinder through which the medium is perfused.

The system has a highly sophisticated computer system controlling pH, oxygen, perfusion rates (constant recirculation, continuous replenishment, and continuous removal of exhausted medium). To initiate the culture 25% of the cells are added and allowed to settle (10 min); the culture is then turned through 90° and another aliquot of cells is added. This is repeated until all four sides of the channels have been seeded. After 1−2 hr perfusion is started at rates up to 1 liter/min. Cell densities of 1.2×10^6 cells/cm (2.5×10^6 cells/ml) have been achieved (36).

This is a highly sophisticated system capable of giving at least a 10-fold increase in process intensity over conventional cultures. Scale-up of the matrix cartridges is limited by the length of the flow path which can be used before an unacceptable drop in nutrient concentration occurs (currently 30 cm). However, the cells in the porous surface are well protected from shear, so that high flow rates can be used.

D. Hollow Fiber Reactors

The use of hollow fiber cartridge systems to grow cells has developed considerably since the pioneer work of Knazek (77) in 1972. The attraction of the system is that ultrafiltration capillary fibers are very porous (with precise molecular cutoffs) and, being very thin (250-μm diameter), thousands can be "potted" together in a small unit to give a huge surface area [100 cm^2/cm^3 of extracapillary space (27)]. Cells are kept in the extracapillary space and medium is perfused through the lumen of the fibers. Nutrients can diffuse through the fibers at one end and the toxic metabolites back into the lumen at the other end (Fig. 10). Cells plus product can be harvested from the extracapillary space. The problem is the pressure drop along the reactor (Fig. 10) (78), which prevents scale-up

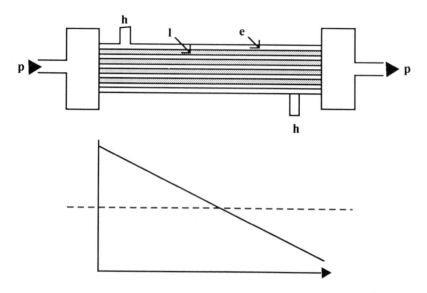

Figure 10   Hollow fiber culture reactor and a diagrammatic representation of the pressure drop/nutrient gradient along the length
of the cartridge. l, lumen of fibers; e, extracapillary space; h,
harvesting port; p, medium perfusion path.

beyond the liter volume size. However, within the system very
high cell densities are achieved [in excess of 2×10^8/ml (27,29)].
Various modifications have been used to overcome this problem.
One is to alternate the fluid cycling processes from within the lumen
to an extracapillary expansion chamber [Acucyst technology (27)].
Other modifications include mixing aeration and medium supply
fibers (29), using filtration grade rather than ultrafiltration grade
membranes (88), and using radial flow techniques which give a
short flow path (Fig. 11). This latter technique has been developed
into designing flat-bed hollow fiber reactors (79–81). The fibers
are used for aeration and as an attachment surface, and three to
six layers are sandwiched between filter membranes, through which
medium is perfused (Fig. 11). This system, which reduced the
need for fast-shearing flow rates, increased the cell yield fourfold
over the cartridge concept owing to the much shorter nutrient flow
path which eliminated concentration gradients. This reactor design
led to the static maintenance reactor (Invitron SMR), which has
cells immobilized in a bulk matrix through which porous tubes perfuse either fresh medium or oxygen (82).

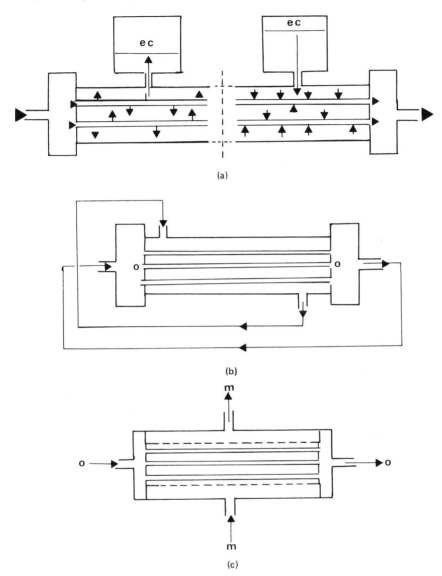

(a)

(b)

(c)

Figure 11   Modifications to increase the efficiency of hollow fiber reactors. (a) Alternate flow paths out of the lumen into an expansion chamber (ec), and reverse flow out of the chamber into the lumen (27). (b) Dual-perfusion system or, alternatively, the use of fibers to supply oxygen (o) with medium perfusion through the extracapillary space (m). (c) Flat-bed radial flow reactor.

Hollow fiber reactors are very successful for suspension cells but the fibers are not ideal for the attachment of adherent cells (especially the cellulose acetate fibers). To solve this problem polypropylene-based fibers can be precoated with poly-D-lysine (29), or serum, fetuin, and poly-D-lysine (27). Cell densities of over 10^7 fibroblast cells/cm^2 have been reported (29). Another approach has been to pack the extracapillary space with microcarriers (83).

Hollow fiber systems have large surface area-to-volume ratios, high cell densities, continuous removal of waste products, separation of the cells from the medium flow thus eliminating shear effects, and the possibility of concentrating the product in a small volume (depending on the molecular weight cutoff of the fibers). Disadvantages of the system are limitations to homogeneity and scale-up due to diffusion problems, nonsteam sterilization, possibility of fiber blockage and membrane leakage, complexity of process control, and monitoring growth.

E. Membroferm Reactor (24,25,34)

The membroferm is a modular reactor consisting of layers of flat membranes separated by a fluorocarbon matrix. Cells can either be grown on the surface of the matrix fibers or as suspension cells in the openings in the fabric. The membranes form reactor chambers (28 × 32 cm × 0.6 mm high) with 500 cm^2 of surface area. The matrix provides between 1400 and 7000 cm^2 of surface area for substrate cells per chamber depending on the material used. It is possible to stack 30−400 layers of chambers giving up to 20 m^2 membrane surface and between 7 and 35 m^2 of matrix surface area in a 4-liter unit. Different types of membrane spacers can be used to form medium, cell, and product chambers, and by changing the combinations different operating modes are possible. The two-chamber system (Fig. 12a) uses microfiltration membranes which retain the cells but allow the medium to perfuse through the chambers. The three-chamber system (Fig. 12b) also uses an ultrafiltration grade membrane which retains the product. It is thus analogous to a flat-bed hollow fiber reactor with short medium flow paths through the cell mass. The membrane unit is supplied from a reservoir in which environmental monitoring and control is carried out.

The unit is capable of supporting cell densities in the region of 10^8/ml (cell chamber). There are many advantages to this system including in situ steam sterilization, accumulation of product within the unit, scale-up to a 35-m^2 surface area, flexibility in operation and control, and a good geometric configuration for medium perfusion (25). The unit is available from a commercial supplier (84).

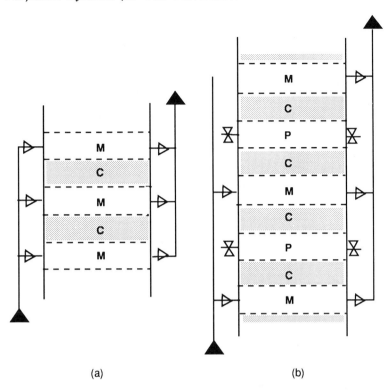

(a)   (b)

Figure 12   The membroferm bioreactor (24,25,34). (a) Two-
chamber system. (b) Three-chamber system to separate product
from the medium. M, medium; C, cell; P, product, compartments.

F. Tubular Biological Film Reactor

The roller bottle is a tubular film reactor since when rolling the
cells are permanently covered with a nutrient liquid film which is
renewed during the submerged phase in the bulk medium. During
the time out of the bulk medium gas diffusion through the thin
liquid film is very efficient. The advantages of this type of culture
are no stress from gas bubbles, a reasonable homogeneity, and
suitability for both attached and suspension cells. As explained
earlier, the roller bottle is considered a simple and reliable multiple-
batch process. Process intensity has been increased by perfusion
techniques (43). The tubular biological film reactor (37) is another

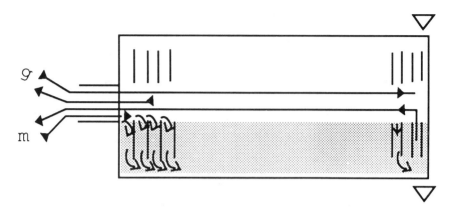

Figure 13   Tubular film reactor (37). A roller bottle with short
vertical plates that allow the medium to perfuse around each element. The elements are alternatively in a gas phase (retaining a
liquid film over the cells and elements which allows rapid gas exchange) and liquid phase as the bottle rotates. g, gas; m, medium.

means of intensifying the productivity of the system (Fig. 13). In
this unit the surface area for attachment is increased 10-fold by the
inclusion of a series of film carrier elements. These have to be far
enough apart to prevent liquid bridging. The vessel is rotated at
normal roller bottle speeds (12–20 rph) and continuous perfusion of
medium and gas is carried out. Preliminary results with a laboratory
scale prototype have demonstrated that cell densities of 10^7 cells/cm^2
can be achieved (37). A theoretical, but achievable, scale-up to
a 1 × 1 m cylinder would give over 1000 m^2 of substrate and thus
10^{12} cells in approximately 300 liters of medium (37). It should also
be suitable for suspension cells, which would be immobilized between
the film carrier elements.

VI. CONCLUSIONS

Perfusion permits culture homogeneity rather than a feed-and-fast regime, and has permitted the development of bioreactors which have no gas headspace (thus allowing better utilization of culture volume) and significant (by 2 orders of magnitude) scale-up of cell density. A wide range of culture systems has been reviewed with the emphasis on attached cells. One system not reviewed is the encapsulation technique (31,32,85) of containing the cells in agarose (86), sodium alginate or polylysine spheres (32,33) of up to 500-μm diameter. These spheres allow the diffusion of nutrients (and with some materials product) through the membrane and are kept in suspension in conventional stirred reactors. This is not a true perfusion system as the spheres are in a bulk medium but the main reason for their omission is that they are mainly suitable for suspension cells. Attempts at providing a collagen, or fibrin, matrix within the spheres for cell attachment have not been that successful to date. However, it is one of three systems (hollow fibers and the membroferm being the others) which, with a suitable membrane choice, allow product compartmentalization in a small volume of the total bulk medium (8). This allows a high product concentration to be harvested and thus simplifies the purification procedure.

The range of systems reviewed gives a wide choice from low ($<5 \times 10^6$ cells/ml) to medium (5×10^7 cells/ml) to high ($>10^8$ cells/ml) process intensity values. The choice of system will be influenced by the annual yield of product required, whether the plant is to produce a single product or to be used sequentially for multiple products, and the product expression kinetics from cells (i.e., whether growth- or non-growth-associated). Another factor to consider, especially for industrial use, is the complexity of the process and the ease of handling. Contamination still remains a dire threat to cell culture systems and the more simplistic the culture design the less likelihood there is of losing large product volumes due to contamination or mechanical breakdown (8,87). The use of perfusion techniques carries this risk because of continuous entry and exit of materials from the reactor, often through vulnerable tubing in pumps, etc. Both the bioreactor and the associated equipment should be critically evaluated with this in mind.

REFERENCES

1. Burrows, M. T. A method of furnishing a continuous supply of new medium to a tissue culture in vitro. *Anat. Rec.* 6: 141–144 (1912).

2. Patterson, M. K., Jr. Perfusion and mass culture systems. *TCA Man.* 1:243–249 (1975).
3. Griffiths, J. B. Can cell culture medium costs be reduced? Strategies and possibilities. *Trends Biotechnol.* 4:268–272 (1986).
4. Spier, R. E. and Griffiths, J. B. An examination of the data and concepts germane to the oxygenation of cultured animal cells. *Dev. Biol. Standard.* 55:81–92 (1983).
5. Katinger, H. A tubular biological film reactor concept for the cultivation and treatment of mammalian cells. In *Animal Cell Biotechnology*, Vol. 3 (R. E. Spier and J. B. Griffiths, eds.), Academic Press, London, pp. 239–250 (1980).
6. Murdin, A. D., Kirkby, N. F., Wilson, R., and Spier, R. E. Immobilized hybridomas: Oxygen diffusion. In *Animal Cell Biotechnology*, Vol. 3 (R. E. Spier and J. B. Griffiths, eds.), Academic Press, London, pp. 55–74 (1988).
7. Spier, R. E. and McCullough, K. The large-scale production of monoclonal antibodies in vitro. (In prep.).
8. Griffiths, J. B. Overview of cell culture systems and their scale-up. In *Animal Cell Biotechnology*, Vol. 3 (R. E. Spier and J. B. Griffiths, eds.), Academic Press, London, pp. 179–220 (1988).
9. Scheirer, W. High-density growth of animal cells within cell retention fermenters equipped with membranes. In *Animal Cell Biotechnology*, Vol. 3 (R. E. Spier and J. B. Griffiths, eds.), Academic Press, London, pp. 263–281 (1988).
10. Griffiths, J. B., Cameron, D. R., and Looby, D. A comparison of unit process systems for anchorage dependent cells. *Dev. Biol. Standard.* 66:331–338 (1987).
11. Whiteside, J. P., Farmer, S., and Spier, R. E. The use of caged aeration for the growth of animal cells on microcarriers. *Dev. Biol. Standard.* 60:283–290 (1985).
12. Lehmann, J., Vorlop, J., and Buntemeter, H. Bubble-free reactors and their development for continuous culture with cell recycle. In *Animal Cell Biotechnology*, Vol. 3 (R. E. Spier and J. B. Griffiths, eds.), Academic Press, London, pp. 221–237 (1988).
13. Kuhlmann, W. Optimization of a membrane oxygenation system for cell culture in stirred tank reactors. *Dev. Biol. Standard.* 66:263–268 (1987).
14. Glacken, M. W., Fleischaker, R. J., and Sinskey, A. J. Mammalian cell culture: Engineering principles and scale-up. *Trends Biotechnol.* 1:102–108 (1983).
15. Looby, D. and Griffiths, J. B. A comparison of oxygenation methods in a 40 L stirred bioreactor. In *Modern Approaches*

to *Animal Cell Technology* (R. E. Spier and J. B. Griffiths, eds.), Butterworths, Buildford, U.K., pp. 449–453 (1987).

16. Whiteside, J. P. and Spier, R. E. Factors affecting the productivity of glass sphere propagators. *Dev. Biol. Standard.* 60:305–311 (1985).

17. Burbidge, C. The mass culture of human diploid fibroblasts in packed beds of glass beads. *Dev. Biol. Standard.* 46:169–172 (1980).

18. Griffiths, J. B. The effect of cell population density on nutrient uptake and cell metabolism: A comparative study of human diploid and heteroploid cell lines. *J. Cell Sci.* 10: 515–524 (1972).

19. Spier, R. E. Environmental factors: Medium and growth factors. In *Animal Cell Biotechnology*, Vol. 3 (R. E. Spier and J. B. Griffiths, eds), Academic Press, London, pp. 29–53 (1988).

20. Rose, C. G. The circumfusion system for multipurpose culture chambers. *J. Cell Biol.* 32:89–112 (1967).

21. Graff, S. and McCarty, K. S. Sustained cell culture. *Exp. Cell Res.* 13:348–357 (1957).

22. Gori, G. B. Chemostatic concentrated cultures of heteroploid mammalian cell suspensions in dialyzing fermenters. *Appl. Micobiol.* 13:93–97 (1965).

23. Bioengineering AG, CH 8636 Wald, Switzerland, membrane laboratory fermenter.

24. Klement, G., Scheirer, W., and Katinger H. W. D. Construction of a large scale membrane reactor system with different compartments for cells, medium and product. *Dev. Biol. Standard.* 66:221–226 (1987).

25. Scheirer, W. High-density growth of animal cells with cell retention fermenters equipped with membranes. In *Animal Cell Biotechnology*, Vol. 3 (R. E. Spier and J. B. Griffiths, eds.), Academic Press, London, pp.263–281 (1988).

26. Schonherr, O. T., van Geler, P. T. J. A., van Hees, P. J., van Os, A. M. J. M., and Roelofs, H. W. M. A hollow fibre dailysis system for in vitro production of monoclonal antibodies replacing in vivo production in mice. *Dev. Biol. Standard.* 66: 211–220 (1987).

27. Hopkinson, J. Hollow fiber cell culture systems for economical cell-product manufacturing. *Biotechnology* 3:225–230(1985).

28. Altshuler, G. L., Dziewulski, D. M., Sowek, J. A., and Belfort, G. Continuous hybridoma growth and monoclonal antibody production in hollow fiber reactors-seperators. *Biotechnol. Bioeng.* 28:646–658 (1986).

29. Tharakan, J. P. and Chau, P. C. A radial flow hollow fiber reactor for the large scale culture of mammalian cells. *Biotechnol. Bioeng.* 28:329−342 (1986).

30. Tyo, M. A., Bulbulian, B. J., Menken, B. Z., and Murphy, T. J. Large-scale mammalian cell culture utilizing Acuyst technology. In *Animal Cell Biotechnology*, Vol. 3 (R. E. Spier and J. B. Griffiths, eds.), Academic Press, London, pp. 357−393 (1988).

31. Nillson, K. and Mosbach, K. Preparation of immobilised animal cells. *FEBS Lett.* 18(1):145 (1980).

32. Duff, R. G. Microencapsulation technology: A novel method for monoclonal antibody production. *Trends Biotechnol.* 3: 167−170 (1985).

33. Rupp, R. G. Use of cellular encapsulation in large-scale production of monoclonal antibodies. In *Large-Scale Mammalian Cell Culture* (J. Feder and W. R. Tolbert, eds.), Academic Press, Orlando, pp. 19−38 (1985).

34. Varecka, R. and Scheirer, W. Use of a rotary wire cage for retention of animal cells in a perfusion fermenter. *Dev. Biol. Standard.* 66:269−272 (1987).

35. van Wezel, A. L., van der Velden-de-Groot, C. A. M., de Haan, H. H., van den Heuval, N., and Schasfoort, R. Large-scale animal cell cultivation for production of cellular biolicals. *Dev. Biol. Standard.* 60:229−236 (1985).

36. Berg, G. J. and Bodeker, B. G. D. Employing a ceramic matrix for the immobilization of mammalian cells in culture. In *Animal Cell Biotechnology*, Vol. 3 (R. E. Spier and J. B. Griffiths, eds.), Academic Press, London, pp. 321−335 (1988).

37. Runstadler, P. W. and Cernek, S. R. Large-scale fluidized-bed, immobilized cultivation of animal cells at high densities. In *Animal Cell Biotechnology*, Vol. 3 (R. E. Spier and J. B. Griffiths, eds.), Academic Press, London, pp. 305−320 (1988).

38. Hayman, E. G., Ray, N. G., and Runstadler, P. W., Jr. Production of biomolecules by cells cultured in tri-dimensional collagen microspheres. In *Bioreactors and Biotransformations* (G. W. Moody and P. B. Baker, eds.), Elsevier, London, pp. 99−110 (1987).

39. Larsson, B. and Litwin, J. The growth of polio virus in human diploid fibroblasts grown with cellulose microcarriers in suspension. *Dev. Biol. Standard.* 66:385−390 (1987).

40. McCoy, T. A., Whittle, W., and Conway, E. A glass helix perfusion chamber for massive growth of cells in vitro. *Proc. Soc. Exp. Biol. Med.* 109:235−237 (1962).

41. Looby, D. and Griffiths, J. B. Optimization of glass-sphere immobilized bed cultures. In *Modern Approaches in Animal Cell Technology* (R. E. Spier and J. B. Griffiths, eds.), Butterworths, Guildford, pp. 342–352 (1987).

42. Whiteside, J. P. and Spier, R. E. The scale-up from 0.1 to 100 litre of a unit process system based on 3 mm diameter glass spheres for the production of four strains of FMDV from BHK monolayer cells. *Biotechnol. Bioeng.* 23:551–565 (1981).

43. Kruse, P. F., Keen, L. N., and Whittle, W. L. Some distinctive characteristics of high density perfusion cultures of diverse cell types. *In Vitro* 6:75–88 (1970).

44. Kruse, P. F., Whittle, W. L., and Miedema, E. Mitotic and non-mitotic multiple-layered perfusion cultures. *J. Cell Biol.* 4:113–121 (1969).

45. Perfusion swivel cap roller bottles, New Brunswick Scientific, Edison, NJ 08818.

46. Corbeil, M., Trudel, M., and Payment, P. Production of cells and tissues in a new multiple-tube tissue culture propagator. *J. Clin. Microbiol.* 10:91–95 (1979).

47. Dugre, M., Corbeil, M., and Boulay, M. Improved production of Marek's disease vaccine with the multi-tube technology. In *Modern Approaches to Animal Cell Technology* (R. E. Spier and J. B. Griffiths, eds.), Butterworths, Guildford, pp. 449–453 (1987).

48. Girard, H. C., Sutcu, M., Erdem, H., and Gurhan, I. Monolayer cultures of animal cells with the Gyrogen equipped with tubes. *Biotechnol. Bioeng.* 22:477–493 (1980).

49. Griffiths, J. B. Scaling-up of animal cell cultures. In *Animal Cell Culture: A Practical Approach* (R. I. Freshney, ed.), IRL Press, Oxford, pp. 33–70 (1986).

50. Petricianni, J. C. Changing attitudes and actions governing the use of continuous cell lines for the production of biologicals. In *Animal Cell Biotechnology*, Vol. 3 (R. E. Spier and J. B. Griffiths, eds.), Academic Press, London, pp. 13–25 (1988).

51. Pullen, K. F., Johnson, M. D., Phillips, A. W., Ball, G. D., and Finter, N. B. Very large scale suspension cultures of mammalian cells. *Dev. Biol. Standard.* 60:175–177 (1985).

52. Jensen, M. D. Production of anchorage-dependent cells: Problems and their possible solutions. *Biotechnol. Bioeng.* 23:2703–2716 (1981).

53. Litwin, J. Mass cultivation of mammalian cells. *Process Biochem.* (July):15–17 (1971).

54. Molin, O. and Heden, C. G. Large-scale cultivation of human diploid cells on titanium discs in a special apparatus. *Prog. Immunobiol. Stand.* 3:106–110 (1969).

55. Schleicher, J. B. and Weiss, R. E. Application of a multi-surface tissue culture propagator for the production of cell monolayers, virus and biochemicals. *Biotechnol. Bioeng.* 10: 617–624 (1968).

56. Mann, G. F. Development of a perfusion culture system for production of biologicals using contact dependent cells. *Dev. Biol. Standard.* 37:149–152 (1977).

57. Burbidge, C. and Dacey, I. K. The use of plate heat exchangers in growing human fibroblasts. *Dev. Biol. Standard.* 55:255–259 (1984).

58. Griffiths, J. B., Cameron, D. R., and Looby, D. Bulk production of anchorage-dependent cells: Comparative studies. In *Plant and Animal Cells: Process Possibilities* (C. Webb and F. Mavituna, eds.), Ellis Horwood, Chichester, U.K., pp. 149–161 (1987).

59. Spier, R. E. Monolayer growth systems: Heterogeneous unit processes. In *Animal Cell Biotechnology*, Vol. 1 (R. E. Spier and J. B. Griffiths, eds.), Academic Press, London, pp. 243–263 (1985).

60. Robinson, J. H., Butlin, P. M., and Imrie, R. C. Growth characteristics of human diploid fibroblasts in packed beds of glass beads. *Dev. Biol. Standard.* 49:173–181 (1980).

61. Griffiths, J. B., Thornton, B., and McEntee, I. The development and use of microcarrier and glass sphere culture techniques for the production of Herpes simplex virus. *Dev. Biol. Standard.* 50:103–110 (1982).

62. Brown, P. C., Figueroa, C., Costello, M. A. C., Oakley, R., and Maciukas, S. M. Protein production from mammalian cells grown on glass beads. In *Animal Cell Biotechnology*, Vol. 3 (R. E. Spier and J. B. Griffiths, eds.), Academic Press, London, pp. 251–262 (1987).

63. Reuveny, S. Microcarriers for culturing mammalian cells and their applications. *Adv. Biotechnol. Proc.* 2:2–32 (1983).

64. Butler, M. A comparative review of microcarriers available for the growth of anchor-dependent animal cells. In *Animal Cell Biotechnology*, Vol. 3 (R. E. Spier and J. B. Griffiths, eds.), Academic Press, London, pp. 283–303 (1988).

65. van Wezel, A. L. Microcarrier technology: Present status and prospects. *Dev. Biol. Standard.* 55:3–9 (1984).

66. Himmelfarb, P., Thayer, P. S., and Martin, H. E. Spin filter culture: The propagation of mammalian cells in suspension. *Science* 164:555–557 (1969).

67. Butler, M., Imamura, T., Thomas, J., and Thilly, W. G. High yields from microcarrier cultures by medium diffusion. *J. Cell Sci.* 61:351–363 (1983).

68. Sato, S., Kawamura, K., and Fujiyoshi, N. Animal cell cultivation for production of biological substances with a novel perfusion culture apparatus. *J. Tiss. Cult. Methods* 8: 167–171 (1983).

69. Kearns, M., Comer, M. J., Steegmans, U., and Jungfer, H. Verfahren und vorrichtung zum kultivieren von zellen. Eur. Patent Office EPO 224-800-A2.

70. Emery, A. N., Lavery, M., Williams, B., and Handa, A. Large-scale hybridoma culture. In *Plant and Animal Cells: Process Possibilities* (C. Webb and F. Mavituna, eds.), Ellis Horwood, Chichester, U.K., pp. 137–146 (1987).

71. Karkare, S. B., Phillips, P. G., Burke, D. H., and Dean, R. C. Continuous production of monoclonal antibodies by chemostatic and immobilised hybridoma culture. In *Large-Scale Mammalian Cell Culture* (J. Feder and W. R. Tolbert, eds.), Academic Press, Orlando, pp. 127–148 (1985).

72. Nilsson, K., Buzsaky, F., and Mosbachh, K. Growth of anchorage-dependent cells on macroporous microcarriers. *Biotechnology* 4:989–990 (1986).

73. Looby, D. and Griffiths, J. B. Fixed bed porous glass sphere (porosphere) bioreactors for animal cells. *Cytotechnology* 1, 439–446 (1988).

74. Murdin, A. D., Thorpe, J. S., Kirkby, N., Groves, D. J., and Spier, R. E. Immobilization and growth of hybridomas in packed beds. In *Bioreactors and Biotransformations* (G. W. Moody and P. B. Baker, eds.), Elsevier, London, pp. 99–110 (1987).

75. Pugh, G. G. and Berg, G. J. Two ceramic matrices for the long term growth of adherent or suspension cells. In *Bioreactors and Biotransformations* (G. W. Moody and P. B. Baker, eds.), Elsevier, London, pp. 121–131 (1983).

76. Berg, G. J. An integrated system for large scale cell culture. *Dev. Biol. Standard.* 60:297–307.

77. Knazek, R. A., Guillino, P. M., Kohler, P. O., and Dedrick, R. L. Cell culture on artificial capilliaries: An approach to tissue growth in vitro. *Science* 178:65–67 (1972).

78. Tharakan, J. P. and Chau, P. C. Operation and pressure distribution of immobilized cell hollow fiber bioreactors. *Biotechnol. Bioeng.* 28:1064–1071 (1986).

79. Ku, K., Kuo, M. J., Delenti, J., Wildi, B. S., and Feder, J. Development of a hollow fibre system for large-scale culture of mammalian cells. *Biotechnol. Bioeng.* 23:79–95 (1981).

80. Tolbert, W. R., White, P. J., and Feder, J. Perfusion culture systems for production of mammalian cell biomolecules.

In *Large-Scale Mammalian Cell Culture* (J. Feder and W. R. Tolbert, eds.), Academic Press, Orlando, pp. 1–18 (1985).

81. Feder, J. and Tolbert, W. R.   The large-scale cultivation of mammalian cells. *Sci. Am.* 248:24–31 (1983).

82. Tolbert, W. R., Srigley, W. R., and Prior, C. P.   Perfusion culture systems for large-scale pharmaceutical production. In *Animal Cell Biotechnology*, Vol. 3 (R. E. Spier and J. B. Griffiths, eds.), Academic Press, London, pp. 373–393 (1988).

83. Strand, M., Quarles, J. M., and McConell, S.   A modified matrix perfusion-microcarrier bead cell culture system. *Biotechnol. Bioeng.* 26:503–507 (1984).

84. MBR Bio Reactor Ag, CH-8620 Wetzikon, Switzerland, Memboferm cell culture unit.

85. Lim, F. and Sun, A. M.   Microencapsulated islets as bioartificial endocrine pancreas. *Science* 210:908–910 (1980).

86. Nilsson, K. and Mosbach, K.   Immobilized animal cells. *Dev. Biol. Standard.* 66:183–193 (1987).

87. Arathoon, W. R. and Birch, J.   Large-scale culture in biotechnology. *Science* 232:1390–1395 (1986).

88. van Brunt, J.   Immobilized animal cells: The gentle way to productivity. *Biotechnology* 4:505–510 (1986).

10

Suspension Culture of Mammalian Cells

JOHN R. BIRCH
Celltech Ltd., Slough, England

ROBERT ARATHOON
Genetech, Inc., South San Francisco, California

I. INTRODUCTION

The development of suspension culture methods took place in the
1950s. Owens et al. (1954) grew MBIII mouse lymphoblasts in agitated suspension while Earle et al. (1954) demonstrated growth of
mouse L929 fibroblast cells in Erlenmeyer flasks on a rotary shaker.
By the late 1950s methods had been developed for growing cells in
magnetically stirred "spinner" vessels (Cherry and Hull, 1956;
McLimans et al., 1957), culminating in the use of modified microbial
fermenter vessels (McLimans and Giardinello, 1957; Ziegler et al.,
1958).

   Large-scale suspension culture based on microbial fermentation
technology has clear advantages for the manufacture of mammalian
cell products. The processes are relatively simple to operate and
straightforward to scale up. Homogeneous conditions can be provided in the reactor which allow for precise monitoring and control
of temperature, dissolved oxygen, and pH, and ensure that representative samples of culture can be taken.

   Consequently, by the 1960s suspension culture was the technology chosen for large-scale production of veterinary vaccines, especially foot-and-mouth disease vaccine. In more recent years the

technology has received fresh impetus from the use of continuous
cell lines to produce natural products such as interferon and, more
recently, monoclonal antibodies and proteins produced by recombinant DNA technology.

II. PRODUCTS OF MAMMALIAN CELL SUSPENSION CULTURE

A. Vaccines

The products made in largest volume in animal cell culture are the
foot-and-mouth (FMD) vaccines. Capstick et al. (1962) described a
process for producing FMD vaccine in suspension-adapted baby hamster kidney (BHK) cells. By the 1980s this had formed the basis
of a worldwide industry producing millions of liters of culture fluid
a year based on stirred reactors up to 3000 liters working volume
(Radlett, et al., 1985). Vaccines produced in suspension culture
have been restricted to veterinary applications.

Because it is difficult to retain integrity and immunogenicity of
viruses during rigorous purification procedures, it is usual for
whole-virus vaccines to be relatively impure in comparison with
other cell culture-derived products. Consequently, strict regulations have been laid down governing the use of host cell substrates
because components of the cells (or contaminants) might be incorporated into the vaccines. Because the purity of these vaccines and
their impurities could not be estimated accurately, it was decided
that for human use vaccines should be derived from normal cell substrates with diploid karyotype, finite life spans, and nontumorigenic
characteristics. Such cells cannot be grown in suspension. Suspension culture technology is likely to be useful, however, for the production of vaccines based on viral subunit proteins or glycoproteins.
These can be produced in continuous cell lines which may be grown
at large scale by recombinant DNA techniques and rigorously purified (Bermann et al., 1958; Bermann and Laskey, 1985). This will
be an area of increasing importance, especially where it is impossible to culture the appropriate viruses or to make a safe product
from them.

B. Interferons

The technology used for the large-scale production of FMD vaccines
was adapted during the 1970s for the production of interferon from
human Namalva (lymphoblastoid) cells (Johnston et al., 1979). This
product, now approved for sale in many countries, is made at a

relatively large scale in 8000-liter reactors (Pullen et al., 1985; Phillips et al., 1985). Various other products have been derived from human and primate cell lines grown in suspension at large scales, e.g., interleukins 2 and 3 have been made in this way (Flickinger et al., 1982).

C. Recombinant Products

Recombinant DNA technology allows us to produce in large quantities previously scarce or even completely novel proteins. For many proteins, especially complex eukaryotic proteins, mammalian cells prove to be more appropriate hosts for expression than microbial cells.

A number of expression systems have now been developed for mammalian cells (reviewed by Bebbington and Hentschel, 1985), allowing the production of several proteins of major therapeutic interest. Examples include tissue-type plasminogen activator (t-PA), factors VIII and IX, erythropoietin, and human growth hormone. Various methods for optimizing the production of such molecules have been described (Mather et al., 1986). The continuous line most commonly used for expression of recombinant proteins is the Chinese hamster ovary (CHO) cell. This cell line can be grown in suspension culture, and Fig. 1 shows growth and product synthesis for a CHO cell line expressing human tissue inhibitor of metalloproteinase (TIMP) in an airlift fermenter. Other cell lines which grow readily in suspension and are used to express recombinant proteins include myeloma cells (Rhodes and Birch, 1988) and BHK cells (Pavirani et al., 1987; Wagner and Lehmann, 1988).

D. Monoclonal Antibodies

Monoclonal antibodies already play an important part in the diagnostic industry and are likely to play an increasing role in human therapy. These and other applications, such as immunopurification, have led to the need for efficient production processes. Kilogram-per-year applications already exist (Arathoon and Birch, 1986) and apparently will increase.

Two approaches have been taken to monoclonal antibody production from hybridoma cells: in vivo culture as ascites tumors in rats or mice and cell culture in vitro. For large-scale production, in in vitro culture is gaining favor because it can be scaled up as a unit operation giving significant economies of scale. In addition, the cell culture approach has qualitative advantages which are important for the production of therapeutic antibodies. In particular the risk of contaminating the monoclonal antibody with extraneous

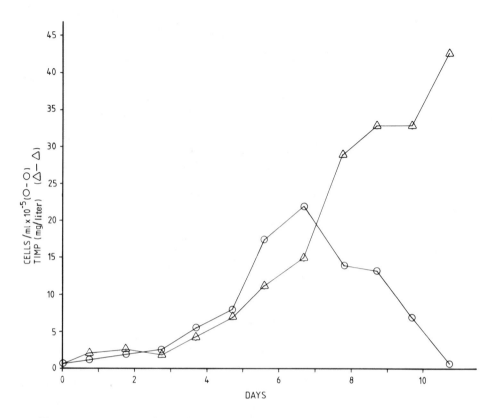

Figure 1   Production of TIMP by CHO cells growing in a 100-liter airlift fermenter in serum-free medium.

mouse antibodies or adventitious agents of rodent origin is greatly reduced.

Hybridoma cells grow in suspension culture and production systems based on airlift reactors (Arathoon and Birch, 1986; Birch et al., 1987) and stirred reactors (Lebherz et al., 1985) have been described.   Antibody production by a mouse hybridoma cell in a fed-batch airlift culture is illustrated in Fig. 2.   In this example production is not growth rate-related and a large proportion of the antibody accumulates during the stationary and decline phases of growth.

The majority of monoclonal antibodies currently in use are of rodent origin.   The in vivo treatment of humans with rodent antibodies may lead to a human anti-mouse antibody (HAMA) response. To avoid this, human antibodies are being developed and in some

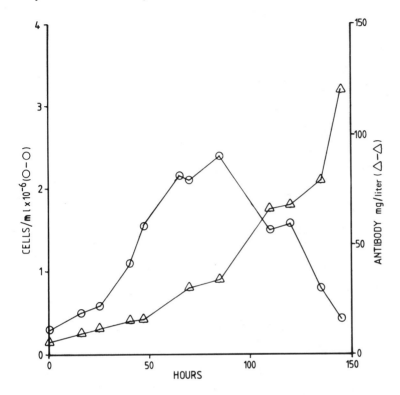

Figure 2   Fed-batch airlift culture of murine hybridoma producing monoclonal antibody.   (From Rhodes and Birch, 1988.   Copyright 1988 *Bio/Technology*.   Used by permission.)

cases it is possible to produce these in Epstein-Barr virus-transformed lymphoblastoid cell lines or in human-human or human-mouse hybridomas.   These human antibody-producing cell lines can be grown in large-scale suspension culture (Birch et al. , 1987).

An alternative approach which may avoid the HAMA response is to use recombinant DNA techniques to generate hybrid antibody genes between mouse variable regions and human constant regions. The hybrid antibody gene is then expressed in a mammalian cell line such as CHO or in myeloma cell line.   Processes are being developed to manufacture these recombinant anbitodies in suspension culture (Rhodes and Birch, 1988).

III. REACTORS FOR SUSPENSION CULTURE

A. Stirred Reactors

The largest suspension processes developed to date in terms of medium volumes and cell numbers use stirred reactors. Vessels of
3000 liters working volume are used for FMD vaccine production
(Radlett et al., 1985) and vessels up to 8000 liters capacity are used
for the production of interferon (Phillips et al., 1985; Mizrahi, 1983).
In recent years stirred reactors have been used successfully for the
culture of CHO cells for the production of recombinant proteins and
of hybridoma cells for monoclonal antibody production (Lebherz
et al., 1985). Many other types of cell have been cultured in suspension at scales up to many hundreds of liters for production of
research materials (see, for example, Moore et al., 1968; Lynn and
Acton, 1975; Zwerner et al., 1981). In conventional systems cells
are grown in stainless steel vessels with height-to-diameter ratios
in the range 1:1 to about 3:1. The cultures are usually mixed with
one or more agitators based on bladed disk (Rushton) or marine
propeller patterns. An agitation system using slowly rotating flexible sheets or sails has also been described (Tolbert and Feder,
1983).

Agitation systems may be driven directly or indirectly by magnetically coupled drives (Lynn and Acton, 1975; Zwerner et al.,
1981; Phillips et al., 1985). Indirect drives reduce the risk of
microbial contamination through seals on stirrer shafts. Both top-
and bottom-driven systems have been used successfully.

B. Vibromixer Reactors

A number of workers (Girard et al., 1973; Tolbert and Feder, 1983)
described the use of vibromixer agitation as an alternative to rotating impellers for animal cell culture. A horizontal disk with con-
ical holes is attached to a shaft which vibrates rapidly in an up-
and-down direction circulating the medium in a vertical direction.
This system has the advantage of reduced contamination risk by
using a static seal in the vessel with vibration transmitted through
a flexible diaphragm. Tolbert and Feder (1983) describe a 100-liter
fermentation system which uses this type of agitation.

C. Airlift Reactors

As with stirred reactors, the principle of the airlift reactor was
first described for microbial culture systems and subsequently
adapted for mammalian cell culture. The reactor relies on a gas
stream both to mix the culture and to supply oxygen. The gas
stream enters the riser section of the fermenter driving circulation.
Gas disengages at the culture surface, causing denser liquid free

of gas bubbles to travel downward in the downcomer section of the
vessel toward the sparger region. Various methods, such as a
draught tube, can be used to separate riser and downcomer sections.
The height-to-diameter ratio (H/D) of airlifts tend to be high (typically as high as 10:1) compared to that of stirred vessels. The design principles are described in reviews by Onken and Weiland,
(1983), Merchuk and Siegel (1988), and Smart (1984). The main
advantage of the airlift-type reactor is its simplicity since there is
no need for the motors and agitators used in stirred culture. The
airlift reactor also has good mass transfer characteristics with respect to oxygen supply but generates relatively low shear forces
(Katinger and Scheirer, 1985; Birch et al., 1987). Katinger et al.
(1979) were the first to use airlift technology for animal cell culture.
They used airlift reactors to culture human lymphoblastoid cells and
BHK21 cells. This type of reactor has subsequently been operated
at scales of up to 2000 liters for the production of monoclonal antibodies from hybridomas and recombinant proteins from CHO cells
(Arathoon and Birch, 1986; Rhodes and Birch, 1988). Figures 3
and 4 show a photograph and schematic layout, respectively, of a
2000-liter manufacturing system.

D. Continuous-Culture Systems

Most large-scale suspension culture processes are operated as batch
or fed-batch processes because these are the most straightforward
to operate and scale up. However, continuous processes based on
chemostat and perfusion principles have attracted attention both as
analytical tools and as potential production methods.

A batch culture is a closed system in which a typical growth
profile is seen (e.g., Figs. 1 and 2). A lag phase is followed by
exponential, stationary, and decline phases. In such a system the
environment is continually changing as nutrients are depleted and
metabolites accumulate. This makes analysis of the factors influencing cell growth and productivity, and hence optimization of the
process, a complex matter. Productivity of a batch process may be
increased by controlled feeding of key nutrients to prolong the
growth cycle. Such a fed-batch process is still a closed system
because cells and waste products are not removed. The technique
is useful for the production of proteins such as monoclonal antibodies which are not growth rate-related.

1. Perfusion Culture

Several systems have been described for continuously perfusing cells
with fresh culture medium. Typically such devices have a filtration
system to retain cells while allowing exchange of culture medium.
One of the earliest perfusion systems was that described by

Figure 3   A 2000-liter airlift fermenter at Celltech Ltd.   (Courtesy of Celltech Ltd.)

Himmelfarb et al. (1969), who used a spin filter in suspension culture of L1210 cells and achieved cell densities approaching 10^8/ml. Rotating the filter helps to prevent its becoming fouled with cells and cell debris.   Tolbert and Feder (1983) and Wagner and Lehmann (1988) described alternative perfusion devices.   Fouling of filters is a problem with perfusion systems and they still represent closed systems in which the cells are retained and the environment changes with time.

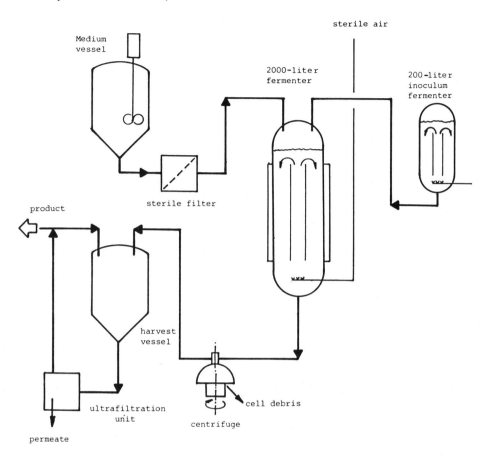

Figure 4   Layout of 2000-liter airlift fermentation system.   (From Rhodes and Birch, 1988.   Copyright 1988 *Bio/Technology*.   Used by permission.)

2. Chemostat Culture

The problems of analyzing the changing environment of the closed system described above can be overcome by using truly open systems in which there is an inflow of medium and an outflow of cells and products. The simplest open system is the chemostat (for principles, see Pirt, 1975).

Culture medium is fed to the reactor at a predetermined and constant rate which maintains the dilution rate of the culture at a value less than the maximum specific growth rate of the cells (to avoid washout of cells from the fermenter).   Culture fluid containing cells and cell products is removed at the same rate.   Steady-state conditions are achieved in the chemostat in which specific growth rate is

determined by the dilution rate and maximum cell density is regulated
by deliberately controlling the concentration of a selected essential
nutrient or inhibitory catabolite. Hence by appropriate selection of
dilution rate and by design of the culture medium one can independently manipulate growth rate, cell density, and the nature of the
limiting nutrient. Steady-state conditions are achieved in which cell
growth is truly exponential and any measured cellular or environmental parameter will remain constant with time, provided genetic stability is maintained. The major use of the chemostat has been to study
the physiology of microbial cells but now there are examples of its
application to mammalian cell studies. Tovey (1985) described the
use of the chemostat to study mouse L1210 cells and Boraston et al.
(1984) described its use with mouse hybridoma cells to investigate
the effect of growth rate on respiration rate. Antibody synthesis
was studied in chemostat culture of a hybridoma cell synthesizing an
IgM by Birch et al. (1985) and was shown to be stable over prolonged culture periods and under a variety of nutrient limitations.

   Apart from its use in the study of cell physiology, chemostat
culture may have some advantages as a production technique. Compared with batch culture the chemostat offers improved utilization of
fermenter capacity due to reduced downtime and the ability to maintain high cell densities and product output for extended periods.
As noted above, it also offers a means of advantageously manipulating cellular physiology. Disadvantages of the method include loss
of flexibility; the reactor is committed to a given cell line for extended periods. The equipment is also more complex than a batch system and therefore more prone to failure. At laboratory scale Fazekas
de St. Groth (1983) used continuous culture to produce antibodies
from 10 different hybridoma cell lines.

   Productivity of continuous cultures may be increased by biomass
feedback, i.e., either to prevent cells from leaving in the effluent
stream or to automatically return a proportion of the cells from the
effluent (see Pirt, 1975 for explanation of principle). This principle has been applied at laboratory scale to a continuous-airlift
culture of hybridoma cells making monoclonal antibody (Birch et al,
1985).

IV. DESIGN CRITERIA AND OPERATING
CONDITIONS FOR REACTORS

A. Aseptic Operation

The relative complexity of mammalian cell processes in terms of equipment and culture media combined with relatively prolonged culture
periods makes them potentially prone to microbial contamination. In
practice asepsis is maintained by ensuring that culture vessels and
associated equipment can be adequately sterilized before use and
that appropriate barriers are present to prevent ingress of micro-

organisms during the fermentation. Small-scale (up to a few liters) suspension culture vessels are typically autoclavable glass vessels. Pilot and production scale reactors are generally stainless steel pressure vessels which can be sterilized in situ using steam at high temperature and pressure. Vessels and pipework are designed to be free-draining so as to avoid buildup of condensate, which could generate cold spots which would not be sterilized. Microporous filters are used to remove microorganisms from gas mixtures entering fermenters and exhaust gases are usually filtered as well to prevent back contamination. Generally culture media are sterilized by microporous filtration although some autoclavable media for suspension culture have been developed (e.g., Nagle, 1968; Keay, 1976). It is common practice in many laboratories to use antibiotics to reduce the risk of microbial contaminations. This procedure is not without risks, however, because it can give rise to antibiotic-resistant microbes and it has been shown that well-designed modern fermenters can be run successfully without antibiotics. Overall failure rates due to contamination of less than 5% at production scale have been achieved (Arathoon and Birch, 1986).

More detailed descriptions of equipment and air sterilization are given by Threlfall and Garland (1985) and Harper (1985).

B. Automation

Increasing use is being made of microprocessor-based computers to automate fermenter operations (Ehsani et al., 1987). In addition to data logging and control of key process parameters, the computer is used to automate valve and pump activation sequences during cleaning, sterilization, and process control (Birch et al., 1987).

C. pH Control

In the absence of pH control it is commonly found that the pH value in a culture drops during growth to an inhibitory value due to the accumulation of lactic acid. It is therefore usual to measure and automatically control the pH value in suspension culture. The pH of most cultures is controlled by a bicarbonate-CO_2 buffering system. Telling and Stone (1964) described a method for controlling pH in stirred suspension culture of BHK 21 cells which relied on the automatic adjustment of air and CO_2 flows to the culture to raise and lower the pH value, respectively. This principle is still widely used. The pH may also be controlled by the addition of sterile base or acid solutions.

The pH value can have a profound effect on cell growth and metabolism. The optimum pH value for growth and product formation will vary with cell type. In the case of a human lymphoblastoid line, for example, the optimum pH for growth was 7.4 (Birch and Edwards, 1979). The growth rate was greatly reduced below pH 6.6 or above

pH 8.0. Effects on maximum population density could be attributed
to the influence of pH on the growth yield values for glucose, which
was shown to be the growth-limiting nutrient.

D. Dissolved Oxygen Concentration

Oxygen is essential for the growth of mammalian cells. The oxygen
uptake rate has been established for several cell lines. For example, a hybridoma cell line and the BHK 21 cell line in suspension
have an oxygen demand of 0.2 mM $O_2/10^9$ cells/hr (Radlett et al.,
1972; Boraston et al., 1984). Oxygen uptake data for other cell
lines are given by Fleischaker and Sinskey (1981). Knowing the
oxygen demand of a given cell line establishes the oxygen transfer
rate which must be achieved in the reactor. Straightforward methods are available for measuring oxygen transfer rate (see, for example, Boraston et al., 1984). Although the oxygen requirement
of animal cells is small in comparison with microbial cultures, it can
nevertheless limit growth in some suspension culture systems, especially those which rely exclusively on surface aeration (Fleischaker
and Sinskey, 1981; Boraston et al., 1984).

Various techniques have been developed to provide adequate
oxygen supply to cultures. The most common method is to bubble
or "sparge" air or oxygen into the culture. This method was originally applied to stirred cultures of BHK 21 cells (Radlett et al.,
1972) and is now used routinely in industrial scale stirred and airlift
reactors. Under some circumstances sparging may damage cells
(Radlett et al., 1972) but in general conditions have been found
empirically which are not harmful. Handa et al. (1987) studied gas/
liquid interfacial effects on the viability of hybridoma cells in bubble
column reactors. They showed that survival of cells in sparged
systems depends on bubble size, cell type, superficial gas velocity,
and the macromolecular composition of the medium. Alternative
means of aeration have been described, and for example, Wagner
and Lehmann (1988) developed a bubble-free aeration system using
hollow fiber membrane aerators.

Some workers (e.g., Radlett et al., 1972; Boraston et al., 1984)
described systems for automatically controlling the concentration of
dissolved oxygen in cultures. Oxygen is typically monitored with a
dissolved oxygen electrode and control is usually based on varying
the supply of air or oxygen to the culture. The optimum oxygen
concentration for growth appears to vary from one cell line to
another. For BHK 21 cells, for example, the optimum for growth
was equivalent to a partial pressure of 80 mm Hg (Radlett et al.,
1972) while for a human lymphoblastoid cell line maximum cell densities were achieved at 166 mm Hg (Mizrahi et al., 1972).

In some laboratories redox electrodes have been used in preference to oxygen electrodes in oxygen control systems because they tend to be more robust. When they are used it is necessary to establish experimentally the redox potential corresponding to the optimum dissolved oxygen concentration. This value has to be established separately for different media and cell types (Arathoon and Telling, 1981; Pullen et al., 1985).

Foaming may occur in sparged cultures, especially when the culture medium contains serum. It can be prevented by the addition of antifoaming agents (Lambert and Birch, 1985). Alternatively, conventional foam-breaking or trapping equipment may be used.

Additional data on the measurement and control of culture parameters including temperature and pressure were reviewed by Harris and Spier (1985).

V. MEDIA FOR SUSPENSION CULTURE

The general nutritional requirements of mammalian cells have been reviewed elsewhere (e.g., Mather, 1984; Lambert and Birch, 1985). Historically media for large-scale culture have contained animal serum. Cost considerations and the drive to improve the quality of protein products has led to the development and widespread use of serum-free media at production scale. Monoclonal antibodies, for example, are produced at 1000-liter scale in serum-free medium (Birch et al., 1987).

Issues which need to be addressed specifically in designing media for suspension culture include the avoidance of cell aggregation, mechanical damage to cells, precipitation of serum protein, and foam formation. Cell aggregation and attachment to surfaces of the culture vessel may be reduced by lowering the calcium concentration in the medium. Mechanical damage to cells in agitated culture is usually not a problem in media containing serum but can cause difficulties in chemically defined, protein-free media or those containing reduced levels of serum protein. Kuchler et al. (1950) demonstrated the protective effect of methylcellulose in agitated serum-free suspension culture. Methylcellulose and other polymers were subsequently used in a number of serum-free medium formulations for suspension culture (Higuchi, 1973; Birch, 1980), although the mode of action of these polymers is not clearly understood. Precipitation of serum protein can occur in some culture systems and this has been avoided by the use of the surfactant polymer Pluronic F68 (Swim and Parker, 1968).

It should be noted that nutritional requirements may change both qualitatively and quantitatively in agitated suspension culture as opposed to static culture. Birch (1980), for example, showed

an increased serum requirement for a human lymphoblastoid cell
line in shake-flask culture compared with static culture. Similarly,
Murakami et al. (1983) showed that a soybean phospholipid fraction
was an essential medium ingredient for a myeloma cell line growing
in serum-free stirred culture but was not required in stationary
culture.

Cell densities achieved in suspension culture (typically <1 mg
dry cell weight/ml) are low compared with those in microbial cultures. Systematic studies of the *quantitative* nutritional requirements of cells demonstrated the potential for improving the productivity of culture media (Birch, 1970; Higuchi, 1973). Standard
media are not designed for the production of high cell densities in
suspension culture. In a study of the BHK 21 cell growing in suspension culture, Arathoon and Telling (1982), for example, showed
that the amino acid composition of a standard culture medium bore
no resemblance to the patterns of utilization by the cells. In some
culture systems maximum population densities may be limited by the
accumulation of inhibitors such as ammonia and lactate (Kimura et al.,
1987). Various strategies have been proposed to overcome this limitation. Glacken et al. (1986), for example, proposed a strategy
for reducing levels of lactate and ammonia based on controlled addition of glucose and glutamine to the culture. Ammonia accumulation
may also be avoided in some cases by replacing glutamine in the culture medium with glutamic acid (Griffiths and Pirt, 1967).

VI. SUMMARY

Mammalian cell suspension culture systems are being used increasingly in the biotechnology industry. This is due to their many advantages including simplicity and homogeneity of culture. Suspension
systems are very adaptable (e.g., for microcarrier, microencapsulation, or other methods of culture). Their engineering is thoroughly
understood and standardized at large scale, and automation and
cleaning procedures are well established. Suspension systems offer
the possibility of quick implementation of production protocols due
to their ability to be scaled easily once the basic culture parameters
are understood.

The only main disadvantage of the suspension culture systems
to date is their inapplicability for the production of human vaccines
from either primary cell lines or from normal human diploid cell
lines (Hayflick et al., 1987 and references therein). One of the
great advantages of suspension culture is the opportunity it provides to study interactions of metabolic and production phenomena
in chemostat or turbidostat steady-state systems. Furthermore, in
suspension culture systems from which cell number and cell mass
measurements are easy to obtain, rigorous and quantitative

estimations of the effects of growth conditions or perturbations of metabolic homeostasis can be made. Such studies can speed up the development of optimal processes.

With our increasing understanding of factors influencing expression in mammalian cells (Cohen and Levinson, 1988; Santoro et al., 1988) and the direct application of new methods in suspension culture (Rhodes and Birch, 1988), its usefulness and importance is likely to increase in the future. In this chapter, we have described some of the potential uses of the various suspension culture systems and have covered most of the established technology and literature. Due to the rapid developments and needs in the biotechnology industry and the versatility of suspension culture systems, it is probable that many more variations on this theme will evolve in the near future at both the pilot and production scales.

REFERENCES

Arathoon, W. R. and Telling, R. C. (1982). Uptake of amino acids and glucose by BHK 21 clone 13 suspension cells during cell growth. *Dev. Biol. Standard.* 50:145–154.

Arathoon, W. R. and Birch, J. R. (1986). Large scale cell culture in biotechnology. *Science* 232:1390–1395.

Bebbington, C. and Hentschel, C. (1985). The expression of recombinant DNA products in mammalian cells. *Trends Biotechnol.* 3:314–317.

Bermann, P. W., Gregory, T., Crase, D., and Laskey, L. A. (1985). Protection from genital herpes simplex virus type 2 infection by vaccination with cloned type 1 glycoprotein D. *Science* 227:1490–1492.

Bermann, P. W. and Laskey, L. A. (1985). Engineering glycoproteins for use as pharmaceuticals. *Trends Biotechnol.* 3:51–53.

Birch, J. R. (1970). Improvements in a chemically defined medium for the growth of mouse cells (strain LS) in suspension. *J. Cell Sci.* 7:661–670.

Birch, J. R. (1980). The role of serum in the culture of a human lymphoblastoid cell line. *Dev. Biol. Standard.* 46:21–27.

Birch, J. R. and Edwards, D. J. (1980). The effect of pH on the growth and carbohydrate metabolism of a lymphoblastoid cell line. *Dev. Biol. Standard.* 46:59–63.

Birch, J. R., Lambert, K., Boraston, R., Thompson, P. W., Garland, S., and Kenney, A. C. (1985). The industrial production of monoclonal antibodies in cell culture. In *Perspectives in Biotechnology* (J. M. C. Duarte, L. J. Archer, A. T. Bull, and G. Holt, eds.), Plenum Press, New York.

Birch, J. R., Lambert, K., Thompson, P. W., Kenney, A. C., and Wood, L. A. (1987). Antibody production with airlift fermenters.

In *Large Scale Cell Culture Technology* (K. Lyderson, ed.),
K. Hanser.

Boraston, R., Thompson, P. W., Garland, S., and Birch, J. R.
(1984). Growth and oxygen requirements of antibody producing mouse hybridoma cells in suspension culture. *Dev.
Biol. Standard.* 55:103–111.

Capstick, P. B., Telling, R. C., Chapman, W. G., and Stewart,
D. L. (1962). Growth of a cloned strain of hamster kidney cells
in suspended cultures and their susceptibility to the virus of
foot-and-mouth disease. *Nature* 195:1163–1164.

Cherry, W. R. and Hull, R. N. (1956). Studies on the growth of
mammalian cells is agitated fluid media. Abstract TCII *Anat.
Rec.* 124:483.

Cohen, J. B. and Levinson, A. D. (1988). A point mutation in the
last intron responsible for increased expression and transforming activity of the c-Ha-ras oncogene. *Nature* 334:119–123.

Earle, W. R., Schilling, E. L., Bryant, J. C., and Evans, V. J.
(1954). The growth of pure strain L cells in fluid suspension
culture. *J. Natl. Cancer Inst.* 14:1159–1171.

Ehsani, N., Green, T., Haskell, S., Polastri, G. D., and Arathoon,
W. R. (1987). Computerised control system for cell culture.
In *Plant and Animal Cells, Process Possibilities* (C. Webb and
F. Mavituna, eds.), Ellis Horwood, pp. 285–290.

Fazekas de St. Groth, S. (1983). Automated production of mono-
clonals in a cytostat. *J. Immunol. Meth.* 57:121–136.

Fleischaker, R. J. and Sinskey, A. J. (1981). Oxygen demand
and supply in cell culture. *Eur. J. Appl. Microbiol. Biotechnol.*
12:193–197.

Flickinger, M. C., Klein, F., Ricketts, R., and Pickle, D. (1982).
Large scale suspension culture production of interleukins 2 and
3 by constitutive-producer cell lines. ACSRAL, Abstr. Pap.
Am. Chem. Soc. 184, Meet., MICR 20.

Girard, H. C., Okay, G., and Kivilcin, Y. (1973). Use of the
vibro fermenter for multiplication of BHK cells in suspension
and for replication of FMD virus. *Bull. Off. Int. Epizoot* 79:
805–822.

Glacken, M. W., Fleischaker, R. J., and Sinskey, A. J. (1986).
Reduction of waste product excretion via nutrient control:
Possible strategies for maximising product and cell yields on
serum in cultures of mammalian cells. *Biotechnol. Bioeng.* 28:
1376–1389.

Griffiths, J. B. and Pirt, S. J. (1967). The uptake of amino acids
by mouse cells (strain LS) during growth in batch culture and
chemostat culture: The influence of cell growth rate. *Proc.
Roy. Soc. B* 168:421–438.

Handa, A., Emery, A. M., and Spier, R. E. (1987). Effect of gas – liquid interfaces on the growth of suspended mammalian cells: Mechanisms of cell damage by bubbles. *Dev. Biol. Standard.* 66:241 – 252.

Harper, G. J. (1985). Air sterilisation. In *Animal Cell Biotechnology*, Vol. 1 (R. E. Spier and J. B. Griffiths, eds.), Academic Press, Orlando.

Harris, L. and Spier, R. E. (1985). Physical and chemical parameters: Measurement and control. In *Animal Cell Biotechnology*, Vol. 1 (R. E. Spier and J. B. Griffiths, eds.), Academic Press, Orlando.

Hayflick, L., Plotkin, S., and Stevenson, R. E. (1987). History of the acceptance of human diploid cell strains as substrates for human virus vaccine manufacture. *Dev. Biol. Standard.* 68:9 – 17.

Higuchi, K. (1973). Cultivation of animal cells in chemically defined media: A review. *Adv. Appl. Microbiol.* 16:111 – 136.

Himmelfarb, P., Thayer, P. S., and Martin, M. E. (1969). Spin filter culture: The propagation of mammalian cells in suspension. *Science* 164:555 – 557.

Johnston, M. D., Christofinis, G., Ball, G. D., Fantes, K. H., and Finter, N. B. (1979). A culture system for producing large amounts of human lymphoblastoid interferon. *Dev. Biol. Standard.* 42:189 – 192.

Katinger, H. and Scheirer, W. (1985). Mass cultivation and production of animal cells. In *Animal Cell Biotechnology*, Vol. 1 (R. E. Spier and J. B. Griffiths, eds.), Academic Press, Orlando.

Katinger, H. W. D., Scheirer, W., and Kromer, E. (1979). Bubble column reactor for mass propagation of animal cells in suspension culture. *Ger. Chem. Eng.* (Engl. Transl.) 2:31 – 38.

Keay, L. (1976). Autoclavable low cost serum-free media: Growth of established cell lines and production of viruses. *Biotechnol. Bioeng.* 18:363 – 382.

Kimura, T., Iijima, S., and Kobayashi, T. (1987). Effects of lactate and ammonium on the oxygen uptake rate of human cells. *J. Ferment. Technol.* 65:341 – 344.

Kuchler, R. J., Marlowe, M. L., and Merchant, D. J. (1960). The mechanism of cell binding and cell-sheet formation in L strain fibroblasts. *Exp. Cell Res.* 20:428 – 437.

Lambert, K. J. and Birch, J. R. (1985). Cell growth media. In *Animal Cell Biotechnology*, Vol. 1 (R. E. Spier and J. B. Griffiths, eds.), Academic Press, Orlando.

Lebherz, W. B., Lee, S. M., Gustafson, M. E., Ricketts, R. T., Lufriu, I. F., Morgan, A. C., and Flickinger, M. C. (1985).

Production of monoclonal antibodies by large scale submerged hybridoma cultures. *In Vitro* 21:16A.

Lynn, J. D. and Acton, R. T. (1975). Design of a large scale mammalian cell suspension culture facility. *Biotechnol. Bioeng.* 17:659–673.

Mather, J. P. (ed.) (1984). *Mammalian Cell Culture, the Use of Serum-Free Hormone-Supplemented Media*, Plenum Press, New York.

Mather, J. P., Tsao, M., and Arathoon, W. R. (1986). Expression of cloned proteins in mammalian cells: Regulation of cell-associated parameters. *Proc. Biofair Tokyo*:127–133.

McLimans, W. F., Davis, E. V., Glover, F. L., and Rake, G. W. (1957). The submerged culture of mammalian cells. *J. Immunol.* 79:428–433.

McLimans, W. F. and Giardinello, F. R. (1957). Submerged culture of mammalian cells: The five liter fermenter. *J. Bacteriol.* 74:768–771.

Merchuk, K. C. and Siegel, M. H. (1988). Airlift reactors in chemical and biological technology. *J. Chem. Tech. Biotechnol.* 41:105–120.

Mizrahi, A., Vosseller, G. Y., Yagi, Y., and Moore, G. E. (1972). The effect of dissolved oxygen partial pressure on growth, metabolism and immunoglobulin production in a permanent human lymphocyte cell line culture. *Proc. Soc. Exp. Biol. Med.* 139: 118–122.

Mizrahi, A. (1983). Production of human interferons: An overview. *Process Biochem.* (August):9–12.

Moore, G. E., Hasenpusch, P., Gerner, R. E., and Burns, A. A. (1968). A pilot plant for mammalian cell culture. *Biotechnol. Bioeng.* 10:625–640.

Murakami, H., Edamoto, T., Shinohara, K., and Omura, H. (1983). Stirred culture of myeoloma MPC-11 cells in serum-free medium. *Agric. Biol. Chem.* 47:1835–1840.

Nagle, S. C., Jr. (1968). Heat-stable chemically defined medium for growth of animal cells in suspension. *Appl. Microbiol.* 16: 53–55.

Onken, U. and Weiland, P. (1983). Airlift fermenters: Construction, behavior and uses. *Adv. Biotechnol. Proc.* 1:67–95.

Owens, O., Gey, M. K., and Gey, G. O. (1954). Growth of cells in agitated fluid medium. *Ann. NY Acad. Sci.* 58:1039–1055.

Pavirani, A., Neulien, P., Harrer, H., Schamber, F., Dott, K., Villeval, D., Cordier, Y., Wiesel, M.-L., Nazurier, C., Van de Pol, H., Piquet, Y., Cazenove, J.-P., and Lecocq, J.-P. (1987). Choosing a host cell for active recombinant factor VIII production using vaccinia virus. *Biotechnology* 5:389–392.

Phillips, A. W., Ball, G. D., Fantes, K. H., Finter, N. B., and Johnston, M. D. (1985). In *Large Scale Mammalian Cell Culture* (J. Feder and W. R. Tolbert, eds.), Academic Press, Orlando.

Pirt, E. J. (1975). *Principles of Microbe and Cell Cultivation.* Blackwell, London.

Pullen, K. F., Johnson, M. D., Phillips, A. W., Ball, G. D., and Finter, N. B. (1985). Very large scale suspension cultures of mammalian cells. *Dev. Biol. Standard.* 60:175−177.

Radlett, P. J., Pay, T. W. F., and Graland, A. H. M. (1985). The use of BHK suspension cells for the commercial production of foot and mouth disease vaccines over a twenty year period. *Dev. Biol. Standard.* 60:163−170.

Radlett, P. J., Telling, R. C., Whiteside, J. P., and Maskell, M. A. (1972). The supply of oxygen to submerged cultures of BHK 21 cells. *Biotechnol. Bioeng.* 14:437−445.

Rhodes, M. and Birch, J. (1988). Large scale production of proteins from mammalian cells. *Biotechnology* 6:518−523.

Santoro, C., Mermod, N., Andrews, P. C., and Tjian, R. (1988). A family of CCAAT-box-binding proteins active in transcription and DNA replication: Cloning and expression of multiple cDNAs. *Nature* 334:218−224.

Smart, N. J. (1984). Gaslift fermenters: Theory and practice. *Lab. Pract.*:9−14.

Swim, H. E. and Parker, R. F. (1968). Effect of pluronic F68 on growth of fibroblasts in suspension on a rotary shaker. *Proc. Soc. Exp. Biol. Med.* 103:252−254.

Telling, R. C. and Ellsworth, R. (1965). Submerged culture of hamster kidney cells in a stainless steel vessel. *Biotechnol. Bioeng.* 7:417−434.

Telling, R. C. and Stone, C. J. (1964). A method of automatic pH control of a bicarbonate-CO_2 buffer system for the submerged culture of hamster kidney cells. *Biotechnol. Bioeng.* 6:147−158.

Threlfall, G. and Garland, S. G. (1985). Equipment sterilisation. In *Animal Cell Biotechnology*, Vol. 1 (R. E. Spier and J. B. Griffiths, eds.), Academic Press, Orlando.

Tolbert, W. R. and Feder, J. (1983). Large scale cell culture technology. *Ann. Rep. Ferment. Proc.* 6:35−74.

Tolbert, W. R., Feder, J., and Kimes, R. C. (1981). Large scale rotating filter perfusion system for high density growth of mammalian suspension cultures. *In Vitro* 17:885−890.

Tovey, M. G. (1985). The cultivation of animal cells in continuous flow culture. In *Animal Cell Biotechnology*, Vol. 1 (R. E. Spier and J. B. Griffiths, eds.), Academic Press, Orlando.

Wagner, R. and Lehmann, J. (1988). The growth and productivity of recombinant animal cells in a bubble free aeration system. *Trends Biotechnol.* 6:1011−1014.

Ziegler, D. W., Davis, E. V., Thomas, W. J., and McLimans, W. F. (1958). The propagation of mammalian cells in a 20 litre stainless steel fermenter. *Appl. Microbiol.* 6:305−310.

Zwerner, R. K., Cox, R. M., Lynn, J. D., and Acton, R. T. (1981). Five-year perspective of the large scale growth of mammalian cells in suspension culture. *Biotechnol. Bioeng.* 23: 2717−2735.

11

Microcarrier Culture Systems

SHAUL REUVENY
Israel Institute for Biological Research, Ness-Ziona, Israel

I. INTRODUCTION

Animal and human cells can be propagated in vitro in two basically different modes: as anchorage-independent cells growing freely in suspension throughout the bulk of the culture, or as anchorage-dependent cells (ADC) requiring attachment to solid substrate for their propagation (monolayer type of cell growth) (Fig. 1).

Cells from continuous established cell lines which have the ability to grow in suspension are the most obvious type to be used for large-scale production of cells and cell products. These cells can be relatively easily cultivated on a large scale in homogeneous culture, products can be produced at a lower cost and a greater bulk, and as they grow indefinitely in vitro, their availability is limited. However, these cells cannot always be used in the production of biologicals because of the following:

1. Cells growing in suspension are considered to have a tumorigenic potential and thus their use as substrate for production of biologicals intended for human and veterinary use is sometimes limited (1,2).
2. Propagation of viruses in suspended culture sometimes causes rapid changes in viral markers, leading to reduction in immunogenicity as compared to viruses propagated on ADC (3).

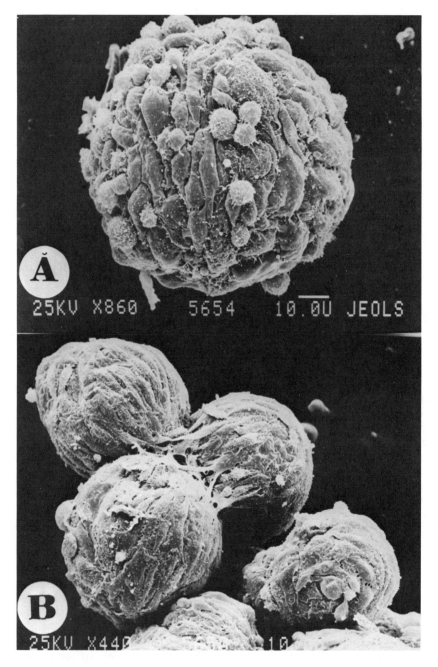

Figure 1   Scanning electron micrograph of anchorage-dependent
cells propagated on beaded microcarriers. (A) Baby hamster kidney
cell line (BHK). (B) Primary chick embryo fibroblasts. Note the
cell bridges between the microcarriers.

3. Sometimes cell lines (including recombinant cell lines) can secrete considerably higher amounts of products when propagated as ADC (attaching to growth surface) as compared with the same cells propagated in suspension (4).

For these reasons, different types of ADCs (human diploid cell strains, primary cells, and cell lines) are used extensively in the production of different biological products, such as interferons, lymphokines, hormones (e.g., growth hormones), growth factors (e.g., platelet-derived growth factor, epithelial growth factor, colony-stimulating factors), tissue plasminogen activator, serum proteins (e.g., factor VIII, factor IX), tumor necrosis factor, erythropoietin, viral vaccines (e.g., poliomyelitis virus, foot-and-mouth disease virus, rabies virus), and viral antigens (e.g., hepatitis, herpes), for human and veterinary applications.

Traditionally, ADCs are propagated on the bottom of small glass or plastic vessels. The restricted surface-to-volume ratio offered by these classical and traditional techniques has created a bottleneck in the production of cells and cell products on a large scale. In an attempt to provide systems that offer large accessible surfaces for cell growth in small-culture volume, a number of techniques have been proposed: the roller bottle system, the stack plates propagator, the spiral film bottles, the artificial capillary propagators, the packed-bed system, the plate exchanger system, and the membrane tubing reels. These systems and their relative advantages and disadvantages were recently reviewed (5,6).

Since these systems are nonhomogeneous in their nature, and are sometimes based on multiple processes, they suffer from the following shortcomings: limited potential for scale-up, difficulties in taking cell samples, limited potential for measuring and controlling the system, and difficulty in maintaining homogeneous environmental conditions throughout the culture. In an effort to overcome these limitations, van Wezel (7) developed the concept of the microcarrier (MC) culturing systems. In this system, cells are propagated on the surface of small solid particles suspended in the growth medium by slow agitation. These cells attach to the MC and grow gradually up to confluency of the MC surface.

In fact, this culture system is a one-unit operation in which both monolayer and suspension culture have been brought together, thus combining the necessary surface on which cells will grow with the advantages of the homogeneous suspension culture known to us from the traditional animal and bacterial cell-submerged cultures.

The purpose of this chapter is to describe the concept of the MC culture, different commercially and noncommercially developed MCs, critical parameters which determine their suitability for supporting cell growth and production of biologicals, the ways in which the MCs are used in the production of biologicals, and the main problems and considerations in the scaling up of MC culture reactors.

II. ADVANTAGES OF THE MICROCARRIER
CULTURING SYSTEMS

As a result of the unique character of the MC culture described above, this technique has the following advantages over other ADC large-scale cultivation methods:

1. High surface-to-volume ratio (which can be varied easily by changing the MC concentration) is achieved. The high surface-to-volume ratio leads to high cell yields per unit volume and the potential for obtaining highly concentrated cell products. Cell yields of up to $1-2 \times 10^7$/ml are achieved using MC-perfused cultures (8,9,11–13).

2. Cell propagation can be carried out in one high-productivity vessel instead of using many small low-productivity units, thus achieving a better utilization and a considerable saving of culture medium. One liter MC culture propagated under batch mode is comparable to 20–50 Roux bottles (14). While in high productivity perfusion culture, a 2-liter MC cell culture using less than 5 liters of medium gave cell and virus yields equivalent to 250 Roux bottles using 25 liters of medium (15). Moreover, cell propagation in a single reactor leads to a reduction in laboratory space and in the number of handling steps required per cell, thus reducing considerably the labor cost and risk of contamination.

3. The well-mixed MC suspension culture, in which cells are homogeneously distributed, makes it possible to monitor and control various environmental conditions (e.g., pH, pO_2, and concentration of medium components), thus leading to more reproducible cell propagation and product recovery.

4. It is possible to take a representative sample for microscopic observations, chemical testing, or enumeration—an option not available with most other techniques.

5. Since the MCs may be settled easily out of suspension, harvesting of cells and cell products can be done relatively easily.

6. The mode of ADC propagation on the MC makes it possible to use this system for new cellular manipulation, such as cell transfer, without the use of proteolytic enzymes (16), cocultivation of cells (17), perfusion of cell culture in columns (18), large-scale mitotic cell recovery (19), studies on cell interactions (20), studies of cell differentiation in a three-dimensional matrix (21), electron microscopy studies (22), and transplantation of cells into animals (23).

7. MC cultures can be relatively easily scaled up using conventional equipment (fermenters) used for the cultivation of bacteria and animal cells in suspension cultures.

Recently, Griffiths et al. (6,24) compared propagation of several

ADCs on three propagation systems: plate heat exchanger system, glass bead reactor, and MC culture. Although the three cultivation systems were efficient in cell propagation, the MC system was found to be superior in its potential for scaling up, its homogeneous natures and its ease of culture monitoring. On the other hand, the MCs used were expensive and could not be reused, and there is a need for the development of critical experimental procedures in order to achieve good cell growth.

III. HISTORY OF MICROCARRIER DEVELOPMENT

van Wezel (7,25,26) was the first researcher to examine the use of a commercial ion exchange resin, diethylaminoethyl (DEAE)-Sephadex A-50 (Pharmacia, Uppsala, Sweden), as MC for cultivation of ADCs. This MC is composed of crosslinked dextran beads, charged with tertiary amine groups (DEAE), having an exchange capacity of 3.5 milliequivalent (mEq/g) dry materials. van Wezel (7) demonstrated that primary cells and cells from a human diploid cell strain can be cultivated on these MCs and that poliomyelitis virus could be propagated. However, a toxic effect on cell growth was observed at bead concentrations exceeding about 1 g of DEAE-Sephadex A-50 per liter culture, as indicated by increased inoculum losses (about 50%), long lag periods, and diminished capacity for cell growth (26). Complete death of the cell inoculum was observed at bead concentrations exceeding 2 g/liter.

Various approaches were employed in order to decrease the toxic effect. The main approach was to coat the MCs with serum proteins (18), nitrocellulose (27), or carboxymethylcellulose (28), but only partial improvement of the MC was achieved. Levine et al. (29–32) found that this toxicity could largely be eliminated by reducing the exchange capacity of the commercial ion exchange resin (DEAE-Sephadex A-50) to 1.5 mEq/g dry materials (about 40% of the exchange capacity of the commercial preparation). Cells from a human diploid cell strain were propagated on the new low-charged DEAE-dextran MCs in concentrations of 5 g/liter, achieving cell concentrations of 4×10^6 cells/ml. These new developments and the increasing demand for large-scale propagation of ADCs have led to the commercial production of the low-charged DEAE-dextran MCs.

The new low-charged MCs were found to be suitable for cultivation of a wide variety of cell types including primary cells, cells from diploid cell strains, and established or transformed cell lines (33). However, several problems connected with cell propagation and product production on these MCs (34) led commercial companies and researchers around the world to develop new MCs with different characteristics which make them more suitable for a specific cell propagation or product production.

IV. GENERAL REQUIREMENTS AND PROPERTIES
OF A MICROCARRIER

The requirements and properties of an optimal MC, described earlier by several groups (26,33,35 – 37,39,40), can be summarized as follows:

1. Functional attachment group. ADCs can grow on the surface of MCs derivatized with several functional groups: positively charged (primary tertiary or quaternary amines), negatively charged (tissue culture treated polystyrene or glass), denaturated collagen and other proteins and amino acids polymerized to the carrier. These functional groups must allow optimal cell attachment and growth under stirred conditions. In the case of using positive or negative charges, the lowest MC charge which allows cell adherence and growth with minimal adsorption of medium ingredients on the MCs should be used.

2. Buoyant density of the MC. The buoyant density of the MC should be slightly above that of the culture medium (between 1.03 and 1.10) in order to allow suspension of the MCs by slow agitation. At a lower buoyant density the MCs will float to the surface, while at a higher buoyant density the stirring speeds needed to keep the MCs in suspension will cause detachment of cells from the MCs due to shearing forces.

3. MC dimensions. In order to maximize the growth surface per unit volume, a high concentration of small beads is required. On the other hand, each MC must have sufficient surface area to support growth of a single viable cell over several generations. Thus, the MC should be about 100 – 250 µm in diameter so that the MC will carry several hundred cells as monolayer and still be easily suspended. By choosing the optimal MC diameter it is possible to reduce the amount of cell inoculum needed for culture inoculation (41).

4. MC size distribution. To guarantee the homogeneity of the culture, the size distribution of the MCs should be as narrow as possible and preferably within ±25 µm (39). Since cells do not move easily from one MC to the other during cultivation, each MC should be inoculated with several cells so that all the MCs reach confluence at the same time. An uneven size distribution of MCs results in an uneven inoculum distribution with selective higher cell attachment to the smaller beads. Moreover, it may result in lower

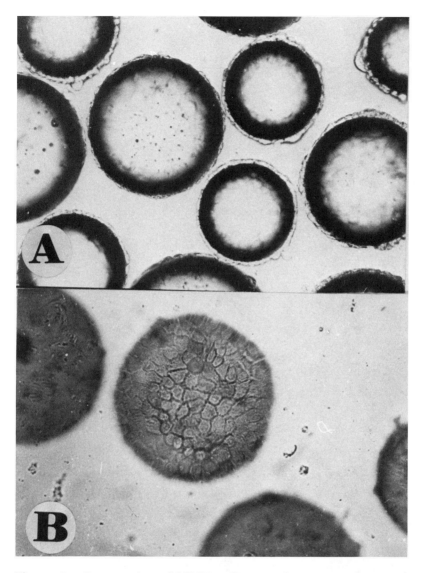

Figure 2 Propagation of MDCK cells on polystyrene microcarriers (light microscopy). (A) Focal plane at the middle of the microcarrier. Note that the cells can be seen on the periphery of the microcarrier. (B) Focal plane at the upper surface of the microcarrier.

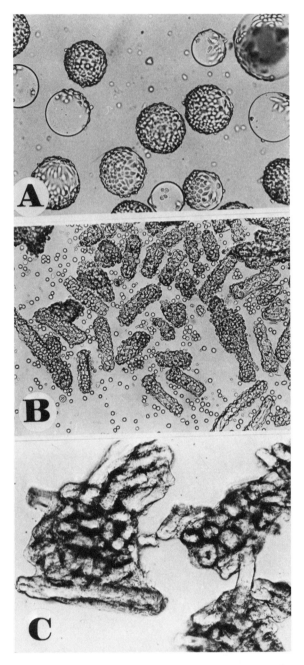

Figure 3   Propagation of anchorage-dependent cells on microcarriers (light microscopy).   (A) Mouse fibroblast cell line (L–929)

filtration velocity during MC bed washing (done mainly during cell harvesting from MC surface).

5. Smooth surface of the MC. The MC surface should be relatively smooth in order to allow good cell spreading and to prevent damage to the cells during MC collisions in stirred cultures.

6. Transparency. Good optical quality of the MC allows microscopic observation of the cells on the MCs with or without staining. Usually porous MCs (e.g, dextran or polyacrylamide) are completely transparent, so that observation of cells attached to the beads is achieved with great clarity (Fig. 3A). With other MCs (e.g, polystyrene, glass, or cellulose) cells can be seen in light microscopy only on the periphery of the MC (Figs. 2,3C) or by applying a special illumination device (42).

7. Toxicity. The MC should be nontoxic to the cells and authorized for use in the production of biologicals for human use (39,43).

8. Rigidity. Maroudas (44) showed that fibroblasts apply mechanical tension on their growth surface and have the ability to bend thin glass rods on which they are growing. Thus a certain degree of rigidity of the growth surface is needed in order to make it possible for cells to spread on the surface (34,44). Various MCs are reported to have different degrees of rigidity. Keese and Giaever (45) described liquid nonrigid microcarriers while others described rigid glass polystyrene and cellulose MCs (35,46−51). There are no available data regarding how this drastic change in the rigidity of MCs affects cell attachment, spreading and growth on their surface. However, there is a possibility that when using rigid MCs, cells may be damaged due to MC collisions during stirring.

9. Porosity. Porous MC can swell in aqueous media and absorb various substances from the growth medium. This is an unwanted phenomenon since it makes it difficult to remove

---

propagated on positively charged dextran-beaded microcarriers. Note that the microcarriers were seeded with low inoculum resulting in uneven distribution of cells on the microcarrier. (B) Mouse fibroblast cell line (L-929) propagated on DEAE-cellulose microcarrier (DE-52). Note that the cells are growing as monolayer and that there is release of rounded cells from the confluent microcarrier. (C) Bovine primary fibroblasts propagated on DEAE-cellulose microcarrier (DE-53). Cells are growing in multilayers in cell-microcarrier aggregates.

medium component from the MC culture. Moreover, porous
MCs can swell or shrink due to changes in pH.

V. TYPES OF MICROCARRIERS

A. Commercial Microcarriers

Several kinds of MCs suitable for cultivation of animal and human
cells are now commercially available from a variety of sources. The
detailed properties of these commercial MCs are described in Table 1.
These MCs can be divided into the following six main groups:

(1) Positively charged tertiary amino derivatized MCs (Biocarrier, Cytodex 1, Dormacel, Superbeads). Animal and human cells
possess a net negative charge on their surface at physiological pH
(52). Thus, these cells are attracted by electrostatic forces to the
positively charged MCs. This strong electrostatic interaction created between the cells and the positively charged MCs is critical in
MC culture since cells must attach rapidly to the MCs and withstand
shearing forces created in the stirred culture without being dis-
lodged from the MCs. Levine et al. (32) demonstrated that there
is a discrete range of charge which is necessary to achieve optimal
cell growth on DEAE-dextran MCs. At an exchange capacity lower
than the optimal range, usually a low degree of cell attachment is
observed and some of the cells which attach to the MCs grow to
some degree and then withdraw from the MC surface, accumulating
in large aggregates. At exchange capacities higher then the optimal, the cells attached to the MC surface and spread; however, a
toxic effect on cell growth is observed. These phenomena can be
explained by the fact that a certain level of cell attraction to the
growth surface is needed in order to achieve cell spreading, DNA
synthesis, and cell growth (34,53). On the other hand, the toxic
affect on the cell growth at high exchange capacity can be attributed to an arrest of cell membrane movement as the result of the tight
cell surface—MC interaction (34).

The need for a discrete range of charge on the MC surface was
demonstrated with several positively charged MCs derivatized with
quaternary, primary, or tertiary amines (32,34,51,54,55). It was
also shown that the hydrophobicity of the MC matrix and of the
charged group affects drastically the rate of cell attachment spreading and growth (54,55), and that the exchange capacity optimal for
cell attachment is not necessarily optimal for cell growth (56). The
commercial positively charged MCs are beaded MCs having a porous
hydrophilic matrix (dextran or polyacrylamide) with positively
charged tertiary amino groups distributed throughout the whole matrix of the MC (Figs. 1,3A). The Dormacel MCs are different from
the other tertiary amine-derivatized MCs in that they have a dimer

DEAE functional group which has a pK of 6.5 instead of monomer DEAE (pK of 9.2), thus reducing the exchange capacity of the MC from 1.5 to 0.3−0.65 mEq/g dry materials (57,58). Moreover, these MCs are available in several degrees of charging.

The main disadvantages of the ionic-charged MCs are that they can swell or shrink due to changes in pH and that serum protein adsorbs strongly to the MCs (36,59). Cellular products obtained from these MC cultures should be further purified from adsorbed proteins. These disadvantages have led to the development of the surface-charged MCs.

(2) Surface-charged MCs (Cytodex 2). Ideally, MCs should be noncharged in order to reduce binding of proteins from the culture medium and to facilitate removal of such proteins from the culture by simple washing. Gebb et al. (60) found that no cell growth was obtained on MCs coated with a layer of quartenary amine positively charged groups and that a certain depth of charged layer is needed to support cell growth.

The only commercial MCs in which the positively charges are located at the outer parts of the MCs are Cytodex 2. These are quaternary amine-derivatized beaded MCs having an exchange capacity of 0.6 mEq/g dry material (as compared with 1.5 mEq/g per dry material with MCs which are charged throughout the matrix of the MC). These surface-charged MCs were shown to adsorb less serum protein than Cytodex 1 MCs (60).

(3) Collagen-coated or gelatin MCs (Cytodex 3, collagen-coated Bioplas, Ventragel, Gelibeads). Biological molecules which play a role in cell attachment (collagen, fibronectin, laminin) were tested as materials for coating or preparing MCs (60,62−64,235). However, only collagen was found to be effective in inducing rapid cell attachment, spreading, and growth under stirring conditions. Commercial MCs are available in two forms: dextran or plastic beads coated with a surface layer of collagen (Cytodex 3, collagen-coated Bioplas), and beads composed entirely of crosslinked gelatin (Gelibeads, Ventragel). The collagen MCs are the most frequently used for cell cultivation for three main reasons:

1. These MCs are effective in supporting growth cells with low plating efficiency (60,65).
2. Cells can be harvested from these MCs by employing a specific protease, collagenase. Using this method, the harvested cells were reported to have maximal retention of the membrane. Gebb et al. (65) reported that in the case of Cytodex 3 MCs the harvested cells have a higher plating efficiency as compared with cells harvested from ionic-charged MCs. In the case of gelatin MC, cells can be harvested by completely dissolving the support material (gelatin) with proteolytic enzymes, thus making the cell transfer process in scaling up easier (66,129).

Table 1   Commercially Available Microcarriers

Type of MC	Registered trade name	Manufacturer	MC matrix composition	Charged group
Tertiary amino positively charged MCs	Biocarrier	Bio-Rad, USA	Polyacryl-amide	Dimethylamino-propyl
	Cytodex 1	Pharmacia, Sweden	Dextran	DEAE
	Dormacell	Pfeifer & Langen, Germany	Dextran	Dimer DEAE
	Mixrodex	Dextran Products, Canada	Dextran	DEAE
	Superbeads	Flow Labs, USA	Dextran	DEAE
Surface-charged MCs	Cytodex 2	Pharmacia, Sweden	Dextran	Trimethyl-2-hydroxyamino-propyl
Collagen-coated MCs	Cytodex 3	Pharmacia Sweden	Dextran	Collagen-coated
	Collagen-coated Bioplas	SoloHill Eng., USA	Crosslinked polystyrene	Collagen-coated
Gelatin MCs	Gelibeads	KC Bio-logical, USA	Crosslinked gelatin	Gelatin

Exchange capacity or equivalent	Specific gravity	Shape	Dimensions (μm)	Surface area (cm^2/g)	Porosity	Transparency	Autoclavable	Ref.
1.4 mEq/g dry materials	1.03	Beads	Diameter 120–180	5000	+	+	+	229
1.5 mEq/g dry materials	1.03	Beads	Diameter 131–220	6000	+	+	+	33, 36
0.3–0.65 mEq/g dry materials	1.05	Beads	Diameter 140–240	7000	+	+	+	57, 58
N.D.	1.03	Beads	Diameter about 150	N.D.	+	+	+	230
2.0 mEq/g dry dextran	N.D.	Beads	Diameter 135–205	5000– 6000	+	+	+	231
0.6 mEq/g dry materials	1.04	Beads	Diameter 114–198	5500	+	+	+	36, 60
60 μg collagen/ cm^2-MC surface	1.04	Beads	Diameter 233–215	4600	+	+	+	36, 60
N.D.	1.02, 1.03, or 1.04	Beads	Diameter 90–150 or 150–212	About 350	–	±	+	232
–	1.03– 1.04	Beads	Diameter 115–235	3300– 4300	+	+	+	129

Table 1   (Continued)

Type of MC	Registered trade name	Manufacturer	MC matrix composition	Charged group
Gelatin MCs (continued)	Ventragel	Ventrex, USA	Crosslinked gelatin	Gelatin
Polystyrene MCs	Biosilon	Nunc., Denmark	Polystyrene	Negative charge (tissue culture of treatment)
	Cytospheres	Lux, USA	Polystyrene	Negative charge (tissue culture of treatment)
	Bioplas	SoloHill Eng., USA	Crosslinked polystyrene	N.D.
Positively charged cellulose MCs	DE-52 or DE-53	Whatman, England	Microgranular cellulose	DEAE
	DEAE, QAE, or TEAE cellulose	Sigma, USA	Fibrous cellulose	DEAE, OAE, or TEAE
Glass MCs	Bioglas	SoloHill Eng., USA	Glass-coated plastic	Glass

N.D., no data available.

Exchange capacity or equivalent	Specific gravity	Shape	Dimensions (μm)	Surface area (cm^2/g)	Porosity	Transparency	Autoclavable	Ref.
–	N.D.	Beads	Diameter 150−250	N.D.	+	+	+	233
Surface charge of $2-10 \times 10^{14}$ charges/cm^2	1.05	Beads	Diameter 160−300	225	−	±	−	32, 45, 110
Surface charge of $2-10 \times 10^{14}$ charges/cm^2	1.04	Beads	Diameter 160−230	250	−	±	−	234
N.D.	1.02, 1.03, or 1.04	Beads	Diameter 90−150 or 150−212	About 350	−	±	+	232
1 or 2 mEq/g dry materials	N.D.	Cylindrical	Diameter 40−50 Length 80−400	N.D.	+	−	+	21, 34, 46, 50, 51, 73 74, 226
About 1 mEq/g dry materials	N.D.	Cylindrical	Diameter 10−20 Length 10−800	N.D.	+	−	+	2, 71
N.D.	1.02, 1.03, or 1.04	Beads	Diameter 90−150 or 150−212	About 350	−	±	−	47−49, 89

3. Since the collagen MCs are noncharged, they absorb less
    serum protein from the culture medium as compared with
    positively charged MCs (60). Sometimes the gelatin MCs
    (not the collagen-coated MCs) have a tendency to aggregate,
    especially in the presence of serum-containing media (67).
    In these cases the specific serum batch should be checked
    before its use in order to prevent aggregation of the gelatin
    MCs (Reuveny, unpublished data).

(4) Negatively charged polystyrene MCs (Biosilon, Cytospheres,
Bioplas). Traditionally, ADCs are propagated on negatively charged
surfaces such as glass or tissue culture-treated polystyrene. It was
demonstrated by Maroudas (68) that optimal cell spreading was obtained at a density of $2-10 \times 10^{14}$ negative charges/cm^2. Cell attachment on a negatively charged polystyrene surface is significantly lower than that on a positively charged one. The reason for this
seems to be an electrical repulsion between the negatively charged
cells and the negatively charged growth surface. The electrostatic
repulsion may be overcome through ionic interaction, protein bridges,
or local concentration of positive charges on the edges of the cell
philopodia (69). Fairman and Jacobson (63) found that porous MCs
in which the negatively charged sulfonated groups are located in a
30-nm layer at the outer surface of the MCs can support cell growth.
However, if these negative charges are distributed throughout the
MC matrix, no cell attachment or growth can be achieved (34), presumably due to excessive electrical repulsion. The commercial polystyrene MCs are composed of hydrophobic nonporous matrix (tissue
culture-treated polystyrene) with a low negative charge on their surface (Fig. 2). Since these MCs are low-charged and nonporous,
there is negligible absorption of medium ingredients into the MC matrix. These MCs do not shrink or expand due to changes in pH
and are relatively resistant to breakage during long-term culturing
under stirring conditions.
     The main disadvantages of these MCs are the low rate of cell at-
tachment on the MC surface (as mentioned above) and the tendency
of the MCs to settle relatively quickly in culture. Because of this
it is difficult to maintain cells on the surface of these MCs in
stirred cultures. The Bioplas MCs are different from other polystyrene MCs in that they are composed of crosslinked polystyrene
(thus they can be autoclaved) and are available with different diameters and densities.
     (5) Glass MCs (Bioglas). Glass is used as a support for cell
attachment and growth in stationary cultures or roller bottles.
However, glass is not suitable as a basic matrix material for MCs
because of its high density (1.5 g/cm^3). Recently glass MCs were
introduced. These MCs are composed of hollow glass beads or
glass-coated plastic beads with a specific density of $1.02-1.04$.

The commercial glass MCs are made of glass-coated plastic beads. They are similar to polystyrene in that they have negative charges on their surface and are nonporous in nature. However, the rate of cell attachment on glass MCs is considerably higher than on the tissue culture-treated polystyrene MCs. Moreover, the glass MCs can be cleaned by applying alcohol and/or acid and thus can be re-used several times (48,49). The attachment mechanism of cells on the glass (and polystyrene) MCs is reported to be different than that on DEAE-dextran MCs (49). Cells attach to the positively charged dextran MCs as a layer, while attachment to glass MCs is done via philopodia. Varani et al. (48,49) reported that, probably due to this phenomenon, cell detachment from glass MCs is accomplished by using a lower concentration of trypsin, thus obtaining better cell viability and plating efficiency (48) as compared with DEAE-dextran MCs.

(6) Positively charged cellulose MCs. Most of the available MCs have a beaded shape. The main reason for this is ease of production since these MCs are usually produced by an emulsion technique (34). The only nonbeaded MCs are the DEAE-cellulose ones. Reuveny et al. (50,51,70) reported that DEAE-cellulose anion exchange resins, mainly used for chromatography, can also be used as MCs for propagating ADCs. Two types of cellulose carriers are used:

1. MCs made of microgranular DEAE-cellulose which are rigid in their nature (DE-52 or DE-53 of Whatman). These MCs support growth of primary cells and cell lines (Figs. 3B, 3C, 4, and 5).
2. MCs made of positively charged fibrous cellulose which are nonrigid in nature and can be bent (QAE, 2-hydroxypropyl aminoethyl, DEAE and TEAE, triethyl aminoethyl, celluloses from Sigma, St. Louis).

These MCs were used for propagating human diploid fibroblasts (2, 71). Both types are cylindrical-shaped, having a cellulose matrix and charged with positive tertiary or quaternary amine groups throughout the matrix. Optimal cell growth was achieved at an exchange capacity of $1-2$ mEq/g dry materials (50). The exchange capacity of the cellulose MCs was sometimes found to affect the pattern of cell growth on the MCs (growth in monolayers or in multilayers; Ref. 226; Fig. 3B, 3C).

These elongated cylindrical cellulose MCs have several advantages over beaded MCs:

1. Cellulose MCs have a higher surface-to-volume ratio which leads to a lower degree of adsorption of ingredients from

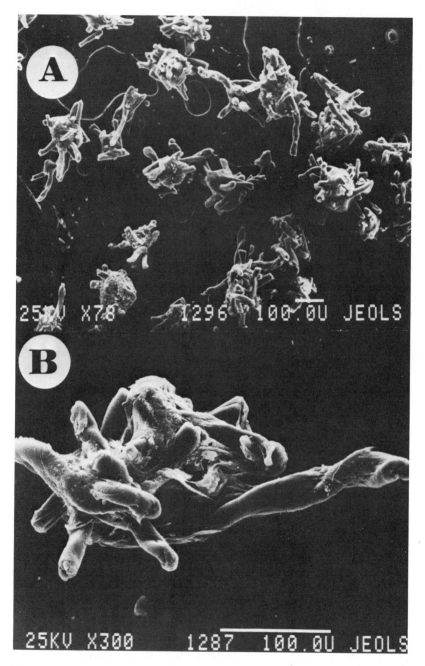

Figure 4   Scanning electron micrograph of primary chick embryo fibroblasts propagated on DEAE-cellulose (DE-53) microcarriers in multilayered cell-microcarrier aggregate culture. (A) Aggregate culture. (B) A single aggregate.

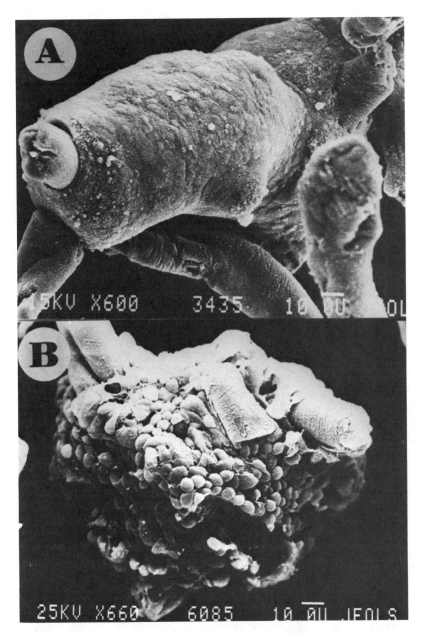

Figure 5   Propagation of human adenocarcinoma cells (A) and *Aedes aegypti* (B) on DEAE-cellulose (DE53) microcarriers. The cells grow in multilayers entrapping the elongated cellulose microcarriers.

the culture medium onto the MC (Reuveny, unpublished data).

2. There is a tendency for the elongated MCs to attach to each other by forming cell bridges (2,50,51; Figs. 3C and 4). As a result of this phenomenon the cells (especially cells with fibroblast morphology) can grow in aggregates in multi-layered fashion (100–200 μm diameter). The cells in the center of the cell-MC aggregate are fed by the flow of medium through the porous MCs.

3. Cells which have a tendency to grow in small aggregates attaching to flat growth surface in Petri dishes, e.g., *Aedes aegypti* cell lines (73) or human adenocarcinoma cells (72), cannot grow on beaded MCs since they peel off very quickly. These cells have the ability to grow in aggregates on the cylindrical cellulose MCs (72,73; Fig. 5).

4. These cylindrical MCs can withstand a higher rate of gas sparging into the culture as compared with beaded dextran MCs (Reuveny, unpublished data).

5. These MCs are significantly cheaper than all the available beaded MCs. High yields of cells from diploid cell strains (2,71), cell lines (51,72–74), and primary cells (21,54,74) were obtained from these cellulose MC cultures; neuronal and mascular cells grow as aggregates on the DEAE-cellulose MCs to a high degree of maturation, i.g., myelination and cross-striation (21), and several viruses and biologicals were produced from cellulose MC cultures (2,71–74).

Recently, a new type of commercial MC (MICA produced by Dr. Muller-Lierheim, KG Biol. Lab., Germany) were reported (75). These MCs are based on Eupergit C beads to which cell adhesion and growth factors are covalently coupled. The nature of these molecules is not specified. Up to now there have been no data in the literature on the use of these MCs in cell cultures. These MCs are not presented in Table 1.

B. Noncommercial Microcarriers

Several researchers have found that commercial MCs do not fit the specific requirements needed by their cell culture system, such as the efficiency of cell harvesting from the MC surface, problems with absorption of proteins by the MCs, etc. (76,77). Therefore, efforts have been made to develop new MCs which will have different characteristics. A list of these noncommercially developed MCs is given in Table 2. Some of these MCs are described below:

1. Keese and Giaever (45) reported that fluorocarbon emulsion droplets coated with polylysine are suitable for use as MCs

Table 2   Noncommercial Microcarriers

Microcarrier	Ref.
Amino acid (glycine)-derivatized polystyrene-divinylbenzene MCs	203
Cell attachment peptide MCs	80
Fibronectin-coated beads	199
Fibrous collagen particles (Fibronectin- and laminin-coated)	198
Fibrous microdisk carriers	78
Liquid MCs	45
Negatively charged polyvinyltoluene MCs	197
Negatively charged (sulfonated), positively charged (polyethyleneimine), or collagen-, BSA-, laminin-, or fibronectin-coated polystyrene-divinylbenzene MCs	76, 63, 204
Polylysine-, gelatin-, or heparin-coated MCs	64
Positively charged (polylysine, DEAE, or diamino-hexane) or gelatin-derivatized agarose-polyacrolein microspheres (Acrobead)	201
Positively charged (teriary amine) polystyrene MCs	200
Primary amino-derivatized polyacrylamide MCs	54, 55
Protein-coated DEAE-Sephadex MCs	202
Serotonin MCs	79

for cell growth. The droplet size distribution is large (100−200 μm) but the system has the advantage of simple cell harvesting by centrifugation, which breaks the emulsion into component phases.
2. Reuveny et al. (54,55) reported that a variety of cells can be propagated on positively charged primary amine-derivatized polyacrylamide MCs. A systematic study was done to determine the optimal degree of charge and hydrophobicity for cell growth. The optimal MC was found to be derivatized with diaminohexane. Its charged density was 0.3 mEq/g dry material, which is considerably lower than that of the tertiary amine-derivatized MCs.

3. Bohak et al. (78) used 0.5- to 1.5-mm-diameter disks (150 μm thick) as MCs for propagating ADCs. These disks are made of thermally bonded fabric sheets made of 10- to 12-μm diameter polyester fibers. The main advantage of these MCs is that low inoculum is needed to initiate the culture.

4. Hannan and Mcauslan (79) showed that several types of cells have the ability to grow on serotonin-coupled agarose beads. They showed that cells growing on these beads expressed a different pattern of polypeptide secretion as compared with cells growing on polystyrene or fibronectin.

5. Ruoslahti et al. (80) coupled synthetic peptides (4-30 amino acids) modeled after the fibronectin cell attachment domain into agarose beads. Good cell growth was obtained using these MCs.

C. Macroporous Microcarriers

Recently, a conceptually new type of MC was described by several researchers (81–83). This is the porous MC in which cell propagation occurs not only on the MC surface but inside as well. The propagation of cells inside macroporous MCs has several advantages. The porous MCs are designed so that cells have access to a large interior surface area on which to grow, while the matrix provides a shell which protects the cells from shear stress. This protection allows propagation of the cells at a higher agitation rate and higher air-sparging rate, thus facilitating the scaling up of the fermentation. Moreover, since the cells are distributed throughout the volume of the MC instead of being restricted to the surface, a much higher cell density can be achieved. Nilsson et al. (82) described a 300- to 400-μm-diameter gelatin macroporous beaded MC in which cells can grow inside. The advantages of these MCs over the conventional ones are reduction of inocula as a result of higher plating efficiency and increased surface area per bead (the ratio of cells to bead was reduced from 5–10 to 2–3), higher cell yield, and mechanical protection to the cells.

Young and Dean (81) described a new fluidized-bed bioreactor for continuous production of biologicals from cell cultures. The basic element of this system is porous carriers. These carriers are approximately 500 μm in diameter and are made of collagen sponge matrix containing heavy particles (specific gravity of 1.5). The highly porous carriers (85% of their volume is open pores) provide space for the propagation of cells and provide a flow of nutrients to the carriers and removal of toxic product and secreted proteins from the carrier to the medium (84). Recombinant CHO (Chinese hamster ovary) cells were propagated in this system to a density of $1-5 \times 10^8$ cells/ml (81,84).

Chan et al. (83) have described porous MCs made of collagen-glycosaminoglycan crosslinked polymer. These MCs have 300- to

500-μm diameters, are highly porous (greater than 98% void volume), and have a high specific surface area of 1 m^2/ml. At the laboratory scale they were able to propagate MRC-5 (cells from a human diploid cell strain) and CHO cells inside these MCs.

A different way to propagate ADCs is by the encapsulation method in which living cells are entrapped inside soft gel beads. Usually, this method is used for propagating hybridoma cells under perfused conditions rather than propagating ADCs. Several systems are described (4,85,86). This encapsulation system is different conceptually from the porous MCs system in that the cells are entrapped inside the bead at the stage of bead formation, and there is no possibility for cells to enter or leave the bead without disrupting it. Nillson et al. (86,87) propagated recombinant CHO cells entrapped inside agarose beads. Although the cells grow inside the agarose beads to a high cell density, a drastic decrease in tissue plasminogen activator production was observed. This encapsulation method will not be described in this chapter.

D. Choosing the Optimal Microcarrier

As described above, nowadays a variety of MCs are available. However, it should be emphasized that there is no MC (commercial or noncommercial) which is optimized for use in all applications. Each MC has its advantages and disadvantages and should be tested in the culturing system in which it is to be used. Factors to be taken into account in choosing MCs are:

1. The efficiency of cell propagation (58,88) and especially product accumulation in culture (4,47,87,89). Varani et al. (47,89) investigated the production of proteolytic enzymes and arachidonic acid metabolites by MRC-5 cells on various MCs (DEAE-dextran, glass, gelatin- and collagen-coated MCs) and found that accumulation of these metabolites in the culture fluid is significantly affected by the types of MCs on which the cells are growing. Similarly, Nilsson et al. (4,87) showed that tissue plasminogen activator and interferon production by recombinant CHO cells is affected by the type of MCs (positively charged and collagen-coated), thus showing clearly that the nature of MC substrate can influence basic cellular properties, the mode of cell propagation, and the efficiency of final product production (87).
2. The ease and efficiency of cell harvesting from the MC surface (48,49; see Section VII.F.1) and the possibility of autoclaving of the MCs. These factors are important in scaling up of the culture.
3. The mode of cell propagation on the MCs in real monolayer form or in multilayered aggregates (Fig. 1 vs. Fig. 4). In general, cell aggregation does not affect the rate of product

production (72-74). However, Schultz et al. (90) reported
a decrease in productivity as a result of cell multilayering,
possibly due to the limited nutrition of the cells located in
the inner layers.

4. The rate of cell detachment from confluent MCs (59,61,91)
   and the mechanical strength of the MCs. These factors are
   very important in long-term perfusion-type cultures.
5. The price of the MCs and the possibility of reusing them
   (48,49). These factors affect the cost of the production
   process.
6. The rate of medium adsorption and especially product adsorption into the MCs. These factors are sometimes important in
   the downstream purification of the final product (60).
7. The rate and efficiency of cell attachment onto the MCs in
   the specific agitated culture in which the MC culture will be
   established. These factors are important in the initiation of
   the MC culture.

VI. APPLICATIONS OF MICROCARRIER CULTURES

A. Cell Propagation

A wide range of cells have been cultivated on various MCs (37,40),
e.g., anchorage-dependent and independent cells; cells of invertebrate, fish, bird, and mammalian origin; transformed and normal
cells; fibroblastic or epithelial cells; and, lately, recombinant cell
lines. A list of cells propagated on MCs was recently published by
Pharmacia (92). A detailed description of a procedure for propagating cells on MCs at the lab oratory scale is described by Lindskog
et al. (93). The most common method for enumerating cells in MC
culture is nuclei count (36,94). In this method a solution of 0.1%
crystal violet and 0.1 M citric acid is added to the cell-covered MCs.
After incubation at 37°C, the nuclei are released from the cells by
vigorous mixing. The crystal violet stains the nuclei, which are
counted in a hemacytometer. A variation of the nuclei count method
is to lyse the cells in a 0.1 M citric acid solution and 1% Triton
X-100. With this method, there is much less background debris and
the nuclei can be counted in an automatic particle counter (67).
Other researchers have trypsinized cells from the MCs surface and
then counted them by the trypan blue exclusion methin in a hemacytometer. There is not always a good correlation between these two
methods. The correlation depends on the viability of the culture and
the efficiency of cell trypsinization (95).

Schon and Wadso (96) used the microcalorimetric method for
monitoring on line the growth of Vero (monkey kidney cell line)
cells on cytodex MC, while Miller et al. (97) evaluated the number

Table 3   Viruses Propagated on Microcarriers

Virus	Ref.
Aujesky virus (pseudorabies)	101
Feline panleukopenia virus	206
Fish viruses	201, 213
Foot-and-mouth disease virus	27, 100, 211, 215
Hepatitis A virus	209
Herpesviruses	113, 126, 214
Influenza	51, 205
Insect viruses	73
Marek virus	74
Oncornaviruses	176
Papovavirus	216
Parvovirus	207, 217
Poliovirus	2, 71, 99, 98, 159, 160, 212, 219
Rabies virus	218, 224
Respiratory syncytial virus	208
Rubella virus	102
Simian virus 40	36
Sindbis virus	51, 98, 103
Vesicular stomatitis virus	98, 174, 220

of attaching cells to MC by using coulter counter as a biomass probe. This cell estimation was done on line while the cells were still attached to the MC.

B. Virus Propagation

Cells cultured on MCs can be used as substrate for the propagation of viruses. A wide range of viruses were propagated in MC cutures (Table 3). In spite of the reduced viral yields per cell seen in some

instances (98), the high cellular productivity appears to be a great
advantage for large-scale virus production. Poliomyelitis virus (99),
foot-and-mouth disease virus (100), and Aujeszky vaccine virus (101)
were propagated in MC cell cultures for vaccine production on an
industrial scale (up to 1000-liter scale). The environmental conditions optimal for replication of viruses in culture sometimes differ
from that for optimal cell propagation (102,103). While some virus
infections destroy cells, others do not cause cell breakdown, and
such viruses (e.g., measles, rubella) are continually produced and
released into the growth medium at a more or less constant rate for
many days. These viruses can be harvested by fully continuous or
semicontinuous perfusion methods (73,102).

Table 4   Biologicals Produced in Microcarrier Culture

Product	Ref.
Bladder cancer-associated membrane protein	238
Carcinoembryonic antigen	72
Growth hormone (recombinant)	61
Human angiogenesis factor	88
Human lysosomal acid lipase	227
Inteferon	
Native (β)	36, 77, 106, 124, 131, 151, 164, 173, 222, 225
Recombinant (β)	9, 58, 90, 108, 146
Recombinant (γ)	87, 91
Interleukin 1	223
Interleukin 2 (recombinant)	195
Nerve growth promoter	228
Steroids	150
Tissue plasminogen activator or fibronolytic enzymes	
Native	10, 104, 105, 156
Recombinant	84, 87, 181, 236

C. Production of Nonviral Biological Products

Several biological products have been reported to be produced in
MC culture (Table 4). These products are obtained either from
"normal," transformed, or recombinant cells. If the product is produced after the induction step (e.g., human beta-interferon), a
batch mode system is adopted. However, if the product is produced
spontaneously, sometimes perfusion culture with a continuous or
semicontinuous mode of product harvesting should be used, e.g.,
carcinoembryonic antigen (72), tissue plasminogen activator (104),
and recombinant tissue plasminogen activator and interferons (86,
90).

The biological products reported to be produced on a relatively
large scale in MC culture are human fibroblast beta-interferon (106,
107), plasminogen activator (104,105), and several recombinant
products (81,84,90,108).

Hu et al. (236) investigated the production of prourokinase in
three cultivation systems: MC culture, roller bottles, and ceramic
matrix. In all three systems they found no significant difference
in the volumetric productivity of the prourokinase. Recently,
Tschopp et al. (109) described the possibility of producing urokinase from MC culture in space under microgravity conditions.

VII. MICROCARRIER GROWTH SYSTEMS

A. General Discussion

At the laboratory scale cells can be cultivated on MCs in Petri dish
stationary cultures inside an incubator in an atmosphere of 95% air
and 5% CO_2. Roller bottles, used initially for cultivation of ADCs on
the inner surface, have also been used for cultivation of ADCs on
MCs (36,110). However, in practice the system employed by most
investigators is the stirred-tank reactor, or fermenter, which can
be used at the smallest as well as the largest scale. The fermenter
is simple in comparison with many of the other devices that have
been used or proposed, and provides a homogeneous, accessible culture that can be sampled, monitored by in situ sensors, and controlled readily. Moreover, much of the technology of microbial fermentation in fermenters can be transferred directly to processes for
the production of animal cells and their products.

Despite the similarity of cultivation of animal cells on MCs in
fermenters to well-established microbial fermentation processes, and
the even greater similarity to cultivation of freely suspended animal
cells, there are peculiarities of MC systems that must be addressed
in designing the vessel as well as in conducting the process (111).
With MC systems two kinds of particles, the cells ($10-35$ μm in
diameter) and the MCs ($100-200$ μm), must be managed separately

and as a complex. During the first 6 hr or so after cell seeding, conditions should favor attachment, with as even a distribution of cells among MCs as possible. During the attachment phase stirring should be gentle. After the attachment phase, the conditions of the process should promote growth of the cells in a monolayer until they are confluent. Shearing forces resulting from rapid agitation must be avoided to prevent detachment or damage to the cells. MC particles, with or without cells attached, tend to settle rather quickly as a result of gravitational force. Special stirring devices and vessel configurations are used to get a homogeneous suspension of MCs without generating high-shear force at any locus in the culture. The gentle mixing systems employed usually do not promote sufficient gas exchange in fermenters of conventional configuration. Therefore changes in dimensions and other design features have been incorporated into new vessels built especially for MC cell culture, and existing vessels have been modified or used in unconventional ways.

The bottom of fermenters designed for MC culture should be fully rounded and the walls should be smooth to avoid opportunities for local accumulation of MCs. As aeration (gas exchange) is sometimes effected solely via the surface, i.e., the liquid-gas interface, the ratio of the height (h) to the diameter (d) should be 1:1, rather than the 2:1 or 3:1 ratio common with fermenters used for microbial fermentations. The inner parts of the vessel that are in contact with the growth medium should be siliconized in order to prevent cell attachment and sticking of the MCs (112).

Special attention should be given to location and configuration of inlet and outlet ports. Although penetration through the bottom of the vessel are undesirable, as they provide stagnant spots where MCs may accumulate, sometimes they are used. On the other hand, too many pipes entering through the head plate and extending below the liquid surface create a baffling effect which may disrupt the flow pattern of the MCs and may increase shearing forces. Furthermore, the pipes may present irregular surfaces on which the MCs may stick. If it is necessary to use a bottom valve for cleaning or emptying the vessel, it should be a ball or ram valve so that as little dead space as possible is created. If penetrations are made through the side wall, they should slope downward within the vessel at 30-45° to prevent accumulation of MCs on the intruding segment and permit working at low volumes. Changing medium or washing cells in batch or perfusion culture can be done easily by withdrawing liquid through an outlet pipe equipped with a 40- to 80-μm stainless steel screen. In order to prevent clogging of the filter, especially in perfused cultures, a cylindrical filter can be mounted on the stirrer shaft. The rotational movement of the shaft-mounted filter prevents its clogging (104,113).

The major problems that usually limit the scale of operation in MC cultures are

1. Sensitivity of cells to mechanical damage during stirring and gas sparging
2. Supply of oxygen to the cells
3. Cell-harvesting from MC surface and inoculum preparation.

These problems and their potential solutions are described below.

B. Sensitivity of Microcarrier Cultures to Mechanical Damage

A major problem encountered in scaling up a MC culture system is the sensitivity of cells to mechanical damage. This sensitivity is the result of lack of a protective cell wall, the relatively large size of the animal cells, and the fact that the cells are anchored to the MC and are not freely suspended. In large volumes this sensitivity may hamper the supply of sufficient oxygen to the cells in the conventional manner (i.e., by sparging of air through the medium and by dispersion of the air bubbles by means of stirring). The sensitivity of ADC to detachment from their growth surface depends on the type of the propagated cells, functional groups on the growth surface, and the type and strength of the shear stress to which the cells are subjected. Weiss (114) found that the mean minimal detachment forces for an established epithelial cell line is $1.06-1.36$ N/m^2 and for rat skin fibroblasts 0.73 N/m^2, thus showing that the sensitivity to shear depends on the type of cell. Moreover, it was shown that the functional groups located on MC growth surface play a major role in the sensitivity of cells to detachment from their growth surface. Treatment of the growth surface with concanavalin A enhanced the adhesiveness of cells. The minimal detachment force was increased from 1.44 to 3.30 N/m^2 (115,116). The effect of shear stress on ADCs (embryonic kidney cells) was studied by Stathopoulos and Helluns (117). They found that shear stress of 0.26 N/m^2 has no effect on cell viability, at intermediate stress levels ($0.65-1.03$ N/m^2) there are morphological changes of the cells and a loss of cell viability ($10-20\%$ detachment), and at higher stress levels (2.6 N/m^2 and higher) marked cell reduction was observed ($75-85\%$ cell detachment). Nerem (118) showed that moderate shear stress, which has no effect on cell viability, induces changes in cell shape and orientation, cytoskeletal structure, and membrane potential.

Tyo and Wang (103) studied the shear forces generated in magnetic bar-type spinner flasks ($0.1-1.0$ liter volumes) and their effects on cell growth and virus production in MC cell cultures. They defined an integrated shear factor as follows:

$$\frac{2\pi NDi}{Dt - Di}$$

where N is rpm, Di is the diameter of the impeller, and Dt is the diameter of the tank. They postulated that the factor should be kept below 40 in order to permit cell attachment and maximum cell yield. If attachment is carried out during a time when no agitation is imposed, the factor may be increased to 80. As virus propagation is more sensitive to shear than is cell growth, the factor should be reduced to below 20 in order to get maximal yield of virus. The maximal allowed integrated shear factor depends on the cell type (67). Schulz et al. (58) found that when using recombinant L cells (producing human interferon), cells are not sheared off the MC up to an integrated shear factor of 30 (70 rpm). This factor can be a useful tool with respect to scaling up of MC cultures. According to estimations made by Tyo and Wang (103), the shear force generated by turbine impellers in large fermenters (1000 liters) is predicted to be significantly less than in small fermenters.

Cherry and Popaustsakis (38,95) studied the hydrodynamic effect on cells in MC cultures. They argued that cell damage in agitated MC cultures attributed to shear stress can be better explained as the effect of turbulence or collisions. Two effects stand out as likely to cause cell damage: (1) interaction between small size-scale turbulence, comparable to the MCs or the spacing between them (turbulent eddies), and the MCs; and (2) collision between the MCs and between MCs and solid objects, particularly the impeller. In a recent publication (95), these authors added an additional factor. During cell propagation MCs tend to aggregate by cell bridges (Fig. 1B). Breaking these bridges during stirring can cause cell destruction. It is not clear which mechanism of cell damage is predominant; that depends on cell and MC densities, intensity of agitation, impeller design, and fluid properties.

Recently, Croughan et al. (119,120) analyzed these hydrodynamic effects in terms of the more fundamental parameters of fluid dynamics which can provide useful principles for the design and scale-up of MC cultures. Moreover, in a recent publication (120), they showed that at mild agitation rates (35 rpm) the hydrodynamic phenomenon (MC collisions and interactions) are insignificant, adding inert MC particles (Sephadex beads) up to a concentration of 30 g/liter has no effect on foreskin diploid fibroblast propagated on Cytodex 1 MCs. At higher agitation levels (150 rpm) hydrodynamic damages occur: adding inert MCs up to concentration of 30 g/liter causes a gradual decrease in growth (exhibiting only 8% of the maximal cell density obtained in the control). This negative effect was especially noticeable during the cell attachment phase. The mechanism of hydrodynamic damage at a high agitation rate is discussed in this work (120).

Animal cells exhibit different degrees of shear sensitivity which depend mainly on cell type. When using shear-sensitive cells, several measures can be taken: using low-shear agitation systems at low agitation rates, using specially designed low-shear reactors, omitting or reducing the rate of air sparging into the culture, and adding special protective additives to the culture medium. Special design features (agitation systems and reactor design) which solve the shear sensitivity problem of MC cultures are described below.

C. Agitation Systems

1. General Requirements

Conventional systems for stirring microbial fermentations are not usually suitable for MC cultures. The following points serve to emphasize the special requirements of MC cultures and the design measures that have been taken to meet these requirements:

1. Stirring should be as gentle as possible, just sufficient to distribute the MCs homogeneously throughout the culture. de Bruyne (121) showed the need for a vertical component in the circulating force.

2. Bearing or hard-moving surfaces in contact with stationary surfaces should be located above the liquid surface to avoid mechanical attrition of the MCs. Thus, a hanging or top-entering agitator shaft is preferred to a bottom-entering type.

3. The stirrer should be driven by a motor that provides smooth and controlled rotation in the range of 10−150 rpm. Stirring speeds for MC cell culture are relatively low, in the range of 10−50 rpm. The stirring speed required for homogeneous suspension of MCs changes according to the volume, vessel dimensions, and impeller size and type.

   During the cell attachment phase stirring should be just sufficient to suspend the particles and may be done intermittently in a timed cycle (122). After the cell attachment phase, stirring speed may be increased to suspend the MCs homogeneously and increase oxygen supply. Later, when confluent growth is approached, stirring speed may be increased further to prevent overgrow of cells from one MC to another and consequent aggregation of MCs.

4. The stirring system should be vibration-free and without erratic motion to avoid trauma to the cells. Changes in speed should be gradual (36,123).

5. The design of the stirrer should be such that a range of culture volumes can be accommodated, since there are

procedures in which cells and MCs are seeded at a low volume and stepwise or gradual increases in volume are made.
6. Baffles, often provided in reactors used for microbial fermentations, should not be used, as turbulence and shearing
forces must be avoided.

2. Types of Impellers

The type of impeller used for MC cell culture deserves detailed discussion in light of the foregoing exposition of requirements and constraints for agitation systems. The spinner flask, developed in the
1950s for growing animal cells in suspension, was used by early investigators for MC cell culture. In this device a Teflon-coated magnetic bar, suspended and free to rotate at the bottom of a fixed
shaft suspended from the top of the vessel, is coupled magnetically
to a motor beneath the vessel. At the 44- and 150-liter scales,
Edy (124) and Scattergood et al. (125) reported the use of conventional turbine impellers. Although these systems support growth of
cells on MCs, they have several disadvantages:

1. Relatively high speeds, with unwanted high-shear forces,
are required to keep the MCs in suspension.
2. There is a stagnant point immediately below the shaft at the
bottom of the vessel where MCs tend to accumulate.
3. There is no prevailing force moving the MCs upward.

One of the first impellers designed especially for MC cell culture
is the paddle and anchor impeller. The rod or turbine is replaced
by a relatively large, flat, vertical blade. With this type of device
it is possible to generate a homogeneous suspension at a relatively
low stirring speed without substantial shearing. Hirtenstein et al.
(123) compared growth of MRC-5 and Vero cells in paddle- and bar-
type spinners and found that the paddle type allowed a 25−35% increase in cell yield. In their hands the stirring speed needed for
obtaining an even suspension of MCs was 20−40 rpm as compared to
50−60 rpm with the bar-type spinner. Although the various
paddle- and anchor-type impellers have been shown to improve cell
yield (10,110,113), they still suffer from the disadvantages that
MCs accumulate in the dead spot located under the shaft, no upward movement of fluid is generated, and adverse shearing effects
are seen when rotational speed is increased to attempt to get homogeneous suspension.

Improvement of the paddle-type impeller was achieved by twist-
ing the paddles at a 30−45° angle from the vertical. This causes
an upward movement of fluid and homogeneous dispersion of MCs at
lower stirring speeds, a condition that favors higher cell yields.
This screw-type impeller was used in several systems (122,123,126).

Hirtenstein et al. (123) found that this impeller is able to generate an even suspension of cell-bearing Cytodex MCs at 15–30 rpm and to accomplish at least a 50% increase in cell yield as compared to that achieved with the traditional bar-type spinner.

A rod with a bulb at the lower end, suspended from the top of the vessel and rotated so that the bulb moves in a circle in the low part of a contoured Pearson flask with the motion of the rod describing a cone, was described by de Bruyne (121). Upward movement of the MCs is generated and there is no accumulation of MCs at the bottom. An even suspension of MCs can be generated at 15–30 rpm. Hirtenstein et al. (123) found that 50–70% greater cell yields can be achieved with this type of spinner as compared with the conventional bar-type spinner.

Feder and Tolbert (127) and Tolbert and Feder (128) described an agitation system that consists of four flexible sheets held vertically, spanning the depth of the culture fluid. These sheets need only to be turned slowly to disperse the MCs adequately.

Reuveny et al. (11) described a new fluid lift agitation system. The basic principle of this impeller is that the culture fluid containing MCs is pumped upward from the bottom of the vessel through the hollow impeller shaft. The fluid leaves the shaft near its top but below the surface of the culture through two short outlet pipes, on opposite sides of the shaft, pointing backward tangentially to the motion of the shaft. Circulation of the culture and an even suspension of the MCs can be obtained with this device at rpm values of 30–60. There is no stagnant spot where MCs might accumulate, there is minimal collision of MCs with the impeller, and the movement of MCs relative to the immediate environment is slow. Reuveny et al. (11) studied propagation of several types of cells on Cytodex 1 MCs using this stirring system and found in each case a significant increase in cell yield as opposed to that with a paddle-type spinner flask. This new stirring system can be operated at a hundred liter scale. The main disadvantage of this stirring system is that it fixes the volume that must be used and thus precludes the useful procedure of "building" the culture in the growth vessel. Lehman et al. (108) described a new agitation system which is composed of a basket impeller. The basket (which is made of coiled hollow fiber tubing) is moving in a nonconcentric way. In this system an even suspension of cells and MCs was achieved in a 22-liter fermenter by an agitation rate of 60 rpm.

3. Reactors Without Impellers

There have been several publications dealing with the growth of cells on MCs in vessels without mechanical stirring devices. Two types of reactors were used: (1) fluidized-bed reactors (81,84,122) used mainly in perfusion cultures and (2) static reactors in which cells are not subjected to shear and are used mainly to maintain the cells in perfused state.

Clark and Hirtenstein (122) described a fluid lift system in which the MCs with cells are retained by filter screens with 100-μm openings. The filtered fluid is circulated through a medium reservoir in which it is aerated. Better cell yields than those with bar spinners were achieved. Using a similar concept, Young and Dean (81) described a large-scale, low-shear, fluidized-bed bioreactor for porous MCs (see Section V.C). The fluidization is accomplished in the reactor by medium flow in a recycle loop. The culture supernatant is separated from the particle slurry in a zone near the top of the reactor and is returned by the recycle loop. The carriers remain suspended in the reactor. Oxygen, pH, and temperature control is done in the recycle loop. Scale-up of the reactor is done by increasing the sectional area of the reactor. The reactor was scaled up to 23-liter volume. Cell concentration of $1-4 \times 10^8$ cells/ml were reported.

Strand et al. (130,131) described a system in which human fibroblasts were propagated on MCs located in the extracapillary space of a hollow fiber matrix. The cells were nourished and induced to produce interferon by nutrients and inducers perfused through the capillaries. High-cell-density interferon production was demonstrated in this system. More recently, Tolbert et al. (12,128,132) described a proprietary device, a "static maintenance reactor," in which cells were grown in a conventional reactor until maximal density was achieved, and then transferred at still higher density (10^8 cells/ml) into the static maintenance reactor in which the MCs are mixed with a semirigid matrix and then perfused while kept in a nonproliferating state. Cells are maintained in this state up to 2 months under low-shear conditions. Fresh medium is added through a relatively low-porosity tube located throughout the reactor, and used medium is removed through a relatively high-porosity tube. Oxygen is supplied through semipermeable silicone tubing running through the matrix and coiled around the medium tubes. In this arrangement, tubes for exchange of medium and gases are in close proximity at any point in the cell matrix. Static maintenance reactors up to 16.5-liter size were reported.

D. Oxygen Measurement and Control

Oxygen is a key nutrient in cell metabolism. Dissolved oxygen (DO) concentration can affect cell yield and thus directly or indirectly affect product expression (133). At low oxygen tension cells tend to grow at a low rate and produce a high concentration of lactic acid (134), while too high an oxygen tension is toxic to the cells (135). The optimal concentration of dissolved oxygen for cell growth has been found to vary from 3 to 20%, 15 to 100% saturation of air

(36,133,135). The oxygen utilization rate (OUR) varies among cell
types in the range of $0.05-0.5$ mmol $O_2/10^9$ cells/hr (133,136).
The OUR of cells in MC culture is sometimes affected by environmental factors. For example, Frame and Hu (137) found that OUR
is affected by glucose concentration.

In microbial fermentations the level of DO in the culture medium
can be controlled over wide limits by changing the agitation speed,
aeration rate, or both. The limit to the speed at which animal cells
or MCs with cells can be agitated and the reasons for this have
been discussed in an earlier section of this chapter (Section VII.B).
Aeration in microbial cultures is usually effected by sparging air
through an open pipe or perforated ring near the bottom of the vessel beneath the impeller. If this type of aeration must be used, it
should be used with caution as sparging air invariably generates
foam, especially with the high protein concentration ($5-10\%$ serum)
common to most cell culture media. Moreover, MCs tend to be entrapped in the rising gas bubbles and thus to float in the foam at
the top of the reactor. This phenomenon is less pronounced with
the elongated cellulose MCs as compared with the beaded dextran
ones. Gas bubbles can actually damage cells (138). Handa et al.
(139) systematically investigated the effect of gas-liquid interface on
the viability of suspended cells. They found that the degree of cell
sensitivity depends on the bubble size, superficial gas velocity, and
cell type. Moreover, they found that the region of highest shear
and turbulence is at the culture surface where bubble collapse
occurs. The effect was found to be less detrimental by adding non-
ionic surfactant such as pluronic polyol F-68, increasing serum concentration, or by ensuring that the bubbles are large (133,139).
Suppressing foam with chemical antifoam agents, a practice common
in microbial fermentations, cannot usually be done with cell cultures
as these agents are toxic to the cells (140). Aunins et al. (140)
investigated the effect of various antifoams on CHO cells MC culture. They found that antifoam must be selected carefully and be
tailored to the cell line in use. Food and medical grade simethicone
antifoams were found to be relatively harmless to cells (67,140).
However, even in the presence of antifoam, foreskin cell MC culture
sparged with air at a mild rate (0.1 cm/sec) without generation of
foam has a negative effect on cell growth. Moreover, they showed
that prolonged aeration of growth medium caused irreversible denaturation of the protein in the growth medium, thus reducing its
effectiveness in supporting cell growth. Hu and Wang (135) reported that with CHO cells, growth was not affected in the presence
of 400 ppm of FG-10 and P-2000 antifoams.

Tyo and Wang (103) found that a 30% decrease in final cell yield
resulted from increasing the culture volume from 100 to 1000 ml.
This decrease was caused by a deficiency in oxygen supply to the

cells. Even at the 1-liter scale, oxygen depletion can be a growth-limiting factor with MC cell culture.

Researchers in many laboratories have taken conventional as well as unconventional approaches to solution of the problems of supplying oxygen to MC cell cultures.

In the case of surface-aerated cultures, oxygen transfer is affected drastically by the rate of stirring and by the ratio of the area of the gas-liquid interface to the culture volume. Fleischaker and Sinskey (136) found that in a 14-liter fermenter in which oxygen was supplied to the cells by surface aeration only, the oxygen mass transfer coefficient (kla) was directly proportional to the stirring speed in the range of 15–100 rpm and inversely proportional to the culture volume according to the expression:

$$kla = \frac{0.414N}{V^{2.05}}$$

where N is the stirring speed and V is the culture volume. Spier and Griffiths (133) found that in an unsparged 4-liter fermenter fitted with an eight-blade turbine impeller the kla was affected even more by stirring speed. The following relationship was proposed:

$$kla = 1.5 \times 10^{-5} \times rpm^{1.85}$$

Realization of the importance of having a high ratio of surface area to volume led most workers to choose vessels with a height-to-diameter ratio of 1:1. As the ratio of surface area (A) to liquid volume (V) is inversely proportional to the fermenter height, maintaining a constant A/V ratio as volume is increased requires maintaining a constant height, and this leads to a highly unconventional vessel. The practical limit of h/d is not necessarily 1:1, but as this ratio becomes lower the vessel becomes increasingly difficult to construct and operate.

Another means of increasing the effectiveness of surface aeration is to enrich the atmosphere in the headspace with oxygen. Glacken et al. (141) calculated that if pure oxygen were used instead of air in the headspace, oxygen limitation would be seen at the 3.5-liter scale. However, several other workers reported that dissolved oxygen can be controlled by surface oxygenation of cultures in 100- to 200-liter fermenters. Hirtenstein and Clark (142) anticipated that oxygen limitation would be seen above the 200- to 300-liter scale. van Wezel (143) saw the effects of oxygen limitation at the 50-liter scale, while Scattergood et al. (125) encountered the problem at 140 liters. These differences in perception of the scale at which surface aeration is no longer effective in maintaining DO at or above the critical level derive from differences in cell concentration, cell type, agitation rate, fermenter configuration, and mixing system.

van Wezel (143) used a special perforated ring, like a sparger, located 2 cm above the liquid to direct a gas mixture (oxygen, nitrogen, and carbon dioxide) downward on the surface and accelerated its rate of solution. Using this device, DO could be controlled in cultures up to 350 liters volume. Hu et al. (144) and others (95) used a two-blade surface aerating impeller in order to induce turbulence in the culture surface and thus increase oxygen transfer rate from the culture surface. This surface aerator is a paddle-type impeller rotating with the agitator shaft at the gas-liquid interface on the culture surface. By introducing the surface aerator, fourfold increase in oxygen transfer rate (OTR) in 8-liter fermenter containing 5-liter culture is achieved (144).

Another means of controlling DO is by increasing the headspace gas pressure in order to increase the solubility of oxygen. However, as mentioned by Spier and Griffiths (133), this measure should be taken only in the latter stages of growth, as the toxic limit of DO may be reached in early stages when oxygen demand is low. A further drawback to elevating headspace pressure is that CO_2 solubility will be increased also and pH will be depressed as a consequence.

Sparging small amounts of air or oxygen into large reactors has been used with MC cell cultures (72,101,145,146). Successful sparging with little or no foaming is usually accomplished with a single-orifice sparger generating large bubbles. Sometimes antifoams are added (67,135), and it is easier to sparge elongated cellulose MC cultures than beaded dextran MC culture.

Spier and Griffiths (133) found kla values of $1-25$/hr with a sparged suspension culture of BHK cells as opposed to $0.1-4$/hr with unsparged cultures. They described the relationship between kla, stirring speed, and air flow rate in a 4-liter vessel with eight turbine blades on the impeller as follows:

$$\text{kla} = 3 \times 10^{-5} \times \text{rpm}^{1.58} \times \text{AFR}^{0.58}$$

when AFR is air flow rate in ml/min. Thus kla can be seen to depend strongly on stirring speed and air flow rate.

Katinger et al. (147) used an airlift fermenter for propagation of animal cells in suspension. They achieved gentle mixing by sparging large bubbles upward through a draft tube, which stabilized liquid flow and minimized foam formation. This system would appear to be applicable to oxygenation in large receptors. Katinger (148) was not able to grow cells on MCs in airlift fermenters.

Circulation systems in which cell-free medium in an external loop or reservoir is aerated in order to supply oxygen to the cells

have been used by several groups (81,105,122,126,149,150).
Griffiths et al. (10,105,126) used a closed-perfusion system for
propagation of MRC-5 cells on Cytodex 1 MCs in reactors up to 20
liters. Dissolved oxygen was monitored and controlled by surface
aeration in a reservoir.

Young and Dean (81) described a system for supplying of oxygen to a perfused fluidized-bed reactor. In this system, the growth
medium is oxygenated outside the reactor by means of a membrane
gas exchanger located on the medium circulation loop. An equation
for calculating the height of the fluidized-bed reactor at which no
anoxic conditions to the cells are generated is presented. This reactor is scaled up by increasing the reactor cross-section and keeping constant bed depth (81). Increasing OTR by recirculation
through an external oxygenator is not common in MC culture. The
main problem with such systems is that high recirculation rates are
required (141,148). Thus, the problem of separating MCs from culture supernatant is difficult (there is a need for large filters in
order to separate the MCs from the culture supernatant or to use
heavier MCs). This problem is aggravated with scaling up as higher circulation rates are required to maintain the critical DO in all
parts of the vessel containing the cells.

Several research groups have found that it is possible to oxygenate cell cultures using thin-walled, oxygen-permeable tubing im-
mersed in the culture medium. Oxygen diffuses through the wall
of the tube without producing bubbles (15,103,108,136,141,142,149,
151–153).

Sinskey et al. (151) found that 5 m of silicon tubing (i.d. 0.147,
o.d. 0.196, wall 0.025 cm) submerged in a 10-liter vessel with pure
oxygen in the tube and air in the headspace, could supply
0.25–0.33 mmol O_2/liter/hr at stirring speeds of 15–120 rpm. The
oxygen transfer rate was less sensitive to changes in agitation than
it was with surface aeration. Eberhard and Schugerl (153) investigated the effect of air pressure, percentage of oxygen in the
gas stream, and agitation speed on OTR in a silicon tubing system.
They found that increasing the pressure from 0.5 to 1.5 bar caused
an increase in OTR by approximately twofold, using pure oxygen in
the silicon tubing increased OTR by fivefold as compared with air,
and increasing the agitation rate from 25 to 100 rpm resulted in a
gradual increase in OTR (153). By using the silicon tubing system they were able to propagate insect cells in 10-liter fermenter to
a concentration of 5×10^6 cells/ml.

Lehman et al. (108) used a polypropylene hydrophobic porous
hollow fiber membrane (0.33 μm pore size) for this purpose. Using
this membrane (2.3 m/liter) they achieved maximum oxygen transfer capacity of 35 mg oxygen/liter/hr using air. An OTR of 0.16
mg/cm^2 tube/hr was achieved with the silicon tubing. The

hydrophobic hollow fibers were coiled around a basket-type agitator
(see Section VII.C.2). The OTR was affected by the tumble speed
of this membrane basket agitator as well as by the gas flow rate in
the tubing. The possibility of supplying oxygen at low stirring
rates is evident. Moreover, since silicon and hollow fiber tubing
are also permeable to CO_2, it is possible to control pH by delivering
CO_2 through the same tubing.

Whiteside et al. (154) and Spier and Whiteside (155) described
a novel caged aeration system. In this system air is sparged into a
closed wire mesh cage attached to and rotating with the stirrer
shaft. The penetrations in the screen are small enough to exclude
the MCs. Bubbles and foam are generated inside the cage only,
while oxygen is delivered to the cells by the medium which flows
freely in and out of the cage. Whiteside et al. (154) reported that
with this system they were able to control DO in a 10-liter baby
hamster kidney (BHK) MC culture. Griffiths et al. (6,156) combined the caged aeration system with the oxygen supply by medium
recirculation in order to produce tissue plasminogen activator in
highly dense perfused MC culture. Recently, the same principle
was introduced into the cell lift agitation system (11). A ring
sparger was introduced into the rotating double-walled draft tube,
the outer wall of which is made of 400-mesh wire cloth. The air
(or gas mixture) is pressurized through the stirring shaft and out
through the ring sparger, generating bubbles inside the double-
walled draft tube and exiting through the two outlets at the top of
the draft tube (11). Kla values in a 5-liter reactor were measured.
By the use of air kla values of 5/hr were obtained as compared with
kla value of 30/hr in the case of pure oxygen. The OTR was found
to be affected drastically by the gas flow rate but is not affected
by agitation rate (11).

Katinger et al. (157) and Karrer et al. (158) described a similar concept in which the bioreactor is divided into two compartments: one for cell cultivation and the other for aeration. The
compartments are separated by a vibrating screen. The vibration
helps to keep the screen clean and improves diffusion of oxygen into
the culture. This system has an OTR of 4 mg O_2/cm^2 cage
surface/hr.

In summary, it should be clear that DO is the most critical
parameter for scaling up MC cell cultures. When designing an
aeration system for MC cell culture, any one or any combination of
the above approaches, namely, control of surface aeration, delivery
of oxygen by sparging, and controlled stirring speed, may be used
to maintain the critical DO concentration. Looby and Griffiths (236)
reported that in a 39-liter reactor stirred at 40 rpm, an OTR of
3 mg/liter/hr is achieved in the case of caged aeration, and an OTR
of 12.6 is achieved in perfused system recirculated at a rate of

1 reactor volume/hr in which oxygenation is done in an external
10-liter sparged vessel. In order to achieve a higher OTR, they
combined both systems (caged aeration and recirculation), increased
agitation rate to 80 rpm, and used oxygen instead of air. By combining these methods OTR of 92 mg/liter/hr is obtained, which is
sufficient for supplying oxygen to cell cultures at a density of
1.5×10^7 cells/ml.

Spier and Griffiths (133) wrote a computer program in BASIC
which permits calculation of the effects of changes in rate of stirring, sparging, and other variables on the oxygen transfer rate in
a specified fermenter, if the specific OUR of the cells is known.

In spite of the interest and research activity in this area there
is still a lack of good systems for supplying oxygen to large-scale
MC cell cultures. This is the main reason that, with few exceptions (159,160), the scale of MC culture is limited to 300–400 liters.

E. Inoculum

On a volume/volume basis the inoculum size in animal cell culture is
usually greater than that used in microbial fermentations. The
range is 5–30% and depends markedly on the plating efficiency of
the cells used (36,161). As a rule, continuous cell lines do not require as much inoculum as diploid cell strains or primary cells. Hu
and Wang (41) found that by determining the optimal size of the MC
and choosing a better growth medium for cell growth, the amount of
inoculum can be decreased. By optimizing these parameters they
were able to increase the multiplication ratio of foreskin diploid cells
to 15–16 as compared with 3–4 found in typical batch culture.
Another means for reducing the inoculum size is to use macroporous
MCs in which cell growth occurs not only on the MC surface but
also inside the MC (78,82). The ratio number of MCs to number of
cells is in the broad range of 1:2 to 1:30, although a ratio of 1:5 to
1:10 is preferred (36,41,161–165). Hu et al. (161) suggested that
a critical number of cells per MC is required for normal cell growth
to occur. Failure to achieve this critical number will decrease cell
growth. These authors presented a system for the estimation of
this critical number (161). The critical number of foreskin cells per
bead was found to be 6. This number can be reduced by using improved growth medium or by choosing optimal MC diameter. Nilsson
et al. (82) showed that in the case of using macroporous MCs the
cell/bead ratio can be reduced to 2–3.

Butler (14) showed that the distribution of inoculated cells on
MCs can be calculated by the Poisson distribution equation. He
showed that with MDCK cells, the cell/bead ratio should be 7:1 in
order for a negligible proportion of uninoculated MCs to be present
in the culture (14).

The concentration of MCs used in batch culture is in the range of 1−5 mg/ml (with Cytodex MCs) which provides about 5−25 cm^2 surface area/ml culture. The MC concentration most commonly used in batch cultures is 3 mg/ml (15 cm^2 of surface/ml). Higher MC concentrations, i.e., higher cell concentrations, can be used but then simple batch culture is no longer possible; frequent medium replenishment is necessary. With MC cell cultures in packed or concentrated perfusion mode still higher concentrations, up to 12 mg/ml, providing about 60 cm^2/ml, can be used (128). Apparently a minimal initial cell concentration of about 5 × 10^4 to 5 × 10^5/ml is needed to initiate rapid cell growth and ensure a relatively short lag period.

In order to ensure even distribution of cells on the MCs and complete utilization of cells in the inoculum, several researchers suggested seeding cells and MCs into a reduced volume, about one-third of the intended operating volume. After the cell attachment phase, medium can be added to bring the volume to that required (36,126,163). These workers agreed that during the attachment phase the culture should be essentially static, with only intermittent stirring. These measures allow more uniform and efficient cell attachment and increase cell yields in small fermenters. However, this procedure seems impractical in larger fermenters as it can create problems in controlling temperature, supplying oxygen to the cells, and distributing the cells evenly among the MCs, which settle at a different rate (124,145). Edy (124) reported that setting up the culture at the final volume and stirring continuously from the beginning had no deleterious effect on cell growth and that the actual cell distribution on the MCs correlated well with the theoretical Poisson distribution. van Wezel (145) also reported that cells become well attached to MCs even though the culture is being stirred continuously during the attachment phase. However, it is generally agreed that during the cell attachment phase stirring should be minimal — only sufficient to keep the cells in suspension (39).

The physiological state of the culture is important if the best utilization of cells in the inoculum is to be achieved. Cells should be harvested for use as inoculum during the late exponential phase and definitely before the stationary phase (166). The seed culture should be fed 24 hr before harvesting (67). It is preferable for the MCs to be placed in the medium prior to its seeding with cells to allow the MCs time to adsorb proteins. Some workers recommend adding amino acids and vitamins to the medium at seeding to increase the plating efficiency of the cells (41,126,164). All of these precautions are more important with cells of low plating efficiency, e.g., diploid cell strains, or when cells must be seeded at low density for pragmatic reasons.

F. Harvesting of Cells

1. Release of Cells from Microcarriers

In a serial transfer procedure to build up volume for the final production stage, the ratio of the size of one fermenter (or its contents) to the next in line may be 1:20 to 1:4. The number of cells
actually transferred depends not only on the volume of the culture
but also on the efficiency with which the cells are harvested. The
harvesting procedure should produce a well-dispersed, single-cell
suspension in order to achieve an even distribution of cells among
the MCs. Cells are usually released from the MCs by treatment with
proteolytic enzymes, e.g., trypsin, collagenase, pronase, or hyaluronidase with or without the addition of a chelating agent such as
EDTA. These treatments cause injury to the cell and affect the integrity of the plasma membrane to some degree (167). Thus, enzymatic treatment should be avoided if possible and controlled carefully and used for short periods, if used at all (166). Tolbert and
Feder (128) reported that cell harvesting with proteolytic enzymes
is more efficient with cell-MC aggregates, which are formed when
cells have overgrown the MCs and bridged between them, than it is
with cells grown in a monolayer on the MCs. Baijot et al. (101)
showed that cell harvesting from MC cultures depends on the quality
of trypsin and that there are batch-to-batch variations. In order
to decrease the damaging effect of trypsinization, Hu et al. (168)
harvested cells from DEAE-dextran MC surface by employing trypsin
and adjusting the pH above the physiological extreme (8.4–9.0) for
a short period. Harvesting cells at pH 8.2 is less effective since a
longer exposure time to trypsin is needed.

RDB, a protein of plan origin with no damaging effect on cells
and no susceptibility to inhibition by serum, was used by Ben Nathan
et al. (169) and by Fiorentini and Mizrahi (74) instead of the former
proteinases with good results. This material was also found to be
capable of dispersing insect cells which cannot be dispersed by
trypsin (73). Lindskog (170) used dextranase to harvest cells from
Cytodex MCs. The cells were harvested as sheets. In order to obtain a single-cell suspension, a combination of trypsin and dextranase
was employed. Wissemann et al. (66) showed that by applying col-
legenase on gelatin beads the MC is entirely digested, leaving cells
free from MCs and suspended in solution while retaining 98% viability.
van Oss et al. (171) detached cultured cells from dextran MCs by
lowering the surface tension of the growth medium.

The efficiency of recovery of cells from MCs is often very low,
ranges between 60 and 95% (26,49,60,65,101,125), and depends on
the cell type, type of growth medium, type of MC (49,60,65), extent
of MC confluency (172), rate of multilayering, quality of trypsin
(101), duration for which cells are exposed to trypsin (166), and

duration for which the cells have been in confluent state (166).
Vertblade et al. (172) and Lindner et al. (166) showed that harvesting of MRC-5 and Vero cells from Cytodex 1 MCs can be
achieved with high yields (100% cell transfer, higher than 95%
viability, and 80-90% cell attachment). Their method employs pre-
washing of the confluent MCs with EDTA solution (pH 8.0), removal
of the EDTA by washing with buffer (EDTA inhibits trypsin activities), and then applying trypsin at pH 8.0. Cell detachment
was obtained in 15 min. By employing that method, they were able
to scale up their culture to 100 liters. Hu et al. (168) obtained a
95% yield by trypsinizing DEAE MCs at high pH. The efficiency of
cell harvesting from the MC surface is reported to be high when
collagen-coated, glass, polystyrene, or denatured gelatin beads are
used (49,60,65).

Harvesting cells from MCs is a risky operation in scaling up MC
cell culture. Different researchers have used different means to
circumvent this problem. One approach is to use primary cells to
avoid cell harvesting from the MC surface. Primary cells, obtained
directly from animal tissue, organs, or embryos by applying a mixture of enzymes and a chelating agent, can be obtained in sufficient
quantity to permit direct seeding of a rather large fermenter.
Scattergood et al. (125) used a cell suspension obtained from 10
chick embryos to seed a 9-liter fermenter and on this basis calculated a need for about 70 embryos to seed a 60-liter fermenter, an
operation that seems feasible. By the same reasoning 1100 embryos,
an impractical number, would be needed to seed a 1000-liter fermenter. At this scale it seems that at least one step of cell harvesting from MCs and transfer to fresh beads in a larger fermenter
would be necessary. The need for an intermediate step is shown
even more dramatically in another process described by van Wezel
et al. (99). They propagated monkey kidney primary cells in MC
culture for production of polio vaccine. Two kidneys taken from one
monkey supplied only enough cells to seed a 10-liter fermenter. Because of the high cost of breeding monkeys and the necessity to adhere to strict and extensive quality control measures, it is more
economical to subculture the cells, even up to 12 generations in culture, than to use primary cells for production of this virus. Except
when inexpensive primary cells can be used, all other systems in
which ADCs are grown on MCs almost invariably require at least one
intermediate harvesting step in the scale-up train.

Another way to avoid the operation of stripping the cells from
the MCs between stages is to use devices other than MCs in the
early stages. Suspensions prepared by trypsinization of cell layers
grown in roller bottles are used to provide the seed for relatively
small fermentations (124,173). However, cell seeding from multiple-
unit propagators involves a high labor cost and, more importantly,

increases significantly the risk of contamination. Consider that
sometimes as many as 100–200 roller bottles are needed to seed a
single 10-liter fermenter. Single-unit propagators (5) can be used
in order to avoid this pooling operation (e.g., multiplate propagator
or glass bead propagator).

Several researchers have shown that under some circumstances
it is possible to scale up MC cell cultures without the conventional
harvesting operations. Crespi and Thilly (174) transferred CHO and
monkey kidney (LLC-MK2) cells directly from MCs to MCs while
stirring in a medium low in calcium content. Similarly, Nahapetion
et al. (175) scaled up culture of human foreskin diploid fibroblast
by bead-to-bead transfer. In several publications (58,104,146,176)
it was reported that scaling up of MC cultures was achieved by
adding fresh MCs to heavily sheathed MC cell cultures in order to
generate cell movement to fresh MCs and a new wave of cell proliferation. The cells which are released from the confluent MCs
(Fig. 3B) attach and grow on the newly added MCs. Kluft et al.
(104) expanded the volume of an MC culture of human melanoma
cells from 3 to 10 to 40 liters by adding fresh medium and MCs in-
crementally. The small fraction of cells free of MCs became attached to the fresh MCs and grew on them. MC-to-MC transfer of
cells, when it can be done, not only saves material and labor costs
but also avoids the risk of contamination entailed in the sequence of
steps in conventional harvesting. Generally cell transfer from MC
to MC is achieved more easily with cell lines than with elongated
fibroblast primary cells or diploid cell strains. The disadvantages
of this method are that the scaling-up process is longer than in cell
harvesting and that uneven distribution of cells on MCs is obtained.

2. Separation of Cells from Microcarriers

Cells are harvested from MCs either in the vessel in which they
have been grown or in a separate vessel. If harvesting is carried
out in the growth vessel, the MCs with adhering cells are washed
by permitting them to settle, decanting the supernatant medium,
adding buffer, and repeating the sequence of steps if desired. Dissociation agents are added in minimal concentration and volume,
stirring is implemented for a brief period, medium containing serum
is added to stop the action of the proteolytic enzyme, and the suspension of cells and MCs (or cells only if stripped MCs are retained
by the filter) is transferred to the next larger vessel. Several disadvantages are evident in this procedure: The minimal volume that
can be used is the smallest working volume of the fermenter; all of
the dissociating agent used is transferred to the next fermenter;
washing by sedimentation and decantation is slow; washing and separation are inefficient unless a filter (60–100 µm to retain MCs but
not cells) is used on the input of the transfer line.

Vertblad et al. (172) and Lindskog et al. (170) showed that it is possible to use such a procedure for scaling up of MC cultures. They have added new MCs and medium to the trypsinized MC culture without separating cells from Cytodex 1 MCs. Residual trypsin remaining in the culture did not affect cell attachment to the new MCs since it was inactivated by inhibitors present in the serum. This procedure cannot be used in serum-free media.

The procedure for harvesting cells from the MCs in a separate vessel is similar in principle but operationally different. A principle advantage is that the volume of the harvest vessel may be chosen to permit the smallest possible operation volume, which is limited by the total mass of MCs. One gram of Cytodex MCs may be expected to give a settled bed volume of 14−18 ml. Scattergood et al. (125) reported that MCs from a 9-liter fermenter gave a bed volume of 350 ml. At minimal volume washing, treating, and separating can all be done more efficiently. van Wezel et al. (26) described a special apparatus for harvesting monkey kidney cells from MCs. Agitation is provided by a vibromixer and the MCs are retained in the vessel by a stainless steel screen with 60-μm openings at the bottom outlet port. This apparatus was used by Scattergood et al. (125) for harvesting chick embryo fibroblasts from MCs. The efficiency of harvesting cells by this apparatus was reported to be 80−90% by van Wezel (39). This apparatus is supplied commercially by Contact, Holland. Spier et al. (177) described the use of a small-bore tube (1.2 mm) for continuous stripping of cells from MCs. Hu et al. (168) detached cells from the MC surface by passing a trypsinized MC suspension through a glass bead column.

In order to obtain an even distribution of cells on MCs in the next (larger) stage, sometimes the freed cells should be separated from the used MCs, which still retain a variable number of viable cells even after the harvesting treatment (26). Used MCs, still retaining viable cells, if transferred to the next stage in effect are sometimes seeded more heavily than the new MCs and thus achieve confluence earlier. In other cases, Hu et al. (168) reported that cell attachment to new MCs is considerably higher than to the old MC carriers from the seed culture.

Cells harvested (released) from the MCs can be separated from them by gravity sedimentation, filtration, or density gradient centrifugation. Billing et al. (178) described testing (a the laboratory scale) these three techniques for cell separation after trypsinization. Gravity sedimentation is based on the difference in settling rates between cells and MCs with no agitation. Filtration was done through an 88-μm pore size nylon screen. Density gradient centrifugation in Ficoll Paque at low speed sent the MCs to the bottom of the tube while the cells were located in an intermediate band. From the standpoint of efficiency of cell recovery the latter two methods

were superior (65 – 75% vs. 35 – 45%). However, the filtration method
of van Wezel et al. (26) is the only one of the three which has been
tested on a large scale.

   MCs separated from cells can be reused only when they are the
ionically charged type. Collagen or gelatin beads cannot of course
be reused after exposure to proteolytic enzymes. Another way to
overcome the need for separating cells from MCs is to apply dextranase on dextran MCs (170) or proteolytic enzymes on gelatin MCs
(66), resulting in the dissolution of the MC.

VIII. MODE OF CELL PROPAGATION
AND PRODUCT PRODUCTION

A. Batch and Modified Batch Mode

Until recently, the most important products obtained from ADC cultures were the viral antigens. The viruses propagated on the cells
are usually cytopathogenic (e.g., poliomyelitis or foot-and-mouth
disease virus), i.e., the virus kills the host cell during its growth.
Thus the most efficient process consists of two batch phases — first
a phase in which maximal cell density is achieved and then a phase
in which virus propagation is effected. Another product produced
in batch mode from ADCs is human fibroblast beta-interferon. After
a first phase in which cells are grown to optimal cell density, induction is carried out for interferon production.

   In the simple batch mode of operation, temperature, pH, and DO
are usually controlled, but nutrients, e.g., glucose and glutamine,
are depleted continuously and inhibitory waste products, e.g.,
lactic acid and ammonia accumulate. Thus cell propagation slows and
stops and cell yield is limited, $1 - 2 \times 10^6$ cells/ml (39,179). Improvement in cell yield is obtained by periodically replacing a portion of the culture supernatant with fresh medium, thereby restoring
nutrients and removing waste products. Clark and Hirtenstein (123)
took another approach, feeding essential depleted nutrients (cystine,
glutamine, inositol, glucose, choline, and pyridoxine) to chick embryo fibroblasts after 3 days growth on MCs. By feeding a concentrate of essential nutrients, the cost of adding whole medium was
avoided. A disadvantage of shot of slug feeding, as pointed out by
van Wezel (145), is that a sudden change in environment can cause
separation of cells from MCs. A more sophisticated approach to
controlling cell propagation is to feed essential nutrients on demand
in order to maintain predetermined concentrations and concomitantly
to limit accumulation of waste products. Fleischaker et al. (149,180)
fed glucose to a MC cell culture on demand to maintain a concentration of 0.5 mM (vs. 20 mM normally batched) and achieved higher
cell density and lower accumulation of lactic acid. Glacken et al.

(141,180) kept the glutamine concentration in a MC cell culture as low as 0.2 mM (vs. 4 mM) by continuous feeding and thus reduced the level of ammonium ion in the culture by over 60%. Although this fed-batch mode of process conduction generally leads to a higher cell density and viability, inevitably cells do begin to die, probably because of waste product accumulation.

An interesting and different approach to maintaining MC cell cultures in a viable state for extended periods was taken by Morandi et al. (107). They propagated MRC-5 cells on Cytodex 1 MCs in 1.2- and 5-liter fermenters while dialyzing the medium against 5 and 20 liters, respectively, of fresh serum-free medium. On the smaller scale the dialysis was done with 150 cm^2 of dialysis tubing; on the larger scale dialysis was done with hollow fibers providing 8000 cm^2 of surface. This system provided low molecular weight nutrients continuously and waste products were diluted continuously. High cell yields were realized.

B. Continuous or Extended Operation

Truly continuous methods for propagating animal cells are not practical in MC cell culture as they would entail continuous or repeated harvesting of cells and replenishment of MCs. However, in recent years a need has arisen for products that are secreted continuously from ADCs, e.g., recombinant cells (84,90,108). An attractive mode of producing such products is to propagate the cells to achieve high density and afterward to establish conditions such that viability is maintained, significant growth does not occur, and product is secreted continuously. Product can be produced by recovering supernatant free of MCs periodically or by perfusing the system continuously.

Manousas et al. (176) propagated oncoviruses in MC culture by intermittent withdrawal of supernatant. Lazar et al. (72) produced carcinoembryonic antigen in an MC culture of adenocarcinoma cells by periodic harvesting of culture supernate. The main advantage of the pseudocontinuous, repeated harvest system is its simplicity. Product is produced in the same vessel in which cells are propagated, and harvest is accomplished by periodically letting the MCs settle and withdrawing supernatant through a closed system. However, in contrast with perfusion systems, relatively low concentrations of cells and products are obtained. Moreover, sudden changes in the growth medium composition occurs, leading to long lag periods and sometimes cell detachment from the MC surface (36).

Wagner et al. (181) investigated the production of prourokinase in MC culture (5 g/liter Cytodex 3) in two continuous systems: a reactor operated in a feed-harvest mode (30% of the culture volume per day) or in a perfusion mode (150% of the culture volume per day). A cell concentration of 3×10^6 cells/ml was achieved in the feed and harvest system and 7×10^6 cells/ml in the perfusion system.

In perfusion cultures there is a constant supply of medium to the cells and a constant removal of waste products. Thus, the environment of the cells remains constant throughout the culture and high cell density is achieved. Cell concentrations of $1-2 \times 10^7$ are achieved in MC cultures having MC concentrations of $7.5-12$ mg/ml (6,12,13,108,128,179).

MC cell cultures, in comparison with microbial or animal cell suspension culture systems, are especially well suited to management in perfusion systems. The MCs are large enough to be separated by relatively large pore size filters or to settle under the force of gravity alone in a reasonable period of time.

A number of special devices for accomplishing perfusion have been described. A stainless steel 100-μm pore size cage mounted on the stirring shaft and rotating with it was used in several cases (24,104,126,145). MCs are excluded from the cage and thus clear culture medium can be pumped out of the vessel from inside the cage while fresh medium is supplied continuously to the culture outside the cage.

Tolbert and Feder (128) added a settling bottle, clarifying vessel, effluent reservoir, and medium reservoir all external to the growth vessel. After multilayer growth was achieved, perfusion was effected by pumping MC-cell aggregates to a settling vessel, where the settling rate exceeded the upward flow rate and supernatant virtually free of MCs was sent to a filtration vessel. From there, the stream was split, some being returned to the medium reservoir and hence to the main vessel, and some being harvested continuously through a spin filter. An additional advantage to the settling bottle is that cells grew not only on the surface of the MCs but also between them, giving a higher cell concentration than would be expected if the cells formed only a single confluent monolayer.

Butler et al. (182) and others (13) used a device conceptually similar to the settling vessel in which culture supernatant is pumped through a gravity column separator located above the liquid level. Fleischaker (67) calculated that in a 100-liter tank perfused at two volumes a day, the minimal diameter of the settling column should be 14.5 cm. An equation for calculating the minimal diameter of the settling column is presented in this publication. More recently, two types of MC perfused cultures were described (see Section VII.C.3): (1) the static maintenance reactor in which cells are maintained under nonproliferating static conditions as described by Tolbert et al. (12,128,132), and (2) fluidized-bed reactors (81,84) in which the porous carriers are suspended by the medium flow in the recycled loop. Both systems were scaled up to production scale and high cell densities in the range of 10^8 cells/ml were reported.

In the foregoing paragraphs perfusion systems are discussed mainly for their advantages in the manufacture of products secreted during a nonproliferating stage. However, several groups (81,84,105,126) have also used closed perfusion systems as a means of obtaining efficient oxygen transfer by incorporating external oxygenation system on the medium circulation loop (see Section VII.D).

Among the characteristics of perfusion systems that have been cited as advantageous for economical, efficient production of biologicals are the following:

1. Propagating cells at high-density perfusion cultures can potentially reduce production costs by increasing volumetric productivity, increasing concentration of the desired product and sometimes increasing the specific production rate of the product.

2. Saving in growth medium and especially serum. Tolbert and Feder (128) showed that in perfusion systems medium is utilized four times as efficiently as in roller bottles. Kluft et al. (104) and others (146) reduced the serum content of the perfusion medium to 0.5−1% during the maintenance period. Delzer et al. (146) suggested reducing the serum concentration by adding BSA.

    Perfused MC cultures can be fed by either an open or a closed recirculated system. In an open system, the efficiency of utilization of medium is sometimes reduced since there is continuous removal of nonutilized growth medium. In closed perfusion systems in which medium is circulated through the culture into a reservoir where it is optimally reequilibrated and returned to the culture, the efficiency of medium utilization is increased and the product can be accumulated in relatively small volume, but accumulation of toxic metabolic products will occur as well. Griffiths (183) reported a three- to fourfold increase in efficiency of medium utilization in closed MC perfusion culture as compared with open systems.

3. The concentration of nutrients and waste products in contact with the cells can be changed by varying either the flow rate or the concentrations in the feed stream (126,165, 179), thus achieving good cell nutrition and efficient removal of toxic product.

4. The fact that cells are maintained at steady-state conditions and the ease with which the nutrient composition of the environment of the cells can be changed in a perfusion

system makes it possible to study the effects of step or
pulse changes on growth and metabolism for basic information as well as to move toward optimal operating conditions.
Moreover, the perfusion system permits continuous calculation of oxygen consumption, CO_2 production, glucose consumption, lactic acid production, etc., by on-line measurements made on the input and output streams.

The foregoing discussion of the characteristics of perfusion systems stresses its advantages. It is worthwhile to review some of the
disadvantages of perfusion systems. Perfusion systems are not
suitable for products produced during active growth or during the
stage of declining metabolic activity and decreasing viability. Wagner
et al. (181) reported a drastic decrease in the prourokinase specific
production rate (10-fold) in maintained confluent cultures as compared with actively growing cells. There is a risk of genetic instability during the maintenance period which may last for months.
Detachment of cells from the MC during long-term cultivation is observed. In general, control of the environment (pH and especially
dissolved oxygen) is more difficult in high-density perfusion systems,
closed or open, than in closed batch systems. The apparatus is
relatively complicated. Pumps (which may fail), filters (which may
clog), level sensors, auxiliary vessels for medium and effluent, and
a requirement for better control instrumentation are all features of
the perfusion system which add to capital costs and may contribute
contamination susceptibility or operating problems.

IX. GROWTH MEDIUM

The media that are used for the growth of cells in stationary monolayer cultures are usually suitable for use in MC cell culture. Several modified formulas or nutrient-feeding regiments have been devised and are discussed in the following paragraphs.

Griffiths and Thornton (126) and Clark et al. (163) suggested
enriching the medium during the initial cell attachment phase to ensure better utilization of the cell inoculum. Several changes related to the higher cell density and consequently more vigorous
metabolic activity have been made. Organic buffers, e.g., HEPES,
have been added to provide pH control to supplement the CO_2 bicarbonate system. Glucose has been replaced by galactose or fructose to reduce lactic acid production (184). High molecular weight
nonprotein polymers have been added to low-serum media to reduce
turbulence, to raise the viscosity and protect cells from mechanical
damage, and to promote cell attachment (35). van Wezel and
van der Welden de Groot (26) and Clark et al. (185) pointed out
that MC cell cultures are more sensitive to the quality of the growth

medium, especially the serum, than are stationary monolayer cultures. Clark et al. (185) proposed that stringent control measures be used in the selection of each batch of serum.

Clark et al. (163) warned that antibiotics, commonly used to protect cell cultures from microbial infection, can affect cell growth rate and saturation density adversely in MC cell culture, especially with cells of low plating efficiency.

Clark and Hirtenstein (122) indicated that reducing the serum concentration in the medium after confluent growth has been attained seems to prevent sloughing off of cells from the MCs. Several researchers (9,88,174,185 – 188,235) demonstrated that it is possible to grow cells in MC cell culture in serum-free medium. Usually the addition of attachment factors and growth factors is needed. Frequently cells are propagated in MC cultures (especially in long-term cultures) in two stages. In the first stage cell spreading and growth on the MC is achieved in the presence of serum; in the second stage it is replaced with serum-free or low-serum medium (61,88,104,105,146). A procedure for testing the suitability of serum-free medium for propagation of ADC on MCs was described by Bodeker et al. (189).

Delzer et al. (146) showed that in certain cases serum-free media used in MC-perfused culture of recombinant mouse fibroblasts resulted in a low yield of interferon. In order to achieve reasonable production, serum levels had to be kept at 2.5% or a combination of 0.2 – 0.4% serum and BSA. In other cases (236,240), it was reported that in long-term perfusion cultures in which recombinant cells were maintained in low-serum growth media, cell growth was not affected by the lower serum concentration; however, lower product yields were obtained.

One of the most critical problems in cell culture generally and MC cell culture especially, because it is aggravated by high cell density, is the inhibitory effect of accumulated products (179). Gaseous products, e.g., CO_2 and to some extent NH_3, can be purged by the gas stream through the system. Highly soluble, nonvolatile products such as lactic acid and NH_4^+ will accumulate in the culture medium (179). Accumulation of lactic acid, which leads to a drastic decrease in pH, can be diminished by changing the carbon source (184) or by feeding glucose slowly, thus keeping glucose at a low level (180,190). The generation of NH_4^+ ion, which seems to have a drastic adverse effect on cell growth, is related to the metabolism of glutamine (179,191). Glacken et al. (141,180) suggested slow feeding of glutamine to cell cultures in order to reduce the amount of NH_4^+ ion that accumulates, while Hassell et al. (239) suggested development of glutamine-free medium (in which glutamine was substituted with glutamate) in order to reduce ammonia generation by over 50%. The possibility that other waste

products may reach growth-limiting concentrations was suggested by
Birch and Cartwright (192). On the other hand, the possibility
that growth slows or ceases because the medium becomes depleted of
essential nutrients has also been explored. Polastri et al. (193) and
Butler (179) and others (149,150) analyzed MC cell cultures for
amino acid utilization in batch and perfusion cultures, and found
some differences related to cell type. Usually, glutamine and
branched amino acids (especially leucine and isoleucine) were the
amino acid consumed most rapidly. Glutamine is needed for anabolic
processes in the cells; however, it is also used as a major carbon
source. Wagner et al. (195) investigated variation in amino acid
concentrations in growth medium of three different recombinant cells
producing human interleukin 2. They found no evidence that amino
acid consumption rate influences product synthesis. Schultz et al.
(90) showed that in high-density MC perfusion cultures which are
limited in their supply of branched amino acids (valine, leucine, and
isoleucine), a drastic decrease in interferon productivity was observed. Supplying these amino acids led to a threefold increase in
productivity.

X. CONCLUSIONS

Five to 10 years ago many believed that the development of the recombinant microbial technology would eliminate the need for large-
scale mammalian cell culture. It was assumed that any product produced by animal cells could be produced by microbial cells into which
mammalian genese has been cloned. However, in recent years it has
become evident that due to problems associated with bacterial systems, such as product instability and a precise folding and glycosylation requirement, sometimes, mammalian cell products cannot be
produced efficiently in microbial cells. Recombinant animal cell lines
are found more efficient for the production of these more complex
products (196). Consequently, we are witnessing a vigorous development of cell culture technology. Efforts are under way for
cloning new biologicals, developing new recombinant cell lines with
amplified expression, developing serum-free media, potentiating productivity by cellular physiology and optimization of the environment,
and scaling up culture size and efficiency.
       Animal cell cultures are of two basic types: those which re-
quire attachment to a surface for their growth (ADC) and those
which grow in suspension. Both of these methods for cell propaga-
ation are used for the production of a wide variety of biologicals.
MC cell culture is the method of choice for producing useful products from ADCs, although cell propagators providing large,

continuous-surface areas may have value in certain applications. It is generally accepted that MC cell culture offers the following principal advantages: a high ratio of growth surface to culture volume is provided; a single large production unit can replace a number of smaller units; mixed homogeneous MC cell cultures can be sampled, monitored, and controlled more easily than inhomogeneous systems; harvesting of cells and extracellular products can be done easily; and MC cell cultures can be scaled up readily using conventional equipment (fermenters) used for microbial processes.

This chapter reviews the situation of MC culture technology: The types of MCs available nowadays, biologicals produced by MC cultures, and the major problems in scaling up MC cultures. In order to further improve the efficiency of MC cultures, research and development in the following areas are needed:

Development of new MCs, especially porous ones, with improved performance.

Development of new reactors for MC cultures. This includes low-shear agitation systems, improved oxygen delivery systems, and new concepts in the management of perfused cultures.

Development of inexpensive serum-free growth media especially suited for MC-perfused cultures.

Improvements in the technology of harvesting cells from the MC surface.

ACKNOWLEDGMENT

The author wishes to express his gratitude to Professor A. Mizrahi for his help.

REFERENCES

1. Petricciani, J. C. Should continuous cell lines be used as substrates for biological products? *Dev. Biol. Standard.* 66:3–12 (1985).
2. Larsson, B. and Litwin, J. The growth of polio virus in human diploid fibroblasts grown with cellulose microcarriers in suspension cultures. *Dev. Biol. Standard.* 66:385–390 (1987).
3. Bahnemann, H. G. Animal cells, viral antigens and vaccines. *Abs. Pap. ACS* 180:5 (1980).
4. Nilsson, K. and Mosbach, K. Immobilized animal cells. *Dev. Biol. Standard.* 66:183–193.

5. Spier, R. E. Monolayer growth systems: Heterogeneous unit process. In *Animal Cell Biotechnology*, Vol. 1 (R. E. Spier and J. B. Griffiths, eds.), Academic Press, Orlando, FL, pp. 243–263 (1985)

6. Griffiths, J. B., Cameron, D. R., and Looby, D. Bulk production of anchorage-dependent animal cells — Comparative studies. In *Plant and Animal Cells: Process Possibilities* (C. Webb and F. Mavituna, eds.), Ellis Horwood Ltd., Chichester, pp. 149–162 (1987).

7. van Wezel, A. L. Growth of cell-strains and primary cells on microcarriers in homogeneous culture. *Nature* 216:64–65 (1967).

8. Butler, M., Imamura, T., and Thilly, W. G. High yields from microcarrier cultures by medium perfusion. *J. Cell Sci.* 61: 351–363 (1983).

9. Fraune, E., Lehmann, J., and Menge, U. Serum free production and purification of beta-interferon. *Proc. 4th Eur. Cong. Biotech.*, p. 594 (1987).

10. Griffiths, B., Atkinson, T., Electricwala, A., Latter, A., Ling, R., McEntee, I., Riley, P. M., and Sutton, P. M. Production of a fibrinolytic enzyme from cultures of guinea pig keratocytes grown on microcarriers. *Dev. Biol. Standard.* 55: 31–36 (1984).

11. Reuveny, S., Zheng, Z. B., and Eppstein, L. Evaluation of a cell culture fermenter. *Am. Biotech. Lab.* 4:28–39 (1986).

12. Feder, J. The development of large scale animal cell perfusion culture systems. *Adv. Biotech. Process* 7:125–152 (1988).

13. Martin, N., Brennan, A., Denome, L., and Shaevitz, J. High productivity in mammalian cell culture. *Biotechnology* 5: 838–840 (1987).

14. Butler, M. Growth limitation in microcarrier cultures. *Adv. Biochem. Eng./Biotech.* 34:57–84 (1987).

15. Griffiths, B., Thornton, B., and McEntee, I. Production of Herpes viruses in microcarrier cultures of human diploid and primary chick fibroblast cells. *Eur. J. Cell Biol.* 22:606 (1980).

16. Ryan, U. S., Mortara, M., and Whitaker, C. Methods for microcarrier culture of bovine pulmonary artery endothelial cells avoiding the use of enzymes. *Tissue Cell* 12:619–636 (1980).

17. Davis, P. F. and Kerr, C. Co-cultivation of vascular endothelial and smooth muscle cells using microcarrier techniques. *Exp. Cell Res.* 141:455–459 (1982).

18. Bone, A. J. and Swenne, I. Microcarriers: A new approach to pancreatic islet cell culture. *In Vitro* 18:141–148 (1982).

19. Crespi, C. L. and Thilly, W. G. Selection of mitotic Chinese hamster ovary cells from microcarriers. Cell cycle-dependent induction of mutation by 5-bromo-2'-deoxyuridine and ethyl methansulfonate. *Mutation Res.* 106:123−125 (1982).
20. Davis, P. E., Ganz, P., and Diehl, P. S. Reversible micro-carrier mediates junctional communication between endothelial and smooth muscle cell monolayer: An in vitro model for vascular interactions. *Lab. Invest.* 85:710−718 (1985).
21. Shahar, A. and Reuveny, S. Nerve and muscle cells on microcarriers in culture. *Adv. Biochem. Eng./Biotechnol.* 34: 33−55 (1987).
22. Hogan, M. E., Hassouna, H. I., and Klomparens, K. L. Sectionable microcarrier beads: An improved method for the preparation of cell monolayer for electron microscopy. *J. Electron Microscopy Techn.* 5:159−169 (1987).
23. Demetriou, A. A., Whiting, J. F., Feldman, D., Levensan, S. M., Chowdhury, N. R., Mascioni, A. D., Kran, M., and Chowdhury, J. R. Replacement of liver function in rats by transplantation of microcarrier attached hepatocytes. *Science* 233:1190−1192 (1986).
24. Griffiths, J. B., Cameron, D. R., and Looby, D. A comparison of unit process systems for anchorage dependent cells. *Dev. Biol. Standard.* 66:331−338 (1987).
25. van Wezel, A. L. The large scale cultivation of diploid cell strains in microcarrier culture: Improvement of microcarriers. *Dev. Biol. Standard.* 37:143−147 (1976).
26. van Wezel, A. L. and van der Velden-de-groot, C. A. M. Large scale cultivation of animal cells in microcarrier culture. *Process Biochem.* 13:6−8 (1978).
27. Spier, R. E. and Whiteside, J. P. The production of foot and mouth disease virus from BHK 21 C13 cells grown on the surface of DEAE-Sephadex A-50 beads. *Biotechnol. Bioeng.* 18: 659−667 (1969).
28. Levine, D. W., Wang, D. I. C., Thilly, W. G. Optimizing parameters for growth of anchorage-dependent mammalian cells in microcarrier culture. In *Cell Culture and Its Applications* (R. Action, ed.), Academic Press, New York, pp. 191−216 (1977).
29. Levine, D. W. Production of anchorage-dependent cells on microcarriers. Ph.D. thesis, MIT, Cambridge (1979).
30. Levine, D. W., Wong, J. S., Wang, D. I. C., and Thilly, W. G. Microcarrier cell culture: New methods for research scale application. *Somatic Cell Genet.* 3:149−155 (1977).
31. Levine, D. W., Thilly, W. G., and Wang, D. I. C. Parameters affecting cell growth on reduced charge microcarriers. *Dev. Biol. Standard.* 42:159−164 (1979).

32. Levine, D. W., Wang, D. I. C., and Thilly, W. G. Optimiza-
    tion of growth surface parameters in microcarrier cell culture.
    Biotechnol. Bioeng. 21:821−845 (1979).
33. Hirtenstein, M., Clark, J., Lindgren, G., and Vretbald, P.
    Microcarrier of animal cell culture: A brief review of theory
    and practice. *Dev. Biol. Standard.* 46:109−116 (1980).
34. Reuveny, S. Research and development of animal cell micro-
    carrier cultures. Ph.D. thesis, The Hebrew University,
    Jerusalem (1983).
35. Nunc, Biosilon Bulletin No. 1. Cultivation principles and
    working procedures (1981).
36. Pharmacia Fine Chemicals. Microcarrier cell culture: Principles
    and methods (1982).
37. Reuveny, S. Microcarriers in cell culture: Structure and
    applications. *Adv. Cell Culture* 4:213−247 (1985).
38. Cherry, R. S. and Papoutsakis, E. T. Hydrodynamic effect
    on cells in agitated tissue culture reactors. *Bioprocess Eng.*
    1:29−41 (1986).
39. van Wezel, A. L. Monolayer growth systems: Homogeneous
    unit processes. In *Animal Cell Biotechnology*, Vol. 1 (R. E.
    Spier and J. B. Griffiths, eds.), Academic Press, New York,
    pp. 265−282 (1985).
40. Reuveny, S. Microcarriers for culturing mammalian cells and
    their applications. In *Advances in Biotechnological Processes*,
    Vol. 2 (A. Mizrahi and A. L. van Wezel, eds.), Alan R. Liss,
    New York, pp. 1−32 (1983).
41. Hu, W. S. and Wang, D. I. C. Selection of microcarrier diam-
    eter for the cultivation of mammalian cells on microcarriers.
    Biotechnol. Bioeng. 30:548−557 (1987).
42. Johansson, A. and Nielsen, V. Biosilon: A new microcarrier.
    Dev. Biol. Standard. 46:125−129 (1980).
43. Windig, W., Haverkamp, J., and van Wezel, A. L. Control
    on the absence of DEAE-polysaccarides in DEAE-Sephadex
    purified poliovirus suspensions by pyrolysis mass spectrometry.
    Dev. Biol. Standard. 47:169−177 (1981).
44. Maroudas, N. G. Chemical and mechanical requirements for
    fibroblast adhesion. *Nature* 244:353−354 (1973).
45. Keese, C. R. and Giaever, I. Cell growth on liuqid micro-
    carriers. *Science* 219:1448−1449 (1982).
46. Reuveny, S., Bino, T., Rosenberg, H., and Mizrahi, A. A
    new cellulose based microcarrier culturing system. *Dev. Biol.
    Standard.* 46:137−145 (1979).
47. Varani, J., Dame, M., Rediske, J., Beals, T. F., and
    Hillegas, W. Substrate-dependent differences in growth and

biological properties of fibroblasts and epithelial cells grown in microcarrier culture. *J. Biol. Standard.* 13:67–76 (1985).

48. Varani, J., Bendelow, M. J., Chun, J. H., and Hillegas, W. A. Cell growth on microcarriers. Comparison of proliferation on and recovery from various substrates. *J. Biol. Standard.* 14:331–336 (1986).

49. Varani, J., Dame, M., Beals, T. F., and Wass, J. A. Growth of three established cell lines on glass microcarriers. *Biotech. Bioeng.* 25:1359–1372 (1983).

50. Reuveny, S., Silberstein, L., Shahar, A., Freeman, E., and Mizrahi, A. DE-52 and DE-53 cellulose microcarriers. I. Growth of primary and established anchorage-dependent cells. *In Vitro* 18:92–98 (1982).

51. Reuveny, S., Silberstein, L., Shahar, A., Freeman, E., and Mizrahi, A. Cell and virus propagation on cylindrical cellulose-based microcarriers. *Dev. Biol. Standard.* 50:115–124 (1982).

52. Borysenko, J. S. and Wood, W. Density, distribution and mobility of surface anions on a normal/transformed cell pair. *Exp. Cell Res.* 118:215–227 (1979).

53. Folkman, J. and Moscona, A. Role of cell shape in growth control. *Nature* 273:345–349 (1978).

54. Reuveny, S., Mizrahi, A., Kotler, M., and Freeman, A. Factors affecting cell attachment, spreading and growth on derivatized microcarriers. I. Establishment of working system and effect of the type of the amino-charged groups. *Biotechnol. Bioeng.* 25:469–480 (1983).

55. Reuveny, S., Mizrahi, A., Kotler, M., and Freeman, A. Factors effecting cell attachment spreading and growth on derivatized microcarriers. II. Introduction of hydrophobic elements. *Biotechnol. Bioeng.* 25:2969–2980 (1983).

56. Himes, V. B. and Hu, W. S. Attachment and growth of mammalian cells on microcarriers with different ion exchange capacities. *Biotechnol. Bioeng.* 29:1155–1163 (1987).

57. Dormacell microcarriers for cell cultures Pfeifer and Langen Dormagen (1985).

58. Shulz, R., Krafft, H., and Lehmann, J. Experiences with a new type of microcarrier. *Biotechnol. Lett.* 8:557–560 (1986).

59. van Wezel, A. L. Cultivation of anchorage-dependent cells and their applications. *J. Chem. Tech. Biotechnol.* 32:318–323 (1982).

60. Gebb, C., Clark, J. M., Hirtenstein, M. D., Lindgren, G., Lindgren, U., Lindskog, U., Lundgren, B., and Vertbald, P. Alternative surfaces for microcarrier culture of animal cells. *Dev. Biol. Standard.* 50:93–102 (1982).

61. Gray, P., Morsden, W., Ponnowitz, D., Crowley, J., and Pirhonen, J. The production of human growth hormone from recombinant mammalian cells. Paper presented at the Eng. Foundation Conf. on Cell Culture Engineering, Palm Coast, Florida, Jan. 30, 1988.

62. Nilsson, K. and Mosbach, K. Preparation of immobilized animal cells. *FEBS Lett.* 118:145–150 (1986).

63. Fairman, K. and Jacobson, B. S. Unique morphology of HeLa cell attachment, spreading and detachment from microcarrier beads covalently coated with specific and non-specific substratum. *Tissue Cell* 15:167–180 (1983).

64. Obernovitch, A., Sene, C., Maintier, C., Boschetti, E., and Monsigny, M. New microcarriers suitable for the culture of fibroblastic cells. *Biol. Cell* 45:28–35 (1982).

65. Gebb, C., Lundgren, B., Clark, J., and Lindskog, U. Harvesting and subculturing cells growing on denatured-collagen coated microcarriers (Cytodex 3). *Dev. Biol. Standard.* 55:57–65 (1984).

66. Wissemann, K. W. and Jacobson, B. S. Pure gelatin microcarriers: Synthesis and use in cell attachment and growth of fibroblast and endothelial cells in vitro cell. *Dev. Biol.* 21: 391–401 (1985).

67. Fleischaker, R. Microcarrier cell culture. In *Large Scale Cell Culture Technology* (B. K. Lydersen, ed.), Hanser, New York, pp. 59–79 (1987).

68. Maroudas, N. G. Sulfonated polystyrene as an optimal substratum for the adhesion and spreading of mesenchymal cells in monovalent and divalent saline solutions. *J. Cell. Physiol.* 90: 511–520 (1977).

69. Grinnell, F. Cellular adhesiveness and extracellular substrata. *Int. Rev. Cytol.* 53:65–144 (1978).

70. Reuveny, S., Mercado, A., Silberstein, L., and Mizrahi, A. Growth of foot and mouth disease virus in BHK cells propagated on DEAE cellulose microcarriers. *In Vitro* 16:262–263 (1980).

71. Litwin, J. Preliminary results on the use of human diploid fibroblasts for inactivated polio virus vaccine production. *Dev. Biol. Standard.* 60:237–242 (1985).

72. Lazar, A., Reuveny, S., Geva, J., Marcus, D., Silberstein, L., Ariel, N., Epstein, N., Altbaum, Z., Sinai, J., and Mizrahi, A. Production of carcinoembryonic antigen from a human colon adenocarcinoma cell line. I. Large-scale cultivation of carcinoembryonic antigen-producing cells on cylindric cellulose-based microcarriers. *Dev. Biol. Standard.* 66:423–428 (1987).

73. Lazar, A., Silberstein, L., Reuveny, S., and Mizrahi, A. Microcarriers as a culturing system of insect cells and insect viruses. *Dev. Biol. Standard.* 66:315–323 (1987).

74. Fiorentini, D., Shahar, A., and Mizrahi, A. Production of herpesvirus of turkeys in microcarrier culturing system: A new method for production of vaccine against Marek's disease. *Dev. Biol. Standard.* 60:421–430 (1985).

75. Muller Lierheim Biol. Lab., In vivo in vitro 3.

76. Jacobson, B. S. and Rayan, U. S. Growth of endothelial and HeLa cells on a new multipurpose microcarrier that is positive negative or collagen coated. *Tissue Cell* 14:69–84 (1982).

77. Damme, J. V. and Billiau, A. Large-scale production of human fibroblast interferon. *Methods Enzymol.* 78:101–111 (1981).

78. Bohak, Z., Kadouri, A., Sussman, M. V., and Feldman, A. F. Novel anchorage matrices for suspension culture of mammalian cells. *Biopolymers* 26:S205–S213 (1987).

79. Hannan, G. N. and Mcauslan, B. R. Immobilized serotonin: A novel substrate for cell culture. *Exp. Cell Res.* 71:153–163 (1987).

80. Ruoslahti, E., Pierschbacher, M. D., Oldberg, A., and Hayman, E. G. Synthetic peptides causing cellular adhesion to surfaces. *BioTechniques* (Jan–Feb):38–41 (1984).

81. Young, M. W. and Dean, R. C., Jr. Optimization of mammalian-cell bioreactors. *Biotechnology* 5:835–837 (1987).

82. Nilsson, K., Birnbaum, S., Buzaky, F., and Mosbach, K. Growth of anchorage-dependent cells on macroporous micro-carriers. In *Modern Approaches to Animal Cell Technology* (R. E. Spier and J. B. Griffiths, eds.), Butterworths, pp. 492–503 (1987).

83. Chan, F., Yannos, I. V., and Steuer, A. F. Porous micro-carrier particles for mammalian cell culture. Paper presented at ACS meeting, Anaheim, Sept. 11 (1986).

84. Hayman, E. G., Ray, N. G., Tung, A. S., Holland, J. E., and DeLucia, D. E. Production of monoclonal antibody and recombinant t-PA on grams per day scale. Proceedings of the 1988 Miami Biotechnology Winter Symposium, p. 54 (1988).

85. Boag, A. H. and Sefton, M. V. Microencapsulation of human fibroblasts in a water soluble polyacrylate. *Biotechnol. Bioeng.* 30:954–962 (1987).

86. Nilsson, K. Mammalian cell culture. In *Methods in Enzymology*, Vol. 135 (K. Mosbach, ed.), Academic Press, New York, pp. 387–393 (1987).

87. Nilsson, K., Birnbaum, S., and Mosbach, K. Microcarrier culture of recombinant Chinese hamster ovary cells for

production of human immune interferon and human tissue-type plasminogen activator. *Appl. Microbiol. Biotechnol.* 27: 366–371 (1988).

88. Fantini, J., Galans, J. P., Abadie, A., Canioni, P., Gazzone, P. J., Marvaldi, J., and Tirard, A. Growth in serum free medium of human colonic adenocarcinoma cell lines on microcarriers: A two step method allowing optimal cell spreading and growth. *In Vitro* 23:641–646 (1987).

89. Varani, J., Hasday, J. D., Sitrin, R. G., Brubaker, R. G., and Hillegas, W. A. Proteolytic enzymes and arachidonic acid metabolites produced by MRC-5 cells on various microcarrier substrates. *In Vitro* 22:575–582 (1986).

90. Schultz, R., Krafft, H., Piehl, G. W., and Lehmann, J. Production of human β-interferon in mouse L-cells. *Dev. Biol. Standard.* 66:489–493 (1987).

91. Tan, B., Jobses, I., and Rousseau, I. The effect of cation concentration and medium composition on cell adhesion. *Proc. 4th Eur. Cong. Biotechnol.*, p. 599 (1987).

92. Pharmacia, Inc. Cytodex Reference List (1984).

93. Lindskog, U., Lundgren, B., Wergeland, I., and Billig, D. Microcarrier cell culture: Vero cells on Cytodex 1. *J. Tissue Cult. Method.* 9:205–210 (1985).

94. Sanford, K. K., Earle, W. R., and Evans, V. J. The measurement of proliferation in tissue culture by enumeration of cell nuclei. *J. Natl. Cancer Inst.* 11:773–795 (1951).

95. Cherry, R. S. and Papoutsakis, E. T. Physical mechanisms of cell damage in microcarrier cell culture bioreactor. *Biotechnol. Bioeng.* 32:1001–1014 (1988).

96. Schon, A. and Wadso, I. Microcalorimetric measurements on tissue cells attached to microcarriers in stirred suspension. *J. Biochem. Biophys. Method.* 13:135–143 (1986).

97. Miller, S. J. O., Henratte, M., and Miller, A. O. A. Growth of animal cells on microbeads. I. In situ estimation of numbers. *Biotechnol. Bioeng.* 28:1466–1473 (1986).

98. Giard, D. J., Thilly, W. G., Wang, D. I. C., and Levine, D. W. Virus production with newly developed microcarrier system. *Appl. Environ. Microbiol.* 34:668–672 (1977).

99. van Wezel, A. L., van der Velden-de-Groot, C. A. M., and van Herwaarden, J. A. M. The production of inactivated p polio vaccine on serially cultivated kidney cells from captive-bred monkeys. *Dev. Biol. Standard.* 46:151–158 (1980).

100. Meignier, B. Cell culture on beads used for the industrial production of foot and mouth disease virus. *Dev. Biol. Standard.* 42:141–145 (1979).

101. Baijot, M., Duchene, M., and Stephenne, J. Production of aujesky vaccine by the microcarrier technology "from the ampoule to the 500 litre fermentor. *Dev. Biol. Standard.* 66: 523−530 (1987).
102. van Hemert, P., Kilburn, D. G., and van Wezel, A. L. Homogeneous cultivation of animal cells for the production of virus and virus products. *Biotechnol. Bioeng.* 11:875−885 (1969).
103. Tyo, M. A. and Wang, D. I. C. Engineering characterization of animal cell and virus production using controlled charge microcarriers. In *Advances in Biotechnology*, Vol. 1 (E. M. Young, C. W. Robinson, and C. Venzina, eds.), Pergamon Press, New York, pp. 141−146 (1981).
104. Kluft, C., van Wezel, A. L., van der Velden, C. A. M., Emeis, J., Verheijen, J. H., and Wijngoards, A. Large scale production of extrinsic (tissue type) plasminogen activator from human melanoma cells. In *Advances in Biotechnological Processes*, Vol. 2 (A. Mizrahi and A. L. van Wezel, eds.), Alan R. Liss, New York, pp. 97−100 (1983).
105. Griffiths, J. B., McEntee, I. D., Electricwala, A., Atkinson, A., Sutton, P. M., Naish, S., and Riley, P. A. The production and properties of a tissue plasminogen activator from normal epithelial cells grown in microcarrier culture. *Dev. Biol. Standard.* 60:439−446 (1985).
106. Edy, V. G., Augenstein, D. C., Edwards, C. R., Gruttenden, V. F., and Lubiniecki, A. S. Large scale tissue culture for human IFN-b production. *Tex. Rep. Biol. Med.* 41:169−174 (1982).
107. Morandi, M., and Valeri, A. Continuous dialysis of cultures of human diploid cells (MRC-5) grown on microcarriers. *Biotechnol. Lett.* 4:465−468 (1982).
108. Lehmann, J., Piehl, G. W., and Schulz, R. Bubble free cell culture aeration with porous moving membranes. *Dev. Biol. Standard.* 66:227−240 (1985).
109. Tschopp, A., Cogoli, A., Lewis, M. L., and Morrison, D. R. Bioprocessing in space: Human cells attach to beads in microgravity. *J. Biotechnol.* 1:287−293 (1984).
110. Nielsen, V. and Johannsson, A. Biosilon-optimal culture conditions and various research scale culture techniques. *Dev. Biol. Standard.* 46:131−136 (1980).
111. Reuveny, S. and Thoma, R. W. Apparatus and methodology for microcarrier cell culture. *Adv. Appl. Microbiol.* 31: 139−179 (1986).
112. Noteboom, W. D. and Will, P. C. Siliconizing glassware to be used for suspension cell culture. *J. Tissue Cult. Methods* 7: 9−11 (1982).

113. Griffiths, J. B., Thornton, B., and McEntee, I. The development and use of microcarrier and glass sphere culture techniques for the production of Herpes simplex viruses. *Dev. Biol. Standard.* 50:103–110 (1982).

114. Weiss, O. L. The measurement of cell adhesion. *Exp. Cell Res.* 8:141–153 (1961).

115. Grinnell, F. Concanavalin A increases the strength of BHK cell attachment to substratum. *J. Cell Biol.* 58:602–607 (1973).

116. Rees, D. A., Lloyd, C. W., and Thorn, D. Control of grip and stick in cell adhesion through lateral relationships of membrane glycoprotein. *Nature* 276:124–128 (1977).

117. Stathopoulos, N. A. and Helluns, J. D. Shear effects on human embryonic kidney cells in vitro. *Biotechnol. Bioeng.* 27:1021–1026 (1985).

118. Nerem, R. M. Shear stress effects on anchorage dependent mammalian cells. Paper presented at the Eng. Foundation Conf. on Cell Culture Engineering, Palm Coast, Florida, Jan. 31 (1988).

119. Croughan, M. S., Hamel, J. F., and Wang, D. I. C. Hydrodynamic effects on animal cells grown in microcarrier cultures. *Biotechnol. Bioeng.* 29:130–141 (1987).

120. Croughan, M. S., Hamel, J. F. P., and Wang, D. I. C. Effect of microcarrier concentration in animal cell culture. *Biotechnol. Bioeng.* 32:975–982 (1988).

121. de Bruyne, N. A. A high efficiency stirrer for suspension cell culture with or without microcarriers. *Adv. Exp. Med. Biol.* 172:139–149 (1984).

122. Clark, J. M. and Hirtenstein, M. D. Optimizing culture conditions for the production of animal cells in MC culture. *Ann. NY Acad. Sci.* 369:33–46 (1981).

123. Hirtenstein, M. D., Clark, J. M., and Gebb, C. A comparison of various laboratory scale culture configurations for microcarrier culture of animal cells. *Dev. Biol. Standard.* 50:73–80 (1982).

124. Edy, V. G. Interferon production in microcarrier culture of human fibroblast cells. *Adv. Exp. Med. Biol.* 172:169–178 (1984).

125. Scattergood, E. M., Schlabach, A. J., Mcleer, W. J., and Hilleman, M. R. Scale-up of chick cell growth on microcarriers in fermentors for vaccine production. *Ann. NY Acad. Sci.* 413:332–339 (1983).

126. Griffiths, J. B. and Thornton, B. Use of microcarrier culture for the production of herpes simplex virus (type 2) in

MRC-5 cells. *J. Chem. Technol. Biotechnol.* 32:324–329 (1982).

127. Feder, J. and Tolbert, W. R. The large scale cultivation of mammalian cells. *Sci. Am.* 248:36–43 (1983).
128. Tolbert, W. R. and Feder, J. Large scale cell culture technology. *Annu. Rep. Ferm. Process* 6:35–74 (1984).
129. Paris, M. S., Eaton, D. L., Sempolinski, D. E., and Sharma, B. P. A gelatin microcarrier for cell culture. Presented at the 34th Annual Meeting of the Tissue Culture Association, Orlando, June 12–16 (1983).
130. Strand, M. S., Quarles, J. M., and McConnel, S. A modified matrix perfusion-microcarrier bead cell culture system. 1. Adaptation of matrix perfusion system for growth of human foreskin fibroblasts. *Biotechnol. Bioeng.* 26:503–507 (1984).
131. Strand, M. S., Quarles, J. M., and McConnel, S. A modified matrix perfusion-microcarrier bead cell culture system. 2. Production of human (β) interferon in matrix perfusion system. *Biotechnol. Bioeng.* 26:508–512 (1984).
132. Tolbert, W. R., Lewis, C., Jr., White, P. J., and Feder, J. Perfusion culture systems for production of mammalian cell biomolecules. In *Large-Scale Mammalian Cell Culture* (J. Feder and W. R. Tolbert, eds.), Academic Press, New York, pp. 97–123 (1985).
133. Spier, R. E. and Griffiths, B. An examination of the data and concept germane to the oxygenation of cultured animal cells. *Dev. Biol. Standard.* 55:81–92 (1984).
134. Sardonini, C. A. and Barngrover, D. Oxygen bicarbonate and acid metabolism determination in mouse cell microcarrier culture. *Biotechnol. Bioeng.* 32:1073–1078 (1988).
135. Hu, W. S. and Wang, D. I. C. Mammalian cell culture technology: A review from an engineering perspective. In *Mammalian Cell Technology* (W. G. Thilly, ed.), Butterworths, pp. 167–198 (1986).
136. Fleischaker, R. J. and Sinskey, A. J. Oxygen demand and supply in cell culture. *Eur. J. Appl. Microbiol. Biotechnol.* 12:193–197 (1981).
137. Frame, K. K. and Hu, W. S. Oxygen uptake of mammalian cells in microcarrier culture-response to changes in glucose concentration. *Biotechnol. Lett.* 7:147–152 (1985).
138. Kilburn, D. G. and Webb, F. C. The cultivation of animal cells at controlled dissolved oxygen partial pressure. *Biotechnol. Bioeng.* 10:801–814 (1968).
139. Handa, A., Emery, A. N., and Spier, R. E. On the evaluation of gas-liquid interfacial effects on hybridoma viability

in bubble column bioreactors. *Biol. Standard* 66:241–253 (1987).

140. Aunins, J. G., Croughan, M. S., Wang, D. I. C., and Goldstein, J. M. Engineering developments in homogeneous culture of animal cells: Oxygenation of reactors and scaleup. *Biotechnol. Bioeng. Symp.* 17:699–723 (1986).

141. Glacken, M. W., Fleischaker, R. J., and Sinkey, A. J. Large scale production of mammalian cells and their products: Engineering principles and barriers to scale up. *Ann. NY Acad. Sci.* 413:355–372 (1983).

142. Hirtenstein, M. D. and Clark, J. M. Further comments on the physical characteristics of microcarriers. *ESACT Newslett.* 6:9–17 (1981).

143. van Wezel, A. L., Cultivation of anchorage dependent cells and their applications. *J. Chem. Technol. Biotechnol.* 32:318–323 (1982).

144. Hu, W. S., Meier, J., and Wang, D. I. C. Use of surface aerator to improve oxygen transfer and cell growth in cell culture. *Biotechnol. Bioeng.* 28:122–125 (1986).

145. van Wezel, A. L. Microcarrier technology, present status and prospects. *Deve. Biol. Standard.* 55:3–9 (1984).

146. Delzer, J., Hauser, H., and Lehmann, J. Production of human interferon-β in mouse L-cells. *Dev. Biol. Standard.* 60:413–419 (1985).

147. Katinger, H. W. D., Scheirer, W., and Kramer, E. Bubble column reactor for mass propagation of animal cells in suspension. *Ger. Chem. Eng.* 2:31–38 (1979).

148. Katinger, H. W. D., Scheirer, W., Frenzi, H., and Bliem, R. F. Proposal for the improvement of microcarrier culture. *ESCAT Newslett.* 5:5–8 (1980).

149. Fleischaker, R. J., Weaver, J. C., and Sinskey, A. J. Instrumentation for process control in cell culture. *Adv. Appl. Microbiol.* 27:137–167 (1981).

150. Dhainant, F., Gerbert-Gaillard, B., and Maume, B. F. Culture of mouse adrenal cortex y-1 cell line on microcarriers in cell reactor. Growth and steroidogenesis. *J. Biotechnol.* 5:131–138 (1987).

151. Sinskey, A. J., Fleischaker, R. J., Tyo, M. O., Giard, D. J., and Wang, D. I. C. Production of cell derived products: Virus and interferon. *Ann. NY Acad. Sci.* 369:47–59 (1981).

152. Kuhlmann, W. Optimization of a membrane oxygenation system for cell culture in stirred tank reactors. *Dev. Biol. Standard.* 66:263–268 (1987).

153. Eberhard, U. and Schugerl, K. Investigations of reactors for insect cell culture. *Dev. Biol. Standard.* 66:325–339 (1985).

154. Whiteside, J. P., Farmer, S., and Spier, R. E. The use of caged aeration for the growth of animal cells on microcarriers. *Dev. Biol. Standard.* 60:283–290 (1985).

155. Spier, R. E. and Whiteside, J. P. The description of a device which facilitates the oxygenation of microcarrier cultures. *Dev. Biol. Standard.* 51:151–152 (1984).

156. Griffiths, J. B. and Electricwala, A. Production of tissue culture plasminogen activators from animal cells. *Adv. Biochem. Eng./Biotechnol.* 34:147–166 (1987).

157. Katinger, H., Reiter, M., Weignang, F., Ernst, W., Dohlhoff-Dier, O., Broth, N., and Steindl, F. A scaleable modular vibrating cage system for enhanced gas/liquid diffusion and cell immobilization in continuous perfused culture of mammalian cells. Paper presented at the Eng. Foundation Conf. on Cell Culture Engineering, Palm Coast, Florida, Jan. 31 (1988).

158. Karrer, D. Indirect aeration/perfusion system for cell culture. Paper presented at the Eng. Foundation Conf. on Cell Culture Engineering, Palm Coast, Florida, Jan. 31 (1988).

159. Montagnon, B. J., Vincent-Falquez, J. C., and Fanget, B. Thousand liters scale microcarrier culture of vero cells for killed polio virus vaccine-promising results. *Dev. Biol. Standard.* 55:37–42 (1984).

160. Montagnon, B. J., Fanget, B., and Vincent-Falquet, J. C. Industrial-scale production of inactivated poliovirus vaccine prepared by culture of Vero cells on microcarrier. *Rev. Infec. Dis.* 6 Sup2:341–343 (1984).

161. Hu, W. S., Meier, J., and Wang, D. I. C. A mechanistic analysis of the inoculum requirement for the cultivation of mammalian cells on microcarriers. *Biotechnol. Bioeng.* 27:585–595 (1985).

162. Butler, M. and Thilly, W. G. MDCK microcarrier cultures: Seeding density affects and amino acid utilization. *In Vitro* 18:213–219 (1982).

163. Clark, J., Hirtenstein, H., and Gebb, C. Critical parameters in the microcarrier culture of animal cells. *Dev. Biol. Standard.* 46:117–124 (1980).

164. Clark, J. M. and Hirtenstein, M. D. High yield culture of human fibroblasts on microcarrier: A first step in production of fibroblast derived interferon (human beta interferon). *J. Interferon Res.* 1:391–400 (1981).

165. Butler, M., Hassel, T., and Rowley, A. The use of micro-
     carriers in animal cell culture. In *Plant and Animal Cells:
     Process Possibilities* (C. Webb and F. Mavituna, eds.),
     Horwood, Chichester, U.K., pp. 64–74 (1986).
166. Lindner, E., Arvidsson, A. C., Wergeland, I., and
     Billig, D. Subpassaging cells on microcarriers: The im-
     portance for scaling up the production. *Dev. Biol. Standard.*
     66:299–305 (1987).
167. Anghilieri, L. J. and Dermietzel, P. Cell coat in tumor
     cells —Effect of trypsin and EDTA: A biochemical and
     morphological study. *Oncology* 33:11–23 (1976).
168. Hu, W. S., Giard, D. J., and Wang, D. I. C. Serial propa-
     gation of mammalian cells on microcarriers. *Biotechnol.
     Bioeng.* 27:1466–1476 (1985).
169. Ben Nathan D., Barzilai, R., Lazar, A., and Shahar, A.
     RDB —A new product of plant origin for cell culture disper-
     sion. *Dev. Biol. Standard.* 60:467–473 (1985).
170. Lindskog, U., Lundgren, B., Billig, D., and Lindner, E.
     Alternatives for harvesting cells grown on microcarriers:
     Effects on subsequent attachment and growth. *Dev. Biol.
     Standard.* 66:307–313 (1987).
171. van Oss, C. J., Charny, C. K., Absolom, D. R., and
     Flanagan, T. D. Detachment of cultured cells from micro-
     carrier particles and other surfaces by repulsive van der
     Waals forces. *Biotechniques* (Nov–Dec):194–196 (1983).
172. Vretblad, P., Wergeland, I., Lindner, E., and Billig, D.
     Pilot and production scale microcarrier cultivation of
     mammalian cells on Cytodex. *Proc. Biotechnol.* 85 (Europe)
     Meeting, Online Publications, pp. 97–105 (1985).
173. Morandi, M., Stanghellini, L., and Valeri, A. Problems
     involved in the large scale production of biological products,
     such as β-interferon, using diploid fibroblast cells as sub-
     strate. *Dev. Biol. Standard.* 60:405–412 (1985).
174. Crespi, C. L. and Thilly, W. G. Continuous cell propagation
     using low-charge microcarriers. *Biotechnol. Bioeng.* 23:
     983–994 (1981).
175. Nohapetian, A. T. Expansion of human foreskin diploid
     fibroblast cells in a microcarrier culture by bead-to-bead
     transfer. *Fed. Proc.* 46:1329 (1987).
176. Manousos, M., Ahmed, M., Torchio, C., Wolfi, J.,
     Shibley, G., Stephens, R., and Mayyasi, S. Feasibility
     studies of oncornavirus production in microcarrier culture.
     In Vitro 16:507–515 (1980).
177. Spier, R. E., Whiteside, J. P., and Bolt, K. Trypsiniza-
     tion of BHK-21 monolayer cells grown in two large scale

unit process system. *Biotechnol. Bioeng.* 19:1735–1738 (1977).

178. Billig, D., Clark, J. M., Ewell, A. J., Carter, C. M., and Gebb, C. The separation of harvested cells from microcarriers: A comparison of methods. *Dev. Biol. Standard.* 55: 67–75 (1984).

179. Butler, M. Growth limitations in high density microcarrier cultures. *Dev. Biol. Standard.* 60:269–280 (1985).

180. Glacken, M. W., Fleischaker, R. J., and Sinskey, A. J. Reduction of waste product excretion via nutrient control: Possible strategies for maximizing product and cell yields on serum in cultures of mammalian cells. *Biotechnol. Bioeng.* 28: 1376–1389 (1986).

181. Wagner, A., Villernaux, S., Marc, A., Engasser, J. M., and Einsele, E. Prourokinase production by human kidney tumor cells on microcarriers. Paper presented at the Eng. Foundation Conf. on Cell Culture Engineering, Palm Coast, Florida, Jan. 31 (1988).

182. Butler, M., Imamura, T., Thomas, J., and Thilly, W. G. High yields from microcarrier cultures by medium perfusion. *J. Cell Sci.* 61:351–363 (1983).

183. Griffiths, B. Can cell culture medium costs be reduced? Strategies and possibilities. *Trends Biotechnol.* 4:268–272 (1986).

184. Imamura, T., Grespi, C. L., Thilly, W. G., and Brunengraber, H. Fructose as a carbohydrate source yields stable pH and redox parameters in microcarrier cell culture. *Anal. Biochem.* 124:353–358 (1982).

185. Clark, J., Gebb, C., and Hirtenstein, M. D. Serum supplements and serum-free media: Applicability for microcarrier culture of animal cells. *Dev. Biol. Standard.* 50:81–92 (1982).

186. Mignot, G., Dhainaut, F., Ibanez, M. D., Chessebeuf-Padieu, M., and P. Padieu. Culture de lignees de cellules epitheliales dans des milieux chimiquement definis sur microsuppors en bioenerateurs. *C. Roy. Soc. Biol.* 180:460–472 (1986).

187. Deaver, G., Price, P. J., and Vass, H. F. Serum free growth of attached and/or non-attached CHO cells. *In Vitro* 23:37A (1987).

188. Clark, J. Reducing the requirement for serum supplements in high yield microcarrier cell culture. *Proc. Eur. Conf. on Serum Free Cell Culture* (Fisher, ed.), pp. 6–15 (1983).

189. Bodeker, B. G. D., Berg, G. J., Hewlett, G. T., and Schulmberger, H. D. Procedures to test for the serum-free

cultivation of adherent cell lines on microcarriers. *J. Tissue Cult. Method.* 8:161–165 (1983).

190. Hu, W. S., Dodge, T. C., Frame, K. K., and Himes, V. B. Effect of glucose on the cultivation of mammalian cells. *Dev. Biol. Standard.* 66:279–290 (1987).

191. Butler, M. and Spier, R. E. The effect of glutamine utilization and ammonia production on the growth of BHK cells in microcarrier cultures. *J. Biotechnol.* 1:187–196 (1984).

192. Birch, J. R. and Cartwright, T. Environmental factors influencing the growth of animal cells in culture. *J. Chem. Technol. Biotechnol.* 32:313–317 (1982).

193. Polastri, G. D., Friesen, H. J., and Mauler, R. Amino acid utilization by Vero cells in microcarrier culture. *Dev. Biol. Standard.* 55:53–56 (1984).

194. Thilly, W. G., Barngrover, D., and Thomas, J. N. Microcarriers and the problem of high density cell culture. In *From Gene to Protein: Translation into Biotechnology.* Academic Press, New York, pp. 75–103 (1982).

195. Wagner, R., Krafft, H., Ryll, T., and Lehmann, J. Variation of amino acid concentrations in the medium of human interleukin 2 producing cell lines. *Proc. 4th Eur. Cong. Biotechnol.* 576–579 (1987).

196. Arathoon, W. R. and Birch, J. R. Large-scale cell culture in biotechnology. *Science* 232:1390–1395 (1986).

197. van Meel, F. C. M., Steenbakkers, P. G. A., Schonherr, O. T., Haefs, C. A. M., and Magre, E. P. Negatively charged polyvinyltoluene microcarriers evaluated in tissue culture. In *Innovation in Biotechnology* (E. H. Houwink and R. R. van der Meer, eds.), Elsevier, Amsterdam (1984).

198. Denomme, L. and Qu, G. Y. Propagation of mammalian cells in a Celligen bioreactor using collagen microcarriers. Paper presented at the ASM Meeting, Atlanta, 1–6 March (1987).

199. Schwarz, M. A. and Juliano, R. L. Interaction of fibronectin-coated beads with CHO cells. *Exp. Cell Res.* 152:302–312 (1984).

200. Reuveny, S., Corett, R., Freeman, A., Kotler, M., and Mizrahi, A. Newly developed microcarrier culturing systems — An overview. *Dev. Biol. Standard.* 60:243–253 (1985).

201. Lazar, A., Silverstein, L., Margel, S., and Mizrahi, A. Agarose-polyacrolein microsphere beads: A new microcarrier culturing system. *Dev. Biol. Standard.* 60:457–465 (1985).

202. David, A., Segard, E., Braun, G., Berneman, A., and Harodniceanu, F. An improved microcarrier for mass culture

of human diploid cells. *Ann. NY Acad. Sci.* 369:61–64 (1981).

203. Kuo, M. J., Lewis, R. A., Jr., Martin, A., Miller, R. E., Schoenfeld, R. A., Schuck, J. M., and Wildi, B. S. Growth of anchorage-dependent mammalian cells on glycine-derivatized polystyrene in suspension culture. *In Vitro* 17:901–912 (1981).

204. Burke, D., Brown, M. J., and Jacobson, B. S. HeLa cell adhesion to microcarrier in high shear conditions: Evidence for membrane receptors for collagen but not laminin or fibronectin. *Tissue Cell* 15:181–191 (1983).

205. Gusky and Bergstram, J. High yield growth and purification of human Parainfluenza type 3 virus. *Gen. Virol.* 54:115–123 (1981).

206. Rivera, E., Karlsson, K. A., and Bergman, R. The propagation of Feline Panleukopenia virus in microcarrier cell culture and use of the inactivated virus in protection of mink against viral enteritis. *Vet. Microbiol.* 13:371–381 (1987).

207. Rivera, E., Sjosten, C. G., Bergman, R., and Karlsson, K. A. Parcin parvovirus: Propagation in microcarrier cell culture and immunogenic evaluation in pregnant gilts. *Res. Vet. Sci.* 41:391–396 (1986).

208. Hayle, A. J. Culture of respiratory syncytial virus infected diploid bovine nasal mucosa cells on Cytodex 3 microcarriers. *Arch. Virol.* 84:81–88 (1986).

209. Widell, A., Hansson, B. G., and Nardenfelt, E. A microcarrier cell culture system for large scale production of hepatitis a virus. *J. Virol. Method.* 8:63–71 (1984).

210. Colyer, T. E. and Boyle, J. A. Optimization of conditions for production of channel catfish virus DNA. *Appl. Environ. Microbiol.* 49:1025–1028 (1985).

211. Butchaial, G. and Roo, B. U. Growth of BHK 21 c13 monolayer cells on microcarriers for production of foot and mouth disease virus vaccines. *Curr. Sci.* 56:346–349 (1987).

212. von Seefried, A., Chun, J. H., Grant, J. A., Letvenuk, L., and Pearson, E. W. Inactivated poliovirus vaccine and test development at Connaught Laboratories, Ltd. *Rev. Infect. Dis.* 6:345–349 (1984).

213. Nicholson, B. L. Growth of fish cell lines on microcarriers. *Appl. Environ. Microbiol.* 39:394–397 (1980).

214. Bektemirov, T. A. and Nagieva, F. G. Propagation of herpes virus in suspension culture on microcarrier. *Vopr. Virusol.* 5:615–618 (1980).

215. Meignier, B., Maugeot, H., and Favre, H. Foot and mouth disease virus production on microcarrier growth cells. *Dev. Biol. Standard.* 46:249–256 (1980).

216. Fohring, B., Tjia, S. T., Zenke, W. M., Sauer, G., and
      Doerfler, W. Propagation of mammalian cells and of virus in
      a self regulating fermenter (40852). *Proc. Soc. Exp. Biol.
      Med.* 164:222–228 (1980).

217. Bergman, R. and Straman, L. Production of a feline parvo-
      virusvaccine using monolayer cell systems in roller fleasks and
      microcarriers. *Dev. Biol. Standard.* 55:77–78 (1984).

218. Nagieva, F. G., Bektemirov, M. S., Matevosyan, K. S.,
      Bektemirov, T. A., and Pille, E. R. Production of fixed
      rabies virus in Quail fibroblast culture growing in suspension.
      Vopr. Virusol. 4:429–431 (1980).

219. Mered, B., Albrecht, P., Hopps, H. E., Petricciani, J. C.,
      and Salk, J. Monkey kidney cell growth optimization and
      poliovirus propagation in microcarrier culture. *Dev. Biol.
      Standard.* 47:41–54 (1981).

220. Frappa, J., Beaudry, Y., Quillan, J. P., and Fantages, R.
      Multiplication of normal and virus-infected diploid cells cul-
      tivated on microcarriers. *Dev. Biol. Standard.* 42:153–158
      (1979).

221. Crespi, L. C., Imamura, T., Leong, P. M., Fleischaker,
      R. J., Brunenbraber, H., and Thilly, W. G. Microcarrier
      culture: Applications in biological production and cell biology.
      Biotechnol. Bioeng. 23:2673–2689 (1981).

222. Giard, D. J., Fleischaker, R. J., Sinskey, A. J., and Wang,
      D. I. C. Large-scale production of human fibroblast inter-
      feron. *Dev. Indust. Microbiol.* 22:299–309 (1982).

223. Prestidge, R. L., Sandlin, G. M., Koopman, W. J., and
      Bennett, J. C. Interleukin 1: Production by P388D1 cells
      attached to microcarrier beads. *J. Immunol. Method.* 46:
      197–204 (1981).

224. van Wezel, A. L. and van Steenis, G. Production of an in-
      activated rabies vaccine in primary dog kidney cells. *Dev.
      Biol. Standard.* 40:69–75 (1978).

225. Giard, D. J., Loeb, D. H., Thilly, W. G., Wang, D. I. C.,
      and Levine, D. W. Human interferon production with diploid
      fibroblast cells grown on microcarriers. *Biotechnol. Bioeng.*
      21:433–442 (1979).

226. Kotler, M., Reuveny, S., Mizrahi, A., and Shahar, A.
      Ion exchange capacity of DEAE microcarriers determined the
      growth pattern of cells in culture. *Dev. Biol. Standard.* 60:
      255–261 (1985).

227. Sando, G. N. and Rosenbaum, L. M. Human lysosomal acid
      lipase/Cholesteryl ester hydrolase. Purification and properties
      of the form secreted by fibroblasts in microcarrier culture.
      J. Biol. Chem. 260:15186–15193 (1985).

228. Norrgren, G., Ebendal, T., and Wikstrom, H. Production of nerve growth-stimulation factor(s) from chick embryo heart cells. *Exp. Cell Res.* 152:427–435 (1984).
229. Bio-Rad Laboratories. Biocarrier: The new high strength matrix for mammalian cell culture. *Bioradiation* 31:1 (1979).
230. Dextran Products. Microdex microcarriers for cell culture.
231. Flow Laboratories. Superbeads microcarriers (1978).
232. SoloHill Engineering. Collagen coated beads (1987).
233. Ventrex Laboratories. The gelatin microcarrier (1984).
234. Lux. Cytospheres (1980).
235. Sayer, T. E., Butler, M., and MacLeod, A. J. The attachment of MDCK cells to three types of microcarriers in different serum free media. In *Modern Approaches to Animal Cell Technology* (R. E. Spier and J. B. Griffiths, eds.), Butterworths, pp. 264–279 (1987).
236. Hu, W. S., Tao, T. A., Bohn, M. A., and Ji, G. Y. Kinetics of prourokinase production by mammalian cells in culture. In *Modern Approaches to Animal Cell Technology* (R. E. Spier and J. B. Griffiths, eds.), Butterworths, pp. 365–380 (1987).
237. Looby, D. and Griffiths, J. B. Comparison of oxygenation methods in a 39L stirred bioreactor. In *Modern Approaches to Animal Cell Technology* (R. E. Spier and J. B. Griffiths, eds.), Butterworths, pp. 449–453 (1987).
238. Lundgren, B., Wergeland, I., Braesch-Anderssen, S., Paulie, P., Koho, H., and Perlmann, P. Production and purification of a bladder cancer associated membrane protein. In *Modern Approaches to Animal Cell Technology* (R. E. Spier and J. B. Griffiths, eds.), Butterworths, pp. 738–744 (1987).
239. Hassell, T. E., Allen, I. C., Rowley, A. J., and Butler, M. The use of glutamine-free media for the growth of three cell lines in microcarrier culture. In *Modern Approaches to Animal Cell Technology* (R. E. Spier and J. B. Griffiths, eds.), Butterworths, pp. 245–263 (1987).
240. Lundgren, B. and Nilosson, S. A model system to study gene expression and scale up of microcarrier cultures or recombinant DNA manipulated cells. In *Modern Approaches to Animal Cell Technology* (R. E. Spier and J. B. Griffiths, eds.), Butterworths, pp. 397–405 (1987).

12

Microencapsulation of Mammalian Cells

JOSEPH E. TYLER
Abbott Biotech, Inc., Needham Heights, Massachusetts

I. INTRODUCTION

Numerous methods of immobilizing mammalian cells have been described in the literature (1). The practical benefits of immobilization include reuse of the biological catalysts (cells), increased product concentration, and protection of the cells from hostile environments. Systems such as microcarriers, gel entrapment, and microencapsulation in which the cells are retained, associate with, or adhere to small particles have been extensively studied for more than 10 years. One method which has shown to be particularly useful for culturing mammalian cells is microencapsulation. The mammalian cells are retained inside of a semipermeable hydrogel membrane. A porous membrane is formed around the cells permitting the exchange of nutrients and metabolic products with the bulk medium surrounding the capsules. The first practical method of microencapsulating viable mammalian cells was described by Lim (2) in 1978 and patented in 1982 (3). Since that time microencapsulation has been utilized for large-scale mammalian cell culture (4) and alternative methods of encapsulation have been developed.

II. METHODS

As summarized by Lim (5), the process of encapsulation and the subsequent microcapsules must have the following properties:

1. The process must be rapid and gentle in order to minimize cell damage.
2. The reagents and the membrane material must be nontoxic.
3. The membrane should be permeable to nutrients and waste products.
4. The porosity of the membrane should be controllable.
5. The membrane should have sufficient mechanical strength to withstand stirring, pumping, shaking, and other disruptive forces occurring during the culturing process.

A. Polyamino Acid/Alginate Microcapsules

The procedure described by Lim (2,3) in 1978 was the first to meet the above criteria and successfully permit the encapsulation of viable mammalian cells. This microencapsulation procedure was commercialized by Damon Biotech as the Encapcel process. The steps for forming the microcapsules around the cells are as follows:

1. The mammalian cells are grown in a suitable culture system and aseptically harvested using centrifugation or membrane separation devices.
2. The cell concentrate is washed with saline and combined with sodium alginate solution to a final alginate concentration of 0.6–1.2%.
3. The alginate/cell suspension is formed into droplets of uniform size and shape. The droplets are formed by forcing the suspension through a small orifice and then breaking up the stream into droplets.
4. The droplets fall into a calcium chloride solution which gels the alginate.
5. The gelled droplets are washed with saline to remove any residual calcium chloride.
6. A semipermeable membrane is cast on the gelled droplets by reacting polyamino acids with alginate. Membrane formation occurs due to ionic bonding between the polyamino acid amino groups and the alginate carboxyl groups. Several polyamino acids have been used including poly-L-lysine, poly-L-orinithine, and poly-L-glutamate.
7. The gelled alginate support is then liquified using a chelating agent to remove the calcium ion from the gel.
8. The microcapsules are now suitable for culturing. The entire process can be completed in 1–4 hr, even on a large scale. The process is shown in descriptive form in Fig. 1. For more detailed descriptions of the capsule formation process, see Refs. 2–7.

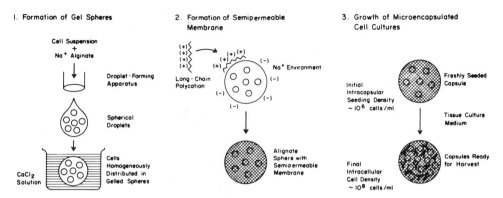

Figure 1 The microencapsulation process. (From Ref. 4.)

Considerable flexibility with regard to permeability, properties, and size of the capsules has been reported. King et al. (6) were able to vary the molecular weight cutoff of the membrane from 20,000 to 300,000 by varying the intrinsic molecular weight of poly-L-lysine (PLL) from 14,000 to 525,000 and the alginate/PLL reaction time from 6 to 40 min. The effects of varying the PLL molecular weight and reaction time on diffusion of marker proteins into the capsules are shown in Figs. 2–4. The same group was also able to adjust the amount of retained alginate by first casting a highly porous membrane, allowing the alginate to diffuse out, and then casting a second less permeable membrane. Varying the size and membrane of the capsules was also studied by Goosen et al. (8,9) and Edmunds et al. (10).

B. Calcium Alginate Hollow Spheres

An alternative method to polyamino acid/alginate hydrogels was developed by Vorlop et al. (11). Here the cell concentrate is mixed with a 1.1% methylcellulose/1.3% calcium chloride solution in a ratio of 1:3. The resulting solution is dropped into a 0.75% sodium alginate solution. A layer of calcium alginate is formed around the droplet during a reaction period of 15 min. Reported capsule diameters are approximately 2.5 mm. This method is simpler than the polyamino acid/alginate membrane process.

C. Bipolymer Membranes

Kim and Rha (12,13) recently reported on chitosan-alginate bipolymer microcapsules. A droplet containing cells to be encapsulated is

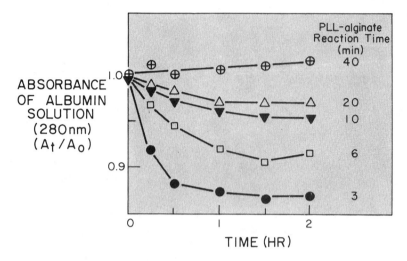

Figure 2   Effect of increasing the alginate-PLL reaction time from
3 to 40 min on microcapsule membrane molecular weight cutoff. The
molecular weight of the PLL was 21,000 and albumin (MW = 66,000)
was the diffusing protein. (From Ref. 6.)

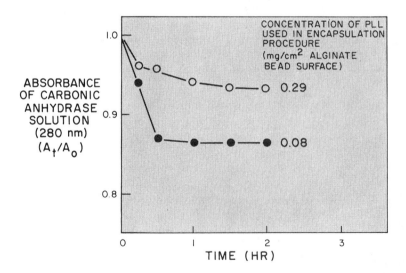

Figure 3   Dependence of microcapsule membrane molecular weight
cutoff on PLL concentration. The molecular weight of the PLL
was 65,000 and the alginate PLL reaction time was 6 min. (From
Ref. 6.)

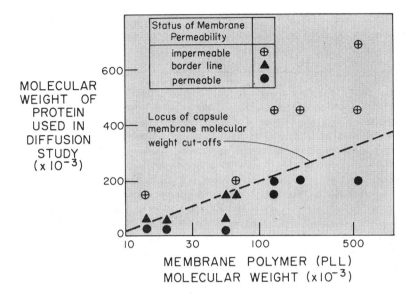

Figure 4   Effect of increasing the molecular weight of PLL from 14,000 to 525,000 on the membrane molecular weight cutoff. Reaction time was 6 min and PLL concentration 0.13 mg/cm². (From Ref. 6.)

layered with a chitosan solution and dropped into sodium alginate solution. The alginate bonds to the chitosan layer. The porosity is controlled by the ionic strength of the chitosan solution which affects the hydrodynamic volume of the chitosan molecule. The membranes formed by this process had a thickness of 5 μm and a strength of 3−5 N.

III. GROWTH OF ENCAPSULATED MAMMALIAN CELLS

A. Culturing Conditions

After formation of the membrane, the capsules can be used in typical cell culture systems. Laboratory scale methods include spinners and t flasks. Medium can be exchanged by allowing the capsules to settle, decanting the spent medium, and refilling with fresh medium. For large-scale culturing, both stirred tanks and airlift fermentors have been used (4,6). Rupp (14) described an inexpensive and simple stirred-tank design used at Damon Biotech. The polyamino acid/alginate microcapsules typically are 500−1500 μm in diameter and are easily retained in the reactor using a 200-300-μm screen. The ratio of capsule volume to total working volume normally is in the

range of 1:10 to 1:2. The effect of microcapsule volume to media
volume ratio with erythroleukemic cell line 745 is shown in Fig. 5.
Maximum cell densities were observed at a capsule total volume ratio
of 1:33. The other ratios studied were 1:11 and 1:6. Additional
medium was not added during the culturing period. A variety of
cell culture media listed in Table 1 such as DMEM, RPMI, and EMA,
both with and without serum, have been used successfully. Maximum
cell densities of greater than 1×10^8 cells/ml of capsules occurs when
the capsules are perfused with fresh medium continuously. The capsules offer some protection from mechanical shear so the culture can
be agitated and direct gas sparged without damage to the cells. An
example of encapsulated cell growth for a hybridoma cell line is shown
in Fig. 6. On day zero individual cells can be observed dispersed
throughout the capsule. As the culture ages the cellular mass increases and the cells begin to clump together. By day 20 very
dense masses of cells practically fill the capsules. In some cases
cell density becomes so great that capsules break open due to over-
crowding of the cells. Figure 7 presents the cell counts and the
monoclonal antibody production from the hybridoma culture of Fig. 6.

B. Cell Types

Table 1 shows a sampling of the many different types of cells that
have been successfully grown in microcapsules. The maximum cell
density achieved was 5×10^8 cells/ml of capsule with a recombinant
murine myeloma (18). In this case the cells were maintained in culture for 150 days. In the majority of examples in Table 1 the peak
cell densities per unit volume are significantly higher than can be
achieved in suspension or attachment cultures. Shown in Figs. 5,
7, and 8 are the time courses of cell growth for some of the examples in Table 1.

C. Advantages and Limitations

A number of advantages over other culture methods are apparent
with microencapsulated cells. The advantages can be summarized as
follows:

1. The capsules provide protection from the deleterious effects
   of shear stresses which occur from gas sparging and agitation. Membrane strengths of 3−5 N have been reported.
2. Cell retention in the bioreactors is easily accomplished using
   relatively coarse mesh screens which do not plug as readily
   as separation devices required with free suspension cultures.
3. The effective bioreactor is the capsule; consequently the
   scale-up is less difficult than that in culture systems such

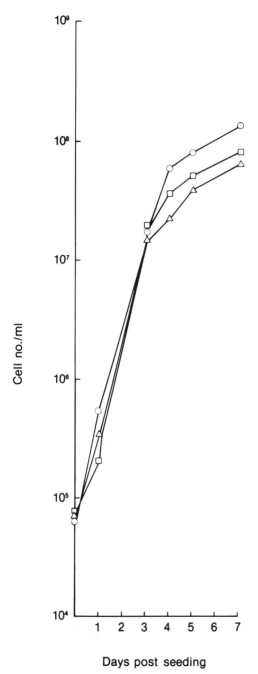

Days post seeding

Figure 5   Effect of microcapsule media ratio on the final cell density of culture of murine erythroleukemic cells clone 745.   The ratios were 1:33○ ; 1:11□ ; 1:6△.   (From Ref. 16.)

Table 1 Cell Types Propagated in Microcapsules

Cell type	Culture Period (days)	Peak cell density cells/ml of capsules	Medium	Ref.
Rat islets	27	NS	CMRL 1969	5
Hepatocytes	14	NS	NS	5
Rat hybridoma 187.1	27	1.5×10^7	DMEM + 5% FBS	15
Rat hybridoma M1/93	27	1.3×10^7	DMEM + 5% FBS	15
Murine erythroleukemia 745	8	1.6×10^8	RPMI + 10% FBS	16
Namalva	7	4.6×10^6	RPMI 1640 + 10% FBS	7
FS-4	NS	NS	DMEM + 10% FBS	7
Murine hybridoma	24	$2 \ 10^8$	EMA + 5% serum	4

Various murine hybridomas	20 – 30	$1.45 \times 10^6 - 3 \times 10^8$	EMA + 5% serum	14
Murine hybridoma AcV$_I$-II$_{20}$	18	5.4×10^7	DMEM	6
Murine hybridoma ATCC CRL1606	NS	8.2×10^7	NS	13
Murine hybridoma ATCC HB8852	NS	2.8×10^7	NS	13
Murine hybridoma 126/38	7	NS	HGDM DIF	11
Human tumor cells	12 – 20	$0.8 - 12 \times 10^5$	RPMI 1640 + 10% FBS	17
Mouse myeloma J558L (recombinant)	150	5×10^8	EMA	18
Insect cells	NS	2.5×10^7	NS	19

NS = Not stated.

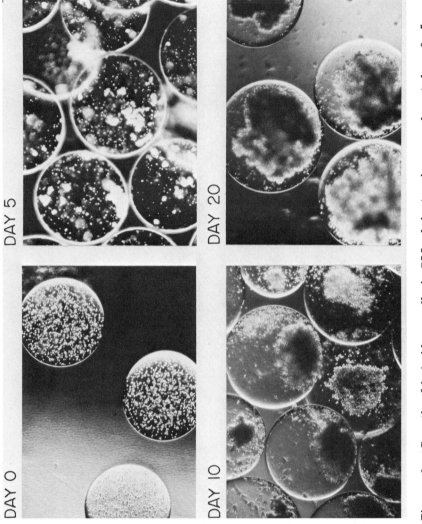

Figure 6   Growth of hybridoma cells in PLL–alginate microcapsules at days 0, 5, 10, 20. (From Ref. 4.)

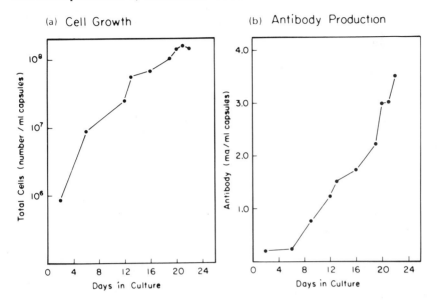

Figure 7   Hybridoma cell growth and antibody production in microcapsules.   (From Ref. 4.)

as large-scale suspension cultures or hollow fiber reactors.
Maintaining similar substrate concentrations in the bulk fluid
and minimizing liquid film resistances are the most important
factors in scale-up.

4.   By adjusting the porosity of the membrane, expressed products can either be retained in the capsule affording a degree
of concentration or allowed to diffuse into the medium.

There are some limitations with microcapsules, however.   Heath
and Belfort (20) and Glacken et al. analyzed diffusional limitations
resulting from the membrane and the intracapsular cell mass, and
determined that under certain circumstances of high cell densities
and reduced nutrient concentrations severe concentration gradients
can exist.   Glacken et al. suggested that to prevent oxygen limitation the maximum capsule diameter should be limited to 170 μm.
Edmunds et al. were able to increase the rate of antibody formation
and cell proliferation by increasing the agitator speed (reducing external liquid film resistance) and by reducing the capsule diameter to
350 μm.   The range of capsule diameters studied was 350−900 μm.

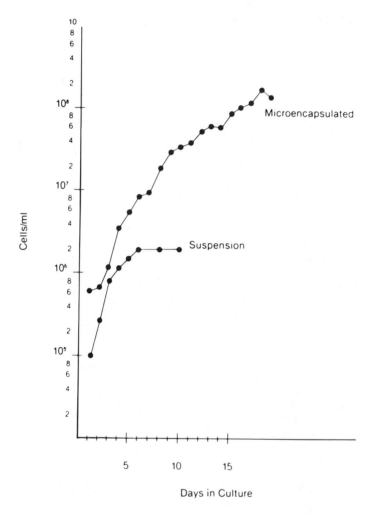

Figure 8   Comparison of cell growth for a hybridoma in suspension and PLL-alginate microcapsules.   (From Ref. 14.)

They independently reached the conclusion that 200 μm was the maximum diameter required to prevent intracapsule mass transfer limitations. The structure of the membrane can also affect mass transfer in and out of the capsule. The same group observed that capsules with double coats had lower rates of cell growth and MAb formation than single-coated capsules.

A second limitation is the encapsulation process itself. The process of forming the capsules is relatively benign and rapid at small scale but becomes increasingly hazardous to the cells with scale-up. Processing times increase with increasing scale for batch encapsulation methods and greater loss of cellular viability occurs during exposure to nutrient starvation and nonphysiological conditions. Fortunately, because of the high cell densities achieved in the capsules, large culture volumes are frequently not needed. Finally, the process involves many steps thereby increasing the risk of contamination by foreign organisms.

IV. APPLICATIONS OF ENCAPSULATED MAMMALIAN CELLS

A. Hybridomas and Production of Monoclonal Antibodies

The major application of encapsulated mammalian cells is the production of monoclonal antibodies. The Encapel process has been used to produce multiple gram quantities of antibodies from over 100 different hybridoma cell lines (4,14,21). For the production of antibodies the capsule porosity is adjusted to a molecular weight cutoff of about 90,000, which retains the immunoglobulin inside the capsule. The capsules are harvested at the end of the culture period, broken open, and the antibody purified from the resulting crude solution. Table 2 and Figs. 7 – 9 are examples of antibody production by encapsulated hybridoma cells. All examples were either batch-fed or perfused with media. The concentration of intracapsular antibody ranged from 1250 to 5300 mg/liter in 7 – 27 days. Three of the examples include suspension culture data, which enables one to compare the two systems on the basis of several criteria. The comparison of the three examples can be summarized as follows:

	Antibody (mg/liter)	Days	Liters of fermenter capacity required to produce 1 g MAb	Liters of medium required to produce 1 g MAb
Encapsulation	2135	29	2.29	22.4
Suspension	131	8	9.35	9.5

Table 2   Monoclonal Antibody Production by Encapsulated Hybridomas

Hybridoma	Capsules				Suspension				Ref.
	Antibody (mg/L capsule)	Days	Fermentor capacity Liters/g antibody	Medium used Liters/g antibody	Antibody (mg/L suspension)	Days	Fermentor capacity Liters/g antibody	Medium used Liters/g antibody	
Not reported	3500	22	NA	NA	NA	NA	NA	NA	2
Mouse/mouse AcV$_I$-II$_{20}$	1250	7	NA	NA	NA	NA	NA	NA	6[a]
	5300	7	NA	NA	NA	NA	NA	NA	6[b]
Mouse/mouse 100	1500	18	2.67	40.0	65	7	15.38	15.4	14
Rat/mouse 187.1	2855	27	1.75	11.4	117	10	8.55	8.5	15
Rat/mouse MI/9.3	2050	27	2.44	15.9	212	6	4.72	4.7	15

[a]Single membrane.
[b]Double membrane.
NA = not available.

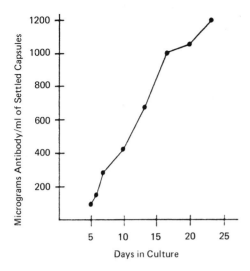

Figure 9   Accumulation of intracapsular monoclonal antibody.   (From Ref. 14.)

Final antibody concentration is much higher using encapsulated cells. This translates into a lower requirement of culture capacity to produce the same amount of antibody. More medium is required to produce the same amount of antibody with the perfused capsule system than with batch suspension culture.

Since the monoclonal antibody is retained in the capsules, considerable purification and concentration is achieved during the culturing process. Reports indicate the antibody purity at the end of the culture period ranges from 45 to 80% (4,14,21). Foreign immunoglobulin in the medium is excluded from the capsule by the membrane; thus subsequent purification is simplified.

The other methods of encapsulation have also been used to culture monoclonal antibody-secreting hybridomas. For the calcium alginate hollow spheres equivalent antibody concentration was achieved in both suspension and encapsulated cultures. The hollow spheres had a 60% lower initial cell density (11). Peak concentration of antibody were 20 times greater with chitosan-alginate membrane encapsulation cultures than with corresponding suspension controls in two examples (13).

B. Other Applications

Encapsulated cells have been used for other purposes. An interesting application has been to inject encapsulated islet of Langerhan

cells into an experimental animal suffering from diabetes mellitus
(5,22). Normal blood sugar levels were observed in diabetic rats
tested with encapsulated islets for periods of up to 52 weeks. Capsules recovered from the rats at the end of 52 weeks contained viable
insulin-secreting islet cells.

Jarvis and Grdina (7) used microencapsulated FS-4 and Namalwa
cells to produce interferon by induction with chemical and viral

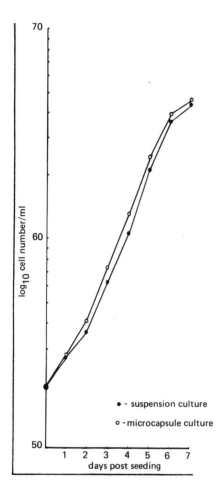

Figure 10   Comparison of Namalwa cell growth in suspension and
PLL-alginate microcapsules. (From Ref. 7.)

agents. Both cell lines grew well in capsules compared to suspension or monolayer cultures for the specific cells. Total and specific interferon production for both cell lines in microcapsules were comparable to suspension or monolayer cultures. The growth curve for the Namalwa experiment is shown in Fig. 10.

Microencapsulated tumor cells have been used to assess the effect of various anticancer drugs in vivo (17). Human tumor cells were encapsulated and injected intraperitoneally into mice. The antitumor drugs to be tested were administered and the response of the encapsulated tumor cells measured. The drugs were found to cross the capsule membrane and act on the tumor cells inside the capsules. The level of inhibition and/or killing corresponded to other in vitro and in vivo determinations of drug activity.

Recombinant cell lines expressing proteins of commercial interest are currently being cultured in microcapsules (18). A recombinant protein of about 65,000 molecular weight expressed in a myeloma cell line has been maintained productively in perfusion culture for periods greater than 150 days. In this case a relatively permeable membrane was prepared and the expressed product diffused from the capsule into the surrounding medium. The conditioned medium containing the product was harvested continuously. During the protein production period low-cost serum-free medium was used. Oxygen uptake rates of as high as 5 mM/liter/hr were observed with the high cell densities maintained in the capsules.

V. SUMMARY AND FUTURE PROSPECTS

As previously described, microencapsulation has been successfully used to grow a large number of mammalian cell types. The encapsulation process is gentle enough to permit a high degree of survivability and flexible enough to allow control over the porosity of the capsular membrane. The microencapsulated cells have been used to produce large quantities of monoclonal antibodies and recombinant proteins. Other applications include both in vivo and in vitro uses.

For the future simpler methods of encapsulation are being developed. This will permit easier scale-up of the process of preparing the capsule membranes. Limited analysis and optimization of the capsule system has been completed. Mass transfer capabilities, diffusional resistances, and mechanical strength of microcapsules need to be fully characterized to allow the optimal use of microcapsules in culturing mammalian cells.

ACKNOWLEDGMENT

My special thanks to Sheila Horsley-Ladd for her assistance in preparing and typing the manuscript.

REFERENCES

1. Glacken, M. W., Fleischaker, R. J., and Sinskey, A. J. Large-scale production of mammalian cells and their products: Engineering principles and barriers to scale-up. *Ann. NY Acad. Sci.* 413:355–372 (1983).
2. Lim, F. Research Report to Damon Corp. (1978).
3. Lim, F. Encapsulation of biological materials, U.S. Patent 4,352,883, October 5, 1982.
4. Posillico, E. G. Microencapsulation technology for large-scale antibody production. *Biotechnology* 4:114–117 (1986).
5. Lim, F. Microencapsulation of living cells and tissues: Theory and practice. In *Biomedical Application of Microencapsulation* (F. Lim, ed.), CRC Press, Boca Raton, pp. 127–154 (1984).
6. King, G. A., Daugulis, A. J., Faulkner, P., and Goosen, M. F. A. Alginate-polylysine microcapsules of controlled membrane molecular weight cutoff for mammalian cell culture engineering. *Biotechnol. Prog.* 3:231–240 (1987).
7. Jarvis, A. P. and Grdina, T. A., Production of biologicals from microencapsulated living cells. *Biotechniques* 1:22–27 (1983).
8. Goosen, M. F. A., Oshea, G. M., Gharapetian, H., Chou, S., and Sun, A. M. Optimization of microencapsulation parameters: Semipermeable microcapsules as a bioartificial pancreas. *Biotechnol. Bioeng.* 27:146–150 (1985).
9. Goosen, M. F. A. *Critical Reviews in Biocompatibility* (D. F. Williams, ed.), CRC Press, Boca Raton, pp. 1–24 (1987).
10. Edmunds, W. W., Kargi, F., and Sorenson, C. Mass transfer effects in microencapsulated hybridoma cells producing monoclonal antibodies. *Appl. Biochem. Biotechnol.* 20/21:1–27 (1989).
11. Vorlop, K. D., Spiekermann, P., and Klein, J. Production of monoclonal antibodies with CA-alginate hollow spheres. Presensented at Cell Culture Engineering, Palm Coast, Florida (1988).
12. Rha, D. K. Biomaterials for cell culture processing. Presented at Cell Culture Engineering, Palm Coast, Florida (1988).
13. Kim, S. K. and Rha, C. K. Encapsulation for mammalian cell culture. Presented at Cell Culture Engineering, Palm Coast, Florida (1988).
14. Rupp, R. G. Use of cellular microencapsulation in large-scale production monoclonal antibodies. In *Large-Scale Mammalian Cell Culture* (J. Feder and W. R. Tolbert, eds.), Academic Press, Orlando, Florida, pp. 19–38 (1985).
15. Gilligan, K. J., Littlefield, S., and Jarvis, A. P., Jr., Production of rat monoclonal antibody from rat x mouse hybridoma cell lines using microencapsulation technology. *In Vitro Cell. Dev. Biol.* 24:35–41 (1988).

16. Jarvis, A. P., Grdina, T. A., and Sullivan, M. F. Cell growth and hemoglobin synthesis in murine erythroleukimic cells propagated in high density microcapsule cultures. *In Vitro Cell. Dev. Biol.* 22:589−596 (1986).

17. Gorelik, E., Quejera, A., Shoemaker, R., Jarvis, A., Alley, M., Duff, R., Mayo, J., Herberman, R., and Boyd, M. Microencapsulated tumor assay: New short-term assay for in vivo evaluation of the effects of anticancer drugs on human tumor cell lines. *Cancer Res.* 47:5739−5747 (1987).

18. Rayner, J. and Gilooly, A. Damon Biotech, pers. Commun. (1988).

19. Goosen, M. F. A., King, G. A., Faulkner, P., Daugulis, A. J., and Bugarski, B. Culturing of animal and insect cells in multiple membrane micricapsules for the enhanced production and recovery of proteins and baculoviruses. Presented at Cell Culture Engineering, Palm Coast, Florida (1988).

20. Heath, C. and Belfort, G. Immobilization of suspended mammalian cells: Analysis of hollow fibers and microcapsule bioreactors. *Adv. Biochem. Eng.* 34:1−30 (1987).

21. Duff, R. G. Microencapsulation technology: A novel method for monoclonal antibody production. *Trends Biotechnol.* 3:167−170 (1985).

22. Goosen, M. F. A., Oshea, G. M., and Sun, A. M. F. Microencapsulation of living tissue and cells, Canadian Patent 1,196,862, November 19, 1985.

13

Continuous Culture with Macroporous Matrix, Fluidized Bed Systems

PETER W. RUNSTADLER, JR., AMAR S. TUNG, EDWARD G. HAYMAN,*
NITYA G. RAY, JOHANNA v. G. SAMPLE, and DAVID E. DeLUCIA
Verax Corporation, Lebanon, New Hampshire

I. INTRODUCTION

In recent years, researchers have identified a large number of naturally produced proteins that hold promise for the diagnosis, prevention, or treatment of human diseases (1,2). Among these proteins are potential products that will be produced by biotechnology and pharmaceutical companies, such as monoclonal antibodies, vaccines, blood chemistry compounds, anticancer agents, hormones and endocrines, growth factors, and other potential therapeutic molecules such as the interleukins and interferons. While these proteins are produced naturally by naturally occurring, unaltered cell lines, the productivity of these cell lines is low compared to the output of these products by cells altered through recombinant DNA technology. Recombinant DNA engineering has been successfully used to develop expression systems in selected mammalian cell lines. This has allowed the synthesis of these complex proteins at relatively high rates with the proper three-dimensional configurations and modifications, including posttranslational glycosylation, which are required to produce the desired activity found in naturally occuring proteins (3-10). As a result, mammalian cell culture is now generally ac-

Present affiliation: T-Cell Sciences, Cambridge, Massachusetts.

knowledged as the preferred approach to producing efficacious quantities required for the commercialization of these products (6–8).
In the early research and development phases of these products, mammalian cell culture, as developed for laboratory use, is used.
In general terms, the cell culture, generally in petri dishes or t flasks, is kept in incubators while the cells grow throughout the culture vessel. This process is usually repeated using multiple vessels, roller bottles, or cell factories, to obtain enough cells to produce the desired protein in quantities needed for laboratory purposes or other preclinical uses.

This type of laboratory process, however, is not economically viable for the production of the quantities of product that are required for commercial introduction of these proteins (11–17). Consequently, a number of technologies have been pursued for the purpose of developing large-scale, commercial cell culture manufacturing systems that address the complexity and expense of producing these proteins in large quantities by mammalian cell culture (18–22). The biotechnology and pharmaceutical industry is currently using traditional fermentation technologies for the introduction of the first wave of these products (23,24). However, the competitive markets that will be created by the large number of companies that are involved in this technology will emphasize the need for effective, large-scale mammalian cell-culturing processes that provide ease of process optimization, have maximum cost-effective product yield, and give predictable scale-up (Ref. 20,introduction;25,26).

Several years ago we started the development of a unique cell-culturing technology that would achieve the low-cost manufacture of large-scale, commercial amounts of therapeutic proteins produced by mammalian cells (25,26). This commercial process technology utilizes continuous culture with immobilized cells inside a special matrix.
The matrix is fluidized in a fluidized bed bioreactor.

II. OPTIMIZED CELL CULTURE

A system for large-scale commercial mammalian cell culture should have the following characteristics. It should embody the ability to

1.   Retain high-producing cells for long periods of time under continuous, steady-state operation
2.   Immobilize all (or most) of these cells and grow them to high cell densities
3.   Have a homogeneous cell microenvironment throughout the bioreactor
4.   Be applicable to both anchorage-dependent and suspension cells

5. Provide culture stability
6. Be easily scaled up, and
7. Have flexibility for process optimization and be easy to operate

These characteristics are important for the following reasons.

A. Continuous Culture with Cell Retention

A continuous process with cell retention, in contrast to a batch process, allows steady-state operating conditions and permits a true optimization of the culture environment (27). Under steady-state conditions, i.e., where all process variables such as pH, temperature, cell density and viability, etc., have reached a constant, steady-state value, one parameter at a time can be changed to effectively define the best process operation. Cell retention permits an additional degree of freedom in this optimization process in that the nutrient feed rate to the bioreactor can be varied independently (26,29); this cannot be done in a chemostat or continuous-suspension culture that operates without cell retention (30).

B. Immobilized Cells at High Density

Immobilization of cells enables and encourages the cells to grow to high densities (25,27,31). It is also believed that immobilized cells can alter their metabolism and under the proper conditions can redirect their metabolic energy from growth to product formation (27,28).

Most cells in vivo grow to very high densities, exceeding 10^9 cells/ml. Evidence points to the fact that at high densities cells form an altered microenvironment. It is hypothesized that this is an enriched environment in which nutrients, hormones, growth factors, enzymes, other cell factors, and the cells themselves interact in ways not observed at lower cell densities. It is believed that cell surface receptor proteins, autocrines, attachment factors, and secreted secondary metabolites act to alter the extracellular environment. Although not well understood, the feedback mechanisms of the cells are believed to respond to the altered extracellular environment by changing their intracellular processes.

In an immobilized cell, continuous, steady-state process, the cell culture microenvironment can be optimized by controlling the rate of addition and the relative concentrations of nutrient substrates and other medium components fed to the culture. It is also possible to control the concentration of inhibitory culture metabolites (waste or toxic products produced by the cells) by the rate at which they are removed.

C. Medium Substrates, the Immobilization Matrix, and the Microenvironment

Figure 1 is a display of the complex factors that interplay in establishing the microenvironment of the cell. The development of serum-free media for animal cell culture in recent years has highlighted the importance of the many nutritional and hormonal contributions of sera to the extracellular environment (28). Nutritional substrate requirements vary between cell types. The density of cells also appears to influence the type and concentration of nutrients and other factors on which the growth and productivity of the cell depend. Continuous immobilized culture is a convenient method to independently vary and thus be able to optimize the medium substrates in a single culture. These substrates include sugars, amino acids, vitamins and related compounds, lipoproteins, fatty acids, cholesterol, and trace elements, such as iron, selenium, etc.

Theoretically, other factors and components in the extracellular milieu can also be varied to condition the cells and change their growth and productivity. These potentially include hormones, such as growth factors, and binding proteins that recognize hormones or nutrients and thus modulate their action. In addition, there are attachment factors that mediate the cell anchorage to a culture

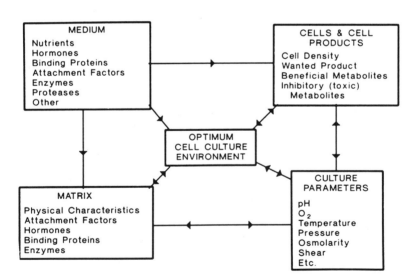

Figure 1   Factors affecting culture optimization.

substrate or to other cells. There are also extracellular enzymes or proteases that can influence the stability of the cell-expressed product.

The substrate or matrix used to immobilize the cells may offer an additional ability to influence the extracellular microenvironment. The immobilization matrix can be selected and altered to provide some of the beneficial factors, hormones, and components discussed above. This may lead to better cell attachment, provide faster cell growth, encourage high cell density, and enhance per-cell productivity.

Finally, physical parameters also influence the optimization of the culture. These include chemical parameters such as pH and osmolarity as well as physical parameters such as oxygen tension, temperature, pressure, and shear rate. The task of optimizing a culture for obtaining maximum productivity is in part one of selecting the best combination of these parameters. The continuous, immobilized culturing process is a way to optimize these cultures that is impossible, or at the least very difficult, to do using transient, batch culture.

D. A Homogeneous Extracellular Environment

The best productivity from a given culturing system will ensue if all of the cells in the system see the same optimized extracellular environment. This is usually the case, for example, in a well-mixed suspension culture. However, suspension cultures usually have reactor volume productivities well below the best values achievable from high-density perfusion systems. They may also have per-cell productivities less than in high-density perfusion cultures because of the important microenvironmental conditioning factors discussed above. But high-cell-density perfusion systems commonly suffer from nonhomogeneity; they often include necrotic regions of low cell viability and productivity. A high-cell-density perfusion system should attempt to ensure bulk reactor homogeneity, often a difficult task if it is designed to operate continuously over long periods of time.

E. Anchorage-Dependent and Suspension Cells

Ideally, the culturing process should work with all cell types, whether they grow in suspension or are anchorage-dependent. Many of the new biochemicals to be produced have been recombinantly engineered into a variety of cells. To be generally useful and applicable, a system should accept and work well with all types of cells.

F. Culture Stability

Genetic mutations or other factors present in a living cell culture
pose the potential problem of culture instability. If nonproducer,
or low-producer, cells grow faster than high-producer cells, the former may soon become the predominant population in the growing culture. This can lead to culture instability (a decline in the space-
time productivity of the culture). It is also possible that genetic
mutations may alter the biochemical makeup of the culture liquor over
time, leading to culture instability in the sense of variability in the
consistency of the harvest composition, product purity, or product
stability. Culture instability can adversely affect downstream processing costs and/or cause concerns about regulatory compliance.
A continuous process which inherently possesses culture stability
thus has distinct advantages over processes which do not.

G. Ease of Scale-up

There is inevitably a need to bring to market commercial products
on a timely schedule. Confidence in the reliability of the process,
as it is put into commercial production, is a prime reason for want-
ing a process that is easily scaled. The development process for recombinantly engineered products usually proceeds with the culturing
of cells at small scale (for the production of research and preclinical
or early clinical quantities of materials) and evolves to large-scale
production as clinical and market demands dictate. Reliability is one
of the most important factors in evaluating process scale-up. Reliability encompasses the ability to know that the process will function
well at large scale as production increases and that properly sized
equipment will be in place when needed.

III. THE CONTINUOUS, IMMOBILIZED CELL, FLUIDIZED BED, CULTURING PROCESS

During the past several years we have developed technology and
large-scale hardware to achieve a culturing process with the characteristics described in Section II (25,26). Cells are cultured by first
immobilizing them in weighted, porous, essentially spongelike beads
called microspheres. A vertically oriented flow through the bioreact-
or vessel suspends the microspheres as a fluidized bed, while a separate flow stream continuously adds medium and removes harvest
liquor containing the cell product. The excellent mass transfer characteristics of the bioreactor fluidized bed, and of the membrane gas
exchanger in a recycle loop, provide oxygen supply and removal
of carbon dioxide at the very high rates demanded by the cell densities created inside the microspheres. The fluid velocities in the

bioreactor are sufficient to suspend the cell-containing microspheres in the culture liquor but do not damage the fragile mammalian cells which are protected and live inside the sponge microspheres.

A. Microsphere Technology

The basic element of this technology is the porous, weighted microsphere. The spongelike microsphere is a tridimensional matrix of natural collagen. The collagen is in the form of a sphere with a diameter of 500−600 μm and contains interconnected pores and channels on the order of 20−40 μm wide. These pores and channels allow cells to enter and populate the internal volume of each sphere. Figure 2 is a scanning electron micrograph of microspheres. Visible in the micrograph are both collagen and the weighting material used to provide the spheres with a high specific gravity. The collagen in the microspheres has a leaflike morphology, providing large surface areas for cell attachment, or mechanical entrapment for suspension type cells, and proliferation. The microsphere's internal volume,

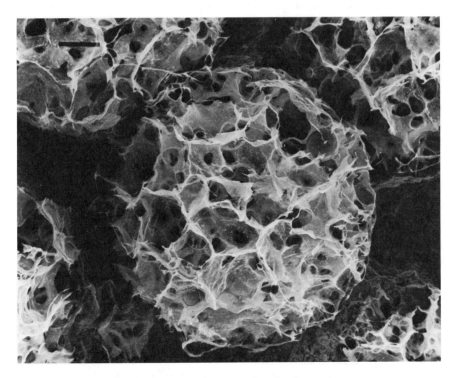

Figure 2   Scanning electron micrograph of microspheres.

which is nearly 85% open, enables cell population at high densities throughout. Because cells can be entrapped as well as anchored to the leaflike surface, both anchorage-dependent and suspension cells populate the microspheres to cell densities in excess of 10^8 viable cells/ml of microsphere volume. This provides a high-viability cell mass, approaching tissue cell densities, throughout the volume of the porous matrix in the fluidized bed.

The spongelike microspheres are derived from native, type 1, bovine collagen, using a proprietary process. The microspheres are classified by their terminal velocity to establish acceptable fluidization characteristics. They are strongly crosslinked to ensure durability and long life within the bioreactor environment. The collagen morphology, pore size, and specific gravity are controlled during manufacture and microspheres can be engineered to meet the needs of various types of cells.

The open-sponge structure of the microspheres allows cells to enter and populate the microspheres after they are put into a bioreactor. Microspheres are therefore made prior to cell inoculation and are stored in a ready-to-use state.

This collagen matrix is nontoxic to animal cells (32,33). Collagen is the major constituent of the extracellular structural matrix in higher animals and it is immunologically benign. The collagen used in Verax microspheres is a natural substrate for cell adhesion for anchorage-dependent cells (34−37).

Specifications for the porous, collagen microspheres are given in Table 1.

Table 1   Porous Collagen Microspheres

Average diameter	$500-600$ μm
Pore size	$20-40$ μm
Wet specific gravity	$1.6-1.7$
Typical fluidization velocity	75 cm/min
Effective microsphere number density in reactor (25% solids)	3.8×10^6 microspheres/liter
Microsphere outer surface area based on expanded fluidized bed (25% solids)	3 m^2/liter
Estimated microsphere surface area (internal)	> 30 m^2/liter
Biocompatible	
Postinoculate with cells	

B. The Fluidized Bed Bioreactor and Recycle System

The central feature of the bioreactor system is the fluidized bed of microspheres. A fluidized bed occurs when particles are put in a vertical vessel and a fluid of lower density is caused to flow upward through them. When the velocity of fluid is sufficiently high, the particles no longer rest on one another, but are suspended in the fluid and are free to move about. In the fluidized bed bioreactor, the particles are the microspheres with the cells inside them. The upward flowing fluid is the recycled culture liquor that surrounds each microsphere. The advantages of a fluidized bed bioreactor are excellent scalability, high mass transfer capability needed to provide oxygen and removal of carbon dioxide from the cells, high rates of mixing, and low hydrodynamic shear that otherwise might damage fragile mammalian cells.

This system is shown in Fig. 3. On the left of the figure is the fluidized bed bioreactor. A recycle flow of culture liquor surrounds the microspheres in the bioreactor vessel. This provides nutrients by diffusion through the microsphere pores to the cells living inside the microspheres. Cell products secreted by the cells diffuse out of the microsphere and into the culture liquor.

The microspheres are fluidized as a thick slurry in the bioreactor vessel. The cells are then inoculated into the bioreactor. Cells rapidly enter the microspheres and each microsphere is populated with thousands of cells. The microspheres circulate freely throughout

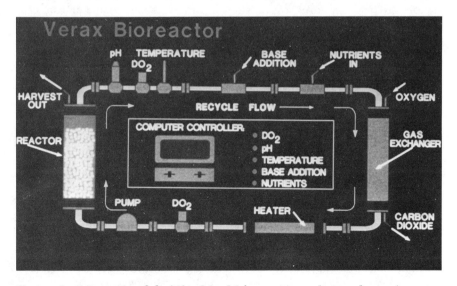

Figure 3 Schematic of fluidized bed bioreactor and recycle system.

the bioreactor vessel. This provides a uniform, homogeneous, highly viable cell concentration with high productivity of the cells throughout the reactor volume.

We hypothesize that a feature of the technology *yet unproven* is the stability of cultures in the fluidized bed system. We believe that in addition to a stabilization of the cells, as a result of immobilization in the collagen matrix microspheres, these reactor systems also become essentially immune to genetic mutations affecting the total culture. We hypothesize that this occurs as a result of isolating such mutations to one or, at the very most, a few microspheres within the millions of microspheres that compose the reactor bed. This effect is enhanced by the very high dilution rates at which medium is perfused through the fluidized bed. These high perfusion rates tend to rapidly remove free cells from the reactor before they can exchange from one microsphere to another.

C. The Recycle Flow

The slurry of microspheres in the bioreactor vessel is fluidized and suspended using a recycle flow. The other piping components shown in Fig. 3 illustrate the recycle portions of the system. Other flow paths are provided for the addition of nutrients and the extraction of harvest from the continuously operating system. The specific gravity of the microspheres establishes the fluidization velocities required in the bioreactor. The liquid/solid fluidized bed of culture liquor and microspheres establishes a natural separation horizon near the top of the bioreactor vessel. This provides a clear zone between the recycling culture liquor and the slurry of microspheres. The recycle flow removes culture liquor from the zone above the separation horizon at the top of the reactor. The culture liquor is conditioned in the recycle loop and returned to the bottom of the reactor vessel. The microspheres, containing most of the cells in the bioreactor vessel, never enter the recycle loop but remain in the suspended slurry in the bioreactor. The loop consists of a recycle pump; a membrane gas exchanger (to add oxygen and extract carbon dioxide); sensors for the measurement and control of pH, dissolved oxygen, and temperature; a heater to control temperature of the culture liquor; and a flow meter to measure the recycle flow rate.

Metering pumps continuously supply medium to the bioreactor from a chilled-medium reservoir. Because there is no gas in the bioreactor, medium addition and product harvest occur at the same rate.

The all-liquid culture fluid —oxygen and carbon dioxide are present in the culture liquor only in the dissolved gas state— eliminates all gas-liquid interfaces and any resultant foaming. Foaming is found in some reactor systems and has the potential to damage fragile proteins and mammalian cells. The fluid dynamics of the bioreactor and recycle loop flow must be designed to minimize shear forces that can otherwise lyse cells or damage the biomolecules made by the cells (38,39).

D. System Monitoring and Control

Automated monitoring and control of the bioreactor system is done using a microcomputer and analog-digital interface. This provides measurement, recording, and control of the reactor system and culturing process. Control is maintained against operator set points to provide continuous, steady-state operation for long-term cultures.

E. Summary

The development of the weighted, collagen microsphere, fluidized bed bioreactor system has produced an effective mammalian cell-culturing process. This system has the following capabilities:

Predictable scale-up
Homogeneous reactor environment
High productivity per cell
High space-time productivity
Consistent product output with good purity and activity
Cost-effective product yield

The hardware embodiments of this system incorporate designs to obtain highly reliable aseptic operation. This enables continuous culture operation for long periods of time (measured in months). The systems are also simple to use and easy to operate.

The following sections give the details of several production scale hardware configurations that use this technology. Several examples are presented showing data from runs using recombinant cells and hybridomas. These examples are given to illustrate scalability and productivity. Finally, an example is given to illustrate the production economics of this technology for an important pharmaceutical protein, tissue plasminogen activator (tPA).

IV. HARDWARE SYSTEMS

A. System 2000 Production Bioreactor

The System 2000 is a large-scale, continuous-culture bioreactor designed to use the fluidized bed/microsphere technology to produce kilogram quantities per year of medical proteins on a production scale. This system is engineered for reliable aseptic operation using industrial grade components.

Figure 4 is a photograph of the System 2000 skidmounted reactor
and process components. The entire system is 1.4 m wide by 3.25 m
long and 2.3 m high and occupies 4.6 m^2 of floor space. Two 1500-
liter, sanitary jacketed vessels are used for daily medium feed and
harvest withdrawal. The system has auxiliary equipment for the
aseptic sampling of culture liquor and microspheres and for the initial loading of microspheres and cell inoculum at the start of a production run.

The system is designed in accordance with GMP/FDA guidelines
and is operated as clean-in-place (CIP) equipment and as a sterilize-
in-place (SIP) system. Computer instrumentation controls the environment within the reactor and logs and displays process information
during startup and operation. The system controls temperature, pH,
dissolved oxygen concentration, medium flow rate, recycle flow rate,
and medium temperature. Measured system variables include reactor
pressure, liquid level in the storage vessels (medium, harvest, and
base), medium temperature, recycle flow rate, change in dissolved

(A)

Figure 4   Verax System 2000 fluidized bed bioreactor. (A) System
in operation. (B) Side view.

oxygen across the reactor, and reactor temperature. In addition, the system is alarmed to alert operators when measured system variables exceed specified limits around set point operation.

The system has a 24-liter fluidized bed volume. When cells in the matrix grow to a density of 3×10^8 cells/ml of microspheres and the fluidized bed is operated with 25% of the fluidized bed volume occupied by microspheres, the total number of cells in the 24-liter bioreactor is 1.8×10^{12}. The total number of viable cells contained in a bioreactor is the primary variable on which to base reactor performance; for suspension cells the above numbers make the System 2000 equivalent to over 1000 liters of conventional stirred-tank culture capacity. For attachment-dependent cells in stirred-tank culture using microcarriers, the equivalent capacity is 300–450 liters.

(B)

Figure 4   (Continued)

Table 2   Culture Parameters for the System 2000 and System 200

Parameter	System 200	System 2000
Fluidized bed height (m)	1.35	1.35
Bed diameter (m)	0.038	0.15
Fluidized bed volume (liters)	1.5	23.9
Typical equivalent batch reactor volume (liters)	40–110	600–1800
Typical total number of cells[a]	1.2×10^{11}	1.8×10^{12}
Typical oxygen consumption rate (mg-mol/min)	0.19	3.0
Typical fluidization velocity (cm/min)	68	68
Typical recycle flow rate (liters/min)	0.77	12
Typical medium (harvest) flow rate (liters/day)	20–30	300–500

[a]At 3×10^8 cells per ml of matrix with matrix occupying 25% of fluidized bed volume.

Culture parameters for the System 2000 are given in Table 2. Typically, 400–800 liters of harvest will be obtained on a daily basis under steady-state operation. The recycle rate is usually set at 12 liters/min and depends on a number of physical parameters of the bed itself. The operating parameters of the bioreactor are adjusted for each cell line and product in order to obtain the best yield.

Figure 4B is a side view of the System 2000 and shows the two system skids: the process skid on the right contains the bioreactor, the recycle loop, the oxygenator, and the computer; the utility skid on the left contains the base addition system, controls for gases, filtration apparatus for sterilizing medium, and the medium fill and harvest connections.

B. System 200 Pilot Plant/Optimization Bioreactor

The System 200 fluidized bed bioreactor (Fig. 5) is designed for pilot plant use and/or process development and optimization. The system 200 has many of the same design features as the System 2000, but with about one-fifteenth the production capacity. The System 200 is a 1.6-liter fluidized bed bioreactor and is also a CIP and SIP system.

Figure 5 Verax System 200 fluidized bed bioreactor (designed for pilot plant and process optimization).

The System 200 is typically operated at a recycle flow rate of 0.8 liters/min. At a cell density in the microspheres of 3×10^8 cells/ml and 25% microsphere solids in the fluidized bed, the total number of cells in the system is 1.2×10^{11} cells. Using a typical hybridoma cell line, the System 200 is capable of producing $1-2$ g of monoclonal antibody per day.

Like the System 2000, this system is designed for GMP manufacture of biological therapeutics and has fully documented and validated hardware. Computer instrumentation controls the conditions within the reactor and records, displays, and alarms process information during startup and operation.

V. CULTURE PERFORMANCE

A variety of cell types, both anchorage-preferred and suspension cell types, have been successfully cultured using this technology. Among the anchorage-preferred, genetically engineered cells that have been cultured are:

    Chinese hamster ovary cells
    Baby hamster kidney cells
    African green monkey kidney cells
    Human embryonic kidney cells
    Mouse mammary tumor cells

Nonengineered cells that have been cultured include:

    Rat kidney cells
    Transformed rat kidney cells
    Human hepatoma cells

Over 30 hybridomas have been cultured successfully, including

    Mouse/mouse
    Mouse/rat
    Mouse/human
    Human/human

These hybridomas have made a wide range of immunoglobulin products.

A. Monoclonal Antibody Production

A hybridoma cell was cultured in a small, 150-ml, fluidized bed reactor system for 56 days using a proprietary, low-protein, serum-free medium. The monoclonal antibody yield was approximately

80 mg/liter at steady state, twice as high as was achieved with the same cell line in a controlled-suspension culture. This small system produced 200 mg/day of monoclonal antibody at a continuous perfusion rate (medium feed rate) of 2.5 liters/day.

This same hybridoma was cultured in serum-free medium in a System 2000 and the per-cell productivity was the same as in the laboratory scale system. The System 2000 scale-up experiment continued for 20 days, with a daily perfusion rate of 230 liters achieved by day 14. At that point the reactor produced an average of 14 g/day. Since the system was not operated at full capacity during this run, it has been calculated that this cell, at a perfusion rate of 400 liters/day, will produce 24 g/day of monoclonal antibody in a System 2000 production reactor.

Figure 6 shows the medium feed rate and monoclonal antibody production during this run. Near the 2-week mark the culture approaches steady state. Total production on a per-day basis follows the day-by-day changes in medium feed rated until both level off at about the beginning of the third week into the culture.

Figure 7 displays the scalability characteristics of the fluidized bed process. Monoclonal antibody production data are given for the System 2000, the small 150-ml fluidized bed laboratory size reactor, and an intermediate size system. Data are presented on the basis of volume productivity of monoclonal antibody (g/liter of microspheres per day).

The monoclonal antibody produced from these systems was greater than 50% pure (based on total protein) at the output of the reactor. Reduced samples were applied to an SDS gel and stained with Coomassie blue. Figure 8 is a photograph of the result. Lanes 1, 6, and 9 are a series of molecular weight markers as indicated in the figure. Lane 7 is the antibody in the raw harvest. The heavy and light chains are shown as the predominant bands. Lane 8 is the unconditioned serum-free medium used as control.

B. Tissue Plasminogen Activator Production

A recombinant Chinese hamster ovary (CHO) cell line producing tPA was cultured in a small, 150-ml, fluidized bed bioreactor continuously for more than 60 days. Initially, medium containing 1% fetal bovine serum was utilized, but changeover to a serum-free medium was accomplished by day 30. The tPA yield from this small fluidized bed system was 60 mg/liter during the steady-state portion of the run. Forty to fifty milligrams per day of tPA was produced at a perfusion rate of 850 ml of medium per day. Product yield in this small system was sevenfold higher than was obtained in a conventional controlled-chemostat cell culture.

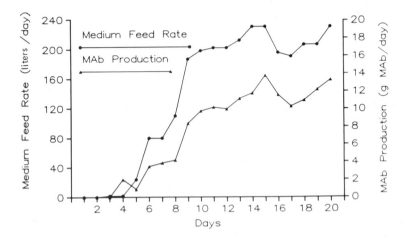

Figure 6   Medium feed rate and monoclonal antibody production in a System 2000.

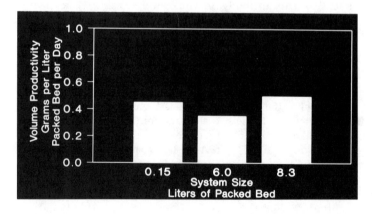

Figure 7   Scalability characteristics of the fluidized bed process.

SERUM FREE HARVEST SAMPLES OF HYBRIDOMA CELL LINES

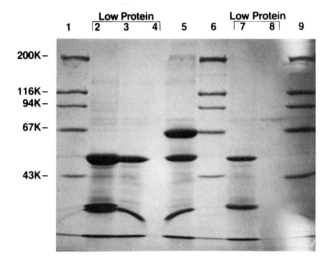

7% SDS PAGE (reduced)

Lanes 1, 6, 9: Molecular weight standards.
Lane 2:   Unselected parent, IgG double light chain.
Lane 3   Selected clonal variant, IgG single light chain.
Lane 5   Single producer, serum free medium with BSA.
Lane 7   Single producer, serum free, low protein medium.
Lane 8   Control, unconditional serum free, low protein medium.

Figure 8   Reduced sample of monoclonal antibody produced from the System 2000 (SDS gel stained with Coomassie blue).

These same recombinant CHO cells were also inoculated into a System 2000 bioreactor, which was operated continuously for 26 days (40). Table 3 gives the culture parameters and other details of this tPA run in the System 2000 bioreactor. The proprietary serum-free medium was supplemented with 1% fetal bovine serum for only the first 7 days. By day 14, the System 2000 was processing 330 liters/day. Figure 9 shows the daily medium consumption and harvest concentration of tPA. Figure 10 displays the daily output of tPA during the 26-day run. For the last 2 weeks of the run, the yield of tPA was 65 mg/liter.

Table 3   System 2000 Operation and Culture Parameters

A. Operation

Day −6[a]   7 liters of microspheres (without cells) loaded

Day −5 to −1   System washed with 600 liters of serum-free me-
                       dium

Day −1   Injected fetal bovine serum in the system to
                       final concentration of 5%

Day 0   Transferred 400 ml of wet microspheres from
                       inoculum reactor to System 2000

Day 0 to 7   Fed with 700 liters of continuous culture medium
                       supplemented with 1% serum

Day 8 to 27   Fed with serum-free continuous culture medium;
                       total protein content of culture medium 54 µg/m

Day 8 to 14   Medium feed rate 270 liters/day

Day 15 to 26   Medium feed rate 330 liters/day

B. Operation Parameters

pH   = 7.2 ± 0.02

T   = 37 ± 0.1°C

Dissolved O_2 = 100 ± 5 mm Hg oxygen partial pressure at the
                       inlet of the gas exchanger (see schematic in
                       Fig. 1)

Reactor   = 10 ± 1 psi
pressure

[a]Day −6 denotes 6 days before the System 2000 was inoculated with
microspheres containing cells.
Source: Adapted from Ref. 40.

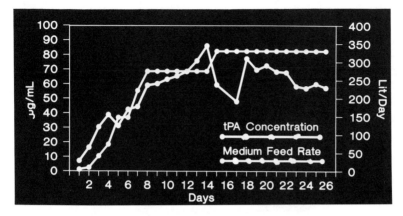

Figure 9 Daily medium consumption and harvest concentration of tPA in System 2000.

The steady-state production of tPA in the System 2000 at a medium (serum-free) feed rate of 330 liters/day was 21–23 g/day.

The scale-up properties of the Verax process were clearly demonstrated with these experiments. Figure 11 shows the scale-up data between the 150-ml benchtop system and the 24,000-ml production System 2000. The gram per liter yield was approximately the same in both systems, and when the perfusion rate was scaled from 800 ml/day to 330 liters/day, the productivity in grams scaled similarly.

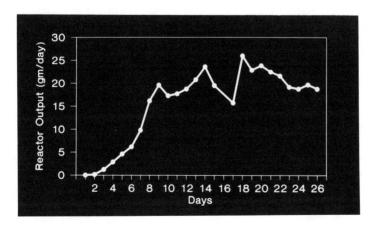

Figure 10 Daily output of tPA during 27-day System 2000 run.

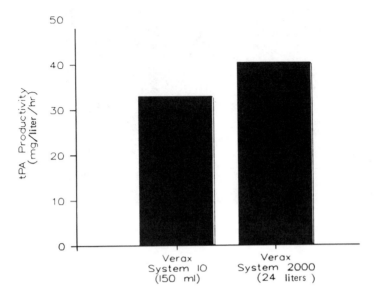

Figure 11   Grams per liter scalability between 150-ml benchtop system and 24,000-ml production System 2000.

These runs also demonstrated that the fluidized bed, continuous culture system produced high-quality, single-chain tPA on a large scale. tPA samples from these systems were analyzed by SDS-PAGE and subjected to anti-tPA overlay (Western blot) after protein transfer to nitrocellulose membranes. The proteins which exhibited immunospecific response to anti-tPA antibodies were of molecular weight 63 and 66 kDa. Figure 12 shows these results. Lane A is a sample of harvest liquor which was reduced and separated on SDS-PAGE and stained with Coomassie blue. Lanes B and C show the Western blot of the same SDS gel. Lane B shows both 63- to 66-kDa and 33- to 38-kDa proteins. This lane was from the run while in 1% fetal bovine serum. Lane C, taken while the reactor was running on serum-free medium, shows only high-quality tPA in the form of a doublet single-chain form. The calculated molecular weights of the doublet are similar to those reported by Vehar et al. (41) and Ranby et al. (42) for purified tissue plasminogen activator synthesized by a human melanoma cell line.

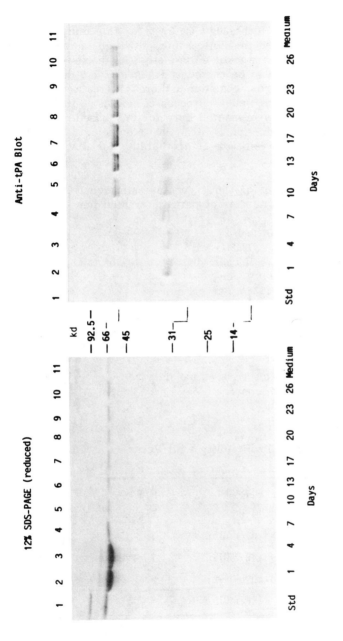

Figure 12 tPA samples analyzed by SDS-PAGE and Western blot in kilodaltons.

VI. PROCESS ECONOMICS

As commercial process technology matures for this industry, minimum
production costs become the bottom line, although factors related to
timing and reliability of market entry, etc., as discussed in the preceding section, must also be considered. A decision on selection of
a process technology must consider a thorough treatment of market
and technical factors as well as process economics. An example of
this type of analysis is presented here for tPA based on the data
obtained from the production System 2000 bioreactor.

Table 4 shows our estimate of profitability for tPA based upon
an analysis of

> *Total market size* of 55,800 g/year as estimated from dose per
>   treatment and number of patients treated per year (U.S.
>   market)
> *Selling price* of $14,000/g
> *Market potential* of $781M/year
> *Market share* of $148M/year (based on 19% of total market)
> *Cost of goods* of $10M per year
> *Gross margin** of $138M per year or 93%

The cost of goods estimate is obtained from an analysis of the
production process using the tPA data presented in the preceding
section and includes details of

> Cell culture specifics from actual data
> Estimate of medium costs

Table 4   Estimate of Profitability with Verax
Process (tPA)

Market size	55,800	g/year
Selling price	14,000	$/gram
Market potential	781 million	$
Market share	148 million	$ revenue
Cost of goods	10 million	$
Gross margin	138 million	$ (93%)

---

*Gross margin = gross revenues − cost of goods.   %Gross margin =
(gross margin/gross revenues) × 100%.

Table 5 Cell Culture Specifics Using Verax Process (tPA)

Productivity	1.9	$\mu g/10^6$ viable cells/hr
Density	1×10^8	Viable cell number/ml of Matrix
Viability	80	%
Generation time	19 and 30	Growth and production/hr
Medium feed rate	0.02	$ml/10^6$ cells/hr
Yield	94	mg/liter

Estimate of costs for labor and production plant overhead at 100% of labor costs

Material and supply costs

Bioreactor costs (including microsphere costs, technology license, equipment depreciation, and maintenance costs)

An estimate of downstream separation and purification costs and yields

Table 5 shows the essential information on the culture specifics used for this cost analysis. These data are typical for the large-scale performance obtained from the fluidized bed bioreactor process using the collagen macroporous microspheres.

Using this information, the cost of tPA produced by the culturing process (prior to any downstream processing) is less than $100/g. With an annual requirement of 10.6 kg of purified tPA per year, four fluidized bed bioractor systems of the size of the System 2000 would meet the annual production needs in this example.

VII. CONCLUSION

The fluidized bed bioreactor using the collagen microsphere matrix has been shown to be useful for many, if not all, cell types, both anchorage-dependent and suspension cell lines. Operation under continuous culture conditions allows optimization of the culture environment through the understanding of the kinetics of growth and product formation. This process under the proper operating conditions yields a product with excellent consistency, purity, and biological activity. These factors have important implications for downstream purification that can lead to reduced costs, if a total system approach is used to optimize the overall production process.

The fluidized bed bioreactor technology is inherently scalable. Properly operating systems are and have been easily scaled up over many orders of magnitude in reactor volume size in many other applications of fluidized bed technology. The smallest reactors run to date are 10-ml fluidized beds, while the largest reactors have a bed volume of 24 liters. The compatibility of cells with native collagen and the morphology and structure of the microspheres ensures maximum cell densities. All of these factors lead to an extremely cost-effective bioreactor system.

This technology has been demonstrated at production scale by making tPA and monoclonal antibodies with excellent results as measured by daily output, product yield, product quality (43), and production cost.

REFERENCES

1. Arathoon, W. R. and Birch, J. R. Large-scale cell culture in biotechnology. *Science* 232:1390−1395 (1986).
2. McCormack, D. Pharmaceutical Market for the 1990's. *Biotechnology* 5:27 (1987).
3. Collen, D., Stassen, J. M., Marafino, B. J., Jr., Builder, S., Cock, F. D., Ogez, J., Ta, D., Pennica, D., Bennett, W. F., Salur, J., and Heyng, C. F. Biological properties of human tissue-type plasminogen activator obtained by expression of recombinant DNA in mammalian cells. *J. Pharmacol. Exp. Ther.* 231:146−152 (1984).
4. Wood, W. I., Capon, D. J., and Simsen, C. C. Expression of active factor VIII from recombinant DNA clones. *Nature* 312:330−337 (1984).
5. Toole, J. J., Knopf, J. L., Wozney, J. M., et al. Molecular cloning of DNA encoding human antiharmophilic factor. *Nature* 312:342−347 (1984).
6. Kaufman, R. J., Wasley, L. C., Furie, B. C., Furie, B., and Shoemaker, C. B. Expression, purification, and characterization of recombinant gamma-carboxylated factor IX synthesized in Chinese hamster ovary cells. *J. Biol. Chem.* 261:9622−9628 (1986).
7. Grinnell, B. W., Beng, D. T., Walk, J., and Yan, S. B. Trans-activated expression of fully gamma-carboxylated recombinant human protein C, an antithrombotic factor. *Biotechnology* 5:1189−1192 (1987).
8. Pavarini, A., Meulein, P., Harrer, H., et al. Choosing a host cell for active recombinant factor VIII production using vaccina virus. *Biotechnology* 5:389−392 (1987).

9. Lin, F. K., Suggs, S., Lin, C. H., et al. Cloning and expression of the human erythropoietin gene. *Proc. Natl Acad. Sci. USA* 82:7580−7584 (1985).
10. Cate, R. L., Mattaliano, R. J., Hession, C., et al. Isolation of the bovine and human genes for Mullerian inhibiting substance and expression of the human gene in animal cells. *Cell* 45:685−698 (1986).
11. Margaritis, A. and Wallace, J. B. Novel bioreactor systems and their applications. *Biotechnology* 2:447−453 (1984).
12. Hu, W. S. and Dodge, T. Cultivation of mammalian cells in bioreactors. *Biotechnol. Progr.* 1:209−215 (1985).
13. Nilsson, K., Scheirer, W., Merten, O., et al. Entrapment of animal cells for production of monoclonal antibodies and other molecules. *Nature* 302:629−630 (1983).
14. Feder, J. and Tolbert, W. R. (eds.). *Large-Scale Mammalian Cell Culture.* Academic Press, New York (1985).
15. Pugh, G. G., Berg, G. J., and Sear, C. H. J. Two ceramic matrices for the long-term growth of adherent or suspension cells: Scalability. In *Bioreactors and Biotransformations* (G. W. Moody and P. B. Baker, eds.), Elsevier, New York, pp. 121−131 (1987).
16. Wilkinson, P. J. The development of a large-scale production process for tissue culture products. In *Bioreactors and Biotransformations* (G. W. Moody and P. B. Baker, eds.), Elsevier, New York, pp. 111−120 (1987).
17. Brunt, J. V. A closer look at fermenters. *Biotechnology* 5:1133−1138 (1987).
18. Levine, D. W., Wang, D. I. C., and Thilly, W. G. *Biotechnol. Bioeng.* 21:821 (1979).
19. Whiteside, J. P. and Spier, R. E. The scale-up from 0.1 to 100 litres of a unit process system for the production of four strains of FMDV from BHK monolayer cells. *Biotechnol. Bioeng.*:551−565 (1981).
20. Lydersen, B. K. (ed.). *Large-Scale Cell Culture Technology.* Hanser, New York (1987).
21. Feder, J. and Tolbert, W. R. The large-scale cultivation of mammalian cell culture. *Sci. Am.* 248(1):36−43 (1983).
22. Spier, R. E. and Griffiths, J. B. (eds.). *Animal Cell Biotechnology, Vol. 3.* Academic Press, London (1987).
23. Birch, J. R., Lambert, K., Thompson, P. W., Kenney, A. C., and Wood, L. A. Antibody production with airlift fermenters. In *Large-Scale Cell Culture Technology* (B. K. Lydersen, ed.), Hanser, New York, pp. 1−20 (1987).
24. Fleischaker, R. Microcarrier culture. In *Large-Scale Cell Culture Technology* (B. K. Lydersen, ed.), Hanser, New York, pp. 59−79 (1987).

25. Runstadler, P. W., Jr. and Cernek, S. R. Large-scale, fluidized-bed, immobilized cultivation of animal cells at high densities. In *Animal Cell Biotechnology, Vol. 3.* Academic Press, London (1987).

26. Dean, R. C., Jr., Karkare, S. B., Phillips, P. G., Ray, N. G., and Runstadler, P. W., Jr. Continuous cell culture with fluidized sponge beads. In *Large-Scale Cell Culture Technology* (B. K. Lydersen, ed.), Hanser, New York, pp. 145–167 (1987).

27. Vieth, W. R. and Venkatasubramanian, K. Immobilized microbial cells in complex biocatalysis. In *Immobilized Microbial Cells* (K. Venkatasubramanian, ed.), American Chemical Society Symposium Series 106, American Chemical Society, pp. 1–11 (1979).

28. Barnes, D. Serum-free animal cell culture. *Biotechniques* 5(6):534–542 (1987).

29. Messing, R. A., Oppermann, R. A., and Kolot, F. B. Pore dimensions for accumulating biomass. In *Immobilized Microbial Cells* (K. Venkatasubramanian, ed.), American Chemical Society Symposium Series 106, American Chemical Society, pp. 13–28 (1979).

30. Pert, S. J. *Principles of Microbe and Cell Cultivation.* John Wiley and Sons, New York (1975).

31. Van Brunt, J. Immobilized mammalian cells: The gentle way to productivity. *Biotechnology* 4:505–510 (1985).

32. Yang, J. and Nandi, S. Growth of cultured cells using collagen as substrate. *Int. Rev. Cytol.*:249–286 (1983).

33. Sabelman, E. Biology, biotechnology, and biocompatibility of collagen. In *Biocompatibility of Tissue Analogs.* CRC Press, Boca Raton, pp. 26–66 (1985).

34. Hayman, E. G., Pierschbacher, M. D., Suyuki, S., and Ruoslahti, E. Vitronectin-A major cell attachment-promoting protein in fetal bovine serum. *Exp. Cell Res.* 160:245–258 (1985).

35. Kleinman, H., Klebe, R. J., and Martin, G. Role of collagen matrices in the adhesion and growth of cells. *J. Cell Biol.* 88:473–485 (1981).

36. Schor, S. L., Schor, A. M., Winn, B., and Rushton, G. The use of three-dimensional collagen gels for the study of tumor cell invasion in vitro: Experimental parameters influencing cell migration into the cell matrix. *Int. J. Cancer* 29:57–62 (1982).

37. Hall, H. G., Farson, D., and Bissell, M. Lumen formation by epithelial cell lines in response to collagen overlay: A morphogenetic model in culture. *Proc. Natl Acad. Sci. USA* 19:4672–4676 (1986).

38. Seirshey, A. J., Fleischaker, R. J., Tyo, M. A., and Wang, D. I. C. *Ann. NY Acad. Sci.* 369:47 (1981).

39. Hu, W. S., Ph.D. thesis, Massachusetts Institute of Technology (1983).
40. Tung, A. S., Sample, J. vG. S., Brown, T. A., Ray, N. G., Hayman, E. G., and Runstadler, P. W., Jr. Mammalian cell culture technology: Production of tissue-plasminogen activator through mass culture of genetically-engineered Chinese hamster ovary cells. *Biopharm. Manufact.* 2:50−55 (1988).
41. Vehar, G. A., Kohr, W. J., et al. Characterization studies of human melanoma cell tissue plasminogen activator. *Biotechnology:* 1051−1057 (1984).
42. Ranby, M., Bergsborf, N., et al. Isolation of two variants of native one-chain tissue plasminogen activator. *FEBS Lett.*: 289−292 (1982).
43. Finkelman, F. D., Katona, I. M., Urban, J. F., Jr., Holmes, J., Ohara, J., Tung, A. S., Sample, J. vG., and Paul, W. E. Interleukin-4 is required to generate and sustain in vivo IgE responses. *J. Immunol.* 141:2335−2341 (1988).

14

Downstream Processing of Proteins from Mammalian Cells

JOHN R. OGEZ and STUART E. BUILDER
Genentech, Inc., South San Francisco, California

I. INTRODUCTION

Mammalian proteins first came into widespread use as therapeutic agents with the introduction of insulin in 1925. However, it was not until the development of the blood fractionation industry, around the time of World War II, that a large number of partially purified mammalian proteins became available. Today, animal tissues still supply the bulk of medicinal proteins on a mass basis (Table 1). However, the recent discovery of techniques for creating hybridomas and recombinant organisms has opened up the field for the production of a virtually limitless variety of protein therapeutics. These new technologies have removed the limitations characteristic of tissue sources such as low titer, limited supply, and variable quality. Given the concerns about the contamination of the human blood and tissue supply by infectious agents such as hepatitis virus, human immunodeficiency virus (HIV), and the causative agent of Creutzfeld-Jacob disease, there is an increasing trend toward the use of recombinant hosts for therapeutics which were previously derived from blood or tissue sources, such as factor VIII, hepatitis vaccine, and human growth hormone.

Recombinant DNA technology creates new opportunities for therapy by allowing the production of commercial quantities of otherwise rare proteins, such as tissue plasminogen activator (t-PA) and

Table 1   Mammalian Protein Therapeutics

Source		Name	Use
I. Animal tissue and fluid	1.	Albumin	Hypoproteinemia; plasma extender for shock, burns
	2.	Human IgG	Augmentation of immune system capacity, especially during hepatitis or tetanus infection
	3.	Rh IgG	
	4.	Equine antivenins (snake, house ant, black widow spider, etc.)	
	5.	Human factor VIII	
	6.	Human factor IX	
	7.	Bovine thrombin	Topical for control of oozing after injury or surgery
	8.	Bovine plasmin	
	9.	Human Cohn fraction fibrinolysin	
	10.	Hepatitis vaccine	
	11.	Porcine lipase and amylase	Pancreatitis; to increase absorption of dietary fats
	12.	Pepsin	
	13.	Trypsin (and chymotrypsin)	Cataract surgery; to reduce inflammation and edema after trauma

14.	Insulin	Control of glucose metabolism
15.	Bovine DNAse	
16.	Glucagon	To terminate severe hypoglycemia as a result of insulin overdose
17.	Human growth hormone	Induction of bone growth in hypopituitary dwarfism
18.	Trypsin inhibitor (Trasylol)	
19.	Bovine collagen	
20.	Hormones (FSH and LH) from postmenopausal women	Induction of ovulation; stimulation of spermatogenesis
21.	Chorionic gonadotropin from pregnant women	Cyptorchidism in males; stimulation of spermatogenesis
22.	Hyaluronidase (bovine testicular)	

II. Conventional fermentation

| 1. | Urokinase | Dissolution of blood clots |
| 2. | Vaccines: DPT, MMR, polio, rabies, yellow fever, etc. | Confer specific immunity |

III. Recombinant fermentation

1.	t-PA	Myocardial infarction
2.	Factor VIII	Blood clotting
3.	Erythropoietin	Anemia in end-stage renal disease
4.	OKT-3	Graft-vs-host rejection

Table 1 (Continued)

Source	Name	Use
III. Recombinant fermentation (continued)	5. Superoxide dismutase	Reperfusion injury
	6. CD4	HIV infection
	7. Other monoclonal antibodies	Cancer, septic shock, platelet aggregation

General references: *AMA Drug Evaluations*, first through sixth editions; American Hospital Formulary Service Drug Information; *Physicians Desk Reference*; PMA "Biotechnology Medicines in Development" annual surveys.
Source: Ref. 35.

erythropoietin (EPO). It also provides a way to produce novel protein sequences that exhibit improved activity compared to that of the natural molecule. Such improvements may be manifested through greater stability to high temperature and oxidizing agents as in the case of subtilisin mutants (1), or by altered pharmacokinetics to increase the circulating half-life, as demonstrated in the construction of the new CD4 "immunoadhesin" hybrid molecules (2). Genetic engineering can be used to create hybrids which combine the activities of two or more proteins into a single molecule to give a new therapeutic effect, such as the fusion of CD4 with bacterial exotoxins for the treatment of AIDS (3) and multistrain immunogens for vaccination of cattle against foot-and-mouth disease (4). Site-directed mutagenesis (5) and cassette mutagenesis (6) techniques are powerful tools for rapidly generating a series of mutants which can be used to probe the structure-function relationships within a protein.

This chapter focuses on the downstream processing of therapeutic proteins produced in mammalian cells. Mammalian cells provide special opportunities for controlling product composition and quality, particularly with regard to glycosylation and subunit assembly. Special considerations of process design and operation are necessary to maximize yield, quality, and consistency in the production of these complex therapeutic molecules.

II. PRODUCTION SYSTEMS

A. Bacteria vs. Mammalian Cells as Hosts

Initially, most recombinant polypeptides were cloned and expressed in *E. coli* because its genetics were well understood, restriction enzymes were available, and it could be grown cheaply. The first generation of recombinant therapeutics, including human insulin, human growth hormone, and alpha-interferon, were produced in this host. As larger, more complex proteins became targets for production, it became apparent that mammalian cells could carry out some aspects of molecular processing more efficiently than bacteria or the protein chemist. Today, many new recombinant therapeutics including t-PA, EPO, monoclonal antibodies (e.g., OKT3), factor VIII, and CD4 are being produced in mammalian cells.

Mammalian cells grow much more slowly, and subsequently have much lower volumetric productivity than bacterial fermentations. For the same reason, mammalian cell cultures can be rapidly overgrown by contaminating bacteria, thereby requiring stringent precautions to ensure the sterility of cell culture media and equipment. In addition, mammalian cells have very complex nutritional requirements, and therefore the media are much more complex and

expensive than those for bacteria. Considering the substantially
higher costs associated with producing recombinant proteins in mammalian cells, it is worth reviewing when and why they would be the
system of choice.

B. Factors Affecting the Choice of Mammalian
Expression Systems

The desire to produce a naturally glycosylated protein is probably
the most important factor dictating the use of mammalian cells. The
presence of carbohydrate residues can affect the activity, stability,
solubility, and pharmacokinetics of a polypeptide. For example, the
solubility of rt-PA decreases by about an order of magnitude when its
carbohydrate residues are removed by glycosidase treatment. Clearance of proteins from the circulation is often mediated through the
action of liver cell receptors specific for certain types of attached
carbohydrates. When the terminal sialic acid residues are removed
from complex-type carbohydrate side chains to expose the penultimate
galactose moieties, the protein is rapidly removed by hepatocyte receptors (7). Therefore, the type as well as the presence or absence
of carbohydrate can be important. Yeasts are capable of glycosylation, although they often produce hypermannosylated forms of the
protein which may not have the properties expected of the wild-type
carbohydrate moieties and may be immunogenic. There are no practical ways to achieve "proper" glycosylation either in bacteria or by
chemical modification. Thus, there is currently no satisfactory alternative to mammalian cells for the proper glycosylation of proteins.
    The next two related reasons to use mammalian cells are that they
are much more likely to produce cloned mammalian proteins in their
native, properly folded state with correct disulfide bonds, such as
factor VIII or immunoglobulins. For smaller or moderately sized peptides like proinsulin or growth hormone, a high level of productivity
along with high quality and acceptable yield can be achieved in bacteria. However, the production of many mammalian proteins in pro-
karyotes leads to the formation of improperly folded and insoluble
aggregated product. Under conditions which promote rapid expression,
this nonnative protein aggregates to form dense, insoluble particles
in the bacterial cytoplasm. These particles are known as "inclusion"
or "refractile" bodies because they are visible by transmission electron
microscopy or phase-contrast light microscopy, respectively. In traditional biochemistry, product in such a denatured state would normally have been discarded. However, at the time these feedstocks
were first produced no other starting material was available, so recovery methods were developed which could selectively extract and
purify the small percentage of properly folded material (8). Methods
were later developed which allowed the recovery of molecules that had
initially been improperly folded or had mispaired disulfide bonds

thereby dramatically increasing the process yields (9,10). Improved expression systems have recently been developed which lead to higher yields of properly folded product in the fermenter. These systems generally employ secretion of the polypeptide into the more oxidizing environment of the periplasmic space or the extracellular medium, where disulfide bond formation can take place.

The fourth advantage of the use of mammalian cells is that the unwanted addition of methionine at the N-terminus is not a problem. Recombinant proteins which are expressed intracellularly in bacteria such as *E. coli* usually possess methionine as the N-terminal amino acid. For proteins which have this amino acid as the natural N-terminal residue, this is obviously not an issue. For others, the structure, activity, and safety of the resulting polypeptide must be evaluated on a case-by-case basis. Several approaches have been used to avoid an undesired N-terminal methionine. Cyanogen bromide cleavage is one method for removing methionine after synthesis, especially when the product must be separated from a peptide leader sequence (11), although it is limited to molecules with no internal methionine residues. Also, with certain N-terminal sequences, an endogenous aminopeptidase may be used to process away the terminal methionyl residue (12). Another approach is to use an expression sequence that allows secretion into either the periplasmic space or the extracellular medium. With this system the leader sequence, or "signal peptide," may be cleaved off by the organism as the product passes through the cell membrane or wall, yielding the mature protein with the correct N-terminus (13). As an alternative, the fusion protein can be engineered to remain intact, with the leader serving as an aid in the purification of the product. For example, in the EcoSec expression system the leader has homology with Staphylococcal protein A, which binds tightly to the Fc region of antibodies. This allows rapid affinity purification of the fusion protein on an immobilized IgG column. The desired product is then separated from the leader sequence following specific cleavage using hydroxylamine (14).

C. Principles Directing the Selection and Sequence of Downstream Operations

The previous discussions call attention to important differences in the feedstocks from mammalian vs. bacterial fermentations as they become available for downstream processing. Products from mammalian cell culture will in general be outside the cell, glycosylated, and be properly folded with the correct disulfide bonding. These properties, and others discussed below, have substantial impact on the choice and order of subsequent separation and purification operations.

Since the protein produced in a mammalian cell culture is already in the proper conformation, the purification process is designed to *maintain* the existing molecular integrity and conformation rather

than *create* it via the use of chaotropes and reducing agents usually
required for protein refolding. This may place narrower limits on
temperature, pH, and solvent composition compared to the ranges permitted in the recovery of bacterial products.

Gram-negative bacteria such as *E. coli* produce significant amounts
of pyrogenic endotoxins as a component of their cell walls. One expects that bacterially derived starting material will be highly pyrogenic; thus the introduction of endotoxin by, for example, fermentation raw materials is of little consequence. The downstream operations must be capable of reducing the pyrogen levels by six to nine
orders of magnitude to nearly undetectable levels. In contrast, a
mammalian fermentation may contain little or no pyrogen, and the
recovery process may not need to incorporate steps to remove pyrogen. The process strategy thus becomes oriented more toward keeping pyrogens out rather than reducing their levels, and it is much
more important to keep raw materials and equipment pyrogen-free.
At any point in the process, the presence of pyrogen then becomes
a useful way to monitor process cleanliness and sanitation, and it
serves as a marker for the presence of contaminating (gram-negative)
organisms. It is critical to minimize the introduction of adventitious
bioburden because in addition to pyrogen, microorganisms may contaminate the process stream with glycosidases and proteases that irreversibly alter or inactivate the product and which may adversely
affect product stability after vialing.

The glycosylation of proteins by mammalian cells creates special
challenges in downstream processing. Figure 1 is a 2D electrophoretogram (15). It shows the separation of *E. coli* proteins labeled with
[^{14}C]amino acids in vivo. Under optimal conditions, it is possible
to count 1000 spots in the original autoradiogram. It is important to
note that most of the spots on the 2D gel are the result of a single
gene product and each gene in *E. coli* is usually represented by a
single spot. Given high-resolution purification procedures which
separate on the basis of size and charge, one can get high purity
along with high yield using just these two modes of separation. This
is indeed the case for many products when purified from *E. coli*, even
when the product is a recombinant human protein. However, Fig. 2
shows a very different pattern (16). It is a two-dimensional map,
but this time of plasma proteins. No longer is there the one-to-one
correlation between individual mammalian genes and spots on the gel.
Many proteins are glycosylated in a complex fashion and genes can
be represented by families of spots representing families of glycopeptides. The typical pattern is observed as a group of spots moving
up and to the left (in other words, becoming more acidic and higher
in molecular weight) with increasing glycosylation. Not only is an
individual gene product represented by a collection of spots rather
than a single one, but there can be significant overlap in charge and
size with neighboring families of glycopeptides. In this case it is no

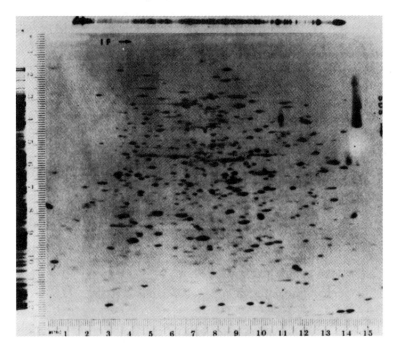

Figure 1 Two-dimensional electrophoretogram of *E. coli* proteins. Proteins were labeled by growing cells in the presence of [14C]amino acids. Isoelectric focusing was carried out in the first (horizontal) dimension, then the SDS gel having a 9–14% polyacrylamide gradient was run in the second dimension. A total of 180,000 cpm of radioactivity was loaded, and the autoradiogram was exposed for 825 hr. Over 1000 spots can be visualized in the original autoradiogram. (From Ref. 15.)

longer possible to achieve both high yield and high purity using only size and charge as the modes of separation. At least one additional mode of separation is required to accomplish this goal. Only a few years ago biospecific adsorption, or affinity chromatography, was seen as a major new purification opportunity. Today, at least for glycoproteins, it may be a necessity and should be considered as a likely step in the purification of pharmaceutical proteins from mammalian cells.

Monoclonal antibodies (MAbs) are good examples of the use of affinity chromatography in protein purification. Because of their tremendous specificity, they are useful tools for affinity purification of protein therapeutics such as leukocyte interferon (17) and factor VIII (18). MAbs are also useful as therapeutics, and it is interesting to note the use

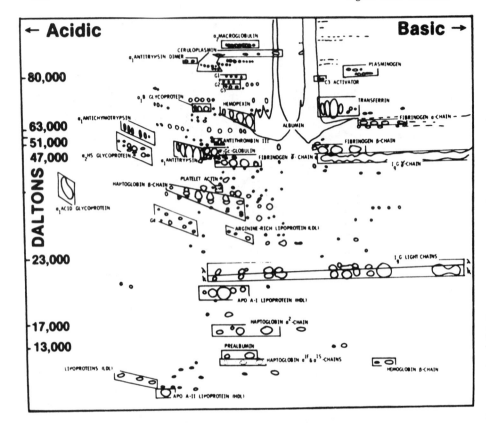

Figure 2   Map drawn from a stained two-dimensional electrophoretogram of mammalian plasma proteins and labeled to indicate the positions of known proteins.   (From Ref. 16.)

of affinity chromatography for purifying the antibody itself: tremendous purification can be achieved in a single step by passing the
conditioned ascites or hybridoma fluid over a column of immobilized
protein A (19).   Protein A is itself a protein "ligand," isolated from
the cell wall of *Staphylococcus aureus*, which binds to the Fc region
of the antibody.   Thus it is possible to use affinity chromatography
to purify a protein which will subsequently be used as the affinity
ligand for purifying yet another protein.

III. DOWNSTREAM PROCESSING OPERATIONS

A. Integration with Upstream Operations

In the narrowest definition, downstream processing is the purification of proteins from conditioned media or broths. However, many controllable factors that influence purification occur early in the production process. The integration of downstream processing with upstream operations such as molecular biology and fermentation can provide significant downstream opportunities.

The interaction between molecular biology and recovery can take several forms. With recombinant DNA products one can influence the purification before the starting material is even available. Given the gene sequence, one can begin to predict how the product will behave on size separation media and ion exchange resins, although the actual ionic properties of the protein may be influenced by tertiary structure and sialylation. Even before a clone has been selected and amplified, the parent cell line can be cultured in order to characterize how various host cell impurities will behave in a prospective recovery operation. Leading or trailing segments can be added to impart properties that will make the protein easier to purify. This particular strategy has frequently been used in bacterial production systems which employ fusion proteins to enhance expression, secretion, or the subsequent recovery. In mammalian systems, the tools of the molecular biologist are more commonly used to enhance expression levels or to alter the biological properties of the final product. Higher titers provide a direct benefit to the recovery scientist by increasing the ratio of product to contaminant, thereby reducing the fold purification that is ultimately required and reducing the operation volumes of early steps.

Perhaps the most important examples of process integration occur in the interaction between recovery and fermentation. One of the primary areas of interaction between these disciplines is the development of suitable media for cell growth. In cases where the expression system uses an amplified selectable marker, such as the DHFR (dihydrofolate reductase) gene, it may be necessary to maintain selective pressure with methotrexate during some or all stages of cell culture. Media supplements such as bovine serum may be necessary for cell growth but may also contain substances that release the selective pressure (e.g., glycine, hypoxanthine, thymidine), causing a decrease in expression level. Fractionation of the serum to remove these components will thus allow optimal cell growth in the seed train while maintaining the cells' ability to produce high product titers.

The use of media supplements such as serum may adversely affect the overall recoverability by leading to complex formation and product degradation. This problem can sometimes be overcome by

carrying out a fractionation or partial purification of the serum which
reduces the level of the deleterious components while retaining the
ability to enhance cell growth. A more satisfactory alternative, however, is to remove serum from the production vessel altogether.
van Reis, et al. (20,21) devised a tangential-flow filtration system
which can reduce levels of serum coming from the seed culture by
any factor necessary. The cells of the seed vessel are concentrated
by microfiltration and then diafiltered to remove any deleterious components of the culture medium along with degraded or altered forms
of product that have been synthesized up to that point. The washed
cells are then inoculated into the production vessel which contains a
serum-free medium. Carrying out this type of medium exchange allows one to completely decouple cell growth in the seed train from the
production phase of the culture. This greatly increases the number
of degrees of freedom available to the fermentation scientist because
changes can now be made in order to enhance growth of the seed
train, with minimal qualitative effect on the production vessel.

Determining the optimal time of harvest is an important area of interaction between fermentation and recovery. Often, allowing a culture
to run longer results in an increase in titer (especially by radioimmunoassay), but with a concomitant increase in debris and degraded forms
of the product. Although it may be simple (but not desirable) to overcome the effect of increased cell debris by increasing the capacity of
the downstream equipment, it may be much more difficult to purify
away the slightly altered or degraded forms of the product. Thus, the
selection of the optimal harvest time must include consideration of the
effect on recovery operations and final product yield and quality.

B. Harvest

In general, current mammalian cell production systems secrete the
protein of interest. At the end of the fermentation the cells themselves contain insignificant additional amounts of product, especially
in relation to host impurities, and are therefore discarded.   Cell
breakage is not only unnecessary but undesirable. Cell disintegra-
tion releases membrane fragments that can foul process equipment as
well as undesirable impurities derived from the cell cytoplasm, particularly host cell proteins and DNA. The operation is made difficult
because mammalian cells are much more fragile than bacterial or yeast
cells. The high-speed centrifugation commonly used for bacteria and
yeast is not appropriate for cell-liquid separation in mammalian systems. Centrifugal harvest of mammalian cells must generally be done
with special low-shear, low-centrifugal field equipment. Mammalian
cell harvesting can be effectively and efficiently carried out by using
depth or tangential-flow filtration. A system similar to the one devel-

oped by van Reis et al. for medium exchange can be used to carry out cell harvesting (20,21). The advantages of that technique include the ability to obtain quantitative product yield by washing (diafiltering) the cells and the ability to preserve cell viability. Since the system can also be sterilized in place, it can carry out continuous harvest of a perfusion cell culture.

C. Safety Issues

The number of passages required for cloning, selection, amplification, and banking prior to production clearly necessitates the use of transformed cell lines since primary cells cannot be propagated long enough to give an economically useful life in the production stage. Initially, there was concern about the safety of immortal cell lines since, by definition, they are transformed and are thought to contain oncogenic DNA or proteins. In addition to the issues arising from the transformed nature of the cells, there were also concerns regarding the contamination of cell lines by adventitious agents such as viruses, fungi, and mycoplasma. Many established cell lines were known to exhibit the presence of retrovirus-like particles in electron micrographs. Finally, there were concerns about the immunogenicity of host cell proteins in patients who received drugs which were purified from recombinant sources.

Several safety concerns have been successfully addressed for rt-PA (29) as described in the following sections.

1. Cell Banking

National regulatory agencies and other health organizations have published "points-to-consider" guidelines which outline the recommendations for characterization of cell banks (22,23). In general, the origin and passage history of the cell line should be tested for tumorigenicity. The master cell bank should be shown to be free of adventitious agents such as fungi, bacteria, mycoplasma, and exogenous viruses. Finally, the cells should be probed for expression of endogenous retroviral sequences using conditions that are known to cause their induction. In cases where the process involves the use of a secondary bank (the "working cell bank"), this bank should also be shown to be free of adventitious agents. The exhaustive characterization of cell banks by these diverse methods provides an initial degree of confidence that the resulting products can be safely injected into humans (24).

2. Cells

Concern over the presence of residual host cells has largely been put to rest by relying on well-established techniques for sterile filtration that have been proven in pharmaceutical industry for decades. In addition, it has been possible to establish the benign nature of some cell lines by inoculating them directly into normal animals. In tests on the Chinese hamster ovary (CHO) cell line used to produce t-PA, immunocompetent rodents were injected with large numbers of the host cell. Although nodules formed initially at the site of injection, these growths rapidly disappeared. Thus, there are two lines of evidence for safety of these 'products with respect to cells: (1) the absence of cells as demonstrated by sterility of the product and (2) the relative noninfectivity of the cells in immunocompetent animals.

3. Nucleic acids

When immortalized mammalian cells were first considered as host systems for recombinant protein, there was substantial theoretical concern about the possibility of DNA from recombinant immortal cell lines causing oncogenic events in patients receiving products from these cell substrates. It is possible to directly measure the DNA content of clarified cell culture fluid or early processing steps with a DNA dot blot assay using ^{32}P-labeled DNA derived from the host cell line (25). However, the sensitivity of this type of assay is about 0.1 ng of DNA/ml of fluid. For some products, especially those which are administered in multimilligram quantities, it is necessary to demonstrate a further reduction of two to four logs in order to assure a level of DNA of less than 1 pg per human dose. Further removal can be validated by spiking ^{32}P-labeled DNA into aliquots of process fluid and then purifying the samples on representative scaled down versions of the recovery process operations. For rt-PA, this combined approach using both direct measurement and process validation resulted in a calculated cumulative clearance of greater than nine orders of magnitude, or less than 0.1 pg of DNA per human dose. This was at least two orders of magnitude lower than the level recommended by the FDA as a maximum acceptable level in the points-to-consider. (This document grew out of a meeting in Washington, D.C. in July 1984; see Ref. 26.)

This approach of going far beyond the required specifications has several rationales. First, it is always possible that in the future specifications may be tightened. Also, the sensitivity or specificity of assays may improve over time. Given the possibility of such changes, it is advisable to have a process that will continue to meet future targets (even if they become more stringent) without requiring process changes. In addition, there is always some variability within a normal production operation and by having a margin of safety one can reliably expect that all lots will meet acceptable specifications.

It is useful to obtain today's perspective on concerns about DNA. Relative to a few years ago, the issue is greatly diminished. One of the reasons for this is an experiment by Palladino et al. (27) in which

DNA did not induce any oncogenic events when injected into immuno-suppressed rodents at a level eight orders of magnitude greater than that expected in a dose of t-PA. This is typical of experiments show ing a lack of biological effect of "naked" DNA inoculated into mammals It is most likely degraded very quickly to inactive fragments and nu-cleotides by circulating nucleases. Even though concern about DNA has greatly diminished, there will probably be a tendency to keep the specification the same, at least in part because regulatory agen-cies know it is achievable.

4. Viruses

The presence of retroviruses in continuous mammalian cell lines has received a great deal of attention because of concern that these par ticles can potentially cause oncogenic events in man. In addressing this issue, one can begin by testing the culture fluid directly for the presence of retroviruses. Lubiniecki and May (28) reported on the application of reverse transcriptase assays, electron microscopy and several other methods to assess the presence of retroviruses in conditioned media. Although particles of both the A and C type can be seen in some CHO cell lines by electron microscopy, no technique used has ever demonstrated functional retrovirus. To increase the sensitivity of electron microscopy past the estimated detection limit o 10^6 particles/ml, harvested fluid can be concentrated by ultracentri-fugation. It is important to note that this approach alone, the demc stration of freedom from functional retroviruses in the culture, is usually insufficient to answer regulatory concerns because it is alwa possible that there might be levels of retrovirus just below the sen-sitivity limit of the assays, or that the specificity of the retrovirus assays might not be broad enough to pick up some unusual potentia. contaminant.

Since no amount of additional assay development can ever answe: such hypothetical questions, the direct measurements must again be supplemented by validated process procedures for removal or inacti-vation of putative retroviruses. A significant number of viral clear-ance steps are available from which to choose. They include ultra-filtration, treatment with chaotropes such as urea or guanidine, pH extremes, detergents, heat, chemical derivatization such as formalde hyde, proteases, and organic solvents. The particular steps choser for any given product will depend on the following criteria. First, the product must be stable to treatment while the model virus must obviously be sensitive to the treatment. Second, the procedure should be designed to achieve the largest possible "window of clear ance." The window is defined as the ratio of initial virus titer (spike) in the process fluid prior to the treatment divided by the virus titer after the treatment. Often the measurable titer after treatment is limited by the sensitivity of the assay. This is importa because only that inactivation or removal which can be demonstrated (measured) can be counted in the validation. A typical value for the initial concentration of virus in the spiked fluid is about 10^6 particles/ml. Sensitivity of detection after treatment is often about 10^2 (due to necessary dilutions) which results in a window of 4 logs of clearance for this particular step. If, for example, 12 logs of clearance was desired, three different and independent mechanisms of removal or inactivation, each giving about 4 logs of clearance, would need to be validated. It is not valid to use the same approach, for example, 3 times. Only with steps that are truly independent is it legitimate to determine the total clearance as the product of the clearances from the individual steps. In addition, the use of more than one model virus (e.g., NIH Rauscher leukemia virus and NZB xenotropic virus) and the assay of the virus by more than one technique (e.g., XC plaque assay, mink lung focus assay, reverse transcriptase assay) serve to strengthen the believability and validity of this approach.

In addition to the steps which are specifically incorporated to remove or inactivate potential retroviruses, several other types of processing operations may provide further clearance, such as conventional purification using ion exchange or affinity chromatography. However, it is difficult to validate these steps for viral clearance capability. The logic is that while, for example, heat treatment is likely to be generic for all retroviruses, any particular separation technique such as ion exchange cannot necessarily be relied on to separate all viruses from the product of interest.

The use of different separation methods for the removal and inactivation of model retroviruses in conjunction with the substantial characterization of the culture fluid should ensure the safety of the product beyond any reasonable doubt.

5. Proteins

Because of concern about the safety of proteins from nonhuman sources with respect to the generation of immune responses, recombinant proteins are generally being brought to unprecedented levels of purity. Today it can be as difficult to quantitate and prove the levels of purity as it is to achieve them. For example, whereas the purity of albumin preparations is commonly about 95–99%, the purity of recombinant human growth hormone (Protropin) and recombinant human insulin (Humulin) is greater than 99.99% with respect to host proteins. In order to measure impurities at this level, two major analytical strategies have been developed. The first method, which is uniquely applicable to recombinant products, is the use of a "blank run," i.e., fermentation and recovery using a host cell containing the selectable marker but lacking the gene for the product. In this way, one may be able to specifically prepare and quantitate the host cell-derived impurities. The second approach is the direct measurement of impurities. The most general method uses an immunoassay based on antibodies to the host cell proteins (30). Although this type of assay is complex in both its development and composition, it provides an extremely sensitive way to quantitate protein impurities in each batch of product.

For products made in mammalian cells, there is a particular type of host cell protein impurity that is of special interest, termed the "endogenous homolog." For example, a recombinant hamster cell line producing human factor VIII might also be expected to express small quantities of hamster factor VIII. One of the most useful features of the blank run approach is its ability to detect low-level expression of such endogenous homologs. Depending on the relative expression levels of the human and endogenous proteins, the endogenous impurity may or may not be of concern. In any case, generation of the pool of blank run proteins can facilitate not only detection but also purification and characterization of the endogenous homolog, which can lead to strategies on how to separate it from the product of interest. Two approaches toward the removal of endogenous homologs are described in the patent literature by Anicetti et al. (31) and Morii et al. (32).

IV. PROCESS CONTROL AND MONITORING

A. Process Changes

Molecules such as indomethacin and aspirin are usually classified as drugs, at least in part because it is possible to completely define their chemical nature and to analytically demonstrate purity, potency, and identity. Biologics are usually much more complex compounds (or mixtures of such) and it is generally not possible to either define them as discrete chemical entities or to demonstrate a unique composition. Biologics, as the name implies, were originally derived from biological sources. Blood fractionation products such as albumin, factor VIII (AHF), and viral vaccines, both live and killed, are examples. Most products made in mammalian cells will fall into this category. To compensate for the incomplete analytical capability to define biologics, regulatory agencies have included in the control and monitoring parameters of a biologic the process used to make it. This situation exists because it is possible that if the process changes significantly, it may yield a different product from that previously reviewed and approved. A new product will of course require a new license.

This logic has tremendous implications in terms of the effort, money, and time required to make significant changes in the production process of a licensed biologic. At the very minimum, there is

probably a 2-year delay between the conception of an idea for a better
process and the point at which a license amendment will be granted
to start production by that method. This includes the time needed to
develop the process, produce clinical material, carry out clinical trials,
file the amendment, and finally obtain approval. This effort may cost
millions of dollars. Thus, the usual strategy is to accumulate significant changes into relatively large packages which are submitted infre-
quently, and only when substantial benefit justifies the change. It
should be noted that process changes during the IND phase, or prior
to licensing, are normally expected. Changes at this point will not require the repetition of as much work compared to changes that are re-
quested after a license has been granted. However, while changes at
this point may not cause such lengthy delays, they do delay market
entry, and therefore represent significant income loss. The effect of
lost product sales much be weighed against the perceived benefits of
the change.

B. Analytic vs. Process Chromatography

Both analytical and process scale chromatographic methods are required to manufacture high-purity pharmaceutical proteins (33). An
important criterion for the usefulness of an analytical method is that
it should be distinct from the procedures that are actually used in
the recovery. Thus, at least one method must be reserved for purity analysis and not used in production. Preferably, the analytical
method should be orthogonal (have a different basis of separation)
to the methods used in the recovery process.

It is important to develop high-resolution analytical methods at
the same time that the recovery process steps are being defined.
By codeveloping analytical and process methods, two advantages are
gained: First, by analyzing the intermediate in-process bulks, one
can more easily identify the principal contaminants and characterize
their bahavior relative to the product. Second, having timely feedback on product purity allows the recovery scientist to know when
a single step, or the entire process, has met its purity goals. This
feedback is essential in deciding when to stop developing the recovery process.

C. Control of Raw Materials

A number of raw materials used in the fermentation process can have
significant impact on the subsequent recovery, particularly serum,
albumin, transferrin, and insulin. Because they are derived from
animal sources, they represent potential variable sources of contaminants such as viruses, mycoplasma, or hydrolytic enzymes. Pretreatment of these raw materials by heating, acidification, or sterile
(0.1 μm) filtration is often necessary to avoid contamination of the
cells and product. For example, contamination of the seed train

by serum-borne mycoplasma or virus may irreversibly repress cell growth and product titer. Carrying out a medium exchange prior to the production culture will remove all degraded product forms that have accumulated up to that point. However, the production vessel will still be contaminated, the cells will produce poorly, and the harvest fluid will likely contain degradative enzymes released by the mycoplasma that decrease the quality of the purified product.

A monoclonal antibody may also be a raw material which is used for purification of the product. Currently, regulatory agencies have set standards for the banking, production, and purification of MAbs that are at least as rigorous as those established for other recombinant proteins. The same set of requirements appears to apply whether the monoclonal is used directly in humans or whether it is merely used for the purification of another protein. Therefore, in order to use a MAb in the recovery of a therapeutic protein, one must characterize the hybridoma cell bank, establish that it is free from adventitious agents, assess the purity of the monoclonal, validate its purification for removal of nucleic acids and viruses, and minimize residual levels of MAb in the product.

D. Quality Control of Final Product

For biologics, consistency of manufacture is an essential component in ensuring the product's safety and efficacy in patients, and is at least as important as absolute purity of the product. Many aspects of protein purity such as nucleic acid and virus removal cannot be adequately tested directly in the final product and must rely on process validation studies. It is essential that the process be run as designed, within the specified limits, in order for the process validation studies to retain their significance.

Although no single test can establish the purity and composition of a complex biologic, a combination of analytical methods is useful in giving a fairly complete picture of the purity, composition, and consistency of the final product. For example, the use of SDS-PAGE and HPLC in conjunction with multiantigen ELISAs can provide substantial confidence when measuring host cell protein impurities at the ppm level.

A valuable assay for monitoring product purity and consistency is the technique of peptide mapping. The protein is enzymatically digested and then the fragments are separated by reversed phase HPLC. The resulting peptide map is one of the most sensitive methods available for monitoring small changes in protein structure. These maps are in a sense analogous to an infrared spectrum of an organic compound and are used in a similar way to fingerprint the protein. Figure 3 (34) shows tryptic maps of t-PA and a mutant of t-PA in which glutamic acid 275 has been replaced by an arginine.

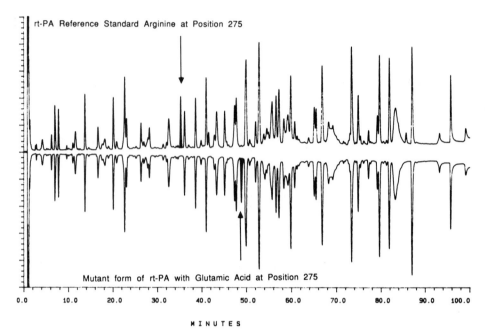

Figure 3   HPLC chromatogram of tryptic digests of native t-PA and
a mutant of t-PA in which the arginine at position 275 in the native
sequence has been replaced by a glutamic acid.   This mutant is resistant to plasmin cleavage at this site, which would otherwise generate the two-chain form of the molecule.   (From Ref. 34.)

It should be noted first that t-PA is quite resistant to trypsinization
and can be cleaved into peptides only after reduction and carboxy-
methylation.   After this initial procedure, one obtains not only a
detailed map but an extremely reproducible one.   The arrows indicate
the shift from one peak to another, demonstrating that even a single
amino acid change in a 527-amino-acid protein can be detected by
this procedure.   It is fair to note that in this particular case 100%
of the molecules are mutated in the same way.   Nevertheless, it is
estimated that this method could detect as little as a 5% contamination
of the product by a mutant polypeptide.   For small peptides, amino
acid composition and sequence are reasonable release criteria.   However, they are unreasonable for such large proteins, so the development of peptide mapping is invaluable in demonstrating the consistency
of a biologic on a lot-to-lot basis.

E. Genetic Stability

As discussed earlier, consistency of the recovery process is important in assuring consistency of the final product. However, it is equally important that the fermentation, including the cells themselves, produce a consistent feedstock. For recombinant products it is important to demonstrate the stability of the product gene throughout the working life of the cells. Perhaps the most likely event with respect to genetic stability is the loss of part of the amplified sequences within subpopulations of recombinant cells. Loss of such sequences can be checked by analysis of specific productivity throughout the manufacturing period. Gross changes in the genetic structure of amplified sequences can also be monitored by fluorescence in situ hybridization of metaphase chromosomes.

The most direct confirmation of genetic stability of the host cell would be to determine the gene sequence at various stages before, during, and after the anticipated production life. However, that is not possible for systems in which multiple gene copies are integrated into the host genome. Instead, it is useful to determine the composition of the protein product as a function of cell age. In this regard, the technique of peptide mapping continues to be invaluable for assessing lot-to-lot consistency.

Additional evidence of genetic stability can be obtained by determining whether the cells retain consistent morphology, growth characteristics, and nutritional status throughout the production lifetime.

V. SUMMARY

Less than a decade ago, the use of continuous mammalian cell lines for the production of cloned proteins was considered strictly a research tool. At that time, few thought it possible to allay the many safety concerns associated with transformed cells. It soon became clear that mammalian expression systems had numerous advantages over bacteria for production of therapeutic proteins, initiating a multidisciplinary effort to address these concerns in a thorough and reliable manner. The success of these efforts is exemplified by the emergence of product molecules into the market. Today, there are seven recombinant human therapeutics that have received FDA approval. Almost half of them (OKT3, t-PA, and EPO) are produced in mammalian cells, with the remainder produced in bacteria (insulin, growth hormone, and alpha-interferon) or yeast (hepatitis vaccine). At least a dozen more recombinant cell culture products are in advanced human clinical trials. With the accumulation of data and experience, continuous mammalian cell lines will no doubt be the preferred hosts for many future products of biotechnology.

REFERENCES

1. Estell, D. A., Graycar, T.P., and Wells, J. A. Engineering
   an enzyme by site-directed mutagenesis to be resistant to chemical oxidation. *J. Biol. Chem.* 260(11):6518−6521 (1985).
2. Capon, D. J., Chamow, S. M., Mordenti, J., Marsters, S. A.,
   Gregory, T., Mitsuya, H., Byrn, R. A., Lucas, C., Wurm,
   F. M., Groopman, J. E., Broder, S., and Smith, D. H. Designing CD4 immonoadhesins for AIDS therapy. *Nature*
   337:525−531 (1989).
3. Chaudhary, V. K., Mizukami, T., Fuerst, T. R., Fitzgerald,
   D. J., Moss, B., Pastan, I., and Berger, E. A. Selective
   killing of HIV-infected cells by recombinant CD4-*Pseudomonas*
   exotoxin hybrid protein. *Nature* 335:369−372 (1988).
4. Kleid, D. G., Yansura, D. G., Dowbenko, D., Weddell, G. N.,
   Hoatlin, M. E., Clayton, N., Shire, S. J., Bock, L. A.,
   Ogez, J., Builder, S., Patzner, E. J., Moore, D. M.,
   Robertson, B. H., Grubman, M. J., and Morgan, D. O. Cloned
   viral protein for control of foot-and-mouth disease. *Dev.
   Indust. Microbiol.* 25:317−325 (1984).
5. Weissmann, C., Taniguchi, T., Domingo, E., Sabo, D., and
   Flavell, R. A. Site-directed mutagenesis as a tool in genetics.
   In *Frontiers in Physicochemical Biology* (B. Pullman, ed.),
   Academic Press, New York, pp. 167−182.
6. Wells, J. A., Vasser, M., and Powers, D. B. Cassette mutagenesis: an efficient method for generation of multiple mutations
   at defined sites. *Gene* 34(2−3):315−323 (1985).
7. Ashwell, G., and Hartford, J. Carbohydrate-specific receptors
   of the liver. *Annu. Rev. Biochem.* 51:531−554 (1982).
8. Olson, K. C., Fenno, J., Lin, N., Harkins, R. N., Snider, C.,
   Kohr, W. H., Ross, M. J., Fodge, D., Prender, G., and
   Stebbing, N. Purified human growth hormone from *E. coli*
   is biologically active. *Nature* 293:408−411 (1981).
9. Builder, S. E. and Ogez, J. R. Purification and activity
   assurance of precipitated heterologous proteins. U.S. Patent
   4,620,948 (1986).
10. Rausch, S. K. Purification and activation of proteins from
    insoluble inclusion bodies. Eur. Patent Application 0,212,960
    (1986).
11. Wetzel, R., Kleid, D. G., Crea, R., Heyneker, H. L., Yansura,
    D. G., Hirose, T., Kraszewski, A., Riggs, A. D., Itakura, K.,
    and Goeddel, D. V. Expression in *Escherichia coli* of a chem-
    ically synthesized gene for a "mini-c" analog of human proin-
    sulin. *Gene* 16:63−71 (1981).
12. Ben-Bassat, A., Bauer, K., Chang, S-Y., Myambo, K.,
    Boosman, A., and Chang, S. Processing of the initiation

methionine from proteins: Properties of the *Escherichia coli* methionine aminopeptidase and its gene structure. *J. Bacteriol.* 169:751–757 (1987).

13. Hsiung, H. M., Mayne, N. G., and Becker, G. W. High-level expression, efficient secretion and folding of human growth hormone in *Escherichia coli*. *Biotechnology* 4:991–995 (1986).

14. Moks, T., Abrahmsen, L., Osterlof, Bjorn, Josephson, S., Ostling, M., Enfors, S-O., Persson, I., Nilsson, B., and Uhlen, M. Large-scale affinity purification of human insulin-like growth factor from culture medium of *Escherichia coli*. *Biotechnology* 5:379–382 (1987).

15. O'Farrell, P. H. High resolution two-dimensional electrophoresis of proteins. *J. Biol. Chem.* 250:4007–4021 (1975).

16. Anderson, N. L. and Anderson, N. G. Two-dimensional analysis of serum and tissue proteins: Multiple gradient-slab gel electrophoresis. *Anal. Biochem.* 85(2):341–354 (1978).

17. Langer, J. A. and Pestka, S. Purification, bacterial expression, and biological activities of the human interferons. *J. Invest. Dermatol.* 83(1 Suppl):128s–136s (1984).

18. Berntorp, E. and Nilsson, I. M. Biochemical and in vivo properties of commercial virus-inactivated factor VIII concentrates. *Eur. J. Haematol.* 40(3):205–214 (1988).

19. Ey, P. L., Prowse, S. J., and Jenkin, C. R. Isolation of pure IgG1, IgG2a and IgG2b immunoglobulins from mouse serum using protein A-Sepharose. *Immunochemistry* 15(7):429–436 (1978).

20. van Reis, R., Arathoon, W. E., Paoni, N. F., Sliwkowski, M. B., Lubiniecki, A. S., and Builder, S. E. A novel separation process for production of recombinant human tissue type plasminogen activator (in prep.).

21. Arathoon, W. R., Builder, S. E., Lubiniecki, A. S., and van Reis, R. D. Process for producing biologically active plasminogen activator. Eur. Patent application 0,248,675.

22. Points to consider in the production and testing of new drugs and biologicals produced by recombinant DNA technology. Office of Biologics Research and Review, U.S. Food and Drug Administration, Bethesda, MD (1985).

23. Points to consider in the manufacture and testing of monoclonal antibody products for human use. Office of Biologics Research and Review, Center for Drugs and Biologics, U.S. Food and Drug Administration, Bethesda, MD (1987).

24. Wiebe and May, Chapter 5, this volume.

25. Kafatos, F. C., Jones, C. W., and Efstratiadis, A. Determination of nucleic acid sequence homologies and relative concentrations by a dot hybridization procedure. *Nucl. Acids Res.* 7:1541–1552 (1979).

26. Petricciani, J. C. and Regan, P. J.   Risk of neoplastic trans-
    formation from cellular DNA: Calculations using the oncogene
    model. *Dev. Biol. Standard* 68:43–49 (1987).
27. Palladino, M. A., Levinson, A. D., Sverdersky, L. P., and
    Obijeski, J. F.   Safety issues related to the use of recombinant
    DNA-derived cell culture products. I. Cellular components.
    From *The 7th General Meeting of ESACT on Advances in Animal
    Cell Technology: Cell Engineering, Evaluation and Exploitation.*
    Dev. Bio. Standard 66:13–22 (1987).
28. Lubiniecki, A. S. and May, L. H.   Cell bank characterization
    for recombinant DNA mammalian cell lines. *Dev. Biol. Standard*
    60:141–146 (1985).
29. Builder, S. E., van Reis, R., Paoni, N. F., and Ogez, J. R.
    Process development and regulatory approval of tissue-type plas-
    minogen activator. In *Proc. 8th International Biotechnology Sym-
    posium* (Paris), Vol. 1:944–959 (1988).
30. Anicetti, V. R., Fehskens, E. F., Reed, B. R., Chen, A. B.,
    Moore, P., Geier, M. D., and Jones, A. J. S.   Immunoassay
    for the detection of *E. coli* proteins in recombinant DNA-derived
    human growth hormone. *J. Immunol. Meth.* 91:213–224 (1986).
31. Anicetti, V., Builder, S., Marks, B., Patzer, E., Ogez, J.,
    and Vetterlein, D.   Method of purifying recombinant proteins
    from corresponding host cell proteins. Patent publication
    PCT W089/04367.
32. Morii, M., Ohoka, M., Kawashima, N., and Suzuki, T.   Method
    of purifying crude tissue plasminogen activator. Eur. Patent
    Application 0,210,870.
33. Builder, S. E. and Hancock, W. S.   Analytical and process
    chromatography in pharmaceutical protein production. *Chem.
    Eng. Prog.* (1988).
34. Garnick, R. L.   Safety aspects in the quality control of recomb-
    inant products from mammalian cell culture. *J. Pharmaceu.
    Biomed. Anal.* (in press).
35. Builder S. E., Garnick, R. L., Hodgdon, J. C., and Ogez,
    J. R.   Proteins and peptides as drugs: Sources and methods
    of purification. In *Proteins and Peptides as Drugs* (in prepa-
    ration).

15

Management of Process Technology

DONALD G. BERGMANN
SmithKline Beecham Pharmaceuticals, King of Prussia, Pennsylvania

I. INTRODUCTION

The measure of success of a large-scale cell culture manufacturing process is the ability of that process to be reliably operated at scale and within the confines of current regulatory environments. The first step of a successful manufacturing process lies in the process design itself. However, poor management of the process technology can cause an otherwise well-designed process to be unreliable in practice.

The key to success in operating a large-scale process plant is the careful attention to detail, particularly to the most important ones. This chapter outlines some of the important considerations and procedures that have been applied to successfully operating cell culture process plants. The primary emphasis is toward the management of large-scale unit processes such as "deep-tank" suspension culture processes. However, many general principles outlined are applicable to smaller multiple-unit process operations such as roller bottles.

A secondary underlying theme is the management of processes intended to produce human pharmaceutical products. Thus, many of the considerations throughout this chapter are from the perspective of compliance with the current regulatory environment related to cell culture and biotechnology manufacturing practices.

II. RAW MATERIAL CONSIDERATIONS

Mammalian cells by nature survive and grow in a relatively narrow
range of environmental conditions and seemingly minor quantities of
inorganic or organic contaminants can have toxic effects on cells.
Cell toxicity by such contaminants can be further exacerbated when
cell are grown under serum-free conditions (1). Microbial contamination is another consideration since cell cultures provide excellent
reservoirs for the propagation of bacteria, mycoplasma, and certain
viruses. All of these can have deleterious effects on the cells, the
product, and/or downstream processing. Another consideration for
human pharmaceuticals is the fact that the current Good Manufacturing Practices (cGMPs) require that the manufacturer establish written
specifications for raw materials used in processing and that they
assure that the raw materials used have met their respective
specifications.

A. Critical Raw Materials

1. Water

Tap water passed through deionizing resins may render water usable
for many types of cell culture media. In practice simple deionization
is not advised. Water that meets or exceeds the USP specifications
for "purified water" (2) is recommended. The method of generating
"purified water" is probably unimportant. However, control over the
levels of bioburden and endotoxins is an important consideration.
Water maintained under conditions that inhibit or minimize microbial
growth is desirable. This can be achieved by maintaining water in
hot circulating loops such as those designed for water-for-injection
(WFI) systems or by routine flushing and sanitization (or sterilization) of any cold circulating loops. Noncirculating use points
should be flushed and sanitized (or sterilized) daily. "Deadlegs"
or static noncirculating systems should be avoided or minimized as
these provide excellent growth reservoirs for microorganisms.

Another consideration is contamination of cell culture processes
by water-borne viruses. Although this probability is low, the
chances of introducing a viral contaminant into the cell culture process can be significantly reduced by heating the water under specified conditions of time and temperature. Ultrafiltration is another
possibility; however, continuous verification of absolute retention can
be difficult to manage compared to other methods available.

2. Cell Culture Media

Cell culture medium often contains 50 or more individual components and may be further supplemented with serum, peptones, and other supplements. Many vendors supply standard media formulations in both liquid and powdered forms and in various lot sizes. Special purchase arrangements are often available which may include the supply of bulk powders in specially sized drums, the production of large lots of powder that produce tens of thousands of liters of reconstituted medium, the manufacture of specialty medium to the user's precise specifications, and yearly production commitments. It is not uncommon for a very large-volume user to produce their own medium from the individual components (3).

Whether media is formulated in-house or by a vendor, detailed specifications need to be established to assure that consistent quality is attained. While chemical purity is a major consideration, the biological quality of medium components is also important. It is not reasonable to expect powdered formulations to be totally free of microorganisms. However, unless rigidly controlled, the methods of manufacturing certain medium components can lead to unacceptably high levels of microorganisms or their byproducts in the final blend.

3. Media Supplements

The most common cell culture medium supplement is animal serum (typically bovine). Serum contains trace factors which promote or enhance cell survival and propagation. Virtually all cells have a requirement for at least some of the factors contained within serum. Great strides have been made in the development of defined media which for some cell types do not require serum supplementation. Despite this, the use of animal serum in the industry has not been broadly replaced.

The growth-promoting qualities and quality of serum in general can vary widely from lot to lot, from season to season, and from year to year. Serum is typically collected at large abattoirs from slaughtered animals. Animals are excellent reservoirs for bacterial, mycoplasmal, or viral infections, all of which can lead to difficulties if introduced into a cell culture process. Thus, the disease state of the animal and the methods of collection and processing are all important considerations. Although filtration can remove bacteria and mycoplasma, small viruses and bacterial byproducts are not readily removed from contaminated serum.

Development of sound biochemical and biological specifications for serum is important. Many serum vendors perform batteries of tests on their serum lots. Vendors typically test for common blood chemistry profiles and for typical biological contaminants found in the source animal. Many vendors also evaluate the growth-promoting qualities of test lots using a standard set of cell lines. Special arrangements can be made which may include collection and processing under custom specifications, additional testing, and yearly production commitments.

The same concerns and considerations regarding serum quality should be applied to any animal-derived biological supplement. The user needs to evaluate the animal source, potential chemical and biological contaminants from the source, methods of manufacturing, and the types of testing performed by the vendor. Detailed specifications need to be established by the user to assure that reliable quality is attained.

Gamma irradiation has been used for a number of years to provide a degree of assurance that any potential biological contaminant in serum or other animal-derived supplement has been inactivated. The necessity of the procedure needs to be evaluated by the user. The animal source, cells employed in the process, and end use of the product all contribute to the risk vs. benefit evaluation of gamma irradiation. The effect of irradiation on trace growth factors in serum cannot always be certain and the situation may be further exacerbated when the serum is used in very low concentrations in cell cultures.

Viral inactivating agents such as binary ethyleneimine are often used by veterinary vaccine manufacturers to treat animal-derived materials such as serum or trypsin. As with irradiation, the risk vs. benefit of the procedure needs to be evaluated on a case-by-case basis.

4. Other Process Additives

In large unit processes quite often the culture environment is controlled by the addition of simple chemicals such as inorganic acids and bases. As with all other raw materials, these chemicals need to have defined specifications that establish their quality standards.

B. Raw Material Testing

Traditionally raw materials (especially media and serum) are tested both by typical chemical analyses and by the performance of cells when cultured with the materials in question. It may be reasoned that simple chemicals should not pose any risk to the process if the material passes typical identity tests. However, when the potential economic loss is high due to the incorporation of an undetected contaminant into a large lot of medium, performance-testing even simple molecules is advised.

1. Defined Media and Chemicals

The largest risk with defined chemicals is usually the chance presence of undetected contaminants that may produce a toxic effect on the cell cultures. Typically one would test for standard chemical identifications for all components and the levels of typical impurities. This information is normally available from the vendor in the form of a vendor certification. The in-house quality control unit should perform a certain subset of tests on the materials as an internal control check of the vendor's performance.

Although the vendor may do cell culture performance testing on each lot of medium and is willing to provide the results to the user, the testing of the test material in a format that faithfully mimics the process is essential. Normally this can only be done in-house. Typically the process is mimicked at small scale using cultures of the production cells under the standard production conditions. The cells should be carried through all cell culture process steps even if the test material is used only at the early seed stages of the process. In this manner any downstream effects of the test material on the production cultures can be evaluated. It is important that the test material be the only variable in the test; all other components should be reference materials that represent the typical quality expected for that material. The parameters that might be evaluated include growth rates, maximum cell densities, productivity in the final production stage, and typical behavior and/or appearance of the cultures. The performance of the test cultures should be evaluated against control cultures carried in parallel and derived from a common cell inoculum. The controls should use only reference materials.

To test the performance of each individual component prior to blending into a lot of medium may prove impractical. A typical strategy is to blend the various test components into a test lot. All other components that are not test components should be from reference lots. If the performance tests on the test lot pass the acceptance criteria, then all test components have passed. If not, then one needs to test each item individually or break the blend into smaller subsets and repeat the tests. Having completed some chemical identity tests on the test materials prior to blending into a test lot helps catch obvious problems before starting the effort.

2. Undefined Growth Factors

Serum is the typical example of an undefined growth supplement.
Often the lack of a critical trace component may not be manifested
in the initial culture. Users will normally test serum performance by
subcultivating cells for several passages (at least three) in the test
serum to assure that their cells will perform consistently under prolonged cultivation with the material. As with defined chemicals the
entire process using production cells should be mimicked. Some
users test serum performance by evaluating the plating or cloning
efficacy of the test material against an established standard. Although this method has value as an initial screening mechanism, it
does not replicate the actual conditions under which the material will
be used in manufacturing.

Vendors will do most chemical and biological tests and provide
test certificates of analysis. The in-house quality control unit
should perform a selected subset of these tests to monitor vendor
performance. Typical tests that a user would perform are standard
tests for bacteria, mycoplasma, possibly some virological tests, some
blood chemistry tests, and one or more tests of identity.

C. Vendor Audits

The assurance that raw material vendors operate within the confines
of good manufacturing and laboratory practices is sound economic
practice and expected by the FDA. During the identification and
qualification of material suppliers the user needs to clearly establish
their quality expectations. Part of the vendor qualification process
should include a visit to the vendor site and a detailed audit of the
vendor's manufacturing and quality practices. Almost all suppliers
of raw materials for the pharmaceutical industry are familiar with
these practices and are willing to entertain audits by potential
customers. Sometimes the initial visit uncovers deficiencies that
need to be corrected before the new vendor can be qualified. Usually, if the new business is attractive enough the vendor will comply
with reasonable requests. Follow-up audits are essential on an annual or biannual basis to assure that the vendor continues to comply
with the user's expectations. In many respects vendor audits by
the customer are similar to initial establishment inspections and routine reinspections by the FDA.

III. EQUIPMENT DESIGN CONSIDERATIONS

The design of process equipment will largely be determined by the
design of the process itself. Several authors have provided detailed
evaluations of various types of large-scale cell culture systems and

considerations related to scale-up of those systems (1,4 – 11). Two
principal systems of cell culture manufacturing predominate to date.
Roller bottle systems are still used extensively in both the human
and veterinary vaccine industry. Roller bottles are one of the easi-
est and least capital-intensive systems to implement.

When the amount of material to be produced becomes very large,
the size of a roller bottle facility and the operating labor can become
prohibitive. "Deep-tank" suspension culture systems encompass the
other major type of large-scale manufacturing system employed today.
Burroughs-Wellcome reported the operation of 8000-liter scale cell
culture fermenters (3) and Genentech recently reported the operation
of 10,000-liter cell culture fermenters (12) used for the production
of proteins from mammalian cells. Very large cultures of cells can
be propagated in fermenters yielding gram to kilogram quantities of
desired proteins. Success rates in excess of 95% have been achieved
in these large-scale systems (9). High success rates can be attributed to appropriate design of vessels, filtration systems, and pipework. Certain general design considerations for equipment apply
regardless of size and degree of sophistication.

A. General Design Considerations

1. Sterilizability

The ability to maintain the sterility of a cell culture process is of
prime importance. It is important that the design of vessels, pipelines, and filtration equipment allow for consistent and reliable sterilization. Pockets and crevices in process equipment should be
avoided or minimized. The design needs to assure that steam vapor
contacts all surfaces being sterilized by steam and that the buildup
of condensate does not occur. Steam entry to vessels needs to
occur such that cold air is adequately purged from the system during
heat-up and that condensate is adequately drained from the vessel
during sterilization. Pipelines connecting equipment need to be constructed such that condensate does not collect in the lines. This is
usually achieved by judicious use of steam traps at low points, slop-
ing lines to traps, and assuring that steam is introduced at high
points in lines. Filtration systems need to be adequately designed
to provide for integrity testing of filters without jeopardizing the
sterility of the system. All welding on process equipment and pipelines should be free of cracks and crevices. Where pipe unions are
required threaded fittings should be avoided. Sanitary fittings
should be used wherever demountable pipe is used. Where a liquid
interface cannot be avoided during sterilization, one needs to ensure
that the liquid phase reaches adequate sterilization temperature and
time.

2. Maintenance of Process Integrity

Maintaining a sterile system is the next major consideration. Wherever possible, welded connections are preferred. Where welded connections cannot be used, sanitary fittings are preferred. All vessel penetrations should be welded and not threaded. Double O-ring seals around flanges are recommended, and steam or sterile condensate blocks between seals should be given serious consideration. Magnetically coupled agitators have been used successfully to eliminate the agitator as a potential source of integrity failure. Due to the availability of very powerful ceramic magnets and the low power input associated with cell culture processes, magnetically coupled agitators have been used in vessels as large as 8000 liters (3). However, recent advances with mechanical seal designs have made mechanically coupled agitators increasingly reliable and attractive due to the ability to impart greater power to the vessel.

3. Ability to Provide Sterile Additions and Transfers

Process equipment and lines should be designed to provide reliable methods of transferring sterile process fluids or inoculum and the introduction of sterile additives as necessary. Ideally this can best be accomplished by welding everything together, sterilizing the system as a whole and maintaining integrity after that. However, this is not always possible or practical. Almost all cell culture systems involve some making and breaking of connections. Numerous procedures, techniques, and designs have been employed to assure that sterility is maintained during these steps. The use of laminar flow cabinets or clean rooms may be practical for some systems and scales of operation. However, when process systems become very large the use of clean rooms or hoods is impractical and not highly reliable. The best system to assure a sterile transfer of material in large-unit process systems is to sterilize or resterilize connections with live steam. Judicious use of sanitary block valves to isolate the sterile parts from nonsterile parts of the system and assurance that steam condensate is adequately purged from the system are all prerequisites of a successful design. Figure 1 schematically depicts a typical design of a steam-sterilizable connection.

A second consideration to assure sterile additions to process vessels is to design for point-of-use filtration of process streams wherever possible. As a general principle in process design, one should always minimize the distance between the final filter and the receiving vessel or system. These considerations apply to both liquids and gaseous materials.

The equipment or system design also needs to provide a method to move process streams from one system or vessel to the next without jeopardizing sterility. The simplest and most reliable method of moving fluids is by sterile compressed air or nitrogen. Peristaltic pumps work well where flexible tubing or hose can be used and very large-capacity pumps are available. The greatest risk lies in tubing or hose failure during pumping, and the history and condition of hoses need to be controlled. Steam-sterilizable sanitary pumps that have very high pumping capacities and pressures are also available. Quite often the problem with sanitary pumps is not pump sterilization but maintaining sterility once in use. Each type of pump has its advantages and drawbacks and the user needs to evaluate each against its intended use.

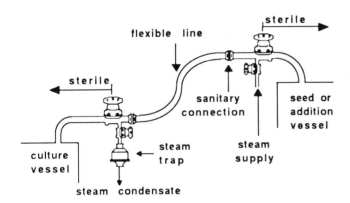

Figure 1 Simplified illustration of a steam-sterilizable connection. The sterile transfer of fluids between two independent vessels can be accomplished by steam-sterilizing the transfer line that interconnects the two vessels. In this illustration the supply (addition) vessel and the culture vessel are both sterile and isolated from the nonsterile external environment by appropriate block valves (double block valve arrangements are often employed). The interconnecting transfer line is steam-sterilized by the addition of steam such that all surfaces between the block valves are exposed to sterilizing conditions. Careful attention to avoid accumulation of condensate at low points or within crevices is essential. The sterilization process should be monitored by appropriate sensing devices. During cool-down applying sterile air pressure (not shown) is recommended to avoid drawing a vacuum within the sterilized line. After cool-down the fluids can be transferred from one vessel to the next through the steam-sterilized connection.

4. Ability to Take Sterile Samples

Cell culture processes require some degree of monitoring which usually involves taking samples from the system. One needs to assure
that the design allows for frequent sampling without compromising
the sterility of the system. Various designs exist on commercially
available production and laboratory scale cell culture fermenters.
The most reliable designs employ the general concepts described for
making sterile transfers (Fig. 1). As an added design feature, the
sample port can be designed to accommodate the attachment of a sterile sample container. This added feature allows for the attainment
of sterile samples and eliminates sampling as a potential source of
contamination.

5. Cleanability

All process equipment and pipelines need to be designed to be cleanable. In very large systems automated clean-in-place systems can
be installed. Both high-pressure and low-pressure/high-flow systems have been successfully applied in the industry. A number of
firms design and fabricate clean-in-place systems for the pharmaceutical and dairy industries. A number of caustic- and acid-based
cleaners for these systems are available that have been proven to
be effective at removing cell culture residuals from equipment. In
smaller systems the cost and effort involved in implementing an automated clean-in-place system is not economical and manual cleaning
makes more sense. Some types of systems or equipment design are
not amenable to clean-in-place systems in which case other methods
must be employed.

B. Instrumentation and Process Control

Virtually all cell culture processes require some degree of process
control. The simplest roller bottle system requires temperature control and bottle rotation rate control. More sophisticated systems
such as a suspension culture system may require temperature, agitation, pH, and dissolved oxygen control to name just a few. Most
mammalian cells are very sensitive to their environment. As such,
one needs to be certain that the monitoring and controlling devices
will accurately control within the process ranges desired, remain
within calibration between calibration or standardization intervals,
and be functionally reliable. Many process control instruments are
presently available which can meet these criteria. Despite the availability of ever-improving instrumentation, one may choose to install
redundant sensors and instrumentation. A very accurate history
should be maintained on the operation and service of critical instrumentation and sensors.

Linking the measurement and control devices to appropriate alarm systems may more than pay for the cost in just one batch saved.

C. System Reliability

An underlying principle to all successfully operated cell culture processes is reliability. Both the process design and the process equipment need to be optimally reliable. Two things usually lead to system reliability. They are (1) simplicity and (2) robustness. The simpler the system, the easier it is to operate, and the more robust (or sturdy) a system or piece of equipment, the less likely it will fail. This is why to date the two predominant cell culture production systems employed in industry are still the roller bottle and some type of stirred reactor. They are both simple in comparison to some other system designs and they both have long histories that have allowed continuous improvements in reliability. Although exploring new systems and equipment has definite merit, those efforts are best left to the development laboratory or pilot operations. Systems or equipment with proven track records are usually preferable for full-scale manufacturing plants.

IV. PROCESS MANAGEMENT

A. General Principles

1. People and Training

There is no substitute for a well-trained and experienced staff. Whether the process manager is responsible for a startup situation or an ongoing operation, that individual should always be certain to maintain a critical mass of experienced staff, and should always be striving to broaden and strengthen the training and experience of the staff. A key attribute of an excellent process operator is genuine interest and enthusiasm for the job. By nature they are very detail-oriented individuals. The importance of attention to details can never be overstated. Training and periodic retraining of staff members is essential. Topics should not only cover process details but also plant safety, cGMPs, scientific and engineering aspects of the process, and should include general round-table discussions on plant improvements. The more involved the staff is with the overall picture, the more interest they will have in the success of the operation.

Training should take the form of classroom-type settings, on-the-job training, and study of written procedures and other appropriate written materials. A training manual for each operator that contains a process description for each step, safety information, references to appropriate written procedures, and so forth can be a very valuable tool. In this manual a written record containing dates of training and level of competency achieved for each operation should be maintained. In this way both the operator and supervisor are aware of one's training status and expectations. The manual also forms part of the written record to satisfy cGMP requirements.

An operator will learn far more by doing than watching or reading. An inexperienced operator should always be accompanied by an experienced operator when performing process operations. In all cases the written procedure for the operation, or standard operating procedure (SOP), should be available at the location where the process is being performed. In operations involving very complex setups, it is advisable that each step of the SOP be checked against the actual operation to assure that no errors occur.

Educational background of the operations staff will vary with the job and the type of process employed. Fairly nontechnical processes do not require operators to have much formal technical background and sufficient training can be obtained on the job. More senior level positions usually require formal technical training. In the emerging biotechnology field and at new startup operations employing sophisticated and technically complex operations, it is not uncommon for many of the operators to have some formal background in a technical discipline. Having a staff of college educated engineers, chemists, and biologists to start up and troubleshoot a new process plant has real advantages initially. However, it can become a challenge for the process manager to motivate the staff as operations become more routine. A balanced mixture of educational and experience backgrounds works best.

2. Documentation

Three types of documentation are normally found in a process plant. The first type is the standard operating procedure (SOP), which normally describe in detail and in a stepwise manner the steps involved and to be followed to operate a piece of machinery, to perform a particular process or operation, to perform particular preventive procedures on equipment, and so forth. SOPs should always be available to the individual performing the operation and followed precisely unless there is reason to deviate.

The second type of documentation is the manufacturing record, often referred to as the batch record or manufacturing formula. These form the record of how a lot of product was produced. As a minimum, batch records contain a record of all ingredients introduced into the process, critical steps performed during the processing, and critical process parameters encountered during the processing (temperatures, pressures, etc.). The cGMPs describe the types of information that should be considered for incorporation into the manufacturing records. Another type of record included in this group is the various logs used to record cleaning and sterilization of equipment, environmental monitoring, and so forth.

The detail in manufacturing records may range from as little detail as described above to containing a complete reiteration of the SOPs that describe the process steps with operator and verifier sign-offs at each step. If the process setup or operation is very complex, it is probably advisable to incorporate those steps into the record along with appropriate sign-offs. This method provides maximum assurance that the proper procedures have been carried out. This also provides a valuable troubleshooting tool in the event that something goes wrong during the processing. Common sense, operating history, and potential impact on product quality are the best determinants of whether to include any given item in the batch record. SOPs and batch records are considered controlled documentation. The formation of and changes to these documents are normally controlled through a quality assurance group and require signatures from appropriate parties that have responsibilities for the process.

The third type of documentation is informal records of process data and other types of information that are not incorporated into batch records but have been determined to be useful in monitoring the performance of the facility and process. This type of information usually has no impact on product quality and is not a critical process parameter. As such it should not be included in the batch record. However, the data provide additional information that allows for process/facility trend analysis and development of data bases. These records are normally not controlled documentation.

The purpose of documentation is threefold. First, the FDA requires appropriate record keeping; and as an ethical manufacturer, one wants to assure that any product that enters the market has been manufactured properly and is safe and effective. Second, if any problems were to occur, the capability to document, trace, or troubleshoot exists. Finally, adequate data collection and documentation allows for the development of data bases for process and facility performance trending. These data bases not only allow one to develop early warning signals of impending problems, but also can lead to the development of improved process methods.

3. Preventive Procedures

Preventive measures are those that prevent batch losses not only due to contamination but also due to mechanical failure and operator error. Perkowski (13) recently detailed some of the major considerations for minimizing batch loss. The most important considerations are as follows:

1. Sound process and plant design
2. Adequate specifications for raw materials and thorough testing

3. Sound SOPs and manufacturing records
4. Well-trained and conscientious staff
5. Sound process and facility validations
6. Comprehensive preventive maintenance program
7. Thorough understanding of process and facility trends, and expected and acceptable ranges of results
8. Thorough record keeping and review systems

A comprehensive preventive maintenance and calibration program is critical to failure-free operation. Thorough SOPs need to be written describing procedures to be performed and their frequency. Complete and accurate documentation of work performed and findings are important for the effectiveness of such a program. Records should be organized in such a manner that a complete history can be developed for each major or critical piece of equipment, system, and instrument. Preventive maintenance should include frequent (batch-to-batch if necessary) inspection and/or replacement of seals, gaskets, valve diaphragms, and so forth. Operations staff can perform many of these routine procedures as part of cleaning or setting up a system. However, a highly trained staff of maintenance mechanics and instrument technicians is essential for the more specialized procedures. A maintenance/instrument staff consisting of experienced persons with specialty skills works best. Again, training is of utmost importance. In new facilities and with new equipment, the frequency and types of preventive procedures performed can usually only be determined arbitrarily. Equipment manuals, vendors, and prior experience can help establish the initial procedures. Preventive procedures can be subsequently optimized according to performance history.

B. Medium Preparation

1. Large Scale vs. Small Scale

The philosophy and approach to cell culture medium preparation will be largely driven by the scale of operation. At very large-volume throughputs a manufacturer may find it economically attractive and necessary to prepare medium from components rather than purchasing preblended medium. Considering the volatility of the bovine serum market, a very large user might also find it attractive to set up their own serum preparation units and donor herds. Following the route described above has a considerable impact on purchasing, quality control, materials management, and plant operations. The financial cost of setting up the procurement, testing, warehousing, and blending capability for such an endeavor may be prohibitive to a smaller startup company.

If medium is prepared from its components, it could be done on a batch-by-batch basis just prior to a batch of medium being

required. However, it is logistically difficult and inconvenient to
attempt to stage, weigh, and blend upward of 50 or more components in a media prep area just prior to the batch of medium being
required. Other strategies have been employed. One approach is
to weigh out the individual components representing a batch of medium and package the components until needed. This approach is only
slightly better than trying to weigh components just prior to use.
It is still very inefficient since only one batch of medium is prepared
at a time. Another approach has been to blend subcomponent parts
of the media mix into groups of concentrates and freeze many aliquots representing batches or partial batches. When a batch of medium is required the appropriate aliquots are then thawed and added
together with other components to make a complete blend. The most
efficient and effective method, however, is to mill large lots of media
into powder where each lot consists of many production batches.

Purchase or preblended medium can usually be done in many
types of packaging configurations. It is best to arrange with vendors a series of packaging configurations that meet various needs
and production volumes of the plant. Whether preparing medium
from components or purchasing as a preblend, having medium plus
other dry components aliquoted into units representing established
batch sizes makes it very convenient when it is time to prepare a
batch of medium. In this manner a batch of medium can be blended
quickly since everything has already been preweighed and all associated documentation completed.

The scale of operation also defines the approach to medium sterility testing. At smaller scale, it is not uncommon to dispense
filter-sterilized medium into containers and to carry out some type of
sterility test before using the batch. At larger scale, holding sterilized medium in containers of any size while on test becomes very
costly and impractical. As a general rule, large batches of filtered
medium used to batch unit process systems, such as fermenters, are
not held on test prior to use. Manufacturers rely on the efficacy of
the sterile filtration design and the integrity of the filters used to
provide consistent batches of sterile media.

2. Medium Preparation Operations

The following are major considerations for reliable medium preparation
operations. First, the equipment and utensils used should be maintained in a clean and sanitary manner. The nature of most cell culture media necessitates that cleaning involve the use of detergents
and hot water to assure that residuals are removed. During or after
cleaning the equipment should be sanitized and then stored dry until
next use. During storage the equipment should be protected from
environmental contamination as much as practical. Many types of
detergents are available that contain sanitizers; hot process water or

live steam can also act as a sanitizer. In some cases it may be difficult to completely dry a large-mix vessel after cleaning. If a vessel has contained residual moisture between uses a rinse with hot process water (80°C) just prior to use can produce a quick resanitization. Occasional steam sterilization of the mix tanks may provide certain preventive benefits; however, routine (each use) sterilization of mix tanks and the nonsterile side of the sterile filtration system is unnecessary provided that the system is properly designed and operated.

Second, very careful attention to component additions to the medium batch is essential. Thorough SOPs and records with appropriate verification and review is important. Accurate measuring devices are important to ensure that appropriate weights and volumes have been added. External analytical methods are useful in providing additional assurances that the medium batch has been prepared properly.

Third, the operators or environment should not contribute any objectionable materials to the medium batch. Hygienic practices and operator health are important considerations. Plant dress should be of the nature that the introduction of dander and so forth from operators is minimized. The methods of introduction of materials into the mixing vessels should be such that little opportunity exists for other environmental contaminants to be introduced into the vessels, and the room should be designed and maintained as a relatively clean environment. It is not unusual for these types of areas to consist of epoxy-painted walls and ceilings, epoxy-sealed floors, and be supplied with Hepa-filtered air. If the blended medium is sterile-filtered and dispensed into sterilized bottles or other types of small containers, then the dispensing operations should be done in a clean room type of environment. Typically medium dispensing into sterilized bottles is done in some form of Hepa-filtered laminar flow hood or room.

Fourth, the sterile filtration design and procedures need to be established to provide the highest assurance of success. Cartridge filters are the most common system today for sterilizing cell culture medium. Pore sizes of 0.22 or 0.1 μm are considered true sterilizing filters. In practice, 0.1-μm filters are recommended as the final sterilizing filter. Prefilters are recommended to prevent final filter blockage. The number of final sterilizing filters used in a series is dependent on the material being filtered. It is not uncommon for freshly collected bovine serum to be processed through a series of three 0.1-μm filters before dispensing. Assuming good control over raw material quality, a single final filter is sufficient to assure sterility of the blended medium.

The design of the filtration system is important. Prefilters can be located just about anywhere before the final sterilizing filter(s).

Normally they are located somewhere adjacent to the mix tank. It
is usually not worthwhile to sterilize these filters. The final sterilizing filter, however, should be located as close to the sterile receiving vessel as possible. Everything from the sterilizing filter
to the final media destination must be sterilized and maintained sterile. The sterilization design and procedures must assure that the
final filter and associated equipment are indeed sterilized and maintained sterile. The system and procedures should be validated to
assure expected performance under standard conditions. Validation
tion procedures should include not only sterilization temperature
profiles and possibly biological challenge but a media challenge as
well. The procedures themselves need to ensure that the filters
are not damaged during sterilization or use.

   Sterile filter integrity should be tested. A number of procedures are available to test filter integrity, the most common of which
is to measure air pressure decay over time of wetted filters. Many
filter vendors supply both the pressure decay data for their filters
and automated machines to test filter integrity. Simple and inexpensive manual test rigs can also be fashioned to perform the same operations. Most decay data are developed on filters wetted with
water. Filters can be purchased that have been factory-tested,
but despite this it is always advisable to test filters after sterilization and just before use, since it is never certain what damage
may have occured during shipping or sterilization. Postuse integrity tests are recommended but may be difficult as the filter is
now wetted with medium which will normally give a different decay
profile from that of water. One approach is to flush the filter
with copious quantities of water, but it may be difficult to assure
that all traces of residual medium have been removed. Another
approach is to have the vendor determine the decay profile of the
medium used. In practice, most problems are identified during
preuse testing and seldom do media filtration operations (if performed properly) ever cause integrity failures. The value of
postuse integrity tests needs to be evaluated against the probable
risks on a case-by-case basis.

   Fifth, medium preparation time should be kept to a minimum.
Cell culture medium is enriched with many nutrients. As such,
the few microorganisms that exist in a nonsterile blend have an
opportunity to multiply. As a general rule, nonsterile medium
should not be held at room temperature for more than 8 – 12 hr and
not more than 24 hr at refrigerated temperatures (even if it contains antibiotics). Logistical timing and staging is very important.
Having all components preweighed and assembled is a benefit. The
right filter selection and filtration design is important. The filtration design should be sized to allow filtration of the entire batch
within a few hours at most.

C. The Culture Process

1. Scale-up to Harvest

The methods of seed culture preparation, scale-up of seed cultures, culture maintenance, inoculation and maintenance of production cultures, and harvest of product are as varied as the myriad of cell culture systems available today. However, certain principles can be applied to most systems.

(1) The cell bank should be maintained in a well-protected area(s) with limited access. More than one site of storage is preferred in which a portion of the bank is stored at each location. Cell bank storage vaults should be maintained under lock and key with limited access to keys. It is a good idea to assign a curator of the banks who is responsible for controlling access and maintaining accurate records of inventories. No expense should be spared on the storage equipment and the systems should be alarmed to assure that the desired temperature is maintained. Equipment should be checked daily.

(2) The seed train must be maximally reliable. Part of this results from optimal protocols which provide cells of dependable growth rate and population density on a very predictable timetable. Wherever practical seed cultures should be maintained with sufficient redundancy to prevent lost production due to the loss of these cultures. At the early stages of scale-up when cultures are in small vessels it is not difficult to maintain redundancy. As the scale increases the cost of maintaining redundancy becomes more difficult and expensive. The degree of redundancy maintained for any particular process will be determined by the manufacturing system, its history, and the assessment of the cost vs. benefit.

One strategy that has been employed with success has been to maintain redundancy at the early seed culture stages, essentially none at the middle stages of scale-up, and then to provide redundancy at the very last stage prior to production scale. In fermentation-type cell culture processes this may take the form of sufficient vessels at the very last stage to allow minimum recovery time if one were to fail. The strategy at this stage sometimes involves seeding cells in these vessels so that the cells are ready to be inoculated into production vessels within a very short time. One technique that has been used has been to utilize only part of a seed vessel to inoculate the production vessel and then to "top off" the seed vessel with fresh medium to allow another growth cycle to occur. Depending on how much culture has been left behind, the cycle time for this type of process can be very short. The process has been termed "solera" by some manufacturers (a word borrowed from a similar practice in the sherry-making business).

Solera vessels maintained for as long as 6 months have been reported (3).

(3) Where feasible seed cultures should be maintained in anti-
biotic-free medium. The stage at which one may choose to begin
to use antibiotics (if at all) in the process is dependent on the
manufacturer's evaluation of the risk vs. benefit and from experience. However, antibiotics should be avoided at the very early
stages of scale-up where greater human manipulation is involved
(i.e., propagation in spinner vessels or T flasks) and the time
span to the production vessel is the greatest. The greatest risk
occurs when antibiotics have masked a bacterial contamination until resistant progeny suddenly appear in the production vessel.
As the scale and proportionate economic loss increases, the use of
antibiotics becomes more attractive.

Details of how one would scale, maintain, produce, and harvest
product will vary with the process design. The most important
factor contributing to the success of a cell culture process from
scale-up to harvest is the careful attention to details in process
plant and equipment design, and in the operation of the process.
The most common loss is usually through contamination. Sterility
must be maintained from cell bank through to harvest. Once control of contamination is in hand, the next most common problem is
usually mechanical failures which can be controlled through adequate equipment design, selection, and preventive maintenance
programs.

2. Process Monitoring

The degree of process monitoring will largely depend on the process
design. Fairly simple roller bottle systems do not require much
other than monitoring of incubator temperatures, roller rack operation, and visual observation of cells in representative selections of
bottles. Both incubators and roller racks can be alarmed to achieve
continuous monitoring. Other types of information can also be collected such as average cell numbers per roller bottle over a representative sample and product yields at harvest. However, between
seeding and harvest of roller bottles not much in-process monitoring
or intervention is practical or necessary.

With more sophisticated systems and/or at larger unit volumes
much more process control and monitoring is practical and usually
necessary. For example, simple deep-tank suspension culture systems require, mixing, temperature, pH, dissolved oxygen control
as a minimum. More sophisticated systems, such as perfusion systems, may have controlled feed rates of nutrients and controlled
harvest rates of cell culture fluids all of which need to be continuously monitored and controlled. Even more sophisticated

monitoring and control schemes have been developed that have fairly
elaborate feed back control loops for multiple parameters all controlled by computer.

These systems should be provided with the capability of alarm-
ing when a process parameter is out of specification. The acceptable ranges of process control need to be determined and then
alarm limits established within those ranges so that the operations
staff has time to respond before the process goes out of specification. Care needs to be taken, however, to avoid nuisance alarms.
Regardless of sophisticated control devices, process operators
should routinely check process equipment on a regular set of rounds
and record pertinent process information on appropriate logs.

Appropriate logs or records should contain the expected and/
or acceptable ranges of process parameters. It is not useful if data
are simply collected and not compared to expected or acceptable
control ranges or process parameters in a timely fashion. The required process control ranges should be indicated directly on the
batch records. Data collected that provide process trend information or other types of data bases but are not control parameters
per se should be assembled elsewhere. This type of information
should be collected and be available to operating staff to refer to
so that they can be aware of expected findings vs. the unusual
ones. These types of data can usually be developed only after
developing some operating history. Figures 2 and 3 represent
examples of data that can be collected and used to determine whether one is operating within expected ranges.

Wherever possible, continuous chart recordings of process measurements should be made. Almost anything that can be monitored
by instruments can be charted on continuous recordings. Records
of observations by operators provide a distinct point in time data
set, but continuous recordings allow the entire process and associated parameters to be viewed and produce a history of the entire process. They also provide an excellent troubleshooting tool.
Process operators should review these recordings each time they
monitor the system to look for any deviations or unusual occur-
rences since their last visit. Depending on the process parameter
monitored, these tracings often become part of the batch record.
For example, a tracing may be the only record of a sterilization
cycle that demonstrates that the equipment has been maintained at
the appropriate temperature for the entire sterilization time.

Certain process parameters can normally only be determined
by periodic sampling of the process vessel. The types of sample
analysis done will largely be dependent on the process. An example of the types of analysis that might be done for suspension
culture process include cell density and viability determinations,
external measurements of pH (to assure that the sensor has not
drifted), product concentration determinations, process sterility

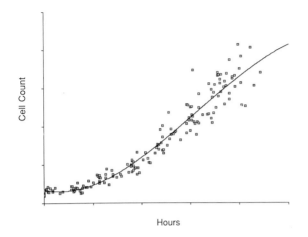

Figure 2   Cell count vs. time in culture. Graph represents 10 successive suspension culture runs under standard conditions. The plot line represents the observed average cell count vs. time of the 10 runs and the scatter plots represent the ranges in cell numbers vs. time (arbitrary units). The data collected over many runs can be used to develop databases that allow one to predict the expected cell densities over time under normal conditions.

determinations, and possibly measurements of certain nutrient concentrations or metabolic byproducts. As with on-line monitoring,
expected results should be indicated on appropriate logs or records.
    Another important aspect of process monitoring is the routine
review and trending of certain critical process parameters by appropriate technical staff and management. It is essential that the
process manager and senior staff have a thorough understanding
of process trends and the possible implications of those trends.
Armed with sufficient information and knowledge the manager should
be able to implement investigations or changes in procedures to
correct undesirable trends.

3. Process Scheduling

The methods of process scheduling will depend on the process involved. In reality scheduling activities in a process plant are more
of an art form than a science and only comes with experience.
Cell culture processes are biological systems and cannot be turned
off to await for the preparation of the next step without possibly
altering the behavior of the culture. As such it is wise to assure

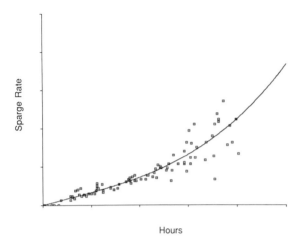

Figure 3   Sparge rate vs. time in culture. Graph represents
the corresponding air sparge rate under standard conditions of the
10 cell culture runs represented in Fig. 2.   The sparge rate is
that rate required to maintain the culture at a predetermined oxygen concentration during the course of the run.   The solid plot
line represents the observed mean sparge rate vs. time for the
10 runs and the scatter plots represent the ranges observed
throughout those runs (arbitrary units).   The data collected over
many runs can be used to develop data bases that allow one to
predict the expected sparge rates under normal conditions.

that the next step in processing is prepared well in advance of
its need.   If possible enough time should be allowed to permit the
entire system to be reset up if the need arises.   In a complex
plant or where throughout closely matches capacity, the scheduler
will normally start at the end of the process and work backward
scheduling each step along the way.
        From the perspective of scheduling, the ideal process is de-
veloped based on time for each step and it is only by time that
complex forward (or backward) schedules can be prepared.   In
reality most processes are developed based on some sort of biological event such as cell density, cell viability, or product concentration.   However, in very well-developed and precisely controlled processes these events can be predicted and controlled to
occur within very well-defined time limits.   Indeed, most product
applications to the FDA not only define the biological events but
the time range in which the events normally occur during
processing.

Ideally, sufficient hold steps should be developed along the product processing scheme to allow adjustments in schedule. In the downstream recovery/purification steps, stable hold steps can usually be found. This is seldom the case in cell culture processing. Some schemes have been employed to lengthen or shorten a process step by varying the cell seeding at that step. This usually works well as long as the seeding densities employed do not have an effect on the overall culture behavior.

Other techniques that have been employed have been to cool the process down to lower temperature to slow up cell metabolism and/or growth. Some cells can be cooled to 4°C to stop cell growth for 24−48 hr and show no effect once rewarmed. Another technique that might be used in suspension processes is to perform a modified solera where a portion of the culture is discarded and the vessel refed with fresh media to allow another 12−24 hr or so of culture time in that vessel. In all cases sufficient data need to be developed to support whatever procedures are employed.

D. Contamination Control

1. Value of Process Sampling

The amount of process sampling done is largely dependent on the process and operating history. In large-unit processes it is not uncommon to sample the sterile filtered batched media prior to introduction of cells, the cell source prior to seeding the next vessel, the newly seeded vessel, the vessel after any major procedure, operation, or additions (exclusive of pH adjustments, air sparging, etc.), the seed vessel after cell transfer, and daily sampling of the vessel while in process. Although extensive sampling of the process may not prevent a current batch from becoming contaminated, it can provide valuable information about the source and time point of the contamination.

2. Environmental Control

The level of environmental control is dependent on the process. Where cultures are openly manipulated, a much more rigid level of control is required than when cultures are contained within closed systems. Environmental monitoring should be done on a routine schedule. Procedures should be in place that describe regular cleaning and sanitization procedures, acceptable bioburden of the air and on surfaces, and action to be taken if the acceptance limits are not met. The impacted areas should always be maintained well within the acceptance limits and detailed documentation should be maintained. Environmental control trends should be maintained and investigations or modifications to procedures implemented if trends are leading toward unacceptable results.

3. Value of Prevention

It is far more desirable to prevent contamination than to attempt
to solve the problem once it has occured.   Part A of this section
provides a brief outline of methods to prevent and/or control
process failures (including contamination).   A detailed treatment
is also provided in a recent paper by Perkowski(13).

4. Troubleshooting

Every incident should undergo a complete evaluation of the cause.
By doing so, new ways will continually be found to improve procedures and plant design.   It is by this painful and time-consuming
process that plant operations can achieve 95% or better success
rates.
    When a contamination occurs the following analyses should be
undertaken (list not all-inclusive):

1. A complete review of all related maintenance, setup, and
    process records, particularly sterilization records
2. A complete review of all related sterility samples
3. A review with operators most recently involved with cri-
    tical operations on the equipment
4. An isolation and identification of the contaminating
    organism(s)
5. A review of system performance during most current opera-
    tion and complete inspection of system for any defects,
    faults, etc.
6. A review of environmental monitoring data

The goal is to find clues or commonalities that lead to the possible
source of the problem.   For example, one might be looking for information that would indicate that the equipment was not properly
prepared or not functioning properly, or possibly a mechanical
failure occurred during operation.   The timing of when the contamination occurred may lead to a specific incident or operation.
The identification of the contaminating organism(s) may provide
clues to the possible source, and a retrospective inspection of the
system(s) involved may lead to undetected mechanical or other types
of failure(s).   The quality of records, attentive staff, and attention
to the slightest detail are important to the success of the investigation.   The investigation and its findings should be documented
for future reference and appropriate corrective action taken.
    If the problem cannot be isolated to a particular cause for cer-
tain, one may choose to revalidate the system or, as a minimum,
repeat sterility trials.   Sterility trials, which normally take the form
of mock runs with bacterial medium, should be combined with an

intensified sampling regimen that has been designed to isolate each and every operation involved with the process. As the troubled system is brought back on-line one will want to do extra sampling and redouble the efforts to ensure that all procedures are performed properly.

V. CONTAINMENT AND BIOSAFETY

A. The Guidelines

The principle containment and biosafety guideline to guide manufacturers has been the NIH guidelines for handling large-scale cultures of recombinant organisms (14). Guidelines do not carry the force of law as statutory regulations. However, some reflection of federal guidelines usually become incorporated into the federal regulations. Both the FDA, EPA, and USDA have taken an active role in sorting out how the biotechnology industry will be regulated. During 1986, the White House Office of Science and Technology Policy formulated a set of written guidelines which call for assigning a single agency to approve a particular product. Where approval or oversight of a product must be performed by more than one agency, the lead agency would be appointed to coordinate those activities. Since then, these agencies have issued several *Federal Register* notices further elucidating their regulatory positions. These agencies tend to adhere very closely to what is suggested in the NIH/RAC guidelines and in many cases enforce the guidelines as regulations.

Most manufacturers propagating cell with recombinant constructs have elected to comply with the NIH guidelines. For large-scale mammalian cell culture processes, most manufacturers follow the guidelines for biosafety level 1 (large-scale) or BL1-LS for processes producing nonpathogenic material. If the process produces material that is pathogenic, then the applicable biosafety guidelines for that material would apply.

B. Containment

1. Physical Containment

Physical containment can be considered the combination of physical design and practices related to the process that minimize or prevent exposure of both the environment and personnel to the biological element. BL1-LS guidelines require that cultures of 10 liters or more be handled in some type of closed system (i.e., fermenter) or other type of primary containment equipment (i.e., biological safety cabinet) designed to reduce the potential for escape of viable organisms. Cultures of less than 10 liters do not

require any special containment features provided that the laboratory practices for BL1 (small-scale) as described by the guidelines
are followed.

Culture fluids should not be removed from a closed system or
other primary containment equipment unless the viable organisms
(cells) containing recombinant DNA molecules have been inactivated
by a validated inactivation procedure effective for that organism.
The collection of samples from and additions to closed systems and
transfers between closed systems should be done in a manner that
minimizes the release of the organism. Exhaust from such systems
should be filtered or treated in a manner that minimizes the release
or organisms to the environment. The closed systems or other type
of primary containment equipment that has contained viable organisms containing recombinant DNA molecules should be sterilized
by a validated procedure shown effective in inactivating that particular organism prior to opening the systems for maintenance or
other purposes. Emergency plans should include methods and procedures for handling large losses of culture on an emergency basis.

2. Biological Containment

Special features of mammalian cells described below provide for some
latitude when applying the NIH guidelines to large-scale cell culture. Mammalian cells have not undergone evolutionary selection
as single-celled organisms, nor are they suited to a free-living
existence in the environment. As unicellular forms they are relatively large ($7-15$ μm) and lack protective cell walls. They have
no ability to survive desiccation or freezing under environmental
conditions. Additionally, mammalian cells grown in culture possess
complex requirements for growth, reflecting the normal homeostatic
mechanisms found in the intact complex animal. Those requirements
include nutritional, hormonal, and environmental factors essential
for growth and survival. Environmental requirements include rather narrow limits for pH, oxygen concentration, temperature, os-
motic pressure, and hydrodynamic shear forces. These requirements are not met simultaneously in air, water, or terrestrial environments. Cells in low serum or serum-free conditions are exquisitely sensitive to many heavy metals, even at levels permissible
in drinking water. And cultures which become excessively diluted,
even with appropriate growth medium, fail to propagate and die.

Unlike microbial organisms, cultivated mammalian cells have
not evolved to live in the external environment. The adverse effects that the environment is likely to pose to the mammalian cell
form a complete biological containment barrier to any potential
survival of these cells in the environment.

3. Exposure of Personnel and the Environment

Many of the biological barriers presented to the mammalian cell by the environment provide protection to the worker. Also, animals including humans possess a number of protective defenses (including immune serveillance) against infection by invading foreign organisms. In general, mammalian cells are not infectious agents and a number of studies have demonstrated the difficulties of transplanting cells into hosts (15,16). Methods used to successfully transplant cells into hosts are not likely to ever be replicated in a process plant.

Another consideration is the possible effect of the product, possible pathogen (i.e., viruses), or recombinant DNA molecules on the worker. The effects of product or pathogen need to be evaluated on an individual basis. Every appropriate measure should be taken to protect the worker from known human pathogens, harmful substances, or immunogens derived from cell culture. The potential danger from recombinant DNA exposure is marginal (15,16).

The protection of the external environment follows the same line of reasoning as the protection of the work force. The permissible levels of release of mammalian cells to the environment will be dependent on the known risks. In principal, despite the perceived safety of the cells and byproducts, the manufacturer should take every reasonable step to minimize the exposure of the environment.

C. Facility and Equipment Design

Several articles have been written covering the topic of facility and equipment design from the perspective of large-scale biotechnological processes. Plant design considerations were recently detailed by Giorgio and McArthur (17,18). Two levels of containment exist. Primary containment are those design and procedures which protect the work force and immediate plant environment. Secondary containment are those designs and procedures which protect the environment and ecosystem outside the plant.

Primary designs range from the process equipment and plant design to protective clothing and vaccinations against infectious agents. Primary procedures include the operating methods and levels of inactivation prior to breaking containment. Most unit processes are designed to protect the process from the outside environment and in doing so protect the external environment from the process.

Secondary containment features protect the external environment. The two containment levels can be thought of as coinciding umbrellas of protection with the secondary containment acting as the outer shell in the event primary systems were to fail. The largest risk of significant environmental exposure is the catastrophic loss of culture from a unit process. In general, floor drains and process equipment drains within the cell culture facility should route to an inactivation/containment system prior to discharge to the sewer system. The common design of an inactivation/containment system consists of a large containment (surge) tank contained within a diked area. The containment tank should drain to a secondary inactivation system where the process wastes are inactivated prior to discharge. Several methods of inactivation are available. Probably the best method is a continuous heat activation system. In the event of a catastrophic loss, the facility floor design should allow the material to be contained within the facility proper or within the inactivation/containment system.

Other secondary containment features may include HEPA-filtered exhaust air from the facility, maintaining the facility at negative pressure with respect to the atmosphere, air locks, and change-in-and-out procedures. The degree of cotainment is dependent on the BL-LS operating level. Levels beyond BL1-LS are usually product- not cell-driven. Finally, an emergency response plan should be in place to handle large volume losses on an emergency basis. This plan should provide the basis of a response whenever the plant is in operation.

VI. SUPPORT REQUIREMENTS

A. Laboratories

Adequate quality control laboratories for the testing of product, raw materials, and the performance of in-process tests are essential. It is also valuable to have these facilities in somewhat close proximity to the process plant. This is particularly important if the process is dependent on rapid turnaround of in-process test results.

The value of "process support laboratories" within the manufacturing unit should not be underestimated. It is within these laboratories that much of the in-process testing can be performed. These laboratories should be within the immediate proximity of the process operation. The types of tests that are usually performed in a process support laboratory range anywhere from cell counting and sterility tests to product concentration analysis. These laboratories are usually manned and operated by the operations staff directly. The advantage of the support laboratory is that these facilities can give immediate response to the operations needs and are logistically positioned to provide rapid responses for in-process tests. Tests that require very sophisticated equipment, are procedurally complex, or require specially trained technicians are probably better preformed within the quality control unit. Those tests that become part of the release specifications for the final product by definition need to be performed by the quality control unit.

Functions that may be performed by a "manufacturing support laboratory" are specific raw material performance tests and certain investigative functions. The types of investigative work that might be performed by these support laboratories are in-process trouble-shooting activities and the development of data on which to develop or improve procedures and so forth. If these types of functions are established within the manufacturing unit they should be established within a laboratory separate from the actual production area. Investigative or raw material performance-testing activities should not interfere with the day-to-day production test activities nor should experiments be performed within a controlled manufacturing area.

B. Vendor Support

Solid support by vendors of equipment is essential. The manufacturer should avoid "fly-by-night" operations as one needs to be able to rely on the quality of the material and the quality of vendor support when needed. The vendors should help guide the selection of equipment to meet one's needs. When necessary they should provide installation, startup, and troubleshooting assistance. Adequate supplies of spare parts should be maintained within the vendors' warehouses and rapid response times available when problems arise. A vendor should be ready to work out equipment problems with the customer and get the troubled system back on-line quickly. Adequate technical support should be readily available. A vendor with local supply houses and technical support is best. Air travel makes distant vendor support less problematic except when help is immediately needed. Foreign vendors of equipment who do not have a regional office with technical support or supplies on the particular continent on which the plant is situated should be avoided.

Good relations with raw material suppliers is also important. Not only is it important in order to perform effective audits but one needs to understand enough of a vendor's operation to help troubleshoot if a problem arises with a particular supply. The manufacturer should always ensure that the vendor has the capacity to supply the required quantities of goods without overextending the operations. This includes assuring that the vendor is financially in good shape, labor relations are good, and the organization is well managed. It is always advisable to have more than one source of all raw materials and supplies if possible.

C. Support Groups

It is important that adequate support by qualified personnel throughout the various disciplines required be available to support plant
operations. These include maintenance and instrument personnel
with appropriate subspecialties, computer personnel where appropriate, engineering personnel in the appropriate disciplines, quality
control and assurance, and process development and validation
support.

The level of support needs to be such that no one discipline
becomes the limiting factor for plant success or throughout. Not
only must the right mix of expertise and manpower be available
but those persons must be available when needed. If the plant
operates 24 hours per day 7 days per week, maintenance and instrument crews are usually available at all times. The same may
be true for the quality control laboratories if in-process test samples are being taken during off-hours that require immediate responses.

D. Plant Technical Services Group

Most large-scale process plants have an internal technical services
group. The technical services personnel are typically responsible
for (but not limited to) the following types of activities:

1. Process and equipment troubleshooting
2. Tracking and evaluation of process trends
3. Coordinating of installation and startup of new equipment
   and systems
4. Coordination of validation activities
5. Leading the development of procedural improvements
6. Leading the development and review of standard procedures
7. Coordinating the initiation of new processes or process
   changes within the plant
8. Leading design reviews of plant expansions, etc.
9. Leading process improvement activities
10. Staffing and managing support laboratories that are not
    associated with day-to-day production testing (i.e.,
    investigational laboratories)

In these roles the technical services personnel become a prime
interface between engineering, maintenance, process development,
quality assurance, validation, and operations. The role of the
technical services group should not diminish the contact and interaction that the production staff has with other groups; in fact,
that contact is essential. However, the production staff cannot

provide the necessary level of attention and detail to the above activities if their daily primary goal is to run a manufacturing operation. The technical services group is often composed of a cadre of technical experts usually with advanced levels of formal technical training in various specialty fields. The ideal is to have personnel with a mixture of backgrounds and experience in both process science and engineering.

VII. SUMMARY

A successful manufacturing scale cell culture process has as its basis a well-designed process. However, even the best designed large-scale process can become unreliable if not managed properly. The primary element to successful management of all large-scale processes is the careful attention to the details, especially the most important ones. The second element that requires serious consideration is the operation of that process within the confines of current regulatory environments. The major considerations are:

1. Adequate raw material specifications and testing
2. Facility and equipment design that maximizes operational success and meets current regulatory requirements
3. Throroughly written procedures covering all aspects of operation and maintenance, and supported by a thorough documentation and review system
4. A dedicated staff with a thorough understanding of systems and the importance of operational details
5. Continual training and retraining programs
6. Sound process and facility validation programs
7. Comprehensive preventive maintenance programs
8. Thorough understanding of process and facility trends, and the expected and acceptable ranges of results
9. Development of sound procedures and practices that minimize batch losses and permit operation within current regulatory requirements
10. Thorough and continual monitoring of all critical process parameters
11. Process scheduling to minimize batch jeopardy or loss while maximizing throughout
12. Maintenance of the appropriate environmental conditions of the plant and general cleanliness of plant and equipment
13. Adequate batch loss prevention programs
14. Adequate troubleshooting programs
15. Adequate support organization and facilities both internally and externally to plant operations.

With the appropriate application of sound process design and management principles, and with operational experience, large-scale cell culture process plants can be operated at success rates that exceed 95%.

ACKNOWLEDGMENTS

The author gratefully acknowledges Genentech colleagues Lidwina Delahaye, Carl Johnson, and Doug Ward for their computer generation of the figures in this text.

REFERENCES

1. Arathoon, W. R. and Birch, J. R.   Large-scale cell culture in biotechnology. *Science* 232:1390−1395 (1986).
2. Purified Water, In USP, Vol. 21, U.S. Pharmacopeial Convention, Inc. Rockville, pp. 1124−1125 (1985).
3. Phillips, A. W., Ball, G. D., Fantes, K. H., Finter, N. B., and Johnston, M. D.   Experience in the cultivation of mammalian cells on the 8000 L scale. In *Large-Scale Mammalian Cell Culture* (J. Feder and W. R. Tolbert, eds.), Academic Press, Orlando, pp. 87−95 (1985).
4. Spier, R. E.   Recent developments in the large scale cultivation. In *Advances in Biochemical Engineering* Vol. 14 (A. Fiechter, ed.), Springer-Verlag, Berlin, pp. 120−162 (1980).
5. Spier, R. E.   Animal cell technology: An overview. *J. Chem. Tech. Biotechnol.* 23:304−312 (1982).
6. Zwerner, R. K., Cox, R. M., Lynn, J. D., and Acton, R. T.   Five-year perspective of the large-scale growth of mammalian cells in suspension culture. *Biotechnol. Bioeng.* 23:2717−2735 (1981).
7. J. Feder and W. R. Tolbert (eds.), *Large-scale mammalian cell culture.* Academic Press, Orlando, (1985).
8. Radlett, P. J. Pay, T. W. F., and Garland, A. J. M.   The use of BHK suspension cells for the commercial production of foot and mouth disease vaccines over a twenty year period. In *Dev. Biol. Standard.* 60:163−170 (1985).
9. Beck, C., Stiefel, H., and Stinnett, T.   Cell-culture bioreactors. *Chem. Eng.* 94:121−129 (1987).
10. Tolbert W. R. and Srigley, W. R.   Manufacture of pharmacologically active proteins by mammalian cell culture. *Biopharm. Manuf.* 1:42−48 (1987).
11. Nelson, K.   Industrial-scale mammalian cell culture. 2. Design and scale-up. *Biopharm. Manuf.* 1:34−41 (1988).
12. Annual Report, Genentech, Inc., South San Francisco (1986).

13. Perkowski, C. A. Controlling contamination in bioprocessing. *Biopharm. Manuf.* 1:62−65 (1987).
14. Guidelines for research involving recombinant DNA molecules. In *Federal Register*, Vol. 51, No. 88, U.S. Government Printing Office, Washington (May 7, 1986).
15. Levinson, A. D., Svedersky, L. P., and Palladino, Jr. Tumorigenic potential of DNA derived from mammalian cell lines. In *Proceedings−Abnormal Cells, New Products and Risk*, Vol. 6 (H. E. Hopps and J. C. Petricciani, eds.), Tissue Culture Association, Gaithersburg, MD, pp. 161−165 (1985).
16. Palladino, M. A., Levinson, A. D., Svedersky, L. P., and Obijeski, J. F. Safety issues related to the use of recombinant DNA-derived cell culture products. I. Cellular components. *Dev. Biol. Standard.* 66:13−22 (1987).
17. Giorgio, R. J. and Wu, J. J. Design of large scale containment facilities for recombinant DNA fermentations. *Trends Biotechnol.* 4:60−65 (1986).
18. McArthur, P. R. Guidelines for pharmaceutical plant design based on recombinant DNA processes. *Pharm. Eng.* 6:33−37 (1986).

16

Monitoring and Control of Animal Cell Bioreactors: Biochemical Engineering Considerations

WEI-SHOU HU and MARK G. OBERG*
University of Minnesota, Minneapolis, Minnesota

I. INTRODUCTION

A typical cell culture medium contains more than 30 species of chemicals of known concentrations plus other growth factors, hormones, and serum proteins whose concentrations are not always defined. During growth a number of metabolites are produced. The complete monitoring and control of the concentrations of all the species, metabolites, and product(s) is impractical. A more realistic approach, which is commonly taken in kinetic studies of animal cell growth in bioreactors, is to monitor the concentrations of "key" nutrients and metabolites. The compounds most often studied are glucose, lactate, glutamine, and ammonium. They are the species present in much higher concentrations in the growth medium than almost all the other nutrients or metabolites. Besides these compounds, analyses are also often made for the other amino acids in the conditioned culture medium. The measurement of amino acid concentrations is primarily for the prevention of their depletion. At times the concentrations

Present affiliation: Exxon Chemical–Baytown Olefins Plant, Baytown, Texas.

of a large number of nutrients and metabolites are measured; however, these measurements are not always clear. The metabolisms of glucose, glutamine, and other amino acids are highly interactive (1–3); a complete depiction of their interactions and the effects of different metabolic states on cell growth and production has not emerged. However, it is sometimes felt that the monitoring of these species in the bioreactors is desirable. Unfortunately, reliable sensors for on-line measurement of most of these compounds are not available and the monitoring at the present stage of biochemical engineering research is mostly done off-line. Therefore, a major roadblock to the on-line monitoring of compounds of interest to cell culture processing is in the unavailability of sensors. This area was thoroughly reviewed recently and will not be discussed in this chapter (4). An alternative to the approach of direct measurement is to use the indirect line measurement of other physical and chemical parameter. Two of the well-established and most widely used on-line sensors are pH and dissolved oxygen probes. The focus of this chapter will be on the manipulation of cell metabolism using the measurement of these two sensors. We will also discuss the unique features of monitoring and controlling pH and dissolved oxygen in cell culture processes, since the accurate measurement and control of these parameters is essential for the estimation of process variables.

II. CHARACTERISTICS OF ANIMAL CELL PROCESSES

The control of chemical and physical parameters in animal cell reactors is similar to that in microbial fermentations. The physical and chemical variables which are usually measured and manipulated on-line are temperature, pH, and dissolved oxygen concentration. Animal cell processes, despite their close resemblance to conventional microbial fermentation, have some unique features; three are probably most directly related to the development of instrumentation for on-line monitoring and control: (1) a large time constant for growth and metabolism, (2) a slow response to control action, (3) a more stringent requirement for the chemical environment. Because of these features, the control actions usually taken to control physical and chemical variables also show some variation from those for microbial processes.

   In an immobilized cell system, such as hollow fiber and ceramic module, the concentration of cells can possibly be as high as 10^8 cells/ml. In other system, the cell concentration typically varies between 10^5 and $2-3 \times 10^6$ cells/ml. With cell retention or cell recycle in a continuous flow or perfusion system, cell concentration of 10^7 cells/ml (5,6). However, even at the high concentration of 10^7 cells/ml, the dry biomass is only in the range of $3-4$ g/liter. The specific growth rate for animal cells is in the range of $0.06 - 0.01$ hr^{-1} (doubling time $12-60$ hr). Both the biomass concentration and the specific growth rate of animal cell processes are thus almost one order of magnitude lower than that of conventional microbial fermentation. The process load (the amount of control agent required to control the variables at a desired value) typically increases with both cell mass and the specific growth rate. It is thus not surprising that the process load for animal cell culture is far smaller than that for microbial fermentation. Except for continuous operations at steady state, load change is an inherent characteristic of animal cell or microbial culture. The load change has to be taken into consideration in selecting the controller and the final control elements. For a typical suspension batch culture, such as that for hybridoma cells, the cell concentration increases from 2×10^5/ml to 2×10^6/ml. In a microcarrier culture, it may increase from 4×10^5/ml to as low as 1×10^6/ml, or to as high as 7×10^6/ml. Thus, the load change for a batch cell culture process is in the range or in the lower side of that for microbial processes. Because of the relatively large time constant for growth, the rate of load change is slower than that for conventional microbial fermentation. Therefore, from a viewpoint of system load, the control of physical and chemical variables in a animal cell reactor is similar to that for microbial fermentation.

Once a control action is taken it usually takes some time for the controlled variable to reach a new set point or to return it to the set point from the deviation. Such a delay is affected by the capacitance of the system. For instance, a higher buffer capacity of the culture broth requires a larger amount of control agent to adjust the pH. A variety of buffering agents are used in animal cell culture medium, such as $NaHCO_3$ and HEPES. The concentration used is often in the range of $15-40$ mM. Besides the buffering agent a variety of other compounds present in the basal medium or also provide buffers to pH change. The buffer capacity of media used for microbial fermentation varies widely. Nevertheless, it is mostly in the same order of magnitude, ranging from $10-100$ mM, as that for animal cell medium.

Another factor affecting the slowness or promptness of a system to respond to control action is the resistance. The transfer of gaseous species to cell culture fluid is effected through the gas-liquid interface area in the liquid top through the gas bubbles as in the case of direct sparging, or through a membrane device such as silicone rubber tubing or polypropylene tubing. The oxygen transfer coefficient for cell culture is only in the vicinity of 0.5 hr^{-1} as opposed to that of 100 hr^{-1} for microbial fermentations. The response time to dissolve oxygen control action in animal cell culture will certainly be slower. A similar delay in the response to pH control action also occurs in the case of sodium bicarbonate serving as pH buffer and the pH being controlled by manipulating the CO_2 concentration in the gas phase.

Animal cell culture medium contains proteins ranging from 7 g/ liter in medium supplemented by 10% serum to 0.2 g/liter in some serum-free medium. At a high concentration of protein, foam inevitably occurs when the culture is sparged with air at a high air flow rate. Since the use of large amounts of antifoaming agent is generally undesirable, a higher air flow rate is avoided to prevent possible foaming problem. In order to maintain the dissolved oxygen concentration above a critical level, oxygen-enriched air is sometimes used. Another important aspect of the cell's growth environment is the relatively narrow range of osmolality, approximately from 280 mOsm/kg to 320 mOsm/kg (7). Because of the narrow tolerance to osmolality change, the pH buffer capacity cannot be increased easily. Furthermore, when different methods of maintaining a process variable constant are available, the one which affects the osmolality least is often preferred.

III. MONITORING AND CONTROL OF pH AND DISSOLVED OXYGEN CONCENTRATION

In controlling a process measurement, lag always occurs. This can possibly be caused by the time required for the controlling agent to complete its effect. In a stirred tank for animal cell cultivation, the agitation rate used is low. The mixing time is thus relatively long compared to that of the microbial processes. A significant lag is common for complete mixing and for the manipulated variable to reach a new steady-state value in the reactor. Furthermore, the response of a measuring device to a change in the measured variable is never instantaneous. For the measurement of pH by pH electrode a 90% response takes only a few seconds; however, the 90% response of a dissolved oxygen electrode to a step change of dissolved oxygen concentration from 0 to 7 ppm can take as long as 1−3 min. The sluggish response of the measuring device is of particular concern when the rate of change of the variable is our interest. One example in which the rate of change is to be measured is the on-line oxygen uptake rate measurement as will be discussed later in this chapter.

A. Control of Dissolved Oxygen Concentration

Two types of steam-sterilizable dissolved oxygen electrodes are commonly used: galvanic and polarographic. The reactions taking place at the anode (Pb) and cathode (Ag) of a galvanic probe are as follows:

Anode reaction: $\quad H_2O + \frac{1}{2}O_2 + 2e^- \longrightarrow 2OH^-$

Cathode reaction: $\quad Pb \longrightarrow Pb^{2+} + 2e^-$

The Pb^{2+} generated at the anode precipitates as $Pb(OH)_2$ and deposits on the anode surface. This results in a decreasing signal for a constant dissolved oxygen concentration as more and more of the anode corrodes.

In a polarographic Clark-type electrode, oxygen diffused through the steel mesh-reinforced Si/Teflon membrane is reduced at the cathode. The reactions which take place at the cathode (Pt) and anode (Ag) are as follows:

Anode reaction: $\quad 4Ag + 4Cl^- \longrightarrow 4AgCl + 4e^-$

Cathode reaction: $\quad O_2 + 2H_2O + 4e^- \longrightarrow 4OH^-$

The accuracy of the dissolved oxygen measurement depends on the linearity of the probes response at the particular polarization potential used. In our measurement using 675 mV, the probes response is accurate to within 0.1% over the dissolved oxygen concentration range of 0 up to saturation with pure oxygen (at 760 mm Hg). Linearity makes the electrode easy to calibrate over its entire range of zero to some known reference concentration. For practical purposes the reference concentration is usually saturation with room air. The stability of the probes output can be effected by drifts in the probes calibration, aging electrolyte, and changes in the Teflon membrane. In our experience with a good polargraphic dissolved oxygen electrode, the probes signal usually decreased by approximately 0.5%/week after a 2- or 3-week culture.

The ability to respond quickly to changes in the dissolved oxygen concentration is an essential property of any good dissolved oxygen electrode. A slow response time leads to dynamic errors in dissolved oxygen control, $k_L a$ determinations, and oxygen uptake rate (OUR) measurements. The two principle properties of the probe which effect its response time are the thickness of its membrane and the diffusivity of oxygen through the membrane. The response time is related to these properties by

$$\text{Response time} = K\frac{X^2}{D}$$

where X is the thickness of the membrane, D is the diffusivity of oxygen through the membrane, and K is a constant. Thus, thin membranes that are also highly permeable to oxygen are desired in order to minimize the response time of the probe.

Various methods are used to supply oxygen to the stirred-tank reactor, including direct sparging, surface aeration, and the use of silicone rubber tubing (8). The major resistance to oxygen transfer resides in the gas-liquid interface or in the gas diffusion through the silicone tubing. The rate of transfer is governed by volumetric transfer coefficient $k_L a$ and the driving force $(C^* - C)$. Whether a dissolved oxygen is suitable for a control or measurement task is dependent on the relative magnitude of response time between the probe and the process. The response time of the probe to a change in the dissolved oxygen concentration can be neglected if it is much faster than the response time of the fermenter to a change in the equilibrium dissolved oxygen concentration (C^*). When a step change in the concentration of oxygen is introduced to the gas phase, a mass balance on oxygen in the fermenter gives

$$\frac{dC}{dt} = k_L a(C^* - C)$$

where C is the dissolved oxygen concentration and C^* is the equilibrium dissolved oxygen concentration. If this equation is solved for the case where $C = 0$ at $t = 0$ (for the case that initially the dissolved oxygen concentration is zero), the result is

$$C = C^*[1 - \exp(k_L at)]$$

The time constant for oxygen transfer is the fermenter is simply the reciprocal of the volumetric mass transfer coefficient for oxygen. Thus, when $t = 1/k_L a$, the dissolved oxygen level is 63% of saturation after inducing a step change in the equilibrium dissolved oxygen concentration (C^*) from 0 to saturation with air. Despite the complexity of oxygen diffusion from bulk liquid through the boundary layer and the membrane to the anode surface, the response of a polarographic probe to a step change in the dissolved oxygen often resembles first-order kinetics. The time constant for the probe (t_p) is thus the time it takes the probe to reach 63% of its final output after being transferred from a nitrogen-sparged, oxygen-depleted vessel to one saturated with air.

The consumption of oxygen or other nutrients by mammalian cells in vitro is generally assumed, and in one case was found (9), to follow Michaelis-Menton-type kinetics. At a high concentration, the uptake of oxygen by mammalian cells is practically zero order. Under such conditions the rate of oxygen uptake is

$$\frac{dC}{dt} = q_o x$$

where q_0 is the specific oxygen consumption rate and x is cell concentration. The time constant in this case is $C^*/q_0 x$. It is cell concentration-dependent.

In a typical cell culture stirred-tank reactor, the time constant for oxygen transfer is $20-40$ min. (7). As a first approximation, the oxygen consumption rate can be assumed to be 0.1 mmol/liter-hr at 10^6 cells/ml and 1 mmol/liter-hr at 10^7 cells/ml (7). The solubility of oxygen in cell culture medium at 37°C is estimated to be 0.18 mmol/liter (10). Thus the time constant for a polarographic probe is in the range of 10 sec. The use of a polarographic probe for rate measurement of either oxygen transfer in fermenter or oxygen consumption by cells poses no serious error due to probe measurement lag.

The measurement of dissolved oxygen is achieved by measuring the rate of reaction involving oxygen in the electrode. The consumption of oxygen by the electrode results in a gradient of oxygen concentration from the bulk liquid to the surface of cathode. In a stagnant solution the dissolved oxygen is replensihed only through diffusion while in agitated solutions the oxygen is replenished not only by diffusion but by convection as well. Thus, the rate of replenishment depends on the degree of agitation. In general, the calibration of probe should be performed under operating conditions. Ideally, one would also like to select a probe and position it in a suitable location in the reactor such that the effect of agitation on its response is minimized. However, for the cultivation of animal cells in stirred-tank reactors, it is not uncommon to have the agitation rate vary with cultivation time. Figure 1 shows the steady-state response of a polarographic probe in a 1-liter cell culture fermenter under different agitation conditions. In the operating range of agitation rate for cell culture ($30-100$ rpm), only a 2% difference in probe output was observed.

In the measurement of rate of change of dissolved oxygen concentration, care should also be taken to avoid errors caused by oxygen consumption by the electrode. Such an error can be significant if the volume of the fluid to be measured is small and the time required for measurement is long. To estimate the consumption rate of oxygen by the dissolved oxygen electrode, one can monitor the dissolved oxygen level as a function of time in a perfectly sealed, agitated vessel which was completely filled with water. The magnitude of the consumption rate of oxygen by the dissolved oxygen electrode varies widely depending on the construction of the electrode. In general, a rate in the order of 0.1 μmol/hr is satisfactory, which is three orders of magnitude lower than the consumption rate of oxygen by cultured cell at 10^9 cells/ml.

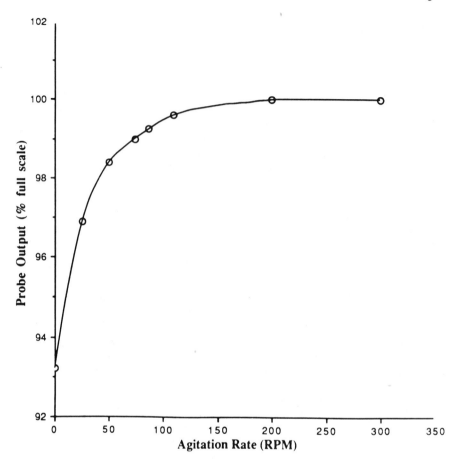

Figure 1   Effect of agitation on the output of dissolved oxygen electrode.

        As mentioned earlier in this chapter, the transfer of oxygen in cell culture bioreactors is achieved by direct gas-liquid contact (sparging, surface aeration) and/or aeration through a membrane device.   Usually the air flow rate used is low and the resistance across the membrane or the gas-liquid interface is large.   In other words, the volumetric oxygen transfer coefficient (k_La) is small. Oxygen-enriched air is often used to supply sufficient oxygen for culture.   The capacitance for oxygen in cell culture fluid is very small. A small increase in the partial pressure of oxygen can result in a large fluctuation in dissolved oxygen concentration. A two-position (on-off) controller seldom gives a satisfactory control. Since a high dissolved oxygen concentration may potentially be growth-inhibitory (11) or have other negative effects (12), over-shoot should always be avoided. A simple PID (proportional-integral-derivative) control strategy can be used to maintain the dissolved oxygen level at a particular set point. In our 1-liter fermenter, air is normally passed through the silicone rubber tubing at a flow rate of 500 cm^3/min and vented into the headspace. It was periodically replaced with a pulse of either nitrogen or oxygen at a flow rate of 1500 cm^3/min. Which gas was pulsed into the fermenter, and for how long, was determined by the PID controller as illustrated in the following equation:

$$\text{Flow rate} = K_c \varepsilon(t) + \frac{K_c}{\tau_I} \int_0^t \varepsilon(t)dt + K_c \tau_D \frac{d\varepsilon(t)}{dt}$$

where K_c is the gain, τ_I is the integral time constant, τ_D is the derivative time constant, and $\varepsilon(t)$ is the error in the dissolved oxygen concentration from its set point. Controller tuning was achieved through a trial-and-error process after using the empirical Cohen-Coon correlations to obtain a set of initial values for K_c, τ_I, and τ_D. When the flow rate was positive, oxygen was pulsed into the fermenter, and when the flow rate was negative, nitrogen was pulsed into the fermenter. The length of each pulse was determined by the magnitude of flow rate, with the maximum time for any pulse being 20 sec.

The Cohen-Coon method of controller tuning is based on the characteristic sigmoidal response of an open-loop real process to a step change in the input. An example is shown in Fig. 2. The sigmoidal response curve was obtained by pulsing oxygen into the 1-liter fermenter continuously. The three required parameters are obtained from this sigmoidal response and form the basis of the Cohen-Coon correlations which are shown in part B of Fig. 2.

The values for K_c, τ_I, and τ_D obtained from the Cohen-Coon correlations, as well as the final working values which were found to provide the best control, are shown in Table 1. As these two sets of values demonstrate, the Cohen-Coon correlations provided a good initial guess for K_c, τ_I, and τ_D. The PID controller was able to maintain the dissolved oxygen level in the fermenter to within 0.2% of saturation with air of the set point (40% of saturation with air). Figure 3 demonstrates the performance of the PID controller in responding to a step change in the set point.

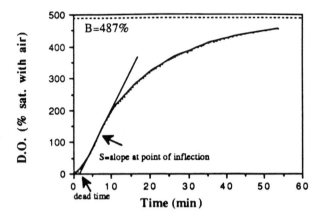

Required parameters: $K = \dfrac{B}{A}$ $\tau = \dfrac{B}{S}$ t_d = dead time

B = output at steady state (478% saturation with air)

A = input at steady state (1500 cm³/min)

(a)

$$K_c = \frac{1}{K}\frac{\tau}{t_d}\left[\frac{4}{3} + \frac{t_d}{4\tau}\right]$$

$$\tau_I = t_d\left[\frac{32 + 6t_d/\tau}{13 + 8t_d/\tau}\right]$$

$$\tau_D = t_d\left[\frac{4}{11 + 2t_d/\tau}\right]$$

(b)

Figure 2   Cohen-Coon correlations for controller tuning.   (a) Sigmoidal response of dissolved oxygen level to a continuous pulse of
oxygen.   The three required parameters for the Cohen-Coon tuning.
The dissolved oxygen level was monitored as a function of time after
first removing all of the oxygen in the fermenter by sparging with
nitrogen and then injecting a continuous pulse of oxygen at time
t = 0.   (b) The Cohen-Coon correlations for PID control.

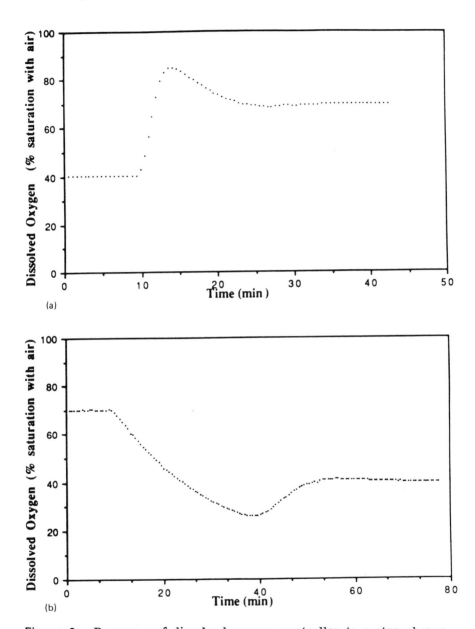

Figure 3 Response of dissolved oxygen controller to a step change in the set point. (a) Step change in the set point from 40 to 70% of saturation with air. (b) Step change in the set point from 70 to 40% of saturation with air.

Table 1   Comparison of Cohen-Coon Tuning and the
Final Controller Values

Variable	Cohen-Coon	Final Value
K_c (cm³/min-% satn.)	52.9	8.0
τ_I (min)	3.6	8.0
τ_D (min)	0.54	1.0

B. pH Control in Cell Culture

Most cell culture media contain $NaHCO_3$ as pH buffer.   The concentration can be as high as 44 mM as in Dulbecco's Modified Eagle's
Medium (DMEM).   Using bicarbonate-buffered medium, cells must
be incubated with a gas mixture containing CO_2. CO_2 dissolves
in medium to act as acid in the buffering system.

$$CO_{2(g)} \rightleftharpoons CO_{2(aq)}$$

The equilibrium relationship is described by Henry's law:

$$P_{CO_2} = H[CO_{2(aq)}]$$

where H is Henry's law constant.   The dissolved CO_2 establishes
equilibrium quickly with biocarbonate:

$$CO_{2(aq)} + H_2O \rightleftharpoons HCO_3^- + H^+$$

$$K_a = \frac{[HCO_3^-][H^+]}{[CO_{2(aq)}]}$$

Thus,

$$pH = pk_a + \log \frac{[HCO_3^-]}{[CO_{2(aq)}]}$$

The pk_a for dissolved CO_2 (or $H_2CO_3^-$) in medium is approximately
6.2−6.3.   It is obvious that the pH of bicarbonate-buffered medium
is affected by the ratio of $[HCO_3^-]/[CO_{2(aq)}]$.   To maintain an
optimal pH in the medium, the concentration of CO_2 in the gas phase
has to be controlled.   The most effective range for any buffer is in
the region where the ratio of the concentrations of salt to acid, in this

case the ratio of $[HCO_3^-]$ to $[CO_{2(aq)}]$, is close to unity. In other words, the most effective region is where pH is close to pk_a. Obviously, bicarbonate is not most effective in the optimum range of pH (6.9–7.3) for cell growth. As cells grow, the production of lactic acid inevitably causes pH to decrease. Two approaches can possibly be taken to control pH at the set point: (1) maintaining a constant CO_2 partial pressure in the gas phase and adding a base solution such as sodium bicarbonate and, (2) using a gas stream to strip CO_2 off the medium and to decrease the dissolved CO_2 concentration. Using the first approach, the concentrations of both $CO_{2(aq)}$ and HCO_3^- were to be kept constant, while in the second approach, the concentrations of both $CO_{2(aq)}$ and HCO_3^- are decreased but their ratio is to be maintained constant. Both approaches are commonly practiced. However, in selecting a pH control mechanism, the effect on osmolality should not be neglected. As mentioned above, the total concentrations of both $CO_{2(aq)}$ and HCO_3^- can be as high as 40 mM in cell culture medium. For some cell lines the optimum range of osmolality for growth spans only 40–50 mOsm/kg. Maintaining the partial pressure of CO_2 and the concentration of HCO_3^- at constant levels by base addition results in controlling the pH at the expense of increasing the osmolality. Thus, between the two approaches the latter one appears to be favorable. The drawback of the latter approach is a longer lag in system response. The direct addition of base to the culture medium results in a relatively fast response of the pH despite the relatively long mixing time for the cell culture reactor. Due to the low gas flow rate and the small gas-liquid mass transfer coefficient, the response to the addition of CO_2 in the gas phase is slower. However, in most cases the rate of response to pH control action is not a major concern in cell culture processing.

Many media developed in the last decade do not contain sodium bicarbonate; instead, other organic compounds are used as pH buffer. These buffers often have pk_a values closer to neutral pH and provide better buffering capacity. The concentration of buffering agent required is, of course, dependent on the pk_a and the process requirement. The most important chemical species in cell culture whose concentration change affects pH is lactic acid produced by cells. The lactic acid production rate of animal cells cultivated in DMEM with 4 g/liter of glucose is in the range of $5-20 \times 10^{-10}$ mmol/cell-hr. At a cell concentration of 10^6/ml, the lactic acid production rate can be as high as 2 mmol/liter-hr. If the pH is to be maintained at the pka of a buffering agent, then the initial ratio of [salt]/[acid] is 1. Using a concentration of 20 mM of lactic acid production causes the ratio to change from 1 to 8/12 = 0.67. This results in a pH change of 0.17 unit. Such a rate of change can be easily corrected by a simple pH controller. The larger the

capacitance, or buffer capacity in the case of pH control, the easier
it is to maintain the controlled variable at the set pointing in the
course of load change. However, the alternative buffers available
for cell culture are often substantially more expensive than bicarbonate. Thus, it may be advantageous in reducing the concentration of buffer and resort the task of maintaining pH to a controller.
As illustrated above, in general when a buffering agent with a pk_a
close to pH set point is to be used, a concentration of 20 mM is more
than sufficient.

A caution which should be taken in using an alternative buffering agent is the requirement of CO_2 for cell growth. CO_2 is required for a number of biosynthetic reactions. At a low cell concentration exogenously added bicarbonate (or CO_2 in the atmosphere)
is required for cell growth (13). Cells grown in bicarbonate-free
medium are often incubated in $1-2\%$ CO_2 atmosphere (14). The
physiological bicarbonate concentration is 25 mM, which equilibrates
with 5% CO_2 at pH 7.2. At times the CO_2 requirement may not be
obvious because that generated by cell metabolism alone is sufficient for cell growth. However, under conditions in which continuous air flow is used to supply O_2, regardless if it is aeration
through headspace, direct sparging, or silicone rubber tubing, the
possibility of stripping off the CO_2 required for cell growth cannot
be ignored. Shown in Fig. 4 is an example of a possible CO_2
requirement for cell growth. Vero cells were cultivated on microcarriers using either DMEM buffered with bicarbonate or DMEM
buffered with tricine (5 mM) with added NaCl to balance the osmolality. Tricine, which has a pk_a of 8.1 and possesses a modest
buffering capacity over the pH range of $7.4-8.8$, exhibits no negative effects on the growth of cells. Vero cells attach to Cytodex
1 microcarriers in both bicarbonate-buffered medium and in tricine-
buffered medium. The gas used for aeration was air/5% CO_2 mixture for the bicarbonate-buffered culture and air only for the
tricine-buffered culture. The CO_2 content of ambient air is only
0.02%. Continuous aeration has the effect of stripping off the
CO_2 produced by cells in excess of that which is in equilibrium with
the 0.02% in air. Vero cells grew normally in the bicarbonate-
buffered culture but were unable to reproduce in the tricine-buffered
medium. But once the air flow through the fermenter was turned
off, presumably allowing the CO_2 produced by the Vero cells to
accumulate, they were able to reproduce and grow normally.

The critical concentration of CO_2 which is required for cell
growth is not well defined. In the cloning of animal cells $1-2\%$
of CO_2 in air is often used. To ensure that cell growth is not
retarded by CO_2 depletion, it is advisable to use a medium with a
low concentration of $NaHCO_3$ and to aerate the culture with a gas
stream which has a constant partial pressure of CO_2. The presence
of $NaHCO_3$ also provides the buffer capacity needed at the initiation

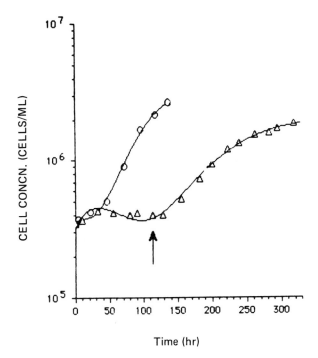

Figure 4 Effect of dissolved CO_2 concentration on cell growth. Both cultures were conducted in the 1-liter fermenter with 5 g/liter of Cytodex 1. Vero cells were harvested from three confluent 150 cm^2 t flasks for each culture and innoculated at a concentration of 3.5 × 10^5 cells/ml. The air used for the bicarbonate-buffered culture (○) contained 5% CO_2, whereas the gases used for the tricine buffered culture (△) did not contain CO_2. At (↑) air flow through the tricine-buffered culture was turned off to allow for the accumulation of CO_2.

of the culture. Since a low concentration of NaHCO3 is used, the subsequent lactic acid production during the cultivation period has to be neutralized by the addition of base.

IV. RATE ESTIMATION FROM ON-LINE MEASUREMENTS

The previous section discusses the control of chemical variables in the cultured environment at the desired value. The ultimate goal of environmental control is to estimate the state of the system (cell concentration, growth stage, etc.) and to manipulate cell growth or metabolism. To achieve this it is necessary to use on-line measurements to estimate various rates. The rates estimated from on-line measurements are volumetric (Q) as opposed to specific (q):

$$Q = qx = \frac{\mu x}{Y}$$

The volumetric rate is the product of specific rate (based on unit cell mass) and cell concentration. Conventional models for microbial growth describe q as a function of specific growth rate (μ), and the yield coefficient (Y) is not necessarily constant. Even under the same specific growth rate (μ), it may vary with the chemical environment. An example is the consumption of glucose by hybridoma cells. The specific consumption is high at a high glucose concentration below 0.1 g/liter (15). Therefore, the volumetric rate is affected by both the rate of change in cell mass (μx) and the metabolic rate (q). Under conditions in which the specific rate of relatively constant, the volumetric rate can be used to infer the rate of increase in biomass. Likewise, under conditions in which biomass is relatively constant (e.g., anchorage-dependent cells in their confluent stage), the volumetric rate is an indication of the cell's metabolic state.

Oxygen consumption rate and lactic acid production rate are the two primary candidates for on-line rate estimation in cell culture. Both oxygen consumption and lactic acid production are related to cell energy metabolism and occur at higher molar rates than other compounds in the culture medium. Both has also been used to correlate to cell growth in bioreactors (7).

A. Lactic Acid Production Rate

A large fraction of glucose consumed by mammalian cells is converted to lactic acid. The ratio of the amount of lactic acid produced to that of glucose consumed ranges from 0.5 to 1.0 on a weight basis. To maintain constant pH in the culture, a stoichiometric amount of base is used to neutralize the lactic acid. This is the basis for lactic acid production rate measurement. The amount of base added at any time can be measured by a load cell. One source of error of such an estimation usually comes from sodium bicarbonate buffer or the bicarbonic acid equilibrated with CO_2 in the aeration gas. In the presence of gaseous CO_2 or bicarbonate, the amount of acid present is strongly affected by CO_2 partial pressure. Obviously when base addition is to be used as a means of estimating lactic acid production, pH of the culture should not be controlled

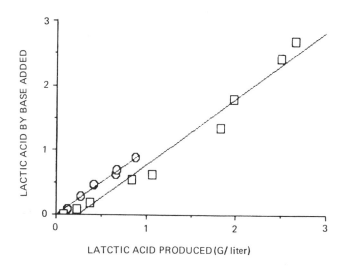

Figure 5 Calculating acid production by measuring base addition using bicarbonate-buffered medium. The actual lactate production of this culture (as determined by enzymatic assay) ($\square$) is compared to the amount of acid production calculated from the addition of base to the fermenter ($\circ$).

by stripping CO_2 off the medium; rather, the partial pressure of CO_2 should be maintained constant to avoid the variation of the capacitance of the system. Caution should also be taken when culture fluid is to be replenished or perfused. The fluid introduced into a reactor should also have a known carbonic acid concentration. In other words, all fluid introduced into the reactor should have been equilibrated with a gas with known CO_2 partial pressure. This is not always a simple task. Thus, it is desirable to use an alternative buffer for pH control. Figure 5 shows the comparison of lactic acid production measured by base addition and that measured by lactate dehydrogenase assay. Data were obtained from two microcarrier cultures of Vero cells using either sodium bicarbonate-buffered (44 mM) or tricine-buffered (5.0 mM) DMEM. In the case of bicarbonate-buffered culture, the calculated amount of acid produced shows a shift from the actual lactate production (as determined by enzymatic assay) ($\square$). There was, however, a reasonable agreement in the incremental increases in both the calculated and the actual acid production data. The accuracy was improved when the bicarbonate-CO_2 buffering system was replaced with tricine.

There was very good agreement between the calculated amount of acid produced and the actual lactate production data. The data presented in Fig. 5 also provide evidence that lactate is the major acidic species (besides CO_2, which is stripped away) produced by the Vero cells.

B. Oxygen Uptake Rate Measurement by Off-Gas Analysis

Analysis of the fractional composition of gases in the inlet and outlet has been used extensively in microbial fermentation. At a quasi-steady state (the concentration of dissolved oxygen of CO_2 in the culture fluid is not changing appreciably in the time period of measurement), the difference in molar oxygen or CO_2 flow rate between inlet and outlet is an estimation of the molar rate of oxygen uptake or CO_2 production.

The volumetric rates of oxygen uptake and CO_2 production in animal cell reactors are two to three orders of magnitude lower than those for microbial fermentation. However, the air flow rate used for animal cell culture, although relatively small, is often not three orders of magnitude lower than that for microbial fermentation. The difference of molar oxygen flow rate between inlet and outlet is therefore very small, rendering accurate measurement difficult. This is especially true for a low cell concentration and a relatively small fermenter.

Another complication in the measurement of exit gas composition is the relatively long holding time of gas stream in the headspace. The smaller the gas flow rate is, the larger the measurable difference between inlet and outlet concentrations can be. This improves the sensitivity of off-gas measurement; however, at a very small gas flow rate, e.g., 0.01 vvm (volume air/volume liquid-min), the holding time in the headspace can possibly render the interpretation of the measurement difficult and poses special problems when the dynamics of the system response to a manipulated variable are to be examined.

C. Oxygen Uptake Rate Measurement by Dynamic Method

An alternative to the off-gas measurement relies on the measurement of the rate of change in the liquid phase. This method entails inducing a transient wherein a gas of known composition is circulated through the fermenter for a certain period of time during which the dissolved oxygen level of the culture is monitored. A mass balance for oxygen in a culture gives

$$\frac{dC}{dt} = k_L a(C^* - C) - OUR$$

The oxygen uptake rate (OUR) of the culture can be solved as

$$OUR = \frac{(C_o - C_f)}{(t_f - t_o)} + \frac{\int_{t_o}^{t_f} k_L a[C^* - C(t)]dt}{(t_f - t_o)}$$

where C_o is the initial dissolved oxygen concentration, C_f the final dissolved oxygen concentration, t_o the initial time, t_f the final time. The trapezoidal method of numerical integration can be used to evaluate the above integral using an on-line microcomputer.

In our 1-liter fermenter measurement was normally carried out using nitrogen gas. Therefore, C^* reduces to zero in the above equation. The nitrogen gas was first used to flush the fermenter of all residual oxygen for a 3-min time period at a flow rate of 1500 cm^3/min. This was immediately followed by a 3-min time period during which the dissolved oxygen level was monitored and the OUR was computed. The flow rate of the nitrogen gas during this time period was also 1500 cm^3/min.

In solving the above equation for OUR, one assumes that the $k_L a$ is known. Without biomass in the reactor, the response of the system to a change in the gas composition can be described as

$$\frac{dC}{dt} = k_L a(C^* - C)$$

The volumetric mass transfer coefficient ($k_L a$) can be determined by aerating nitrogen gas ($C^* - C$) through the fermenter while monitoring the dissolved oxygen level as a function of time. A plot of $\ln(C_o/C)$ vs. time yields a straight line with a slope equal to $k_L a$. The $k_L a$ thus obtained is an overall mass transfer coefficient including mass transfer contributed by aeration through the surface, the silicone rubber tubing, and sparging. An accurate estimation of $k_L a$ affects the accuracy of OUR measurement. If $k_L a$ varies with time during cell growth, the variation should also be known.

The two principal parameters which affect $k_L a$ during a batch culture are agitation rate and the flow of gas in the fermenter. If the agitation rate and gas flow rate change in a batch culture, as in most cell culture operations, their effect on $k_L a$ should be examined. If necessary, the magnitude of $k_L a$ can also be estimated on line.

The two consecutive OUR measurements were performed with two different gas mixtures. Thus, assuming OUR doesn't change significantly during these two measurements, we have:

$$\text{OUR} = \left[\frac{C_o - C_f}{t_f - t_o}\right]_{\text{gas 1}} + k_L a \left[\frac{\displaystyle\int_{t_o}^{t_f} [C^* - C(t)]dt}{t_f - t_o}\right]_{\text{gas 1}}$$

$$\text{OUR} = \left[\frac{C_o - C_f}{t_f - t_o}\right]_{\text{gas 2}} + k_L a \left[\frac{\displaystyle\int_{t_o}^{t_f} [C^* - C(t)]dt}{t_f - t_o}\right]_{\text{gas 2}}$$

Solving for $k_L a$ gives

$$k_L a = \frac{\left[\dfrac{C_o - C_f}{t_f - t_o}\right]_{\text{gas 1}} - \left[\dfrac{C_o - C_f}{t_f - t_o}\right]_{\text{gas 2}}}{\left[\dfrac{\displaystyle\int_{t_o}^{t_f} [C^* - C(t)]dt}{t_f - t_o}\right]_{\text{gas 2}} - \left[\dfrac{\displaystyle\int_{t_o}^{t_f} [C^* - C(t)]dt}{t_f - t_o}\right]_{\text{gas 1}}}$$

In our 1-liter fermenter equipped with 8 ft of silastic tubing and agitated at 40 rpm with a marine impeller, the overall oxygen transfer coefficient (surface aeration and silastic tubing combined) is 3.42 hr^{-1}. The individual contribution from the surface and the silastic tubing are 1.69 and 1.73 hr^{-1}, respectively. Several consecutive $k_L a$ measurements were taken at an agitation rate of 40 rpm and a gas flow rate of 1500 cm^3/min to determine the reproducibility of the measurement. All measurements taken were found to be within 5% of the average value of 3.42 hr^{-1}.

Using the method described above, we measured OUR of Vero cells cultivated on Cytodex 1 microcarriers at a concentration of 5 g/liter. The OUR measurements were conducted during a 6-min time interval. The first 3 min were used to flush the headspace and tubing of the fermenter of all residual oxygen by injecting nitrogen gas at a flow rate of 1500 cm^3/min. The last 3 min were used to monitor the dissolved oxygen level as a function of time (approximately once every 8 sec) while continuing to circulate nitrogen gas through the fermenter at a rate of 1500 cm^3/min. The OUR was computed from the data obtained during these last 3 min. In the range of dissolved oxygen levels studied (35–70% saturation with

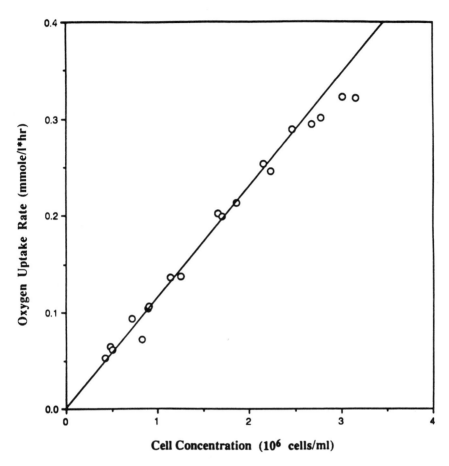

Figure 6   Effect of cell concentration on OUR.   The OUR of a
Vero/Cytodex 1 culture is shown as a function of cell concentration.
The data for this figure were obtained from three fermentation cultures.   All OUR measurements were taken in the presence of lactate
concentrations below 1 g/liter and glucose concentrations above
1 g/liter.

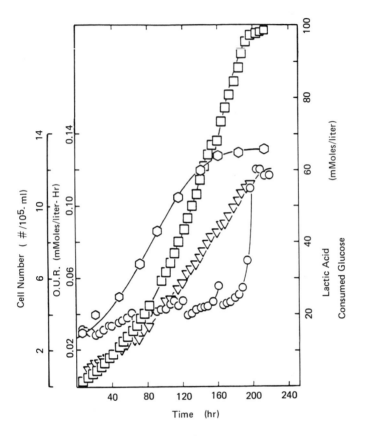

Figure 7 Growth kinetics of FS-4 cells on microcarriers. OUR
was measured on-line using the dynamic method. Lactic acid production was estimated by base addition. (◯) cell number;
(▽) glucose; (○) OUR; (□) lactate.

air), OUR doesn't appear to be affected by the dissolved oxygen
concentration. The OUR increased linearly with cell concentration
over a wide range of cell concentrations encountered in three cultures from which these data points were obtained (Fig. 6). The
specific oxygen consumption rate for Vero cells is 1.1×10^{-10}
mmol/cell-hr cells. The decrease in the growth rate at high cell
concentrations does not appear to change the OUR to any large extent. In these experiments all OUR measurements were taken in
the presence of lactate concentrations below 1 g/liter and glucose

concentrations above 1 g/liter. This was accomplished by measuring the OUR shortly after a medium change. The medium was changed daily for all cultures conducted in the 1-liter fermenter. Thus, for Vero cells cultivated on microcarriers, the oxygen consumption rate can provide a good estimate of cell concentration under defined cultivation conditions. Similar experiments were also previously performed by Fleischaker and Sinskey using FS-4 cells (Fig. 7) (16). In this case both oxygen consumption rate and cumulative lactic acid production parallel cell growth during the exponential growth, but the parallel relationship does not hold after reaching confluence.

V. MONITORING AND MANIPULATION OF CELL METABOLISM IN BIOREACTORS

Animal cells in culture convert the majority of glucose they consume to lactate. With an initial glucose concentration of approximately 4 g/liter, such as that in DMEM, almost 3 g/liter of lactate accumulated at the end of a batch culture (2,7). It is obvious that even if no other nutrient is limiting, the improvement of medium utilization efficiency cannot be achieved by merely increasing the glucose concentration; the resulting increased lactate level will be inhibitory. Thus, a necessary condition for improving medium utilization efficiency is to reduce the accumulation of lactate in the medium. It was demonstrated that in a batch culture of hybridoma cells, the fraction conversion of glucose to lactate as well as the specific consumption rate of glucose could be reduced by maintaining glucose concentration at a level below 0.1 g/liter by programmed feeding of glucose (15). However, in that study glucose was fed merely at an exponential rate. Thus, a constant level of glucose could be maintained only in a period in which both growth rate and the yield coefficient based on glucose are constant. A change in growth rate or metabolism will almost certainly result in deviation of the glucose concentration from the desired level. For a programmed feeding strategy to succeed one needs both a model for developing a feeding scheme and an on-line estimation of metabolic rate for a feedback control. The direct on-line measurement of glucose at a concentration below 0.1 g/liter is not easy. The possibility of using an on-line rate estimation for monitoring cell metabolism thus warrants examination. Following are descriptions of some experiments performed in our laboratory.

To eliminate the effect of changing cell concentration on the OUR, confluent cultures of Vero cells with a constant cell concentration were used for all the studies to be described below. These cells were cultivated in a 1-liter fermenter. After reaching confluence, they were washed once with medium (DMEM supplemented with 10% horse serum but containing no glucose) and then resuspended in medium, transferred to different spinner flasks. Various concentrations of glucose were used in different flasks. The medium was replenished in all flasks as soon as the one with the lowest concentration of glucose decreased almost to depletion. The glucose and lactate concentrations were measured. The oxygen consumption rate was measured separately as described previously (17). The results are summarized in Fig. 8. A Monod model, $q = q_{max} \, s/k(_s + s)$, was fitted to the data using the Marquardt method. This gave a value of 24×10^{-12} g/cell-hr for the maximum specific glucose consumption rate (q_{max}) and a value of 0.1 g/liter for the half saturation rate constant (k_s). The lactate production rate decreased from a value of approximately 14×10^{-12} g/cell-hr at a glucose concentration of 0.8 g/liter to a value of approximately 14×10^{-12} g/cell-hr (indicating that lactate was being consumed) at a glucose concentration of 0.1 g/liter. With decreasing concentration of glucose, not only is the specific lactate production rate reduced, but the amount of glucose consumed and being converted into lactate is also decreased. Although not shown in Fig. 8, the production rate of lactate approaches the consumption rate of glucose at glucose concentrations above 2 g/liter. At high glucose concentrations, essentially all of the glucose consumed by the Vero cells was metabolized via the glycolytic pathway to form lactate. As the glucose concentration decreases, an increasing percentage of the lactate which is produced by glycolysis is further metabolized via the Krebs cycle.

Glucose concentration also affects the oxygen consumption rate. An increase of approximately 40% in the OUR of the culture was observed at low glucose concentrations (0.01 g/liter) compared to its value at high glucose concentrations (> 1.5 g/liter) (Fig. 9). These results were all obtained with culture in which lactate concentration was low. In another set of experiments in which lactate was allowed to accumulate, the OUR appears to be affected by lactate concentration (Fig. 10). At lactate concentrations above 1 g/liter, the OUR of the culture dropped off rapidly at a rate of approximately 3.0×10^{-11} mmol/cell-hr for every 1 g/liter increase in the lactate concentration. The glucose concentrations present at the time of each OUR measurement are also shown in Fig. 10. The glucose concentration was above 1 g/liter for most of the data points. At these concentrations, glucose had little effect on OUR. The decrease in OUR was most likely caused by increasing lactate concentration.

An attempt was made to control the metabolism of the Vero cells by maintaining the glucose concentration of the microcarrier culture within a low range through the controlled feeding of glucose to the fermenter. A feedback control scheme took advantage of the increase

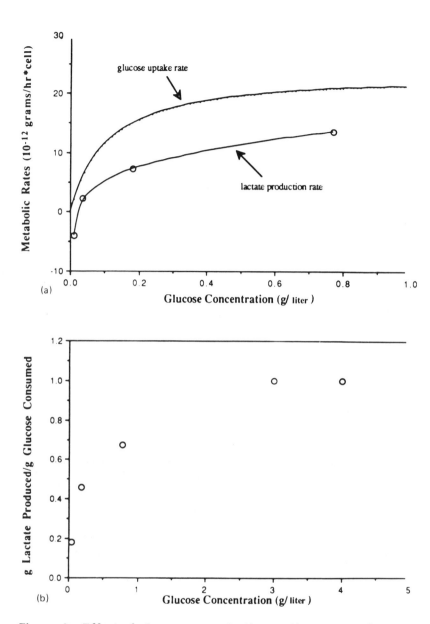

Figure 8   Effect of glucose concentration on the amount of lactate produced per gram of glucose consumed. (a) Comparison of the glucose consumption rate to the lactate production rate at glucose concentrations below 1 g/liter. (b) Effect of glucose concentration on the amount of lactate produced per gram of glucose consumed over a wide range of glucose concentrations.

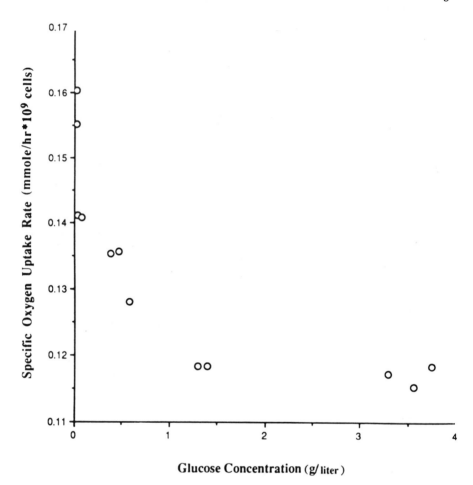

Figure 9   Effect of glucose concentration on OUR. The effect of
glucose concentration on the OUR of a confluent Vero/Cytodex 1
culture with a constant cell concentration is shown. All OUR measurements were taken in the presence of lactate concentrations below
1 g/liter.

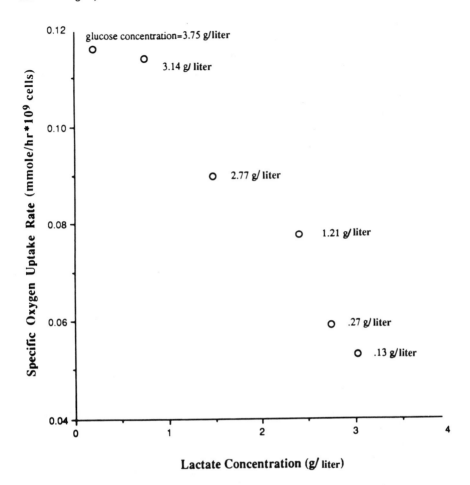

Figure 10 Effect of lactate concentration on OUR. The OUR of a confluent Vero/Cytodex 1 culture with a constant cell concentration is shown as a function of the lactate concentration. The concentration of glucose present at the time of each measurement is also given.

in the OUR as the glucose concentration dropped below 1 g/liter. Since the OUR of the culture is a strong function of the glucose concentration at low glucose concentrations, the OUR of the culture could possibly be used as an indirect measurement of the current glucose concentration in the fermenter (provided that the concentration of lactate was below 1.5 g/liter). For instance, an OUR of 1.4×10^{-10} mmol/cell-hr corresponds to a glucose concentration in the range of 0.05–0.3 g/liter. To test if glucose concentration can possibly be controlled in a range by manipulating OUR at a set point, the OUR of the culture was maintained at its set point by controlling the feed rate of glucose to the fermenter. If the OUR rose above its set point, this indicated that the glucose concentration was too low and the feed rate of glucose was therefore increased. Conversely, if the OUR fell below its set point, this indicated too high a concentration of glucose, and the feed rate of glucose was therefore decreased. A simple proportional control strategy was used to control the feed rate of glucose to the fermenter.

As Fig. 11 demonstrates, by maintaining the OUR of a confluent Vero cell culture at a value of 1.4×10^{-10} mmol/cell-hr, we were able to maintain the glucose concentration in the range of 0.03 to 0.2 g/liter for over 20 hr. Based on the earlier results, the culture exhibited a predictable response to the control strategy. For instance, for OUR values below the OUR set point (t = 12 and 15 hr) the controller responded by decreasing the feed rate of glucose to the fermenter. This resulted in a decrease in the glucose concentration and consequently an increase in the OUR of the culture above its set point (t = 18 hr). The controller then responded to the high OUR value by increasing the feed rate of glucose to the fermenter. This resulted in an increase in the glucose concentration 3 hr later and a drop in the OUR of the culture (t = 21 hr).

The major problem with using this control strategy and operating in a batch mode was the effect that lactate had on the OUR at concentrations above 1.5 g/liter. Even though lower lactate production rates were obtained by operating at low glucose concentrations (approximately 50% lower than those obtained at high glucose concentrations), the lactate still eventually accumulated to a point (approximately 2 g/liter) where it caused the OUR of the culture to decrease sharply (after approximately 27 hr). This caused the control strategy to break down since the OUR was no longer a function of the glucose concentration alone but was also a rather strong function of the lactate concentration.

The data presented in Fig. 11 demonstrates that cell metabolism can be manipulated by controlling the glucose concentration within a low range. The time period in which cell metabolism is manipulated using this control strategy can certainly be extended beyond 20–30 hr in a

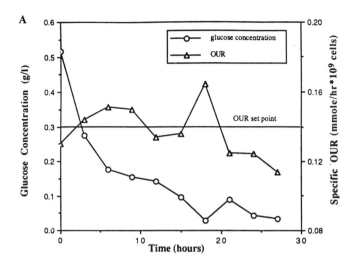

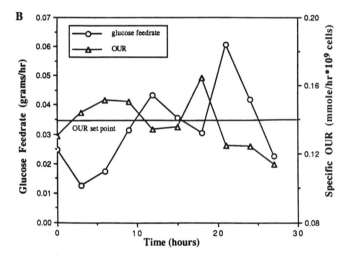

Figure 11 Performance of glucose control strategy. (A) Glucose and OUR profiles as a function of time. (B) Glucose feed rate and OUR profiles as a function of time. The glucose feed rate shown was the feed rate at the time the OUR measurement was taken. The feed rate was then adjusted according to the value of the OUR and maintained at that value until the next OUR measurement was taken 3 hr later.

continuous (perfusion) culture where the lactate concentration could be kept below 1.5 g/liter.

VI. CONCLUDING REMARKS

The ultimate goal of process parameter sensing and control is to direct the reaction or the process toward the optimum path. For biological processes it means manipulating or controlling the gene expression and metabolism of the organism employed. To achieve this goal a means of estimating the process variables of our concern is necessary. Two routes are often taken to obtain the value of the variables: direct measurements when the on-line sensors are available and indirect estimation inferring from other variables for which on-line sensors are available. Only a limited number of on-line sensors are available for cell culture processing. We have discussed the use of two well-established and readily available sensors for on-line parameter control and metabolism manipulation. The use of these indirect measurement can be readily implemented. It is our opinion that the biochemical engineers can exploit their potential in the optimization of animal cell processes.

ACKNOWLEDGMENTS

This work was supported in part by grants from Ecolab, Inc. (St. Paul, MN), Praxis Biologics (Rochester, NY), and the National Science Foundation (Grant ECE-8552670).

REFERENCES

1. Hu, W.-S. and Himes, V. B. Stoichiometric considerations of mammalian cell metabolism in bioreactors. In *Bioproducts and Bioprocesses* (A. Fiechter, H. Okada, and R. D. Tanner, eds.), Springer-Verlag, New York (1989).
2. Zielke, H. R., Zielke, C. L., and Ozand, P. T. Glutamine: A major energy source for culture mammalian cells. *Fed. Proc.* 43:121–125 (1984).
3. Reitzer, L. J., Wice, B. M., and Kennel, D. Evidence that glutamine, not sugar is the major energy source for cultured HeLa cells. *J. Biol. Chem.* 254(8):2669–2676 (1979).
4. Merten, O. W., Palfi, G. E., and Steiner, J. On-line determination of biochemical/physiological parameters in the fermentation of animal cells in a continuous or discontinuous mode. *Adv. Biotechnol. Proc.* 6:111–178 (1986).
5. Kitano, K., Iwamoto, K., Shintani, Y., and Akiyama, S. Effective production of human monoclonal antibody against tetanus

toxoid by selection of high productivity clones of a heterohybridoma. *J. Immunol. Meth.* (in press).

6. Seamens, T. C. Instrumentation of a cell retention reactor for hybridoma. M.S. thesis, University of Minnesota, Minneapolis (1988).

7. Waymouth, C. Major ions, buffer systems, pH, osmolality and water quality. In *The Growth Requirements of Vertebrate Cells in Vitro* C. Waymouth and R. G. Ham, Cambridge University Press, pp. 105–117 (1981).

8. Hu, W.-S., and Wang, D. I. C. Mammalian cell culture technology: A review from an engineering perspective. In *Mammalian Cell Technology* (W. G. Thilly, ed.) Butterworths, Boston (1986).

9. Frame, K. K. and Hu, W.-S. Kinetic study of hybridoma cell growth in continuous culture: I. A model for nonproducing cells. *Biotechnol. Bioeng.* (accepted for publication).

10. Fleischaker, R. L. An experimental study in the use of instrumentation to analyse metabolism and product formation in cell culture. Ph.D. thesis, Massachusetts Institute of Technology, Cambridge (1981).

11. Kilburn, D. G. Lilly, M. D., Self, D. A., and Webb, F. C. The effect of dissolved oxygen partial pressure on the growth and carbohydrate metabolism of mouse LS cells. *J. Cell Sci.* 4:25–37 (1969).

12. Mizrahi, A. Oxygen in human lymphoblastoid cell line cultures and effect of polymers in agitated and aerated cultures. *Dev. Biol. Standard.* 55:93–102 (1984).

13. Ham, R. G. and McKeehan, W. L. Media and growth requirements. In *Methods in Enzymology*, Vol. 58. Academic Press, New York, pp. 44–93 (1979).

14. Bettger, W. J., Boyce, S. T., Walthall, B. J., and Ham, R. G. Rapid clonal growth and serial passage of human diploid fibroblasts in a lipid-enriched synthetic medium supplemented with epidermal growth factor, insulin, and dexamethasone. *Proc. Natl. Acad. Sci. USA* 78:5588–5592 (1981).

15. Hu, W.-S., Dodge, T. C., Frame, K. K., and Himes, V. B. Effect of glucose and oxygen on the cultivation of mammalian cells. *Dev. Biol. Standard.* 66:279–290 (1987).

16. Glacken, M. W., Fleischaker, R. L., and Sinskey, A. J. Reduction of waste product excretion via nutrient control: Possible strategies for maximizing product and cell yield on serum in cultures of mammalian cells. *Biotechnol. Bioeng.* 28:1376–1389 (1986).

17. Frame, K. K. and Hu, W.-S. Oxygen uptake of mammalian cells in microcarrier culture – response to changes of glucose concentration. *Biotechnol. Lett.* 7:147–152 (1985).

17

Methods for the Detection of Adventitious Viruses in Cell Cultures Used in the Production of Biotechnology Products

JUDITH A. POILEY
Hazleton Laboratories America, Inc., Kensington, Maryland

I. INTRODUCTION

The initial use of cell cultures for the growth of virus was reported by Maitland and Maitland (1928). In these early studies, chicken tissue was cultured and used for the production of vaccina virus. Subsequently, Enders et al. (1949) cultured poliovirus type 2 in various human embryonic tissues. This work led to the use of cultured cells as substrates for the growth of viruses. The lack of availability of human embryonic tissue and the expanding base of knowledge of the effects of polio in monkeys led to the use of primary monkey kidney cells as substrates for the production of attenuated poliovirus vaccines. The primary concern of health officials at this time was the possible introduction of altered DNA from continuous cell cultures which might be tumorigenic to vaccine recipients. However, the discovery of other viruses present in primary monkey kidney cells led to additional concerns. Rustigian et al. (1955) reported the occurrence of four viral agents from primary monkey kidney cultures which caused cytopathic effects and Hull et al. (1956) isolated eight viruses from monkey kidney cells. Of most immediate concern, however, was the discovery of Eddy and Coworkers (1961) that SV40 isolated from primary monkey kidney cells by Sweet and Hilleman (1960) induced sarcomas in newborn hamsters. Additional agents found to be present in primary monkey cultures, including herpes virus B and Marburg agent, were shown to be lethal to animal handlers and laboratory personnel.

By 1961 it was determined that sufficient tests were available for detecting and eliminating contaminating viruses to allow continued use of primary monkey cultures (Hopps, 1985). Concern about continuous cell cultures continued to focus on potential tumorigenicity (Committee Report, 1963).

Hayflick and Moorhead (1961) established human diploid fetal lung cultures (WI-38) and characterized them extensively for use in virus vaccine production so that they could be used as an alternative to primary monkey cell cultures. In approximately 1969, a human fetal lung culture (MRC-5) was shown to be suitable for human vaccine production in Britain. In 1972, WI-38 was finally approved for oral poliovaccine production in the United States (Hopps, 1985).

More recent reports dealing with viral contamination of cell cultures include a report of BHK-21 (baby hamster kidney) cells infected with lymphocytic choriomeningitis (LCM) (von der Zeijst et al., 1983a,b). The virus was found to be cell-associated and passed only on cocultivation of infected cells with normal cells or with cell extracts. The virus was not detected in supernatant fluids. The authors felt that the culture could have become infected with LCM after passage in vivo.

Another study by Wognum et al. (1984) described the isolation and characterization of a papovavirus causing cytopathic effects (CPE) in *Cynomolgus* monkey kidney cells after extensive subculturing of the cells. The virus also replicated in BS-C-1, Vero, human embryonic kidney, and calf kidney cells without inducing CPE. After hybridization testing and antibody testing, the virus was determined to be of bovine origin, probably introduced into the culture from calf serum used to supplement the culture medium.

II. SOURCES OF CONTAMINATION OF CELL
LINES BY BOVINE VIRUSES

Fetal bovine serum is most commonly contaminated with infectious bovine rhinotracheitis (IBR), parainfluenza 3 (PI-3), and bovine viral diarrhea (BVD) agents (Molander et al., 1969; Kniazeff, 1968). All commercially produced fetal bovine serum contains BVD and at best the raw serum contains 10^4 infectious BVD virus particles/ml (Hyclone, 1986–87). Routine screening for BVD virus generally yields false negative results because most contaminants are noncytopathic. The serum frequently contains antiviral antibody. Furthermore, BVD is immunosuppressive making it difficult to prepare fluorescent antibody conjugates that have high titers to all three BVD serotypes. Finally, only a small fraction of the total amount of any serum lot is actually tested (Hyclone, 1986–87).

The in vitro host range for BVD includes cells of bovine, porcine, canine, feline, ovine, leporine, and simian origin (Hyclone, 1986–87). Filtering serum through a 0.1-µm filter, as is most commonly done today, will remove IBR and PI-3 but not BVD. Parvovirus has been reported in calf serum only, not fetal bovine serum. Rotaviruses may be present in both fetal bovine and calf serum. Cultures used for the detection of BVD are bovine turbinate cells (BT) which have been continuously grown on horse serum. BVD has not been reported to be present in horse serum.

Recently, a bovine lentivirus was reported which is related to HIV-1 (Gonda et al., 1987). This virus is capable of growing in a wide range of bovine embryonic tissues and is transferred by treatment with infected cell culture supernatant or through the introduction of infected cells into fresh cell cultures (Gonda et al., 1987).

Because of the possible contamination of cell cultures by bovine viruses present in serum, cells which have been in culture for long periods of time should be screened for the possible presence of contaminating bovine viruses.

III. TESTING OF CELL LINES USED TO PRODUCE BIOLOGICS

In the last several years, continuous cell lines (many of which have been shown to be tumorigenic) are coming into use to produce biologicals. Cell lines most commonly used to produce biologicals include Chinese hamster ovary (CHO) cells and mouse cells. More recently, the types of cells used have been expanded to include BHK-21, human, and rabbit cell lines. In many cases, the history of the culture is vague, the cells have been in culture for considerable periods of time, and they have often been passed to a number of different laboratories. Concerns which applied to early cell cultures used for vaccine production also apply to the continuous cell lines, especially the possibility of cell substrate contamination by adventitious viral agents. Many of the cell cultures which were earlier established for vaccine production have been extensively characterized and can be used as sensitive detector cell lines in screening for viruses which might be present in other cell cultures.

The FDA established a series of suggested guidelines for screening cell cultures for the presence of contaminating agents. Among the recommended tests for adventitious viral agents are assays employing both in vivo and in vitro methodologies. Design of in vivo tests to be utilized for detecting adventitious viruses depend on the cell source. For in vivo testing, appropriate animal species are injected with cultured cells and conditioned supernatant by a variety of routes, and then observed for 28 days (Jones and Christenson, 1988).

In the in vivo mouse antibody production test (MAP), the serum is checked for antibodies to ectromelia, GDVII (encephalomyelitis), lactic dehydrogenase, lymphocytic choriomeningitis, Hantaan, minute virus of mouse (MVM), mouse adenovirus, mouse hepatitis, pneumonia virus of mice (PVM), polyoma, Sendai, epizootic diarrhea of infant mice, reovirus type 3, thymic agent, and murine cytomegalovirus.

In vivo testing also includes the hamster antibody production (HAP) test for viruses. Serum from hamsters is tested for antibody to reovirus type 3, PVM, Sendai, SV5, and LCM. For the in vivo rat antibody production test (RAP) the serum is checked for antibodies to Toolans H-1, GDVII, Kilham rat virus, PVM, reovirus type 3, Sendai, rat corona virus, sialodacryoadenitis, and Hantaan virus.

Because patients receiving cell culture products are generally seriously compromised by disease, a normally murine-specific virus may pose serious health risks which would not be a problem in a healthy individual (Jones and Christenson, 1988). However, some rodent viruses are infectious to humans; reovirus type 3 of rodents is indistinguishable from the human isolate, Hantaan virus is infectious to humans, and lymphocytic choriomeningitis is a human pathogen.

In vivo testing for nonrodent viruses includes the inoculation of cells and conditioned medium into fertilized eggs, adult guinea pigs, adult mice, and suckling mice (Table 1). If a rabbit cell culture is used, rabbits must also be included in the in vivo assay. At times the cell culture being evaluated for viruses may be tumorigenic and animals inoculated with tumorigenic cells for the virus test may die from tumor formation. Subsequent pathological evaluation usually demonstrates the inoculated cell culture as the cause of tumorigenicity. In these cases it is recommended that an additional group of animals be inoculated with a frozen/thawed lysate preparation of cells in 10% serum to test for viruses. Other problems that may be encountered in testing are due to components of the medium such as methotrexate or dexamethasone which are often used to stabilize the desired cell type in genetically engineered cells. These additives are often lethal to animals and should

Table 1   Adventitious Viruses Detectable by In Vivo Assay[a]

Animal species	
Embryonated eggs: allantoic and yolk sac routes are less sensitive than chorioallantoic route for detection of poxviruses	Arboviruses Herpes simplex type 1 Herpes simplex type 2 Influenza Mumps Newcastle's disease Parainfluenza 1 (Sendai) Parainfluenza 2 Rabies Vaccinia
Guinea pigs	Arboviruses B virus Ebola Encephalomyocarditis Junin Lassa LCM Marburg Rabies
Adult mice	Arboviruses Encephalomyocarditis Herpes simplex type 1 Herpes simplex type 2 Lassa LCM Rabies
Suckling mice (most sensitive animal for arboviruses)	Arboviruses B virus Coxsackie A virus Coxsackie B virus Foot-and-mouth disease Juni Machupo Variola

[a]References: 1. Lennette, E. H., and Schmidt, N.J., 1979. *Diagnostic Procedures for Viral, Rickettsial and Chlamydial Infections*, 5th Ed. American Public Health Association. 2. Mahy, B. W. J. *Virology: A Practical Approach, 1985.* The Animal Virus Research Institute, Pirbright, Woking Surrey, IRL Press, Oxford. 3. Ballew, H. C. and Forrester, F. T., 1984. *Desk Aid for Isolating and Identifying Viruses.* U.S. Department of Health and Human Services. Center for Disease Control.

Table 2   Cell Cultures Which are Recommended for Detection of Adventitious Viral Contaminants and Viruses Which Have Been Isolated Using These Cell Cultures[a,b]

Virus	HEK	WI-38	MRC-5	Hela
Adenovirus	+	+	+	+
Arbovirus				
BK virus	+	+		
Coronavirus		+	+	+
Cytomegalovirus		+	+	
Coxsackie virus	+	+	+	+
Echovirus	+	+	+	+
Enterovirus	+	+	+	+
Hepatitis A virus		+	+	
Herpes simplex virus	+	+	+	
Human papovavirus	+	+	+	
Influenza	+	+	+	+
Lassa				
Measles	+	+	+	+
Mumps	+	+	+	+
Newcastle's disease virus				+
Parainfluenza virus	+	+	+	+
Paramyxovirus	+	+	+	+
Poliovirus	+	+	+	+
Rabies		+	+	
Reovirus	+	+	+	+
Respiratory syncytial virus	+	+	+	
Rhinovirus	+	+	+	+
Rubella	+	+	+	+
SV40		+	+	
SV-5				

MDBK[c,d]	MEF	Vero	LLC-MK$_2$	BS-C-1	CV-1	AGMK
+		+	+	+	+	+
		+				
				+		+
		+				
	+	+	+	+		+
		+	+	+		+
				+		
+	+	+		+	+	+
+	+	+	+	+		+
		+				+
		+	+		+	+
	+	+	+	+	+	+
+	+	+			+	
+	+	+	+		+	
+		+		+		
		+	+		+	+
	+		+			+
		+	+	+	+	
		+	+	+		
		+	+	+		
		+		+	+	+
		+				+

Table 2   (Continued)

Virus	HEK	WI-38	MRC-5	Hela
Vaccinia virus	+		+	+
Varicella zoster		+	+	
Variola		+	+	+
VSV	+	+	+	+

[a]The lack of a + response may only indicate that the cell line has not been tested with a particular virus.

[b]Refs.: 1. ATCC. 1988. *Catalogue of Animal and Plant Viruses*, 6th ed., ATCC; 2. Lennette, E. H. and Schmidt, N. J. 1979. *Diagnostic Procedures for Viral, Rickettsial and Chlamydial Infections*, 5th ed., American Public Health Association; 3. Lennette, E. H., Balows, A., Hausler, W. I., Jr., and Shadomy, H. J. 1985. *Manual of Clinical Microbiology*, 4th ed., American Society of Microbiology; 4. Mahy, B. W. J. *Virology: A Practical Approach, 1985.* The Animal Virus Research Institute, Pirbright, Woking Surrey, IRL Press, Oxford.

[c]Ballew, H. C. and Forrester, F. T. 1984. *Desk Aid for Isolating and Identifying Viruses*. U.S. Department of Health and Human Services. Center for Disease Control.

[d]Using this culture bovine adenovirus, bovine parvovirus, bovine viral diarrhea, infectious bovine rhinotrachitis, and possible serum contaminants may also be detected.

MDBK[c]	MEF	Vero	LLC-MK$_2$	BS-C-1	CV-1	AGMK
+	+		+		+	+
		+	+		+	
+		+	+	+	+	+

be excluded from the medium used to grow cells for animal inoculation. Final products may also be tested in in vivo assays. It
is suggested that the dose tested be a human dose equivalent.
Stabilizing agents which are used with the final product may cause
problems when inoculated, e.g., by the intracranial route, and
rapid death of the test animals may result. These agents should
be evaluated separately if problems occur. The manufacturer needs
to give consideration as to what approaches constitute appropriate
testing for these products.

In vitro testing for adventitious viral agents as recommended
in the Center for Biologics Evaluation and Research of the U.S.
FDA guidelines includes the use of the test cells as indicator cells,
a normal human embryonic cell culture, and a monkey kidney cell
line. The test cells employed should be cells which were used to
produce the product. Human embryonic cultures commonly used
as detector cells for adventitious viral contamination include human
embryonic kidney (HEK), WI-38, and MRC-5. Monkey kidney cell
lines which can be used as detector cultures include Vero, LLC-
MK_2, BS-C-1, CV-1, and primary cultures of African green monkey kidney (AGMK). A list of potential human viruses which may
be detected with the human embryonic and monkey kidney cells is
given in Table 2. Many contaminating viruses can be detected by
cytopathic effect. Those not exhibiting cytopathic effects can be
detected by hemagglutination and/or hemadsorption utilizing chicken,
guinea pig, and rhesus monkey or human type O erythrocytes.
European regulatory guidelines suggest, in addition to the above-
mentioned cell types, Hela, Madin Darby bovine kidney (MDBK),
and mouse embryo fibroblastic (MEF) cells as detector cells. These
cultures are also monitored for CPE, hemagglutination, and hemadsorption using the same erythrocyte species. A list of possible
human viral contaminants detected by those cells is given in Table
2. The use of a continuous cell line or a well-characterized culture such as Vero or $LLC-MK_2$ is preferable to AGMK primary cultures because AGMK primary cultures may be contaminated with
latent monkey viruses producing similar problems as those encountered in early vaccine production work.

Additional human viruses of concern such as Epstein-Barr virus
(EBV), HIV-1, cytomegalovirus, and hepatitis B virus require different test methodologies. Epstein-Barr virus may be detected by
viral DNA hybridization, nuclear antigen fluorescent staining, or
immortalization of umbilical cord blood lymphocytes (Jones and
Christenson, 1988). More recently, polymerase chain reaction technology can aid in the detection of EBV. Cytomegalovirus testing
as currently recommended by the FDA requires a 6-week tissue
culture assay with a blind passage between the third and fourth

weeks. If cytopathic effects are observed with the test cells, immunoflourescent staining should be performed to confirm that they are due to the presence of cytomegalovirus. Commercially available third-generation kits can be used for hepatitis B screening. The detection of HIV-1, a retrovirus, will be covered elsewhere.

IV. SUMMARY

The possibility for viral contamination exists in established cultures as well as in primary cultures. The use of established genetically engineered cultures in the production of biologicals for human use requires that these cultures be monitored for adventitious viral agents. Among the methods used for this purpose are animal inoculation and in vitro assays which provide a broad-spectrum screen for viral agents.

REFERENCES

1. Committee Report (1963). Continuously cultured cells and viral vaccines. *Science* 139:15−20.
2. Eddy, B. E., Borman, G. S., Berkeley, W. H., and Young, R. D. Tumors induced in hamsters by injection of rhesus monkey kidney cell extracts. *Proc. Soc. Exp. Biol. Med.* 107:191−197 (1961).
3. Enders, J. F., Weller, T. H., and Robbins, F. C. (1949). Cultivation of the Lansing strain of poliomyelitis in cultures of various human embryonic tissue. *Science* 109:85−87.
4. Gonda, M. A., Braun, M. J., Carter, S. G., Kort, T. A., Bess, J. W., Jr., Arthur, L. O., and van der Maaten, M. J. (1987). Characterization and molecular cloning of a bovine lentovirus related to human immunodeficiency virus. *Nature* 330:388−391.
5. Hayflick, L. and Moorhead, P. S. (1961). The serial cultivation of human diploid cell strains. *Exp. Cell. Res.* 25:585−621.
6. Hopps, H. E. (1985). Cell substrate issues: A historical perspective. In *Abnormal Cells, New Products and Risk* (H. E. Hopps and J. C. Petuicciani, eds.), Tissue Culture Association, Gaithersburg, MD.
7. Hull, R. M., Minner, J. R., and Smith J. W. (1956). New viral agents, recovered from tissue cultures of monkey kidney cells. I. Origin and properties of cytopathogenic agents S.V.[1], S.V.[2], S.V.[4], S.V.[5], S.V.[6], S.V.[11], S.V.[12], and S.V.[15]. *Am. J. Hyg.* 63:204−215.

8. HyClone Laboratories, Inc. (1986). BVD virus contamination of fetal bovine serum and infection of cell cultures. *Art to Science* 5:1–2.

9. HyClone Laboratories, Inc. (1986/87). BVD virus contamination of fetal bovine serum and infection of cell cultures. *Art to Science* 5:1–5.

10. Jacobs, J. P., Magrath, D. I., Garrett, A. J., and Schild, G. C. (1981). Guidelines for the acceptability, management and testing of serially propagated human diploid cells for the production of live virus vaccines for use in man. *J. Biol. Standard.* 9:331–342.

11. Joner, E. and Christiansen, G. D. (1988). Hybridoma technology products: Required virus testing. *Biopharmacology*, 50–56.

12. Kniazeff, A. J. (1968). Viruses infecting cattle and their role as endogenous contaminants of cell culture. In *Proceedings: Cell Cultures for Virus Vaccine Production*. Natl. Cancer Inst. Monograph 29:123–132.

13. Maitland, H. B. and Maitland, M. C. (1928). Cultivation of vaccinia virus without tissue culture. *Lancet* 2:596–597.

14. Molander, C. W., Kniazeff, A. J., Paley, A., and Imagawa, D. T. (1969). Further studies on virus isolation from bovine serums. *In Vitro* 6:29.

15. Rustigian, R., Johnston, P., and Reihart, H. (1955). Infection of monkey kidney tissue culture with virus-like agents. *Proc. Soc. Exp. Biol. Med.* 88:8–16.

16. Sweet, B. H. and Hilleman, M. R. (1960). The vacuolating virus, SV_{40}. *Proc. Soc. Exp. Biol. Med.* 108:420–427.

17. van der Ziejst, B. A. M., Noyes, B. E., Mirault, M.-E., Parker, B., Osterhaus, A. D., Swyryd, E. A., Bleumink, N., Horzinelo, M. C., and Stark, G. R. (1983a). Persistent infection of some standard cell lines by lymphocytic choriomeningitis. *J. Virol.* 48:249–261.

18. van der Zeijst, B. A. M., Bleumink, N., Crawford, L. V., Swyryd, E. A., and Stark, G. R. (1983b). Viral proteins and RNAs in BHK cells persistently infected by lymphocytic choriomeningitis virus. *J. Virol.* 48:262–270.

19. Wognum, A. W., Sol, C. J. A., van der Noordaa, J., van Steenis, G., and Osterhaus, A. D. M. E. (1984). Isolation and characterization of a papovavirus from cynomolgus macaque kidney cells. *Virology* 134:254–257.

18

Continuous Cell Substrate Considerations

ANTHONY S. LUBINIECKI
SmithKline Beecham Pharmaceuticals, King of Prussia, Pennsylvania

I. INTRODUCTION

Few issues regarding the development of cell culture biologicals have
been debated as long or as vigorously as the use of continuous
cell lines (CCLs) as substrates. The issue has been discussed since
1954 and is not completely settled in 1989 despite much progress.

To put the debate in perspective it is first necessary to review
two reasons why CCLs are considered useful. First, the cost of
preparing cell culture products is highly dependent on the cost of
preparing the cells for manufacturing. Normal diploid cells (both
primary and serial) grow slowly, require an acceptable surface for
attachment and growth, and attain relatively low cell density at
saturation. In contrast, CCLs frequently divide more rapidly,
usually attain higher saturation density, and may be capable of
growth in suspension. These properties of CCLs translate into
greatly reduced costs per billion substrate inoculum cells prepared
for manufacturing use (1,2). Savings are obtained using CCLs
both in terms of capital outlay (equipment and facilities) and expense items (medium and labor). In the developed world, this
economic issue is a major factor in product development decisions
and in competitive position. In less developed countries the more
favorable economics of CCLs may conceivably mean the difference

between being able to benefit from the availability of a product and not being able to afford the manufacture or purchase of the product.

In addition to economic factors the benefits of some biologicals are only available currently from CCL substrates. Large numbers of population doubling levels (pdl) are consumed in cell cloning. Recombinant and hybridoma cell lines are frequently cloned more than once. Gene amplification techniques may require months of continual cell cultivation. Many cloned lines are further adapted or modified prior to cell bank preparation. Diploid cells enter the senescence state too quickly to be subjected to such procedures, which may require over 100 pdl to complete. Today the technological ability to provide tissue plasminogen activator, erythropoeitin, and monoclonal antibodies for medical use depends exclusively on manufacturing safe product in CCLs.

Thus, there are both economic and technological justifications for the use of CCLs to obtain human health benefits. However, the value of these benefits must be weighed against the putative risk of using products from CCLs. Continuous cell substrate issues basically may be distilled to a single major issue from multiple theoretical causes. That issue is tumorigenicity. The principal question is whether products can be contaminated by preparation in CCLs (which biologically resemble or are derived from tumor tissue) by some substance capable of forming a tumor in product recipients. Historically, this question has been with us for as long as man has employed mammalian cells cultivated in vitro as production substrates (3). Modern science has progressed to the point where four potential sources of tumorigenicity can be hypothesized. These are living substrate cells, transforming proteins, transforming DNA, and infectious viruses (4). The use of multiple filtration steps and other process methods, including lyophilization, effectively precludes contamination of CCL products by intact viable substrate cells. Since the risk of contamination of products from CCLs by intact substrate cells is vanishingly small (5), it will not be discussed in this chapter, while the validation of risk removal is discussed in Chapter 19.

II. CELLULAR PROTEINS

Product contamination by cellular proteins is undesirable from several perspectives. Cells contain hundreds of different proteins (6), some of which may be immunogenic, pharmacologically active, or toxic. Although considerable purification of the protein of interest occurs during downstream processing, the final product cannot be proven to be 100% pure because identification and characterization

of trace amounts of protein impurities is difficult to achieve techno-
logically.

Theoretically, among these contaminating cellular proteins might
be one or more proteins encoded by oncogenes or other oncogenic
DNA sequences. Some tumor cells and continuous cell lines seem
to contain amplified sequences for or to express oncogenes and
activated oncogenes more frequently than diploid cells (7–9). In
some model systems, proteins encoded by viral oncogenic sequences
or activated cellular oncogenes are clearly the means by which the
transformed state is established and/or maintained (9–11). These
considerations form the basis of the theoretical concern regarding
putative contamination of products by "oncogenic proteins" from
continuous cell substrates.

Currently available data suggest several arguments against the
existence of significant risk due to potential oncogenic protein
contamination. First, no mammalian or viral protein is known to
be capable of permanent conversion of diploid cells into continuous
or malignant cells in vivo or in vitro. Microinjection of purified
proteins encoded by oncogenes have only transient, if any, effects
on cell morphology or behavior as has been shown for the v-src
protein (12). Similar observations of transient biological effects
have been reported for cells microinjected with SV40 T antigen
(13). Under unusual circumstances, chronically administered
oncogene-related growth factors such as EGF have been shown
to play a role in the high incidence of progressively growing
mammary tumors in a strain of predisposed inbred mice (14),
possibly analogous to the dependence of some tumors on hormones
such as estrogen (15). While consistent with a negligible risk of
tumorigenesis from protein contaminants, these studies demonstrate
the necessity to achieve adequate purification of the molecule of
interest with respect to biologically active contaminants such as
growth factors.

Second, the continuous synthesis of new molecules of "transforming proteins" is known to be required for maintenance of
morphological alterations. Studies of the effect of metabolic inhibitors and of temperature sensitive mutants of SV40 T antigen
have shown that continual synthesis of active gene product is
necessary for perpetuation of the effect (10). It seems reasonable
to conclude from these data that these types of proteins are rapidly
degraded inside cells. Without the continued presence and active
expression of DNA sequences encoding them, they pose no significant oncogenic risk (16).

In summary, many negative data exist to argue against the
theoretical risks from oncogenic proteins. It can always be speculated that doing additional studies at still larger doses with

lifetime observation periods might produce a positive result. The collective results obtained to date imply that the theoretical possibility of oncogenic events due to contaminating proteins is negligibly small in the absence of DNA-encoding oncogenic sequences, especially in comparison to the real benefits derived from the products of continuous cell lines.

III. CELLULAR DNA

The evidence of tumorigenic potential from contaminating DNA comes from several sources. The DNA of any of several transforming viruses in vivo, such as SV40 (17), polyoma (18), and herpesvirus samiri (19), are capable of tumorigenesis when their complete genome was provided in suitable form to susceptible biological host. DNA from numerous tumor viruses has been shown to "transform" susceptible cells in vitro, including SV40 (10), adenovirus (20), and many others.

Several types of activated oncogene DNAs can "transform" susceptible cells in vitro (11,21); such transfected cell lines are frequently tumorigenic upon implementation in immunosuppressed rodents (22). Fung et al. (23) showed that the v-src oncogene DNA could produce nodules in inoculated chickens which regressed after several weeks. Finally, some oncogenes are capable of causing tumors if introduced transgenically via retroviral vectors (24). This laboratory evidence supports the theoretical possibility of cellular transformation due to contaminating oncogenic DNA but is of uncertain relevance to most examples of product delivery to humans.

These oncogene experiments generally employed calcium phosphate, DEAE dextran, polybrene, or other substances which strongly facilitate DNA transfection in vitro. In the absence of stimulants of DNA uptake (the usual case for protein products), transfection efficiency is greatly reduced. Although activated oncogenes can produce transient nodules in chickens (23), this phenomenon has never been reproduced in any mammal despite considerable effort. The only nonhuman primate study involved the incoulation of 10^8 genome equivalents of T24 bladder carcinoma chromatin into five immunosuppressed rhesus monkeys; after over 3 years of observation, no clinical evidence of tumors has been found (J. Petricciani, pers. commun.).

When the DNA of common CCL substrates for rDNA technology like CHO or DNA of activated oncogenes was inoculated into immunosuppressed rodents, no tumor formation was observed despite the fact that living intact CHO cells were tumorgenic in these animals (25). CHO cell DNA failed to produce tumors when inoculated into newborn Chinese hamsters which were susceptible to tumorigenesis by polyoma viral DNA (26). CHO cell DNA does not seem to transform NIH-373 cells (25). In another experiment, up to 25 μg of a recombinant plasmid containing the hepatitis B virus surface antigen gene plus a selectable marker for neomycin resistance was administered to mice. After up to 56 days, the mice were bled and sacrificed. No evidence of antibody to hepatitis B surface antigen was observed. When spleen cells from the mice were exposed to selective pressure, no evidence of neomycin resistance was observed (27).

These data collectively argue that the risk of tumorigenicity to mammals receiving large quantities of DNA-encoding activated oncogenes must be very small, presumably due to biological barriers such as plasma nucleases, difficulties in transporting DNA across cell membranes, the low efficiency of DNA transfection under ideal conditions, and so forth. When one considers that the vast majority of the minute quantity of DNA-contaminating products from CCLs is not coding for oncogenes, this tiny risk becomes even smaller. Petricciani and Regan estimated that the risk is less than 10^{-9} assuming that a produce dose contains 1 ng of residual cellular DNA from a CCL bearing 100 activated oncogenes per cell genome (28). This is a very small theoretical risk indeed, approaching one per planetary population.

IV. VIRUSES

The role of viruses in human cancer has gradually clarified. The question of whether any human cancers are of viral origin has been answered affirmatively by the discovery of the human T-lymphotrophic virus family as the cause of adult T-cell leukemia/lymphoma (29,30). Agents of the distantly related human immunodeficiency virus family cause AIDS (31,32). Strong statistical associations exist between infection by selected human papilloma viruses and cervical cancer (33). Other examples exist such as the association between Epstein-Barr virus and Burkitt's lymphoma and nasopharyngeal carcinoma (34,35) and between hepatitis B and liver carcinoma (36). While selected viruses may be important in the genesis of specific tumors and more examples may be uncovered by further study, infectious viruses are unlikely to cause all forms of human cancer.

Some of these potential tumor viruses may be detected by assays routinely employed to screen diploid cell substrates for adventitious agents, e.g., adenoviruses, herpesviruses, and papovaviruses (see Chapter 17). However, other types of agents may be missed by such conventional tests, either because they may not be transmitted horizontally (e.g., some endogenous retroviruses) or because they possess a very limited spectrum of in vitro infectivity (e.g., human papilloma and AIDS viruses).

A. Retroviruses

Retroviruses may be transmitted biologically by either of two means:
horizontally from an infected individual to a susceptible one by an
infectious virion (like classical adventitious agents) or vertically
from one generation to the next via a DNA copy of the viral genome
covalently linked to the host genome. Expression of viral genes
may be highly variable for some endogenous retroviruses. A cell
may express retroviral particles which detectably contain viral RNA,
reverse transcriptase, and be infectious. One or more of these
traits might be absent or undetectable despite the clear presence
of the others (37–39). In other cases, assays for all of these
traits may be negative, yet some or all of the viral genes may be
expressed at the cellular level (40). In some cases no evidence
of spontaneous expression exists, but these traits may be induced
(40). Since all vertebrate species examined, including man, possess
such sequences in every cell (41), it must be concluded that each
living vertebrate cell (abnormal or diploid) possesses the potential
to express such genes. Although normal diploid cells may express
some of these sequences only during ontological development (42),
abnormal cells (tumor cells or CCLs) often express endogenous
retroviral particles or genes (40). Expression patterns for retroviral genes vary among cell lines and nonexpressive lines can
become expressive (40,43) and vice versa (44). In some cases,
expression of silent or nearly silent retroviral genes may be enhanced by induction with chemical or viral agents (45,46). As a
result, no single test can define the expression of endogenous
retroviral genes. These complexities make the assessment of risks
due to endogenous retroviruses difficult, time consuming, expensive, and the subject of differing opinions.

   To fully define the putative risk from retroviruses, cell substrates must be subjected to a broad battery of tests looking for
evidence of functional retroviruses. For all assays, cells from
cell banks intended for ultimate manufacturing use must be used in
order for results to be applicable to assessments of product safety.
Transmission electron microscopy (TEM) is useful to detect the presence of retroviral particles. Various morphological and morphogenetic types can be distinguished (47) but this has little value
for qualitative risk assessment purposes. This is because TEM
cannot assess functionality or competence to induce tumors. Defective retrovirus-like particles cannot be distinguished from infectious ones. Also, the method is insensitive. In general, retroviral particles in samples at concentrations below $10^6 – 10^7$ ml
are not detectable by TEM (48). Thus, TEM is useful to generally
screen for the presence of high levels of retroviral particles but
possesses limited power to analyze them or the risks they might
pose.

Reverse transcriptase (RT) is a virion-associated enzyme necessary for infectivity and tumorigenesis (49). Detection of RT activity is accomplished by measuring enzymatic conversion of soluble radiolabeled nucleotide triphosphates to acid-precipitable oligonucleotides when directed by suitable RNA or DNA templates in a hospitable environment (50). To improve sensitivity the virions must be partially disrupted by detergents (50). The pH and cationic optimum for RT activity varies considerably among different retroviruses (50,51). RT activity is generally much greater when directed by synthetic templates than by endogenous virion RNA. RT enzymes of different retroviruses show variable specific activity and relative specificity for RNA over DNA or for different synthetic templates, depending on the morphogenetic retrovirus group (51).

The RT assay depends on the conversion of soluble radioactivity to radioactivity associated with a precipitable particle. Accordingly, it can be effected by a variety of factors including contamination by normal cellular DNA polymerase, gross microbial contamination, entrapment of label by cellular debris, and chemical contaminants, to name just a few. The test article for such assays is usually cell culture fluid concentrated from 10- to 10,000-fold by ultracentrifugation or ultrafiltration (52-54). Greater concentration provides improved sensitivity but potentially lower specificity if cell debris levels are excessive. Independent assessment of enzymatic activity directed by synthetic RNA-DNA hybrid and DNA templates is often necessary to distinguish true RT activity from cellular polymerase contained in concentrated samples. The assay provides qualitative information at best, since the effect of sample dilution on RT activity is often nonlinear and sometimes not dose-responsive. Some samples will appear inactive but actually contain inhibitory substances detectable as a reduction of RT activity in samples of test articles spiked with genuine functional retroviruses. The utility of the assay is further limited by difficulties in providing valid experimental controls and in interpretation. It is not possible to provide a control for a test article which contains the same amount of the same type of cellular debris; therefore, it is difficult to know what "background" levels really are. This makes differentiation of low levels of genuine RT activity from high levels of background difficult. Assay protocol details will affect sensitivity, specificity, and precision, complicating interlaboratory comparison.

In typical assays, incorporation levels should generally exceed background by several thousand counts per minute before the result is termed positive. Lower values should be viewed with considerable skepticism until confirmed frequently, especially with replicate controlled samples. Values below this level for highly

concentrated samples are probably not interpretable, since current
technology cannot meaningfully assess the molecular nature of the
precipitated radioactivity to confirm the presence of RT activity
associated with a novel or unknown retrovirus. Thus, the presence of unequivocal RT activity provides good evidence of the
productive expression of retroviral genes, but the absence of detectable RT levels cannot preclude the possible presence of functional retroviruses. An example of a typical negative result with
appropriate controls is shown in Table 1.

Infectivity assays are widely employed to search for infectious
retroviruses. Commonly employed assays include the S^+L^- focus
formation (55) and XC plaque (56) assays. Genetic restriction of
host range is well known among retroviruses of multiple species
(57,58) and may limit detection of infectious retroviruses by such
assays. A more general type of assay employs exposure of susceptible cells to test fluids or cocultivation with cells from the substrate bank (40). The use of fluids rather than substrate cells is
recommended whenever the substrate cells grow faster than the
indicator cells, precluding mixed cultivation. After two to five
subcultivations, the cultures are examined for RT activity. A
variety of susceptible cell lines from several species may be employed for such assays. Some commonly employed indicator cell
lines include RD, MRC-5, and A549 (human lines), as well as animal lines such as bat lung, mink lung, and dog thymus. A typical
example with negative results and appropriate controls is shown
in Table 2. Cocultivation assay results are subject to the same
interpretation difficulties as RT assay results.

Animal inoculation is not generally useful in cell bank characterization for retroviruses, as many months or even years of observation are generally required to observe retrovirus-induced
pathogenesis in vivo. Formidable biological barriers exist to tumorigenesis, generally requiring large numbers of infectious particles
and host animals of the appropriate genetic background (which is
obviously not knowable in advance when searching for unknown
retroviruses).

Other technologies can be employed for the detection of retroviruses. Evaluation of cloned retrovirus genomes reveals that the
core protein (also known as P30 or interspecies antigen) sequences
are somewhat conserved (59). Evidence of relatedness or serological cross-reactivity is also obtained from competitive radioimmunoassays for this viral antigen, especially between endogenous
retroviruses of the same or closely related species (60). However,
it may prove difficult to distinguish between a generic cross-
reactive P30 antigen and proteolysis of labeled probes associated
with cellular proteases, especially in highly concentrated samples (61).

Table 1 Reverse Transcriptase and DNA Polymerase Activities in Test Article

Sample		Reverse transcriptase[a] (rA·dT)		DNA polymerase[a] (dA·dT)	
		Mn^{2+}	Mg^{2+}	Mn^{2+}	Mg^{2+}
Test article	Undiluted	44	0	39	0
	Diluted twofold in stabilization buffer[b]	9	0	5	4
	Diluted fivefold in stabilization buffer	7	0	0	9
	Diluted twofold in R-MuLV[c]	3,275	413	52	39
	Diluted fivefold in R-MuLV	7,204	682	98	75
	Diluted twofold in DNA polymerase[d]	943	12	19,203	1,531
Positive and negative controls	R-MuLV	12,157	1,780	204	101
	Stabilization buffer diluted twofold in R-MuLV	2,989	343	42	43

Table 1 (Continued)

Sample	Reverse transcriptase[a] (rA·dT)		DNA polymerase[a] (dA·dT)	
	Mn^{2+}	Mg^{2+}	Mn^{2+}	Mg^{2+}
Positive and negative controls (continued)	Stabilization buffer diluted fivefold in R-MuLV	9,440	1,268	95	81
	SMRV[e]	287	7,441	57	82
	DNA polymerase	2,723	22	40,541	3,149
	DNA polymerase diluted twofold in stabilization buffer	953	26	14,329	1,452
	Stabilization buffer	(44)	(44)	(43)	(43)

[a]Expressed as counts per minute [^{3}H]thymidine triphosphate incorporated [mean duplicates minus background (stabilization buffer)].

[b]Stabilization buffer used to dilute test article and positive controls; used undiluted as background for assay.

[c]R-MuLV (Rauscher murine leukemia virus), type C virus positive control.

[d]DNA polymerase, positive control; prepared from mink lung cells.

[e]SMRV (squirrel monkey retrovirus preparation), type D virus positive control.

Table 2 Reverse Transcriptase (RT) Assay Results of 20× Concentrated Supernatant Culture Fluids from Cocultivation of Test Article with MRC-5 Cells for One or Five Passages

Sample	Passage	RT activity (cpm)		DNA polymerase (cpm)	
		Mn^{2+}	Mg^{2+}	Mn^{2+}	Mg^{2+}
Test article	1	92	0	583	84
	5	29	8	56	26
MRC-5 control	1	63	0	512	48
	5	12	0	17	18
R-MuLV-infected controls					
5–10 ffu	1	33	0	106	15
	5	2,145	141	19	18
500–1000 ffu	1	43	0	322	33
	5	3,179	197	31	18
Assay controls					
R-MuLV	—	12,157	1,780	204	101
SMRV	—	287	7,441	57	82
DNA polymerase	—	2,783	22	40,541	3,149
Stabilization buffer	—	(44)	(44)	(43)	(43)

Nucleic acid hybridization and other molecular biology techniques can be an effective tool for the characterization of isolated propagable retroviruses or for quantitation of the presence and expression level of specific known sequences. However, the extreme specificity of these methods makes it impossible to exclude generically the presence of expression of unknown retroviral genes if negative results are obtained. Using cloned type A particle probes from Syrian hamsters, Wurm et al. showed that CHO cells contain many copies of related genetic sequences which are expressed (62). However, the relationship of such sequences to retrovirus-like particles seen by TEM in CHO cells (5) may prove difficult to establish.

Induction of endogenous retroviruses by biological, chemical, and physical agents has been reported by several laboratories (45,46), analogous to the induction of lysogenic bacteriophages (63). These techniques seem rather unreliable and difficult to reproduce. In some reported cases of induction such as CHO cells (64), it is not completely clear that the induced activity represents a functional retrovirus. For example, Manly and associates (65) reported that CHO cells respond to induction with bromodeoxyuridine by producing particles rich in DNA (not RNA as expected for retroviruses).

When large numbers of continuous cell lines are examined for evidence of retroviruses, positive results are found for some lines. Lieber et al. (40) studied 42 cell lines at random and found 10 to be positive. Fenno reported that a series of recombinant CHO cell lines were searched thoroughly for endogenous retroviruses by RT, cocultivation, infectivity, and induction assays. No evidence was found other than the presence of retrovirus-like particles by TEM (66). In contrast, about 30% of murine hybridomas expressed infectious retroviruses.

B. Other Tumorigenic Viruses

Certainly, tumors may be caused by viruses other than retroviruses. In addition to detection by in vitro assays of viral infectivity or morphological transformation, two other methods may be employed. TEM may detect the presence of virions which exhibit narrow host range in vitro, such as papilloma viruses or Epstein-Barr virus. Just as for detection of endogenous retroviral particles, TEM lacks both sensitivity and specificity. This property is occasionally useful in detecting hitherto unknown agents with novel biochemical and biological properties.

Nucleic hybridization is growing in popularity, especially for human cell substrates. This highly sensitive technique is capable of detecting very minute quantities of specific DNA sequences whether they are replicating or infectious or not. With polymerase chain

reaction amplification technology, the methods can reliably detect
a few genomes per million cells (67). While such methods are excellent for detecting the presence of the more fastidious tumor vi-
rises, their extreme specificity requires that a unique probe be
used for each agent of interest. Thus, the method is limited to
detecting known tumor viruses. Its use has been suggested for
the characterization of human cell substrates (68) but has not been
used extensively for nonhuman lines. While clearly powerful, this
type of assay is still under development and has not yet been validated extensively. The broad utility of such methods remains to
be proven.

V. SUMMARY

The debate over the potential risk of tumorigenicity attributable
to the use of CCL substrates for biologicals production has continued
for over 30 years and may continue for some time to come. Manufacturers and regulatory agencies are developing scientifically based
guidelines for such products. It is currently possible to follow
these guidelines to prepare recombinant biologicals and monoclonal
antibodies in CCLs which do not pose unreasonable risks.

This chapter has attempted to describe the scientific tools
available to evaluate the putative risk of tumorigenicity due to
potential virus DNA and protein contaminants. No theoretical or
experimental basis exists to hypothesize that residual cellular protein might present a significant risk of tumorigenicity. The tools
are certainly adequate for characterization of putative risks due to
viruses and DNA but are not sufficiently powerful by themselves
to assure product safety. The subsequent chapter on process
validation describes how adequate assurances of safety ultimately
can be obtained for products of CCLs against theoretical risks
of tumorigenicity due to putative viruses and DNA.

In addition to these safeguards, no evidence of tumorigenicity
has been found in human or livestock animal recipients of the products prepared in CCL substrates. Many patients have received
inoculations of tissue plasminogen activator, erythropoeitin, factor VIII, soluble CD4, GM-CSF, hepatitis B surface antigen vaccine, and various monoclonal antibodies and other recombinant
products of continuous cell lines in clinical trials. For tissue
plasminogen activator, large doses of 100 mg per patient or more
have been used. At the time of writing over 10 kg of CHO-derived
tissue plasminogen activator has been sold since late 1987 for administration to over 100,000 human patients. For recombinant
factor VIII, erythropoeitin, and soluble CD4 proteins, chronic administration has been employed. Millions have received polio and

rabies vaccines prepared in continuous Vero cells. In addition to
this human experience, livestock animals have received annual inoculations of foot-and-mouth virus vaccine prepared in BHK-21 (a
highly tumorigenic CCL) for up to 14 years without effect (69).
No effects have been reported which might be attributed to oncogenic
factors.

Thus, scientific tools of characterization and principles of process
validation are available to protect patients from putative risks of
tumorigenicity associated with products prepared in CCLs. Increasing clinical experience also supports this conclusion.

ACKNOWLEDGMENTS

Retrovirus assay data were provided by Dr. Judy Poilly of Hazel-
ton Laboratories, Kensington, MD. Betty Bullock provided expert
secretarial services for this manuscript. The author is grateful
to these individuals.

REFERENCES

1. Cooper, P. D. Large-scale production of indefinitely propo-
    gated cell lines. *NCI Monogr.* 29:63–69 (1968).
2. Daniels, W. F. Comments on technology of production-contin-
    uous cell lines. *NCI Monogr.* 29:71–72 (1968).
3. Hilleman, M. R. Cell line saga: An argument in favor of
    production of biologics in cancer cells. *Adv. Exp. Biol. Med.*
    118:47–58 (1979).
4. Petricciani, J. C. Regulatory considerations for products
    derived from the new biotechnology. *Pharmaceut. Manuf.*
    5:31–34 (1985).
5. Lubiniecki, A. S. Safety considerations for cell culture de-
    rived biologicals. In *Large Scale Cell Culture*, Carl Hansen,
    Munich, pp. 231–247 (1987).
6. Builder, S. E., Van Reis, R., Paoni, N., Field, M., and
    Ogez, J. R. Process development in the regulatory approval
    of tissue-type plasminogen activators. In *Advances in Animal
    Cell Biology and Technology for Bioprocesses* (R. E. Spier,
    J. B. Griffiths, J. Stephenno, and P. J. Crooy, eds.),
    Butterworth, pp. 452–462, (1989).
7. Dickson, R. B., Kasid, A., Huff, K. K., Bates, S. E.,
    Knabbe, C., Brozzart, D., Glemman, E. P., and Lippmann,
    M. E. Activation of growth factor secretion in tumorigenic
    states of breast cancer induced by 17.B-estrodiol or v-*rasHa*
    oncogene. *Proc. Natl. Acad. Sci. USA* 84:837–841 (1986).
8. Varmus, H. E. The molecular genetics of cellular oncogenes.
    Ann. Rev. Genet. 18:553–612 (1984).

9. Westin, E. H., Wong-Staal, F., Gelmann, E. P., Dela Favero, R., Papas, J. S., Lautenberger, J. A., Eva, A., Reddy, E. P., Tronick, S. A., Aaronson, S. A., and Gallo, R. C. Expression of cellular homologues of retroviral *onc* genes in human hematopoeitic cells. *Proc. Natl. Acad. Sci. USA* 79:2490−2494 (1982).

10. Brockman, W. W. Transformation of BALB/c-3T3 cells by tsA mutants of Simian virus 40: Temperature sensitivity of the transformed phenotype and retransformation by wild-type virus. *J. Virol.* 25:860−870 (1978).

11. Pulciani, S., Santos, E., Lauver, A. V., Long, L. K., Robbins, K. C., and Barbacid, M. Oncogenes in human tumor cell lines: Molecular cloning of a transforming gene from human bladder carcinoma cells. *Proc. Natl. Acad. Sci. USA* 79:2845−2849 (1982).

12. Maness, P. F. and Levy B. T. Highly purified $pp60^{src}$ induces the action transformation in microinjected cells and phosphonylotes selected cytoskeletal proteins in vitro. *Mol. Cell. Biol.* 3:102−111 (1983).

13. Tjian, R., Fey, G., and Graessmann, A. Biological activity of purified simian virus 40 T antigen proteins. *Proc. Natl. Acad. Sci. USA* 75:1279−1283 (1978).

14. Kurachi, H., Okamoto, S., and Oka, T. Evidence for the involvement of the submandibular gland epidermal growth factor in mouse mammary tumorigenesis. *Proc. Natl. Acad. Sci. USA* 82:5940−5943 (1985).

15. Danielpour, D. and Sirbasku, D. A. New perspective in hormone-dependent (responsive) and autonomous mammary tumor growth: Role of autostimulatory factors. *In vitro* 20:975−980 (1984).

16. Lowy, D. R. Potential oncogenic hazards posed by oncogene encoded proteins. *Dev. Biol. Standard.* 68:63−67 (1987).

17. Boiron, M., Levy, J. P., and Thomas, M. Production de tumor chez la hamster par inoculation d'acide desoxyriboni- culeique extrait de cellules infectées par le virus SV40. *Ann. Inst. Pasteur Paris* 108:298−305 (1965).

18. Orth, G., Antanasui, P., Boiron, M., Rebierre, J. P., and Paoletti, C. Infections and oncogenic effects of DNA extracted from cells infected with polyoma virus. *Proc. Soc. Exp. Biol. Med.* 115:1090−1095.

19. Fleckenstein, B., Daniel, M. D., Hunt, R. D., Werner, J., Falk, L. A., and Mulder, C. Tumor induction with DNA of oncogenic primate herpesviruses. *Nature* 274:57−59 (1978).

20. Graham, F. G., Abrahams, P. J., Mulder, C., Heijneker, H. L., Warnaar, S. O., de Vries, F. A. J., Viers, W., and

van der Eb. Studies on in vitro transformation by DNA and DNA fragments of human adenovirus and SV40. *Cold Spring Symp. Quant. Biol.* 39:637−651 (1975).

21. Bishop, J. M. Cellular oncogenes and retroviruses. *Ann. Rev. Biochem.* 52:301−354 (1983).

22. Blair, D. G., Cooper, C. S., Oskarsson, M. K., Ender, E. A., and vande Woude, G. F. New method for detecting cellular transforming genes. *Science* 281:1122−1125 (1982).

23. Fung, Y. K., Crittenden, L. B., Farley, A. A., and Kung, H. J. Tumor induction by direct injection of cloned v. *src* DNA. *Proc. Natl. Acad. Sci. USA* 80:353−357 (1983).

24. Compere, S. J., Baldacci, P., Sharpe, A. H., Thompson, T., Land, H., and Jaenisch, R. The *ras* and *myc* oncogenes cooperate in tumor induction in many tissues when introduced into midgestational mouse embryos by retroviral vectors. *Proc. Natl. Acad. Sci. USA* 86:2224−2228 (1989).

25. Levinson, A., Svedersky, L. P., and Palladino, M. A. Tumorigenic potential of DNA derived from mammalian cell lines. In *Abnormal Cells, New Products, and Risk* (H. E. Hopps and J. P. Petricciani, eds.), Tissue Culture Association, Gaithersburg, MD, pp. 161−165 (1985).

26. Mufson, R. A. and Gesner, T. Lack of tumorigenicity of cellular DNA and oncogene DNA in newborn Chinese hamsters. In *Abnormal Cells, New Products, and Risk* (H. E. Hopps and J. P. Petricciani, eds.), Tissue Culture Association, Gaithersburg, MD, pp. 168−169 (1985).

27. Palladino, M. A., Figari, I. S., Obijeski, J., and Levinson, A. D. Risk Assessment of exogenous DNA uptake in vivo. *In Vitro* 25:40A (1989).

28. Petricciani, J. P. and Regan, P. J. Risk of neoplastic transformation from cellular DNA: Calculations using the oncogene model. *Dev. Biol. Standard.* 68:43−49 (1987).

29. Miyoshi, I., Kubonishi, I., Yoshimoto, D., Akagi, T., Ohtsuki, Y., Shiraishi, Y., Nagata, K., and Hiruma, Y. Type C virus particles in a cold blood T-cell line derived from cultivating normal human and leukocytes and human leukaemic T-cells. *Nature* 294:770−771 (1981).

30. Poiesz, B. J., Ruscetti, F. W., Gazdar, A. F., Bunn, P. A., Minna, J. D., and Gallo, R. C. Detection and isolation of type C retrovirus particles from fresh and cultural lymphocytes of a patient with cutaneous T-cell lymphoma. *Proc. Natl. Acad. Sci. USA* 77:7415−7419 (1980).

31. Barre-Sinoussi, F., Chermann, J. C., Rey, F., Nugeyre, M. T., Chamaret, S., Gruest, J., Danguet, C., Azler-Blin, C., Vezinet-Brun, F., Rouzioux, C., Rouzioux, C., Rosenbaum, W.,

and Montagnier, L. Isolation of T-lymphotropic retrovirus from a patient at risk for acquired immune deficiency syndrome (AIDS). *Science* 220:868–870 (1983).

32. Gallo, R. C., Salahuddin, S. Z., Popovic, M., Shearer, G. M., Kaplan, M., Haynes, B. F., Palker, T. J., Redfield, R., Oleske, J., Safai, B., White, G., Foster, P., and Markham, P. D. Human T-lymphotropic retrovirus, HTLV-III, isolated from AIDS patients and donors at risk for AIDS. *Science* 224:500–503 (1984).

33. Kurman, R. J., Santz, L. E., Janson, A. B., Perry, S., and Lancaster, W. D. Papillomavirus infection of the cervix. I. Correlation of histology with viral structural antigens and DNA sequences. *Int. J. Gynecol. Pathol.* 1:17–28 (1982).

34. Epstein, M. A., Achong, B. D., and Barr, Y. M. Virus particles in cultural lymphoblasts from Burkitt's lymphoma. *Lancet* 1:702–703 (1964).

35. zur Hausen, H., Schulte-Holthausen, H., Klein, G., Henle, W., Henle, G., Clifford, P., and Santesson, L. EB virus-DNA in biopsies of Burkitt tumors and anaplastic carcinomas of the nasopharynx. *Nature* 288:1056–1058 (1970).

36. Nishioka, K., Hirayama, T., Sekine, T., Okochi, K., Mayumi, M., Sung, J. L., Hui, L. C., and Lin, T. M. Australia antigen and hepatocellular carcinoma. *Gann Monogr. Cancer Res.* 14:167–175 (1973).

37. Hanafusa, H., Hanafusa, T., and Rubin, H. The defectiveness of Rous sarcoma virus. *Proc. Natl. Acad. Sci. USA* 49:572–580 (1963).

38. Somers, K. D., May, J. T., Kit, S., McCormick, K. J., Hatch, G. G., Stenback, W. A., and Trentin, J. J. Biochemical properties of a defective hamster C-type oncornavirus. *Interviol.* 1:11–18 (1973).

39. Demsey, A., Collins, F., and Kawka, D. Structure of and alterations to defective murine sarcoma virus particles lacking envelope proteins and core polyprotein cleavage. *J. Virol.* 36:872–877 (1980).

40. Lieber, M. M., Benveniste, R. E., Livingston, D. M., and Todaro, G. J. Mammalian cells in culture frequently release type C viruses. *Science* 182:565–569 (1973).

41. Callahan, R., Chiu, I., Wong, J. F. H., Tronick, S. R., Roe, B. A., Aaronson, S. A., and Schlom, J. A new class of endogenous human retroviral genes. *Science* 228:1208–1210 (1985).

42. Stromberg, K. and Huot, R. I. Preferential expression of endogenous type C viral antigen in rhesus placenta during ontogenesis. *Virology* 112:365–369 (1981).

43. Russell, P., Gregerson, D. S., Alber, P. M., and Reid, T. W.
     Characteristics of retrovirus associated with a hamster melanoma.
     J. Gen. Virol. 43:317–326 (1979).
44. Lubiniecki, A. S. and Szakal, A. K. Induction of virus-
     associated tumors in carcinogen-treated guinea pigs. *Fed.
     Proc.* 38:1450 (1979).
45. Lieber, M. M., Livingston, P. M., and Todero. G. J. Super-
     induction of endogenous type C virus by 5-bromodeoxyuridine
     from transformed mouse clones. *Science* 181:443–444 (1973).
46. Boyd, A., Derge, J. G., and Hampar, B. Activation of
     endogenous type C virus in BALB/c mouse cells by herpes-
     virus DNA. *Proc. Natl. Acad. Sci. USA* 75:4558–4562 (1978).
47. Schildlovsky, G. Structure of RNA tumor viruses. In *Recent
     Advances in Cancer Research: Cell Biology, Molecular Biology,
     and Tumor Virology* (R. C. Gallo, ed.), CRC Press, Cleveland,
     pp. 189–245 (1980).
48. Zeve, V. H., Gonda, M. A., and Lebiedzik, J. Application
     of an automated particle analysis system to the quantitation
     of virus particles. *J. Natl. Cancer Inst.* 53:1099–1102 (1974).
49. Peebles, P. T., Haapala, D. K., and Gazdar, A. F.
     Deficiency of viral ribonucleic acid-dependent deoxyribonucleic
     acid polymerase in noninfectious virus-like particles released
     from murine sarcoma virus-transformed hamster cells. *J. Virol.*
     9:488–493 (1972).
50. Baltimore, D. and Smoler, S. Primer requirement and temp-
     late specificity of the DNA polymerase of RNA tumor viruses.
     Proc. Natl. Acad. Sci. USA 68:1507–1511 (1971).
51. Michaides, R., Schlom, J., Dahlberg, J., and Perk, J.
     Biochemical properties of the bromodeoxyuriding-induced
     guinea pig virus. *J. Virol.* 16:1039–1050 (1975).
52. Smith, R. E. and Bernstein, E. H. Production and purifica-
     tion of large amounts of Rous sarcoma virus. *Appl. Microbiol.*
     25:346–353 (1973).
53. Toplin, I. and Sottong, P. Large-volume purifications of tumor
     viruses by use of zonal centrifuges. *Appl. Microbiol.*
     23:1010–1014 (1972).
54. Mathes, L. E., Yohn, D. S., and Olsen, R. G. Purification
     of infectious feline leukemia virus from large volumes of
     tissue culture fluids. *J. Clin. Microbiol.* 5:372–374 (1977).
55. Bassin, R. H., Simons, P. J., Chesterman, P. C., and
     Harvey, J. J. Murine sarcoma virus (Harvey): Character-
     istics of focus formation in mouse embryo cells cultures and
     virus production by hamster tumor cells. *Int. J. Cancer*
     3:265–272 (1965).
57. Vogt, P. K. and Ishizakii, R. Recriprocal patterns of genetic
     resistance to avian tumor viruses in two lines of chickens.
     Virology 26:664–675 (1965).

58. Odaka, T. and Yamamoto, U. Inheritance of susceptibility to Friend mouse leukemia virus. *Jap. J. Exp. Med.* 32:405–411 (1962).
59. Gilden, R. V. Biology of RNA tumor viruses. In *The Modular Biology of Animal Viruses* (D. Nayak, ed.), Dekker, New York, pp. 435–542 (1978).
60. Charman, H. P., White, M. H., Rahman, R., and Gilder, R. V. Species and interspecies radioimmunoassays for rat type C virus P30: Interviral comparisons and assay of human tumor extracts. *J. Virol.* 17:51–59 (1976).
61. Jackson, M. L., Nakamura, G. R., Lubiniecki, A. S., and Patzer, E. J. Attempts to detect retroviruses in continuous cell lines: Radioimmunoassays for hamster P30 protein. *Dev. Biol. Standard.* 66:541–553 (1987).
62. Wurm, F. M., Williams, S. R., Anderson, K., Dinowitz, M., Obijeski, J., and Arathoon, R. Presence and transcription of endogenous retroviral sequences in CHO cells. In *Advances in Animal Cell Biology and Technology for Bioprocesses* (R. E. Spier, J. B. Griffiths, J. Stephenno, and P. J. Crooy, eds.), Butterworth, pp. 76–81 (1989).
63. Attardi, G., Nasno, S., Rouriere, J., Jacob, F., and Gross F. *Cold Spring Harbor Symp. Quant. Biol.* 28:363–370 (1963).
64. Tihon, C. and Green, M. Cyclic AMP-amplified replication of RNA tumor virus-like particles in Chinese hamster ovary cells. *Nature New Biol.* 244:227–231 (1973).
65. Manly, K. F., Givens, J. F., Taber, R. L., and Ziegel, R. F. Characterization of virus-like particles released from the hamster cell lines CHO-K1 after treatment with 5-bromodeoxyuridine. *J. Gen. Viol.* 39:505–517 (1978).
66. Lubiniecki, A. S. Endogenous retroviruses of continuous cell substrates. *Dev. Biol. Standard.* 70:187–191 (1988).
67. Saiki, R.k, Gelfand, D. H., Stoffel, S., Scharf, S. J., Higuchi, R., Horn, G. T., Mullis, K. B., and Erlich, H. A. Primer directed enzymatic amplification of DNA with a thermo-stable DNA polymerase. *Science* 239:487–491 (1988).
68. Office of Biologics Research and Review. Points to consider in characterization of cell lines used to produce biologicals. Food and Drug Administration, U.S. Public Health Service, Rockville, MD (1987).
69. Regan, P. J. and Petricciani, J. P. The approach used to establish the safety of veterinary vaccines produced in the BHK21 cell line. *Dev. Biol. Standard.* 68:19–25 (1987).

19

Process Validation for Cell Culture-Derived Pharmaceutical Proteins

ANTHONY S. LUBINIECKI
SmithKline Beecham Pharmaceuticals, King of Prussia, Pennsylvania

MICHAEL E. WIEBE and STUART E. BUILDER
Genentech, Inc., South San Francisco, California

I. INTRODUCTION

Decisions by regulatory authorities to permit clinical trials or marketing of cell culture-derived biologicals are based on assessment of benefits and risks associated with use of the proposed product. Risks can be classified into two categories: (1) those associated with potential pharmacological and toxicological effects of the active ingredient(s) and (2) those associated with other components of the product which arise from the method of preparation. One goal of process validation is to demonstrate that real and putative process-associated risks have been reduced to acceptable levels. Another goal is the verification of process consistency, i.e., the ability to produce essentially the same product (with constant safety and efficacy) over time given the normal variability in operating conditions.

This chapter will discuss these concepts and provide examples of their use in the production of cell culture-derived biologicals. Validation is only one of many tools to assure product safety, and it is important to remember that it is best used in combination with other tools such as effective process and facility design. For additional discussion of validation concepts, the reader is referred to recent reviews of this topic (1,2).

II. APPROACH AND RATIONALE

A. Identification and Quantitation of Risk

The first step in any validation effort is the identification of the potential risk factors to be addressed. Cell culture and recovery scientists should begin to develop a list of putative risk factors early in the process development cycle and then update it periodically as additional information and experience are obtained. The process design must include steps that will remove or inactivate potential risk factors in a reliable manner. The biological, biophysical, and biochemical nature of each factor should be assessed to completely define each potential risk in experimental laboratory terms. Part of this information comes from the cell bank characterization program described in Chapters 5, 17, and 18. Table 1 contains a list of typical factors to consider.

In the past, some of these risk factors have been shown to be of actual consequence rather than theoretical concern. For example, in the preparation of inactivated poliovirus vaccine, the risk of serious disease fom residual infectious virus occurred when the inactivation process was inadequately designed and monitored (3). Similarly, some human plasma cryoprecipitate preparations were contaminated by HIV-1 between 1980 and 1985, which resulted in AIDS in many hemophiliacs (4). For purposes of subsequent analysis, development scientists must distinguish between rare but significant real risk factors and frequent theoretical but less significant putative risk factors. Real risks require special consideration and are discussed in more detail in Section II.B.

After the potential risk factors are identified, they must be quantified. For each risk factor it is essential to determine the level present at appropriate process steps (where they might be introduced, concentrated, or reduced), including final product. For example, the levels of residual cellular nucleic acids might be measured in harvested cell culture fluid to determine the initial concentration of this contaminant, in intermediate purification process fluids to determine the extent of concentration or removal, and in the final product to determine the residual concentration. An adequate number of repeated runs should be sampled to provide confidence in estimating the degree of removal or reduction in concentration and to provide sufficient data for estimating the range

Table 1   Putative Risk Factors for Cell Culture-Derived
Products

Putative risk factor	Sources
Intact cells	Cells
Adventitious agents (bacteria, fungi, mycoplasma, viruses)	Cells, raw materials
Endogenous retroviruses	Cells
Residual cellular nucleic acid	Cells
Residual cellular proteins	Cells
Other foreign proteins	Raw materials, antibodies used in purification

Microbial contaminants
  endotoxin
  proteins
	In-process bioburden

Process chemicals
  antibiotics
  ligands
  solvents
  cleaning compounds
  inducers
  nutrients
	Raw materials

of effectiveness of the process. In the event that a given risk factor cannot be detected, the worst case should be assumed, i.e., that the concentration is just below the minimum detectable level.

While the most reliable quantitation of putative risk factors comes from fully scaled processes, much time is lost if quantitation is postponed until full-scale operations are possible. A useful strategy is to begin quantitation and validation work on early laboratory and pilot scale operations. This not only gives important information about how the process must be designed but also can build confidence in the ultimate suitability of a large-scale operation. Often it is useful and sometimes necessary to repeat this validation work on full-scale operations. With the prior knowledge of the methodology and initial results, these tests are more likely to go smoothly and predictably. Safety, regulatory, or practical considerations may also require that validation studies on unit operations be performed outside the manufacturing plant at the laboratory or pilot scale. Thus, readily scalable systems have obvious advantages in risk assessment studies as well as in process development activities.

B. Establishment of Safety Goals

In the language of risk assessment, safety is often defined as the absence of risk. Like absolute purity or sterility, absolute safety is a theoretical concept whose attainment can never be proven. Despite this limitation, reasonable approximations can be developed using established scientific and engineering principles to provide products with negligible or acceptable levels or risk. The evaluation of safety will be made by national control authorities while reviewing the product license application. Therefore, early discussions with regulatory bodies are useful to establish safety goals for the developing product.

Considerable thought must be given to determining what level of *each* putative risk factor is considered to be safe. For example, if a nontoxic natural substance (e.g., an amino acid, salt, sugar) might be present in the final container of a parenterally administered product at a concentration substantially lower than normally found in plasma, this is likely to be considered acceptable and relatively safe.

Other situations are more difficult to assess. Various regulatory and standard-setting organizations may promulgate different advice for the same situation. A recent risk assessment for residual cellular nucleic acid by the World Health Organization suggests that up to 100 pg per dose poses negligible risk to recipients of biologicals from continuous cell substrates (5). The U.S. FDA Center for Biologics Evaluation and Review (CBER) states that products containing less than 10 pg of residual cellular nucleic acid should be considered acceptable for single administration into humans (6). Although these two statements are not in confluct with each other, the potential for confusion exists and the value of international harmonization of such standards is obvious. It is sometimes argued that a more stringent goal is appropriate for a real and present risk than for a putative one. For example, the level of virus inactivation demonstrated for rabies virus in inactivated rabies vaccine should be greater than for putative retrovirus-like particles in a recombinant rabies glycoprotein vaccine made in a continuous hamster cell line like Chinese hamster ovary (CHO). In the former case, real infectious rabies virus is known to be present in the harvested culture medium and will cause a fatal human disease unless completely inactivated. In the latter, the retrovirus-like particles are considered to be defective; therefore an effective biological barrier exists. Process inactivation or removal steps would provide added assurance of safety, but reliance on these measures would be less important since the risk is purely theoretical. Therefore, safety goals for the inactivation steps in the latter case can be substantially less stringent. The establishment of safety goals should also reflect consideration of the intended clinical use, including the route of administration, dose size and frequency, and the total amount of material to be administered. The nature of the host response should also be considered, such as dose response, kinetics, recovery, resistance, and so forth. These issues are complex and require estimation of the possible medical, toxicological, pharmacological, and immunological effects of each substance.

C. Validation Concepts

According to the FDA *Guideline on General Principles of Process Validation* (7), "Process validation is a documented program which provides a high degree of assurance that a specific process will consistently produce a product meeting its predetermined specifications and quality attributes." The major elements of this definition deserve some explanation.

Validation of a process must be a documented program: Validation reports contain a written statement of purpose and goals, study design, protocol, and statistical methodology. Ideally, the study is designed to include worst case operating conditions. The applicability of a validation program will diminish greatly if actual manufacturing conditions are significantly altered from those employed in the validation studies. Validation reports usually contain data collected on process parameters and from analysis of in-process materials and final product(s). As a result, one of the

benefits of validation is that the normal ranges of process parameters become linked to product lots which are shown to be acceptable from a quality control standpoint. Early acceptable product
lots determine these normal ranges; later lots which stay within the
normal ranges provide additional evidence of acceptability. Careful
attention to detail and record keeping is extremely valuable.

Usually, the entire process validation study protocol is given
thorough review and approval by both manufacturing and quality
personnel. Validation programs must be written to comply with
regulations and be accepted by control authorities. Ideally, the
plan is reviewed and agreed beforehand. The results and conclusions should also be reviewed by manufacturing, quality personnel, and the study participants in order for the study to provide its greatest value.

A high degree of assurance must be provided that the process
will routinely achieve the stated goals. A significant part of the
preparation for the study is to establish that the goals are reasonable and possible to achieve, and to assure that the study design is based on solid scientific and engineering principles. End-
points and specifications must be matched to the goals and study
design. While nothing can be proven with statistics alone, it is
difficult to conduct any validation efforts without appropriate statistical analysis of the data.

Demonstration of consistent performance is essential. In practice, this is usually satisfied by a minimum of three distinct studies of identical design in which representative production runs
exhibit comparable and acceptable results. Processes which perform erratically during validation studies usually suffer from inadequate development or lack of control of one or more process
variables. Further developmental investigation is usually warranted
in those cases, followed by additional validation studies. Processes
which cannot be validated are not ready for operation in a regulatory sense and probably are not desirable from the standpoint of
manufacturing reliability.

A validation program is specific for a product manufactured by
a given process. The program is conducted on relevant process
steps in appropriate facilities using calibrated equipment and released lots of raw materials. In some cases where the validation
methods require the use of harmful substances (e.g., radioisotopes,
hazardous chemicals, or infectious agents), equipment and/or facilities other than those normally used for manufacturing may be
employed, providing they faithfully reproduce the manufacturing
process in all key attributes. For any validation study, both the
manufacturing and quality assessment activities (assays,

specifications, and instrument calibrations) should be performed in compliance with current Good Manufacturing Practices (cGMP).

The final point in the regulatory definition of validation pertains to the employment of predetermined specifications. The goals of the validation program are reduced to practice in the establishment of specifications. These must reflect the intended safety, purity, potency, identity, and stability of the product. They must represent fixed targets agreed to by manufacturing and quality personnel in advance of laboratory studies.

Validation is an important tool in the pharmaceutical industry generally, and specifically for manufacture of cell culture biologicals due to their molecular complexity. It provides assurances of risk removal which are vital to proving product quality attributes. National control authorities also recognize this, and some may insist on review of validation programs prior to product licensure. In those cases the validation programs are scrutinized for compliance with principles discussed above. Most validation programs are repeated at least in part at appropriate intervals to verify that the original assurances are still operative, usually every 6−24 months depending on the nature of each program. Any changes to process equipment, operation, or facilities should be carefully reviewed to determine whether repeating validation studies are required.

D. Characterization or Evaluation

Characterization studies resemble validation studies in terms of the underlying science, engineering, and statistical principle employed but have less formal compliance to regulatory guidelines and are repeated less often. Characterization studies are more common in newly emerging scientific fields where validation protocols and guidelines have not yet been established. In some laboratories, these efforts may be known as process evaluation. During characterization studies, established scientific and engineering principles are employed to demonstrate that a process of product achieves one or more specific attributes. Methods are often not validated in the regulatory sense, but they must be qualified scientifically (i.e., have appropriate controls and the ability to yield interpretable data). Many of these studies need to be performed only once or a few times, but with sufficient rigor that their relevance to manufacturing efforts endures. Some of the topics and examples discussed later in the chapter are actually performed as process characterization studies rather than proces validation studies. The characterization studies for mammalian cell banks are described in detail in Chapters 5, 17, and 18.

E. Demonstration of Validation

In previous sections, the need to identify the quantitative risk factors and to establish safety goals has been discussed. Following
these efforts, it is appropriate to obtain relevant data to determine
whether putative risk factors actually are present in the product
at significant levels and to what extent they can be reduced in magnitude when present. The sensitivity of the assay methods employed to determine the concentration of putative risk factors is
an important aspect of the design of studies to measure the removal of unwanted factors. Clearly, more sensitive test methods
will allow proof of greater removal of a putative risk factor. Another important consideration is the limitation in the attainable concentration for analysis. For example, it may not be possible to achieve
material concentrations beyond some desired level (due to their solubility, biological inactivation, or required physical space). Such
limitations often defined the limits of worst case conditions for analysis. Taken together, the sensitivity of the assay and the achievable concentration limits of additives help define a concept often
called the "window" of clearance. This is the difference between
the highest attainable initial concentration of a contaminant or impurity and the lowest detectable concentration of that contaminant.
This difference (on a log scale) represents the amount of impurity
that can be claimed to be measured, hence removed. Many validation studies are constrained to claim only what can be actually measured (i.e., cannot employ dosimetric or extrapolational statistical
methods). In such cases, window size becomes a crucial limiting
factor in the design and usefulness of such validation efforts.

As a hypothetical example, assume that the acceptable level of
an undesirable impurity is assumed to be 10 pg per dose. In addition, assume that the starting concentration of the impurity in harvested culture fluid is 10 µg/ml and one dose can be manufactured
from 1 ml. The sensitivity of the assay is 10 ng/ml. Simple calculation shows that attainment of the goal requires demonstration
of a 1 million-fold removal of the impurity (10 µg ÷ 10 pg), yet
the window size in this example is only three orders of magnitude
(10 µg/ml ÷ 10 ng/ml). Choices available to the validation team
include

1. Resign to failure of the approach (in effect, abandon or
    change the process).
2. Concentrate the sample 1000-fold (which may be prevented
    by some practical limitation such as solubility or result in
    a test article not representative of the manufacturing
    process), or
3. Validate multiple independent process steps for removal of
    the putative risk factor. If the different removal steps

possess fundamentally different mechanisms of action, the individual clearance values can be cumulated.

In the current example, 6 $\log_{10}$ is the goal and the window is only 3 $\log_{10}$. If two different process steps were found to be capable of removing as much of the risk factor as could be measured, then each would contribute 3 $\log_{10}$ of removal for a total of 6 $\log_{10}$. In practice many studies involve the use of radioactively labeled analos of putative risk factors or infectious agents added into process fluids ("spiked") from exogenous sources. A representative process run is then made, or modeled, as appropriate to the study design.

The most reliable data are derived from actual production runs. When this is precluded because of the nature of the reagents or study design, laboratory or pilot facilities are employed. Small-scale equipment is permissible providing that the scalability of the unit operation is demonstrated. Data must be collected under conditions relevant to the full-scale manufacturing process operated under GMP regulations. After appropriate statistical analysis, it should be clear whether the goal has been attained or not. Ideally, the goal should be attainable under worst case conditions.

F. Effects of Validation

The principal goals of process validation are to prove that potential risks have been adequately reduced and to assure that products will consistently possess required quality attributes which have been established. When these goals are achieved, several indirect effects follow. One effect of a properly designed validation program is to reduce the number and scope of quality control tests that are performed on each product lot without sacrificing quality. This has significant impact on the economics of many products. Many of the tests employed in process validation are tedious, time-consuming, and scientifically demanding. While providing useful information, they are often difficult to adapt to or manage in a quality control lot release setting. These tests should be confined to characterization or validation studies whenever possible.

Another benefit of a thorough process validation program can be an improvement in process economics. This may arise from the high level of scrutiny of the development of processes which are to be validated and of the validation process. When done properly, weak points of a manufacturing process may be discovered and strengthened during the validation effort. This has a generally positive impact on process economics, since more reliable processes result in less lost or failed product. This is particularly important for cell culture biologicals which are well known to possess significantly higher manufacturing costs than microbial products.

III. EXAMPLES

A. Risk Reduction

This section will describe some examples of how putative risk factors may be removed or inactivated. The specific implementation of process validation is customized to the chemical, biological, and physical nature of each unwanted agent, to the nature of the product, and to the design of the process.

1. Cells

Recombinant DNA and hybridoma technology (as well as the manufacture of some traditional biologicals such as vaccines) employ continuous/immortal cell substrates. Many of these substrates were originally derived from tumors or have been shown to be tumorigenic when inoculated into immunosuppressed animals. The inoculation of living cells into recipients of these products would obviously be considered unacceptable, and demonstration of their removal from process fluids must be achieved and validated.

Fortunately, mammalian cells are quite large (usually $10-20$ μm) and fragile relative to bacteria. In challenge studies, the use of membrane filters with submicrometer range of nominal porosity is easily shown to be effective in removing many orders of magnitude of bacteria. Since filters remove cells based on size, it is reasonable to conclude that mammalian cells with at least 10-fold larger diameter than bacteria are also removed effectively. This has been confirmed experimentally. Filtration technology is robust, reliable, can be monitored for integrity before and after each use, and is easily validated. Thus, the placement of appropriate filters at strategic points during product recovery operations will not only assure the absence of cells from final product, but also provide a substantial aseptic barrier to bacterial contamination of downstream processing.

Other methods of cell removal including continuous-flow centrifugation, precipitation, floculation entrapment, and sedimentation can also be employed. However, all of these may result in some cell lysis and give rise to higher levels of residual cellular protein and nucleic acid contamination in the starting fluid than would be achieved by use of an appropriately designed filtration process.

2. Endogenous Materials

Despite the fact that cells are easily removed, they may contribute a variety of other contaminants prior to their removal at harvest. One class of these potential contaminants is endogenous retroviruses, whose genetic information resides within the cell substrate genome. Other classes of cellular contaminants include residual cellular

DNA (RC-DNA) and protein. The validation of the removal of each of these contaminants is somewhat different and is presented below.

Retroviruses and Retrovirus-like Particles. Contamination of the bank cells by adventitious (exogenous) agents is an unlikely event following the successful completion of the characterization of the banked cell line (see Chapter 17).

In contrast, the genetic information for endogenous retroviruses has resided in mammalian genomes for tens of millions of years (8) and related sequences have been found in organisms as phylogenetically distant as yeast (9). Their properties are described more fully in Chapter 18. Their expression is highly variable, but their genetic resemblance to infectious or tumorigenic retroviruses requires that they be considered in the worst case as potential pathogens. An outline for a process validation study for their removal is shown in Fig. 1.

The characterization of the master cell bank should include studies to determine the extent of retrovirus expression, which may vary from expression of infectious particles to defective or virus-like particles to absence of detectable particles. All, some, or none of the major endogenous retroviral genes may be expressed. The expression level of viral genes or virus particles should be defined qualitatively and quantitatively. In the event that no particles are observed, it should be assumed that the level just below the limit of detection is a worst case assumption.

Goal	Demonstrate loss of infectivity of model retroviruses when exposed to process conditions
Equipment	Scaled down recovery unit operations as appropriate
Design	Conduct scaled down studies with representative process materials "spiked" with infectious retroviruses; assay residual infectivity
Acceptance criteria	Cumulative inactivation effects leading to less than one putative virus-like particle per million doses

Figure 1   Retrovirus inactivation.

After determining the level of expression, a goal for degree of removal by the process is established. Available evidence for most rDNA expression systems studied so far indicates that endogenous virus-like particles resembling type A particles are devoid of biological activity and generally lack biochemical and biophysical hallmarks of functional retroviruses (10). A target of 10^{-6} residual particles per dose (one particle in 10^6 doses) of final product has been accepted (17) as assuring the safety of the product. The processes of removal and/or inactivation of virus-like particles during product purification must be able to reduce the putative particle concentration in the final product to this point. These efforts at virus inactivation and removal combined with the apparent defectiveness of the virus-like particles result in product usually being deemed acceptable for pharmaceutical applications if they meet the goals set for these risk-benefit analysis.

Let us consider an example where virus-like particles are observed in the cells from the master cell bank but none are detected in harvested cell culture fluids even when a sample is concentrated 5000-fold. If the lower limit of sensitivity of transmission electron microscopy is $10^{7.4}$ particles per ml, then the theoretical worst case level of particles in the harvested fluid is considered to be just below the level of detection divided by the concentration factor; i.e., $(10^{7.4}) \div (10^{3.7}) = 10^{3.7}$ putative particles per ml of harvested cell culture fluids. If, in the worst case, 20 liters of fluid is processed to manufacture one dose of product, then that dose may contain as many as $(10^{3.7}$ particles/ml$) \times (10^{4.3}$ ml/dose$)$ or $10^{8.0}$ putative particles per dose. The minimum clearance required from the worst initial case to the final goal is $\log_{10} (10^{8.0}$ particles per dose$) - \log_{10} (10^{-6}$ particles per dose$)$ or 14.0. As discussed earlier, application of the window concept instructs us that if only $4-5 \log_{10}$ of removal can be demonstrated per step, then at least three process steps with independent mechanisms of action must be validated to achieve the required clearance of $14.0 \log_{10}$. The data are collected under process-relevant conditions. If an infectious retrovirus is present in the cell substrate or process fluidity, the validation exercise is conducted most appropriately with that agent. If none is detected, then the validation studies may be performed using model retroviruses which are available, relatively safe, and biologically measurable. A list of candidate treatments is shown in Table 2. For studies where a surrogate model virus is employed, ion exchange chromatography is not a robust choice as a removal step since the surface physicochemical properties of the putative virus-like particles and the model virus may be different. In constrast, size exclusion chromatography may have more value since retroviruses are generally confined to a size range of $80-130$ nm in diameter. Chemical and/or physical inactivation steps are ideal if they do not damage the protein

Table 2   List of Candidate Retrovirus
Inactivation/Removal Steps

Physical methods	Chemical methods
Irradiation	Solvents
Sonication	Detergents
Heat	Chaotropic agents
Filtration	Extreme pH
	Reactive chemical agents

product, since the properties affected by these methods are common
to multiple members of the virus group.

Each step is then evaluated for effectiveness of removal or inactivation. Appropriate process fluids (preferably samples from
different batches) are spiked with the model retrovirus. The samples are then treated in a process-relevant manner, i.e., they are
exposed to the same conditions as if the actual process were being
conducted. Upon completion of the appropriate processing, samples
are taken for determination of residual infectivity or other measurable property. The results are compared to the titer of un-
treated control samples. After calculation of the efficient of removal
or inactivation for each step, the $\log_{10}$ of clearance steps with different mechanisms of removal are added as described above. In
the example shown in Table 3, a total of at least 15.0 $\log_{10}$ of removal and inactivation was measured by evaluation of four independent process steps. Since the target level was 14.0 $\log_{10}$, the
safety goal was attained.

Once this validation package is completed, care must be exercised to maintain its integrity. The equipment used to carry out
the process must perform consistently over time to achieve the
same removal or inactivation as validated. The facility must be
designed to minimize forward flow of contaminants, i.e., once
endogenous retrovirus removal steps are completed, the process
fluids should not be reexposed to unprocessed culture fluids or
areas or equipment which might recontaminate them.

These types of studies are crucial to assuring the safety of
rDNA and hybridoma products (10,11). They share many similarities with earlier work on human interferon prepared from continuous lymphoblastoid cells (12) and on studies of hepatitis B
surface antigen vaccine prepared from human plasma obtained from
chronic virus carriers (13).

Table 3   Example of Process Validation for Virus Removal/Inactivation

Step	$\log_{10}$ removal inactivation[a]
1	3.5
2	> 4.5
3	> 4.0
4	> 3.0
Cumulative inactivation	>15.0

[a]Minimum value from three
independent experiments.

Nucleic Acids.   The presence of residual cellular nucleic acids
is a potential concern for products of continuous cell substrates since
they may contain genetic information capable of inducing tumorigenesis.   In contrast, biologics prepared from primary or diploid cell substrates may have lower theoretical risk levels since their nucleic acids
are assumed to lack sequences for activated oncogenes.   Removal of
residual cellular DNA (RC-DNA) during processing has been perceived
as an important issue for development of products of continuous cell
origin (14), and the technology to accomplish this has developed during the past decade.   Examples of process steps which have been
successfully validated to remove RC-DNA include ion exchange chromatography and DNase treatment (15–17).   An example of a validation plan for DNA removal is shown in Fig. 2.
    DNA concentrations should be assessed in harvested cell culture
fluids and at appropriate points during recovery and purification.
Sensitive hybridization assays are generally employed for the detection of specific DNA sequences contained in the cell substrate.
    Sensitivity should be maximal without incurring significant non-
specific reactions.   Quantitative evaluation of the ability of the
process to remove total RC-DNA should be performed using labeled
substrate cell DNA as a probe, while detection of removal of specific
sequences (plasmid, promotor, etc.) is best addressed by using
sequence-specific probes.   If process fluids have undetectable levels
of RC-DNA, it may be necessary to perform spiking studies with
labeled cellular DNA spiked into process fluids.

Goal	Demonstrate effective removal of residual cellular DNA (RC-DNA)
Equipment	Scaled down purification unit operations as appropriate
Design	Conduct scaled down studies with representative process materials "spiked" with radiolabeled RC-DNA; assay residual radioactivity
Acceptance criteria	Cumulative reduction of RC-DNA level to less than 10 pg per dose

Figure 2   RC-DNA removal.

One or more downstream processing steps are selected for validation based on the goal that has been set. Probes are labeled by nicktranslation with ^{32}P nucleotides, spiked into process fluids, and treated by process conditions designed to remove DNA. Samples and appropriate controls are evaluated by liquid scintillation counting of radioactive spiked DNA remaining or by hybridization technology (dot blots) to determine initial and final levels of DNA in spiked fluids. The degree of removal is calculated by dividing the initial concentration by the final concentration value. Properly designed and well executed studies have demonstrated removal of 100,000-fold or more RC-DNA by ion exchange chromatography (17). An example showing multiplestep validation used for Activase at Genentech is summarized in Table 4 (17). After initial purification steps, the protein concentration is 0.5 g/liter and the DNA level is 0.11 ng/ml as assessed by direct measurement using dot blot assays. Further removal is measured by experiments spiking ^{32}P-labeled DNA into process fluids and following its removal on representative small columns. The overall clearance is the product of the individual clearances, and the final amount of DNA is the concentration after initial purification divided by the clearance of the ion exchange step times the dose.

Several comments should be made concerning the chemical nature of the residual cellular nucleic acid. First, both RNA and DNA can be substantially removed by ion exchange chromatography. Furthermore most hybridization methods will detect RNA about as well as

Table 4   DNA Clearance During t-PA Purification[a]

	DNA conc. (ng/ml)	rt-PA conc. (ng/ml)	DNA clearance Step	DNA clearance Cumulative
Initial purif.	0.11	0.5	1.9×10^4	1.9×10^4
DE52			1.9×10^5	2.4×10^9

[a]Calculation of residual DNA per dose:

$$\frac{DNA}{Dose} = \frac{0.11 \text{ ng}}{0.5 \text{ mg}} \times \frac{1}{1.3 \times 10^5} \times \frac{100 \text{ mg}}{Dose} = \frac{0.17 \text{ pg}}{Dose}$$

Source: Ref. 17.

DNA. Also, the putative risk of tumorigenesis is considerably less for RNA compared to DNA. All these factors mean the separate removal steps and/or validation studies for RNA are usually unnecessary.

The chemical form of the RC-DNA is also important in process considerations. Inside cell nuclei, DNA exists in both naked and nucleohistone form and presumably both are released by cell substrate as some of the cells due and lyse during production. Care must be taken to ensure the spike is of the appropriate chemical form, i.e., with or without histone for the process step being validated. For example, if the step being validated might contain both naked DNA and nucleohistone, removal of both should be validated independently. The size of the RC-DNA is also important, in that high molecular weight DNA is easier to remove but may carry higher potential risk than low molecular weight DNA. Another consideration in the minimization of RC-DNA levels in final product is to attempt to reduce RC-DNA levels in harvested cell culture fluid by establishing conditions which minimize cell death and disruption during production. Thus, maintaining high cell viability during product collection and harvest is an effective way to minimize contamination of harvested fluid by RC-DNA.

These methods have been very effective in controlling RC-DNA contamination in products of continuous cell substrates. In most cases they are easily implemented. Several reported processes for cell culture biologicals have validated levels of DNA removal which translate to less than 10 pg RC-DNA per dose (15-17). Removal of RC-DNA is frequently a function of how much planning and effort has been incorporated into process design, as described in Chapter 14. Their successes led in part to a growing consensus that products of continuous cell substrates can be rendered acceptably safe for human administration with respect to RC-DNA.

Residual Cellular Protein. Mammalian cells in culture express thousands of proteins, many of which might be immunogenic to a heterologous recipient of the product. A number of sensitive assays have been developed to quantitate residual cellular protein remaining in final product; some of these are discussed in Chapter 20.

An additional approach, which employs principles of validation rather than lot testing, is the so-called blank run characterization (18). This method attempts to estimate how much cellular protein carries through to final product by processing harvested cell culture fluid from the fermentation of a cell line identical to the production line except that the gene encoding the product is missing (18). Care must be taken to guard against potential artifacts leading to selective losses of some proteins in the blank run relative to the normal run, e.g., those which might contaminate final product due to interaction with product during purification or those whose expression might be modified by expression of the product. gene.

The fermentation of the blank cell line and the recovery of protein from it is performed with production raw materials and relevant equipment and methods. As an example, the blank run for recombinant tissue plasminogen activator (rtPA) at Genentech consisted of a CHO host cell identical to that used for production of Activase with the same plasmid, except that the cell did not contain the human tPA gene. All raw materials, equipment, and procedures were identical to those employed for production. Two fermentations and recoveries were performed at the nominal 10,000-liter scale. It was necessary to perform these at full production scale to be able to measure the small amount of cellular protein remaining after purification. The harvested cell culture fluid contained a total of about 1 kg of total protein (residual cellular plus foreign) which yielded only some tens of milligrams when purification was completed. Thus, the recovery process is able to reduce the concentration of unwanted protein in final product by several thousand fold. This estimate of effectiveness of the purification procedure agrees with that calculated by dividing the protein concentration in rtPA-harvested cell culture fluid (less the amount of rtPA) by the amount of protein contaminants in final product measured by immunochemical assays (18). Although costly, this exercise provides an important verification of purity estimates derived from direct immunoassay.

3. Exogenous Materials

It is usually necessary to use many reagents in the fermentation
medium and recovery process which do not belong in the final product. They are treated in the same general way as the endogenous
substances.

Serum and Other Foreign Protein. Animal serum is frequently
employed at some stage of cell growth. The U.S. cGMP for vaccines currently requires that manufacturers remove animal serum
prior to product collection and that the theoretical contamination
of final product by animal serum is less than its original concentration by a factor of 1 million fold. Most manufacturers accomplish
this by extensively washing the cells with protein-free buffer or
medium. The recent advances in protein separation made possible
by powerful protein purification technologies, such as affinity
chromatography, permit substantial removal of animal serum during
downstream processing as well.

Concerns about specific contaminants are usually addressed by
assays specific for those species. For example, the removal of
residual monoclonal antibody leaching from an immunoaffinity purification matrix can be addressed by immunoassays for species-
specific immunoglobulin. Similar assays can be readily developed
for specific serum proteins or other medium proteins of interest.

Adventitious Agents. Adventitious agents include bacteria,
fungi, yeast, mycoplasma, and viruses which are transmitted horizontally from infected individuals or reservoirs to susceptible individuals or cell substrates. Controlling or preventing the possibility that adventitious agents might contaminate the final product
may be addressed in several ways. Validation can represent one
effective method in this effort. The use of validation methods to
ensure freedom from putative contamination by endogenous retroviruses was discussed above. Manufacturers can assure the freedom from contamination by other biological agents using similar
methods. Viruses are especially suited to this approach since direct detection methods often lack sensitivity or may have narrow
specificity. Fortunately, most viruses of a given group usually
have similar structure and possess similar responses to generalized
inactivation methods. Many of the same methods used for putative
retrovirus inactivation will readily inactivate additional virus types
as well, as was recently described for Activase (19).

For example, the use of validated inactivation procedures in
the preparation of urokinase from human embryonic fibroblasts was
accepted by the FDA as an effective substitute for virus detection
assays on each harvested cell culture batch (R. Duff, personal
communication). Even if some virus groups are not convincingly
inactivated during recovery processing, this approach may still be
useful when combined with other control procedures. For example,

poliovirus and, similarly, several other virus families are more resistant to inactivation. However, processes which employ neither
primate cell lines nor unpasteurized human raw materials are
extremely unlikely to serve as a potential source of product contamination by poliovirus. Perhaps the only remotely possible source
of poliovirus is process water. Effective process and facility design
can completely prevent this route of entry (e.g., by exclusive use
of distilled water for medium and buffer preparation). Thus, the
risk of poliovirus contamination of final product is negligible.

   Process Chemicals. There is a long list of low molecular weight
substances used in production which do not belong in the final
product. For most purified proteins there will be several purification process steps where the protein of interest experiences a
change of buffer constituents. These steps are usually performed
by dialysis, precipitation, gel filtration chromatography, or ultrafiltration. Such methods are usually very effective in removal of
low molecular weight components which may have entered during
earlier process steps. Care must be taken to consider the long
list of added chemicals and to defined at what stages they might
be removed. Thus, effective process design will usually provide
adequate assurance of the absence of unwanted small molecules in
the final product.

   Special consideration should be given to removal of toxic or
harmful chemicals. Validation methods should be employed to verify
their removal, in a manner similar to that described above for RC-
DNA. Direct measurement of initial and in-process levels, supplemented by spiking experiments when necessary, will usually provide
adequate assurance of product safety.

   Use of chemicals which are reactive with the protein of interest
should be avoided whenever possible. When not avoidable, their
use should be minimized and tightly controlled (time, pH, temperature, etc.) and their removal accomplished as early in the process
as possible. Their removal should also be validated. Consideration
should also be given to their possible interaction with the protein
of interest and possible resulting modifications to protein structure.

B. Process Performance

1. Genetic Stability

Assurance of product reliability is required by regulation and usually means being able to make the same active ingredient in every
batch, as determined both qualitatively and quantitatively. The
ability to assess qualitative genetic stability (i.e., the existence
of mutations) at the molecular level has been extremely limited or
impossible for traditional biologicals. To partially remedy this,
measurement of the phenotypic and genotypic properties of the host

cell became the surrogate for biochemical evidence of product con-
stancy. The regulations and guidelines of several nations and standard setting organizations request appropriate evidence of genetic
stability for each cell culture production run (20−24).

Quantitative measurement of gene expression is also important.
Abnormal variation in protein product levels in harvested cell culture fluid might reflect mutation or emergence of altered cell populations or infection by adventitious agents such as mycoplasma.
However, traditional biologicals, such as virus vaccine and interferons, have substantial variation in titer from lot to lot, further
complicated by imprecise assays. As a result, the ability to provide meaningful process control by measuring yield is limited for
some types of products or production systems. Thus, earlier
technology was hampered by imprecise or subjective measurements
of the genotypic and phenotypic properties of the cell which were
used as indicators of product consistency.

The development of many analytical technologies that constitute
the rDNA industry has changed this situation. It is now possible
to examine directly the sequence of nucleic acids and the protein
encoded by them. Most regulatory documents, such as U.S. FDA
"Points to Consider" (and similar guidelines from other countries),
suggest that during the characterization of a recombinant cell culture process the DNA plasmid encoding the product gene should
be isolated from the cell substrate, cloned, and sequenced (24).
In systems such as cells transfected by bovine papilloma virus (BPV)
vectors, cloning of the predominantly episomal plasmids is simple and
effective. For other systems in which integration of the plasmid
into host genomic DNA occurs, cDNA cloning and sequencing of produce messenger RNA is often suggested (24). If significant plasmid amplification has occurred, then the use of cloning methods is
hampered by three problems. First, cloning and sequencing techniques can possibly be employed for a limited characterization study
but are too tedious for use in a lot testing mode. Second, no relationship can be constructed between the many protein molecules in
the vial of product and any one of many sequences cloned. Third,
posttranslational modifications can only be detected by analyzing
the protein and not by analyzing the respective nucleic acid. Thus,
although cloning and sequencing methods are mechanically feasible,
no method exists to assess how this information relates to the product in general or to specific product lots.

An alternative to this approach is to directly assess the genetic
stability of the product protein molecules using current methods of
protein analysis. One example is peptide mapping in which controlled enzymatic degradation is followed by HPLC separation to
characterize each digestion fragment (see Chapter 20). These

methods are sensitive, quantitative, and validatable. While they only indirectly assess the genetic stability of the plasmid, they directly assess the genetic stability of the product (which was the original purpose of examining indirect genetic measures of stability).

Another important facet of genetic stability for cell culture biologicals is the establishment of limits on in vitro age for cells used for manufacturing. This is usually measured in terms of population doubling levels (PDL) for diploid cells. It is important to limit the use of diploid cells to ages significantly less than when they begin approaching senescence. Hence, it is necessary to determine at what PDL senescence occurs and to impose strict limits on cell age prior to that point. Currently, rDNA technology primarily utilizes immortal cell lines which in general do not enter senescence. Thus, in vitro cell age need not be defined in terms of PDL, as this has no practical meaning for most immortal lines. Some organizations use the simple concept of time in cultivation (after thawing a working bank vial) to establish limits on cell age in vitro. Once cells are subcultivated for close to the maximum allowable limits, the cultures are destroyed and replaced by cells from a newly thawed vial from the manufacturer's working cell bank (MWCB, see Chapter 5). This procedure also eliminates the imprecision of control by PDL. Time is easy to measure in contrast to PDL which incurs the multiple uncertainties of two cell counts for each determination. Errors (often $10-20\%$ or more per PDL) in PDL measurement are cumulative, e.g., the error in the cell age estmate after 10 passages is about 10 times greater than that after one passage.

We would like to discuss an example of the validation of genetic stability for rDNA cell culture process which combines sensitive protein chemistry methods with more classical measures of genetic stability and a simple temporal measure of cell age in vitro. A summary of the procedure is shown in Fig. 3. The goal of this procedure is to demonstrate stable production throughout the time period for which the progeny of cells from MWCB vial are employed for manufacturing use. As a minimum, this validation study should employ cells of defined age, both below and above the cell age limit, in order to assess the robustness of the manufacturing process biology. The manufacturing process is performed and monitored at various cell age levels: product is recovered and assayed for compliance with quality specification for product release. Much of the data must be collected from laboratory or pilot scale fermentation and recovery systems, since the product collected from cells older than the upper limit established for cell age cannot be marketed. Confirmatory analysis over the permitted range of cell age at full scale is highly desirable. Examples of the parameters that should

Goal	Demonstrate stable production throughout throughout period of manufacturing use
Equipment	Scaled down fermentation and recovery systems
Design	Conduct scaled down operations at various levels of cell age within and beyond normal operating range; assay crude and recovered product
Acceptance criteria	Recovered product passes appropriate valid QC tests for product purity, potency, and safety; productivity per cell remains constant

Figure 3   Genetic stability.

be examined for cell age effects include cell growth rate and maximum cell density, cell viability, titer active ingredient in harvested culture fluid, specific productivity (product mass per cell per unit time), purification yield, product purity, potency, identity, and tryptic fragment profile.

An example of typical data is shown in Table 5. This method was employed to assure the genetic stability of tissue plasminogen activator production by recombinant CHO cells (25). The scaled down systems were operated in this example from 34% of maximum allowable cell age to 156% of the upper limit. The levels of specific productivity are essentially constant throughout the manufacturing period and somewhat beyond the upper limit (through 111%). Despite the possible reduction in specific productivity at advanced cell age, the trend was not statistically significant. These findings were confirmed during full-scale fermentations. No statistically significant effects of cell age during the allowable manufacturing period were observed in any of the parameters evaluated. Thus, the stated goal of the validation program was achieved.

As described here, genetic stability of recombinant processes can be assessed directly, both quantitatively in terms of productivity of the cell population and qualitatively in terms of mutant protein molecules. Using techniques of retrospective validation the data from the full-scale manufacturing batches should be reviewed periodically for effects of cell age. Documentation that review of

Table 5   Results of Cell Age
Studies at Small Scale

Cell age[a]	Specific productivity[b]
34	102 (16)[c]
44	81 (3)
61	96 (8)
94	102 (21)
111	106 (18)
124	70 (3)
156	78 (8)

[a]Percent of maximum permissible age.

[b]Percent of mean for full-scale runs,
as mean of two to three fermentations.

[c]Number in parentheses is 1 SD.

such data reveals no statistically significant cell age effects constitutes periodic revalidation of genetic stability.

2. Proven Acceptable Range Approach

Proven acceptable range (PAR) validation is often used to tie up
all the loose ends. One can identify at least 20 variables in a cell
culture process which might be expected to have some effect on
process performance. Some might be interdependent over some
portion of their range (e.g., concentration of bicarbonate ion, and
fermenter pH and agitation rate). The number of possible interactions is large and lies far beyond the capability of any development organization to define experimentally due to time and economic constraints. The task of the validation team would be impossible unless a mechanism existed to validate the process based on
the manufacture of acceptable product.

PAR validation is essentially the use of principles of retrospective validation to forge a link between the production of acceptable
lots of product from a quality control (QC) standpoint to normal
and permissible values for process operating parameters from a
manufacturing standpoint. For example, a facility has produced
100 batches over a period of time with all batches being acceptable

upon QC testing. Analysis of manufacturing records reveals that bicarbonate ion varied between 1.0 and 3.0 mM, pH varied between 6.8 and 7.5, and agitator speed varied between 90 and 110 rpm. The actual ranges that were recorded for these three process parameters become the acceptable ranges because acceptable products resulted when the process was run within them.

Perhaps at an agitation rate of 200, the pH rises to 7.6 and the bicarbonate concentration falls to 0.5 mM. For the purpose of a process validated by the PAR procedure, these values are beyond the normal range. The reasons that these parameters might be interrelated are scientifically interesting but are not important to the validation exercise. If the product made from this unusual run fails to pass QC tests, then those values for these process parameters are presumed to be unacceptable. If the produce lot passes QC tests and possesses all relevant quality attributes (sometimes including a special stability study), the lot may be accepted if it is clear that the purity, potency, identity, stability, and safety of the produce have not been compromised. These extreme values of those process parameters subsequently may become acceptable in the process validation, especially if the acceptability of final product from these conditions is confirmed by multiple observations. They may not be desirable from a process control viewpoint, i.e., no one intends to deliberately operate in those unusual ranges, but the product might still be acceptable as a rare event from a quality perspective. That is to say, excursion of the parameters to the unusual value will not by itself result in a lot failing. It will usually result in a more careful look to examine the effect of these parameters on product quality (and process control procedures as well).

Once the analysis is completed and the various limits are established, the program is documented and reviewed by manufacturing, development, and quality control personnel. The program is periodically updated by review of production runs performed since the last validation exercise. Revisions are made and documented as necessary. For a more detailed discussion of PAR validation, see the recent review by Chapman (26).

IV. SUMMARY

Principles of process validation are extremely powerful tools in assurance of product quality. They are especially useful for reducing those risks not easily measured routinely during production. When combined with effective process and facility design principles, characterization of cell banks and products, appropriate lot release tests, and adherence to cGMP, safe cell culture biologicals can be prepared in a reliable manner.

REFERENCES

1. Carleton, F. J. and Agalloco, J. P. *Validation of Aseptic Pharmaceutical Processes.* Marcel Dekker, New York (1986).
2. Olson, W. P. and Groves, M. J. *Aseptic Pharmaceutical Manufacturing Technology for the 1990's.* Interpharm Press, Prarie View, Ill. (1987).
3. Nathanson, N. and Langmuir, A. D. The Cutter incident: Poliomyelitis following formaldehyde-inactivated poliovirus vaccination in the United States during the spring of 1955. I. Background. *Am. J. Hyg.* 78:16– 29 (1963).
4. Bloom, A. Acquired immunodeficiency syndrome and other possible immunological disorders in European haemophiliacs. *Lancet* 1:1452–1455 (1984).
5. Petricciani, J. P. and Regan, P. J. Risk of neoplastic transformation from DNA: Calculations using the oncogene model. *Dev. Biol. Standard.* 68:43–49 (1987).
6. Office of Biologics Research and Review. Points to consider in characterization of cell lines used to produce biologicals. Food and Drug Administration, U.S. Public Health Service, Rockville, MD (1987).
7. Center for Drugs and Biologics and Center for Devices and Radiological Health. Guideline on general principles of process validation. Food and Drug Administration, U.S. Public Health Services, Rockville, MD (1984).
8. Aaronson, S. A. and Stephenson, J. R. Endogenous type-cRNA viruses of mammalian cells. *Biochim. Biophys. Acta* 458:323–354 (1976).
9. Garfinkel, D. J., Bocke, J. R., and Fink, G. R. Ty element transposition: Reverse transcription and virus-like particles. *Cell* 42:507–517 (1985).
10. Lubiniecki, A. S. Endogenous retrovirus of continuous cell substrates. *Dev. Biol. Standard.* 70:187–191 (1988).
11. WHO Study Group on Biologicals. Acceptability of cell substrate for production of biologicals. Technical Report Series No. 747, World Health Organization, Geneva (1986).
12. Finter, N. B. and Fantes, K. H. The purity and safety of interferon prepared for clinical use: The case for lymphoblastoid interferon. In *Interferon 1980* (I. Gresser, ed.), Academic Press, New York, pp. 65–80 (1980).

13. Hilleman, M. R., McAleer, W. J., Buynak, E. B., and McLean, A. A. The preparation and safety of hepatitis B vaccine. *J. Infect.* (Suppl) 7:3−8 (1983).

14. Petricciani, J. C., Salk, P., Salk, J., and Noguchi, P. D. Theoretical considerations and practical concerns regarding the use of continuous cell lines in the production of biologics. *Dev. Biol. Standard.* 50:15−25 (1982).

15. Crainic, R., Horodniceanu, F., and Barne, M. Presence and quantification of cell substrate DNA in inactivated poliovirus vaccine. *Dev. Biol. Standard.* 46:275−279 (1980).

16. Wiseman, A., Almazan, M. T., Dale, G. L., Katz, L. R., Lesh, L. S., Lusby, E. L., Marcelletti, S. F., Nonaka, M., O'Hair, C. H., Orida, N. K., and Katz, D. H. Assessment of potential nucleic acid contamination in suppressive factor of allergy purified from a human T Cell hybridoma line. In *Abnormal Cells, New Products, and Risk* (H. S. Hopps and T. C. Petriccerian, eds.), Tissue Culture Association, Gaithersburg, MD, pp. 115−120 (1985).

17. Builder, S. E., Van Reis, R., Paoni, N., Field, M., and Ogez, J. R. Process development in the regulatory approval of tissue-type plasminogen activators. *Dev. Biol. Standard.* (in press).

18. Anicetti, V. R., Fehskens, E. F., Reed, B. R., Chen, A. B., Moore, P., Geier, M. D., and Jones, A. J. S. Immunoassay for the detection of *E. coli* proteins in recombinant DNA derived human growth hormones. *J. Immunol. Meth.* 91:213−224 (1986).

19. Wiebe, M. E., Becker, F., Lazar, R., May, L., Casto, B., Semense, M., Fautz, C., Garnick, R., Miller, C., Masover, G., Bergmann, D., and Lubiniecki, A. S. (1989). A multifaceted approach to assure that recombinant tPA is free of adventitious virus. *Dev. Biol. Standard.* (in press).

20. Food and Drug Administration. Subchapter F − Biologics. Code of Federal Regulations, Title 21, Part 610.18 (1987).

21. Committee for Proprietary Medicinal Products. On the production and quality control of medicinal products derived by recombinant DNA technology. Commission of European Communities, Brussels (1987).

22. National Institute for Biological Standards and Control. Considerations for the standardization and control of the new generation of biological products. London (1987).

23. Ad hoc Committee. Proposed requirement for the hepatitis B surface antigen made by recombinent DNA techniques. *WHO Requirements for Biological Substances*, No. 39, Geneva (1985).

24. Office of Biologics Research and Review. Points to consider in the production and testing of new drugs and biologicals produced by recombinant DNA technology. Food and Drug Administration, U.S. Public Health Service, Rockville, MD, (1985).

25. Lubiniecki, A., Arathoon, R., Polastri, G., Thomas, J., Wiebe, M., Garnick, R., Jones, A., van Reis, R., and Builder, S. Selected strategies for manufacture and control of recombinant tissue plasminogen activator prepared from cell cultures. In *Advances in Animal Cell Biology and Technology for Bioprocesses* (R. E. Spier, J. B. Griffiths, J. Stephenno, and P. J. Crooy, eds.), Butterworth, pp. 442–457 (1989).

26. Chapman, K. G. The PAR approach to process validation. *Pharm. Technol.* 4(12):24–36 (1984).

20

Quality Control of rDNA-Derived Human Tissue-Type Plasminogen Activator

ANDREW J. S. JONES and ROBERT L. GARNICK
Genentech, Inc., South San Francisco, California

I. INTRODUCTION

Nowhere in the development of emerging industries does quality control play as important a role as in biotechnology (1). The uniqueness of this industry is based on the use of genetically altered living organisms for the manufacture of commercial and human health care products. Although it traces its origins to the fermentation industry, biotechnology is best represented by the recombinant DNA (rDNA) and monoclonal antibody industries. Only a decade old, these industries have succeeded in developing a number of products that profoundly affect the quality of human life. An example of a product derived from rDNA technology is the recently approved thrombolytic therapy drug, human tissue-type plasminogen activator (rt-PA) (2). A highly specific enzyme, rt-PA converts plasminogen to plasmin, the active fibrinolytic enzyme that dissolves blood clots causing certain myocardial infarctions.

The development of rt-PA required the codevelopment of quality control systems capable of ensuring the lot-to-lot consistency of a glycosylated biomolecule of approximately 64,000 daltons (Da) manufactured from large-scale mammalian cell culture. The emergence of such a control system resulted from a careful blending of traditional quality control systems long used in the tissue- and serum-derived pharmaceutical industry with many advanced techniques of molecular biology and protein chemistry. This control system was designed to (1) ensure the reproducibility and consistency of rt-PA by careful review of all phases of its manufacture and (2) evaluate a broad spectrum of known and/or suspected product impurities and potential degradation products. Thus, the control system was not relegated to simply ensuring that specifications were met on a routine basis. Some of the assays developed for the in-process control of rt-PA were also used as tools for the validation of the manufacturing process.

The complexity of virtually all quality control systems depends on the size and structural characteristics of the manufactured product. In addition, the intricacies of the manufacturing purification process, which are often directly related to the size of the product, are complicated by a vast number of in-process analyses. It is not surprising, therefore, that the control system developed for rt-PA requires several hundred in-process and final product assays per lot (Genentech, unpublished data).

The analysis of a biomolecule such as rt-PA is only as accurate or complete as the analytical methods used. The consistency of manufacture and the ability to ensure that consistency are the most important means of ensuring the safety of such products (1). This aspect of quality control has great significance because there is little or no information available on the long-term safety of many biotechnology products in humans. Therefore, the extent of "analytical concern" with its resultant cost and time involvement appears more than justified. For example, the efforts of numerous chemists, biochemists, microbiologists, and virologists are required to ensure the quality control of rt-PA. The analytical methods used for testing and validation range from virology to high-performance liquid chromatography (HPLC)-peptide mapping to immunological purity analysis (3,4). Such advanced control systems require automated instrumentation to accurately measure impurities such as DNA, endotoxin, and residual host organism proteins. In addition, a biological, such as rt-PA, is examined and tested by the Center for Biologics Evaluation and Research (CBER) in the United States prior to release by the manufacturer. Thus, unlike other areas of the pharmaceutical industry, the analysis of a biological by a manufacturer is confirmed by a regulatory agency on a lot-by-lot basis.

In this chapter, rt-PA is used to illustrate some of the strategies and analytical methods for the control of recombinant cell culture products. The rationale and limitations of the use of these methods are also discussed.

II. QUALITY CONTROL OF rt-PA AND HOW IT DIFFERS FROM TRADITIONAL PHARMACEUTICAL QUALITY CONTROL

The control of biotechnology products such as rt-PA involves both the individual production process and the product itself. Thus, the quality control strategies developed for human growth hormone (molecular weight of 22,000) produced by mammalian cell culture are considerably different from those developed for rt-PA (molecular weight of 64,000). These strategies may even vary considerably between two separate processes producing the same final product because of significant concerns that must be addressed systematically about end-product purity, safety, and stability (5). Other aspects of biotechnology quality control parallel the methods and protocols developed for more traditional pharmaceutical products. For example, the systems used for raw material testing and release, manufacturing documentation, environmental monitoring, and aseptic processing of rt-PA are virtually identical to those used for the manufacture of small organic drugs (see Fig. 1). The quality control of cell culture products like rt-PA, however, differs from traditional control systems in three major areas: (1) the development and characterization of the production cells, (2) cell culture processes, and (3) the recovery/purification process and final products obtained from cell culture.

The first of these topics is discussed in detail in Chapters 5, 17, and 18. Examples from the second and third areas are discussed in the following sections.

III. SIGNIFICANT ISSUES RELATING TO THE CONTROL OF THE CELL CULTURE PROCESS USED FOR rt-PA

The production of protein products such as rt-PA using transformed cells such as Chinese hamster ovary cells (CHO-K1) has been examined by numerous regulatory agencies and industry, and by consensus is acceptable if adequate documentation exists verifying the suitability and safety of such cells lines. In addition, issues that relate directly to the control of the cell culture process must be examined closely. The most relevant production concerns are listed in Table 1. Examples of these issues are discussed in the following sections.

The absence of nonhost organisms in the CHO-K1 cell cultures used to produce rt-PA is of major importance. In addition to demonstrating that bacteria, yeast, and molds are not present in

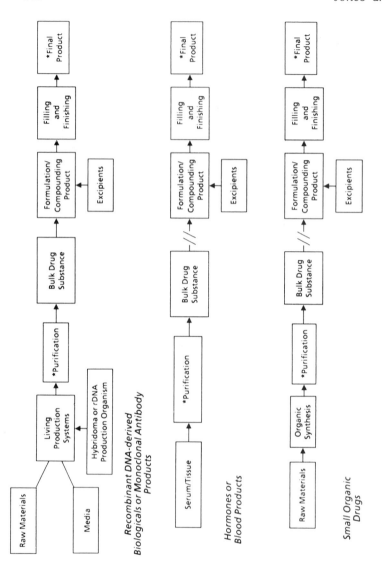

Recombinant DNA-derived
Biologicals or Monoclonal Antibody
Products

Hormones or
Blood Products

Small Organic
Drugs

* Purification process complexity and analytical in-process and final product
testing are often a direct function of the molecular size of the product

Figure 1 Examples of typical manufacturing processes for the production of pharmaceuticals.

Table 1   Major Safety Concerns in the Use
of Mammalian Cells for Producing
Recombinant DNA Pharmaceuticals

Absence of retroviruses or exogenous viruses

Host cell removal

Residual cellular protein

Residual cellular DNA

Biochemical purity

Genetic stability

Molecular stability

Toxicity and/or antigenicity

---

these cultures, one must show that mycoplasmas and viruses are
also absent (6). This necessitates that valid and highly reliable
testing methods be developed to ensure the lack of such contamination. For rt-PA, a viral screening method using three indicator cell lines and three positive control human viruses was developed and validated.

Consistency in the extent of glycosylation is an important consideration in the design of cell culture conditions for the manufacture of proteins such as rt-PA. The degree of glycosylation
may affect the half-life of the product in vivo as well as its potency.
Although difficult to assess, a uniform glycosylation level for a
cell culture product such as rt-PA can be verified after demonstration that reproducible culture conditions exist. Carbohydrate
analysis, although very complicated and difficult to perform, can
serve as a viable method to ensure the consistency of the product.

CHO-K1 cells such as that used for rt-PA possess sophisticated
growth requirements. Thus, they are affected by multiple growth
factors derived from different components in the culture medium.
Alteration of the delicate balance of these components can produce
major changes in the cell culture, resulting in poor consistency in
cell growth, cell viability, and/or product yields. The quality control of each lot of basal medium and its components by means of
rigid release criteria is critical. In testing the basal medium and
its components used for rt-PA production, the test systems involve
a scaled down mimic of the production processes. In each test system, the CHO-K1 cell culture is grown simultaneously in two sets
of appropriate basal medium with components. One of the sets

becomes the reference material and the other set becomes the test culture in which the component under investigation is varied. For example, a test serum lot would be substituted for a reference material serum lot in a completed medium with all other components remaining unchanged. Under testing conditions, both the reference material and the test lots are similarly prepared at the same time. Cell growth performance is monitored simultaneously in both media sets over an appropriate period and then verified by quantitative product yield determinations.

IV. THE IN-PROCESS, BULK DRUG SUBSTANCE, FINAL PRODUCT, AND STABILITY TESTING OF rt-PA

The recovery of rt-PA produced in CHO-K1 cells relies on an effective combination of protein separation techniques such as molecular sizing, ion exchange, and affinity chromatography. The recovery process begins with isolation of the rt-PA, in a highly impure state, from the cell culture medium. An obvious advantage of the cell culture process is that rt-PA is expressed directly into the medium. Thus, harvesting requires only the separation of the cells to achieve a significant purification.

The recovery process developed for rt-PA was designed to purify the final product to a level greater than 99%. The rationale for this level of purity is discussed in more depth later in this chapter. Although the absolute purity requirement for a given product depends on many factors, products intended for chronic human use are usually required to be of much higher purity than those intended for single-use purposes (7). Many of the safety concerns over rt-PA pertained to certain unique impurities such as trace amounts of DNA, residual host cell proteins, pyrogens, and residual media proteins that required considerable planning and effort in the recovery processes to eliminate or minimize them. The most common impurities of concern for mammalian cell culture products like rt-PA and suitable methods of detecting them are listed in Table 2.

The rt-PA recovery process was designed to be readily sanitized and required extensive process validations to demonstrate that the final product made using this process is consistent from lot to lot. In addition, extensive in-process testing is required to monitor the protein recovery. Tests such as protein content, endotoxin by *Limulus* amebocyte lysate (LAL), bioburden, and potency or activity are performed at key steps in the recovery process. Because these tests are performed at multiple points, development of automated analytical methods aids in the data generation. Examples of such automated methods for rt-PA include the

Table 2   Potential Impurities of Cell Culture Products

Impurities	Detection method
Endotoxin	LAL,[a] rabbit pyrogen
Host cell proteins	SDS-PAGE,[b] immunoassays
Other protein impurities (media)	SDS-PAGE, HPLC,[c] immunoassays
DNA	DNA hybridization, UV spectro-photometry
Protein mutants	HPLC-tryptic mapping
Oxidized methionines	Amino acid analysis, HPLC-tryptic mapping, Edman degradation analysis
Proteolytic clips	IEF,[d] SDS-PAGE (reduced), HPLC, Edman degradation analysis
Deamidation	IEF (standard comparison), HPLC
Monoclonal antibodies	SDS-PAGE, immunoassays
Amino acid substitutions	Amino acid analysis, HPLC-tryptic mapping
Viruses (endogenous)	CPE,[e] HAd,[f] electron microscopy, reverse transcriptase activity

[a]*Limulus* amebocyte lysate.

[b]Sodium dodecyl sulfate–polyacrylamide gel electrophoresis.

[c]High-performance liquid chromatography.

[d]Isoelectric focusing.

[e]Cytopathic effect.

[f]Hemadsorption.

use of robotics for endotoxin analysis (8) and a microcentrifugal analyzer for potency determination (9), which is described in more detail later in this chapter.

The result of the rt-PA recovery process is a bulk drug product that is suitable for further processing at a later stage. The quality control of this material is specifically designed to thoroughly address its purity. Testing at this stage typically consists of assays for residual host proteins and endotoxins and sensitive methods to ensure that molecular changes have not occurred in the product during recovery. These assays include electrophoresis, Edman sequence analysis, peptide mapping, amino acid analysis, HPLC purity profiles, host cell protein enzyme-linked immunosorbent assays (ELISAs), and LALs to confirm the consistency of manufacture. Total protein content and potency measurements are also performed at this stage of production to ensure that the final processing of the rt-PA bulk will yield a product that meets specifications.

The rt-PA final product is produced in various dosage sizes after sterile filtration, filling, and lyophilization. The quality control assessments at this point therefore concern primarily those factors that may result from final processing operations. Particulate testing, content uniformity, pyrogenicity, sterility, potency, protein content, identity, and excipient chemical content are examples of final product tests performed for rt-PA. In addition, the analysis of the moisture content of a lyophilized product is a key concern for biologicals.

The stability of biologicals traditionally has been determined by evaluating product potency or activity in a whole-animal bioassay. Measurement and/or identification of the degradation products of most biologicals have proved extremely difficult because of their complexity and relative purity levels. The modes of protein degradation, however, are reasonably well understood and include proteolysis, dimerization, aggregation, denaturation, deamidation, oxidation, photolysis, and chemical reduction by excipients. In the case of rt-PA, a number of analytical methods such as sodium dodecyl sulfate–polyacrylamide gel electrophoresis (SDS-PAGE), HPLC –size exclusion chromatography (SEC), native molecular sizing, and potency by in vitro clot lysis are used to screen for evidence of these degradation modes in order to demonstrate the stability of the product.

V. DETAILS OF THE CONCEPTS AND ANALYTICAL METHODS USED FOR DETERMINING THE PURITY, IDENTITY, POTENCY, AND SAFETY OF rt-PA

A. Purity

The purity of a biological product can be defined as the measurement of the active ingredient in relation to the total substances (excluding

additives) present in the final product (7). Potential contaminants
include adventitious chemicals and microorganisms in the raw materials and in the manufacturing process. The impurities evaluated
by quality control procedures are primarily those associated with
the cell culture itself and possibly materials used during manufacturing. As noted above, contaminating microorganisms are controlled by in-process testing for bacteria, mycoplasmas, and viruses.
Two of the major purity concerns raised by the use of rDNA methods
for the production of biologicals are residual cellular protein and
residual cellular DNA.

1. Residual Cellular Protein

The traditional view that biologicals do not need to be rigorously
purified to levels of greater than 99% arose from several factors.
First, the impurities in the past were generally of human origin
because the source of the biological was human tissue or blood,
and their presence did not constitute a great medical or health
concern. Second, many of the therapeutics obtained from such
sources were present in very low concentrations and rigorous purification often led to loss of potency and yield. Much of the pioneering development of biologicals as therapeutics preceded the
development of purification methods that only recently became
available.
    These views have now been tempered by the transmission of
infectious agents (such as viruses associated with hepatitis, AIDS,
and possibly Creutzfeldt-Jakob disease) through human blood and
tissue-derived products. These new findings have led to stricter
controls in the use of human tissue as the starting material for biologicals and the addition of inactivation steps during the manufacturing process to minimize the risks of disease transmission (10).
    The original concern over the use of nonhuman sources for
the production of pharmaceuticals – in particular, the use of rDNA
methods and continuous cell lines – was that these sources might
contain nonhuman proteins. This could result in the production
of antibodies to the foreign impurities leading to anaphylactic reactions and allergic manifestations similar to those observed in serum
sickness. Products from rDNA technology using cell culture have
therefore been expected to be of high purity to alleviate this concern. Residual cellular DNA is usually expressed in picograms per
dose (see the following section) while protein purity is presented
in terms of percent purity or relative impurity levels (11). Different degrees of purity may be acceptable if the useage of the therapeutic varies. Acceptable levels of residual cellular protein also
should be considered on a per-dose basis (12). The approach developed for rt-PA included preparation of the product at a high
level of purity to minimize the possibility of any adverse reaction to
potential impurities.
    Immunological Assays for Residual Cellular Protein. The develop-
ment of immunological assays for residual cellular protein is complex

and the principles and criteria for validity are still evolving (12, 13, 31). With the exception of bioassays for extremely potent bioactive proteins, the most sensitive methods for measuring protein concentrations are generally immunoassays. The major benefit of immunoassays is that they can be performed in the presence of large concentrations of nonreactive proteins. Thus, one can assay for picogram amounts of the impurity protein in the presence of mg amounts of the product protein (in this case, rt-PA). Traditional analytical methods based on protein separation techniques do not possess this dynamic range and cannot quantitate the impurities unless these undergo separation from the product (see below). Many proteins must be detected and quantitated in products from cell culture. These proteins include those actively secreted along with rt-PA, any proteins carried over from additions of serum to the growth medium, and any proteins present in the medium as a result of cell disruption. An immunoassay must be capable of reacting to all of the most likely protein impurities. The development of a multiantigen immunoassay for a product such as rt-PA is thus a major challenge. Such extremely complex assays require rigorous validation at several levels before the results can be accepted. For a more detailed discussion of these issues, the reader is directed to Refs. 12 and 13.

One of the prime requirements for an impurity assay is obviously a source of impurities both as an assay standard and for generation of the antibodies to be used. These impurities should not contain any product, which would result in the generation of anti-product antibodies. In principle, it might be possible to obtain a sample of starting material from a production run and remove the product by immunoaffinity methods, leaving the impurities behind. The pitfalls associated with this approach are significant: the antibody used to remove the product cannot contain any antibodies to the impurities; it is difficult to remove absolutely all of the product; and it is very difficult to demonstrate that no impurities are removed at the same time as the product, either adventitiously or by an interaction with the product. It is advantageous to use an approach unique to rDNA methodology: to perform the manufacturing process exactly as in a normal production run but using a cell that lacks the gene for the product. This procedure is known as a "blank run." This technique results in a preparation of "pure impurities" that are totally lacking in the product but otherwise identical to those present in a production run. It is theoretically possible that the absence of the product gene and its expression have an effect on the proteins produced by the cells. This possibility has been tested, e.g., with rt-PA by examining the proteins produced in the presence and absence of the rt-PA gene. Such comparisons indicate that, except for the product, there is little detectable difference.

A source of antigens then makes it possible to raise antibodies to the mixture. To ensure a wide spectrum of antibody reactivity,

serum from a number of individual animals immunized for many
months is pooled. Although such a total serum pool contains antibodies to the impurities described, the levels of the individual antibodies vary widely depending on relative concentrations and immu-
nogenicities of the individual components. An assay could be developed using this pool but it might respond well to some of the
more abundant or immunogenic components and only poorly, if at
all, to those that are rarely found or only weakly immunogenic.
To overcome this serious potential problem, an antigen-selected
immunoassay (ASIA) has been developed (31). In this assay,
the antigen mixture is immobilized on a column matrix and used
to affinity-purify the antibodies. When the serum pool passes
over such a column, the antigens themselves select the antibodies:
those proteins that are rare and/or weakly immunogenic will be
enriched and those that are abundant or highly immunogenic will
be saturated and excess antibody will flow through the column.
The eluted antibodies will be (as closely as possible) present in
a stoichiometric ratio to the immobilized antigens. The spectrum
of reactivity of these antibodies is assessed by conventional methods such as SDS-PAGE followed by Western blotting (14). These
analyses show that there is generation of a pool of purified antibodies that recognize at least those antigens detectable by the
most sensitive nonimmunological assay of silver staining (13).

With the impurities from a blank run and an appropriate preparation of antibodies to them, an ELISA could be developed in a
conventional way, using microtitration plates and a double-antibody
sandwich format. This format is essential for sensitivity for the
following reason. The aim of this assay is to detect ng/ml levels
of contaminants in the presence of mg/ml levels of rt-PA. Without
the coat (or "capture") antibody, introduction of the product sample into the well results in most of the binding sites on the plastic
being occupied by the rt-PA itself and only a miniscule amount of
contaminant is bound. This leads to unacceptably low signals
when the product is highly purified. The plates are therefore
initially coated with the affinity-purified antibody described above.
The introduction of the sample allows maximal amounts of impurities
to bind to the plate with minimal hindrance from the product. After
the sample is washed from the plate, the bound impurities are detected by an enzyme-antibody conjugate (prepared from the same
affinity-purified antibody described above) as in a conventional
single-antigen ELISA (see Fig. 2). In this format, the assay
achieves adequate sensitivity in the ng/ml range. The use of the
sandwich format, however, requires that an impurity bind to two
antibodies (i.e., the coat and conjugate) simultaneously. Comparison of the direct protein determination and the ELISA-derived
concentration for fractions of standard separated by gel filtration
chromatography shows that the proteins in the mixture can bind

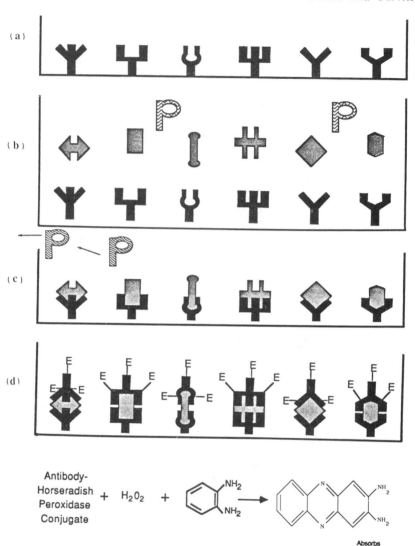

to two antibody molecules and that the two values are in good
agreement.

Even with such rigorous control of the specificity of the antibody preparation, the validity of the concentrations of impurities
determined has not been addressed. In a single-antigen immunoassay, the antibody concentration and standard curve range are
optimized to yield an assay that has a signal proportional to the
analyte concentration. For a multiantigen assay, there is a similar situation with a mixture of antigens in constant proportion.
When an unknown sample (such as final product in which the distribution of antigens is unknown) undergoes analysis, it is essential to demonstrate that the signal is proportional to the impurities
present. The apparent concentration of impurities may depend on
the dilution of the sample because of a lack of antibody excess.
If there is more antigen present than antibody, then the signal will
vary only slightly, if at all, as a function of sample dilution (i.e.,
the signal will not show linear dilution). The magnitude of the signal
will not be proportional to the analytic concentration and an incorrect
value will result. It is therefore necessary to show that the apparent
impurity level (percentage or ppm) is independent of sample concentration for every unknown sample. This requires that each sample be
analyzed in a dilution series to demonstrate the condition of antibody
excess before the result can be considered valid.

The use of the blank run to generate pure impurity preparations, the use of immobilized antigens to "select" the appropriate
assay antibodies, and the demonstration of antibody excess by
assaying dilution series for unknown samples have resulted in a
high degree of confidence in the validity of these immunoassays.
As noted above, however, some methodological constraints and
assumptions that appear to be involved can only be partially verified experimentally. Thus, such assays are used in conjunction
with the results from conventional methods of purity analysis as
part of the integrated testing and control system for evaluating
the quality of rt-PA.

Figure 2   The various steps involved in the performance of multiantigen ELISA. (a) Coat 96-well microtitration plate with process-
specific polyclonal antibodies, (b) Add sample containing product
(P) and impurities (antigens) of various types, (c) Incubate the
mixture and wash the plate to remove product, (d) Add the enzyme-
labeled second antibody and incubate. This completes the double-
antibody sandwich, (e) Add substrate to provide color development.

Nonimmunological Assays for Assessing Product Purity: SDS-PAGE and HPLC-SEC. The complexity and variability of the properties of proteins means that no single assay can be used to assess the purity of a protein preparation. Rather, a group of methods must be used that are capable of detecting a broad spectrum of impurities. The overall picture obtained by combining the information from these methods provides the assurance of product consistency and purity. The most common method for assessing the purity of polypeptides is SDS-PAGE coupled with silver-staining techniques (15,16). As with all methods that rely on physical properties of proteins, this method cannot detect impurities with properties similar to those of the product. SDS-PAGE separates polypeptides on the basis of their molecular weight and, using silver stain, can readily detect proteins with molecular weights different from that of the product at a sensitivity of 25–50 ppm (16). rt-PA can be resolved by SDS-PAGE because it is a mixture of several forms (17). SDS-PAGE provides a determination of the consistency of the ratios of these forms as well as a routine evaluation of the absence of impurities.

Like non-rDNA-derived t-PA, rt-PA exists in two forms that differ in the extent of glycosylation at one particular site in the polypeptide chain. These two forms are known as type 1 and type 2 (18). In addition to this glycosylation heterogeneity, rt-PA exists as equally active one-chain and two-chain forms (19) because of cleavage between residues 275 and 276 of the protein (20). Traces of other proteolytic processing events are also evident when a heavily loaded SDS gel is stained with silver (see Fig. 3). These trace bands are derived from rt-PA and form a highly consistent pattern. Although silver-stained gels cannot be readily quantitated, the use of a control preparation allows the method to be used as a sensitive measure of reproducibility of polypeptide distribution from batch to batch of product. It also provides a means of detecting new bands if the process failed to remove impurities. SDS-PAGE allows an evaluation of the consistency of the ratios of type 1 and type 2, a feature independent of the proteolytic processing events that occur during production and purification. Accurate measures of the cleavage of rt-PA from the one-chain to the two-chain form can be obtained by a size exclusion HPLC method in the presence of SDS and a reducing agent (see Fig. 4). This method not only yields an accurate value for the ratio of the two forms but can be used as an indicator for batch-to-batch consistency of proteolysis during manufacture. Thus, immunological assays for impurities, silver-stained SDS-PAGE, and HPLC together provide a powerful source of data on purity, proteolytic processing, and product consistency.

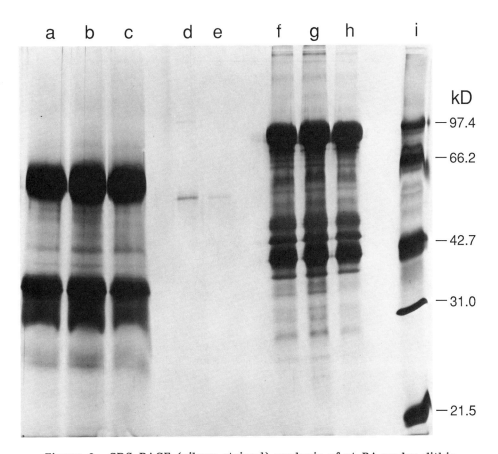

Figure 3 SDS-PAGE (silver-stained) analysis of rt-PA under dithiothreitol (DTT)-reduced and DTT-reduced, carboxymethylated conditions. (a) rt-PA reference material: 5 μg, DTT-reduced. (b)
rt-PA reference material: 7.5 μg, DTT-reduced. (c) rt-PA sample:
5 μg, DTT-reduced. (d) bovine serum albumin control: 10 ng.
(e) bovine serum albumin control: 2.5 ng. (f) rt-PA reference material: 5 μg, DTT-reduced, carboxymethylated. (g) rt-PA reference material: 7.5 μg, DTT-reduced, carboxymethylated. (h)
rt-PA sample: 5 μg, DTT-reduced, carboxymethylated. (i) Molecular weight standards (× 1000).

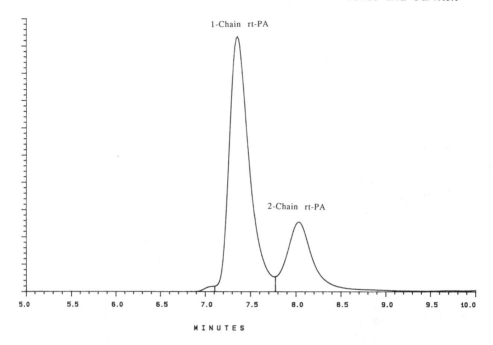

M I N U T E S

Figure 4   Chromatogram of rt-PA after reduction showing the one-
and two-chain forms obtained using a Dupont GF-250 column in a
mobile phase containing 0.1% SDS.

2. Residual Cellular DNA

As discussed in the introduction, the control of the quality of the
final product is complex and is achieved by combining many aspects
of testing throughout the manufacturing process. A good example
of this integration is the testing for residual cellular DNA. The
doses of rt-PA that are administered in the treatment of myocardial
infarction are about 100 mg per patient given over several hours
(21). The concern over residual DNA in products manufactured
from continuous cell lines, whether generated by rDNA or other
techniques, stems from the theoretical possibility of DNA with transforming potential (i.e., oncogenic DNA) being present in the final
product and causing an undesirable effect in the recipient of the
drug. This issue has been the subject of much discussion (22)
and has led to some agreement on the order of magnitude of DNA
levels that may be of concern. A consensus now exists as to what
levels of DNA represent minimal concern based on a combination
of theoretical treatments and actual data on the fate of DNA if it

is introduced into circulation. Levels of 10 – 100 pg of DNA per
dose are considered acceptable from the point of view of associated
risk to the patient (23). This translates to an acceptable level of
10 – 100 pg DNA per 100-mg dose of rt-PA.

The most sensitive assays for DNA are currently based on dot
blot hybridization methods (24). The sensitivity of these methods
depends on the complexity of the DNA used as the probe. Genomic
DNA is usually the least sensitive in terms of the amount of DNA
detectable but is the most sensitive in the terms of the types of
DNA detected. Although data have been obtained by using specific probes, routine testing of products from cell culture should
include the final product for the presence of any kind of DNA because it is not known exactly which type may be of concern. The
control of rt-PA employs cellular DNA isolated from CHO cells as
the probe to ensure that all potential DNA fragments would be detected if present. To achieve the desired sensitivity, it would be
necessary to show in the assay that all of the DNA from a single
dose was quantitatively transferred to the "dot" because the sensitivity routinely achieved with genomic DNA in this assay is of
the order of 100 pg. This has proven technically very difficult;
the problem has been addressed by the use of in-process testing
and process validation (discussed in more detail elsewhere in this
volume). In principle, a sample of rt-PA is obtained from an intermediate stage of the process. This point in the process is carefully chosen such that DNA is barely detectable. This point, however, is positioned prior to several steps in the process that additionally will remove any DNA that might have been present at levels
below the sensitivity of the assay. These latter steps are then
thoroughly validated to demonstrate that the DNA level would be
reduced by several orders of magnitude. This process validation,
coupled with the in-process specification, routinely provides the
assurance that the DNA level in the final product is well within
acceptable limits.

B. Identity and Genetic Stability

The identity of a protein such as rt-PA can be readily established
by its reactivity in a clot lysis assay or in an immunoassay. A
more complex issue is the possibility of a mutation occurring at some
stage during the cell culture process leading to a subtle change
in the molecule. However, the cell culture stage of the manufacturing process is rigidly controlled in terms of the number of generations the cells have divided after growth from the master cell
bank. If mutations occur with any significant frequency, they
cannot accumulate over more than a small, limited number of cell
divisions. (See other chapters in this volume for a detailed discussion of this aspect of product control.)

For a small polypeptide, it is sometimes possible to detect a
mutation in which one amino acid has been changed to another by
using amino acid analysis. For a protein the size of rt-PA, such
an approach is limited to the rarest amino acids because a change
of a single amino acid is close to or below the precision of the method.
In fact, when as much as 10% of bovine serum albumin (w/w) was
deliberately added to rt-PA during an evaluation of amino acid
analysis as a test method, the expected difference in amino acid
composition could not be differentiated within the experimental error of the method. Clearly, amino acid analysis is a low-resolution
method for detecting small changes in structure or purity.

The most powerful method available for evaluating the fidelity
and consistency of the primary structure of proteins the size of
rt-PA is peptide mapping. This method has long been used in
conjunction with Edman degradation (25) to determine the sequence
of amino acids in unknown proteins. Peptide mapping is accomplished
by initially reducing and carboxymethylating any disulfide bonds
present in the protein. This is followed by cleaving the protein
into smaller peptides of usually 25 residues or less by either an
enzymatic (typically trypsin) or, occasionally, a chemical method
that cleaves the protein at specific sites (see Fig. 5). The resultant peptides are then easily separated by reversed-phase (RP)
HPLC, ion-exchange chromatography, two-dimensional thin-layer
chromatography (TLC), or gel electrophoresis. A map or "fingerprint" is obtained, which allows for differentiation between proteins
of similar, but not identical, primary structure.

A tryptic map of rt-PA was developed that is extremely reproducible and informative. The elution times for the peaks may vary
on an absolute basis from day to day and from instrument to instrument, but there is high reproducibility within a day. Thus,
the intraday precision of the method coupled with the analysis of
a control (and of a mixture of the control and sample digests)
yields a very discriminating test for the presence of mutants.
Glycosylation of the molecule gives rise to clusters of peaks with
the same polypeptide sequence because of the heterogeneity of
the glycosylation pathways. The exact correspondence of the test
sample and the control sample provides assurance that the primary
sequence is correct and that the carbohydrate chains are consistent from batch to batch (32). It is estimated that as little as 5%
of the molecules with a single substitution can be detected in most
cases. For example, a mutant form of rt-PA was produced with
a glutamic acid residue at position 275 instead of an arginine residue. The resultant tryptic map shown in Fig. 6 clearly shows
the power of this method to detect substitutions. Peptide mapping
also provides the ability to detect the presence of proteolytic cleavages in the molecule, provided they are not at sites where trypsin

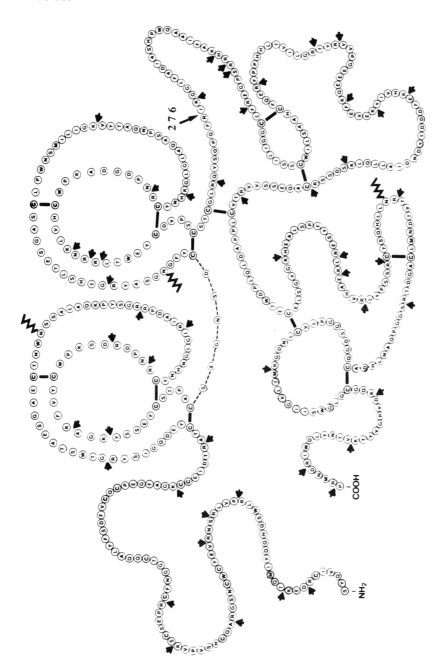

Figure 5   Structure of rt PA.   Arrows indicate the tryptic sites on the molecule giving rise to some 57 tryptic peptides.   The solid lines indicate the position of disulfide bonds.   Zigzag lines indicate positions of attachment for carbohydrates.

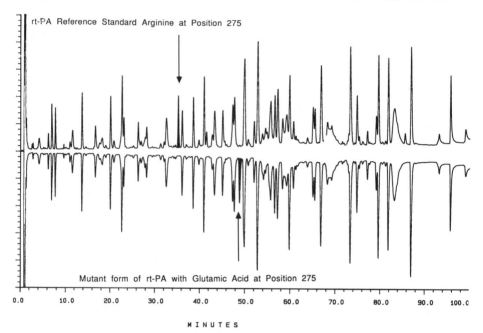

Figure 6   Tryptic map chromatograms of the rt-PA reference standard and a mutant form of rt-PA with a glutamic acid residue in place of the normal arginine residue at position 275. Arrows illustrate the differences in the two chromatograms caused by the substitution.

(or similar enzymes) can cleave. However, the presence of new proteolytic cleavages by trypsin-like enzymes would be detected by SDS-PAGE with silver stain as described above. Thus, a combination of methods is needed to obtain an overall analysis of the properties of the molecule on a batch-to-batch basis.

C. Potency

Clinical studies demonstrate the efficacy of a product in a controlled environment (21) but a test method is required to evaluate the potency of the molecule to ensure its consistency from batch to batch and its stability over its shelf life. Plasminogen is the natural substrate for rt-PA and assays have been developed using small chromogenic substrates that rt-PA will cleave (e.g., S-2288, see Ref. 26). However, the activity of rt-PA on this substrate is not

a true reflection of its ability to activate plasminogen for clot dissolution (27). More sophisticated assays have been developed that
use plasminogen and a chromogenic substrate for plasmin, the product of plasminogen activation (28). Although these assays use
the appropriate substrate and represent a situation closer to that
of an actual clot, they require the use of a clot substitute such
as fragments of fibrin or polylysine to mimic the stimulation of
plasminogen activation known to occur during fibrinolysis in vivo.

An in vitro clot lysis assay was developed that uses purified
components of human origin to examine the potency of rt-PA (9).
Briefly, rt-PA is mixed with thrombin, and fibrinogen is mixed
with plasminogen. These two solutions are then rapidly mixed at
the start of the reaction. Thrombin acts on fibrinogen to generate
a turbid clot, which is monitored in a microcentrifugal analyzer.
The rt-PA acts on the plasminogen to produce plasmin, which dissolves the fibrin, and the turbidity decreases. When the turbidity
has decreased to a predefined level, the clot is said to be lysed,
and the time at which this occurs is called the end point. A log-
log plot of end point vs. rt-PA concentration is linear and allows
interpolation of the activity of the test sample. Thus, rt-PA activity is measured in a setting that closely approximates the physiological setting because the standard curve is linear from the
concentrations expected to be incorporated into a clot as it is formed
naturally up to those concentrations obtained during therapy.

The clot lysis assay has been automated. With an accuracy of
99.5% and a coefficient of variation of 5%, the assay has been invaluable in the development and testing of rt-PA (9). The assay
has also been adapted to microtitration plate format (29) and is
very similar to that used by the World Health Organization to develop a reference material for rt-PA (30).

D. Safety

The safety of rt-PA is ensured by a number of traditional biological tests performed on the final product. These methods include
sterility, pyrogenicity using both the U.S. Pharmacopeia (USP)
rabbit pyrogen test and the LAL assay, and a general safety test
in mice and guinea pigs. In addition, reconstituted vials are assayed for particulate content according to USP guidelines. These
tests are performed on each lot of rt-PA by the manufacturer and
are reviewed and confirmed by the CBER before release.

VI. CONCLUSION

The quality control of rt-PA occupies a special place in the annals
of pharmaceutical product development. The techniques used to

ensure the quality of such a complex molecule as rt-PA require a
blending of those control methods common to the parenteral pharmaceutical industry with advanced immunological analysis methods,
molecular biology techniques, and modern analytical chemistry. The
control system described is an excellent example of what is both
necessary and important in the control of complex molecules. It
is hoped that as new analytical techniques are developed, they will
enhance the control of such products as rt-PA.

REFERENCES

1. Smith, J. W. G. Aspects of regulating biological medicinal
    products of biotechnology. In *The World Biotech Report 1984,
    Vol. 1. Europe.* Online Publications, Pinner, United Kingdom,
    pp. 69–76 (1984).
2. Collen D. Biological properties of plasminogen activators. In
    Tissue Plasminogen Activator in Thrombolytic Therapy (B. E.
    Sobel, D. Collen, and E. B. Grossbard, eds.), Marcel Dekker,
    New York, pp. 3–24 (1987).
3. Jones, A. J. S. and O'Connor, J. V. Chemical characteriza-
    tion of methionyl human growth hormone. In *Hormone Drugs,
    Proceedings of the FDA-USP Workshop on Drug and Reference
    Standards for Insulins, Somatropins, and Thyroid-axis Hormones,
    May 19–21, 1982, Bethesda, Maryland,* U.S. Pharmacopeial
    Convention, Rockville, MD, pp. 335–351 (1982).
4. Jones, A. J. S. and O'Connor J. V. Control of recombinant
    DNA produced pharmaceuticals by a combination of process
    validation and final product specifications. In *Standardization
    and Control of Biologicals Produced by Recombinant DNA Tech-
    nology,* Developments in Biological Standardization, Vol. 59,
    Karger, Basel, pp. 175–180 (1985).
5. Bogdansky, F. M. *Pharm. Technol.* (September 1987), p. 72.
6. *Code of Federal Regulations,* Food and Drug Administration,
    Department of Health and Human Services, Title 21, Subchapter
    F – Biologics, Part 610.12 (Sterility), Part 610.30 (Mycoplasma)
    (Revised April 1, 1989).
7. American Society for Testing and Materials (ASTM), *Standard
    Guide for Determination of Purity, Impurities, and Contaminants
    in Biological Drug Products,* 1989 Annual Book of ASTM
    Standards, Vol. 11.04, Designation E 1298-89, pp. 898-900,
    Philadelphia, PA (1989).
8. Carlson, R. H., Garnick, R. L., Stephan, M. M., Sinicropi, D.,
    du Mée, C. P., and Miller, C. Laboratory robotics applied to
    turbidimetric endotoxin analysis of recombinant DNA-derived
    pharmaceuticals. In *Advances in Laboratory Automation*

Robotics, Vol. 4 (J. R. Strimaitis and G. L. Hawk, eds.), Zymark Corporation, Hopkinton, MA, pp. 235−248 (1988).

9. Carlson, R. H., Garnick, R. L., Jones, A. J. S., and Meunier, A. M. *Anal. Biochem.* 168:428−435 (1988).

10. Taylor, D. M., Dickinson, A. G., Fraser, H., Robertson, P. A., Salacinski, P. R., and Lowry, P. J. *Lancet* ii:260−262 (1985).

11. Liu, D. T., Gates, F. T., III, and Goldman, N. D. Quality control of biologicals produced by r-DNA technology. In *Standardization and Control of Biologicals Produced by Recombinant DNA Technology*, Developments in Biological Standardization, Vol. 59, Karger, Basel, pp. 161−166 (1985).

12. Jones, A. J. S. Sensitive detection and quantitation of protein contaminants in rDNA products. In *The Impact of Chemistry on Biotechnology: Multidisciplinary Discussions* (M. Phillips, S. P. Shoemaker, R. D. Middlekauff, and R. M. Ottenbrite, eds.), ACS Symposium Series, Number 362, American Chemical Society, Washington, D.C., pp. 193−201 (1988).

13. Anicetti, V. R., Fehskens, E. F., Reed, B. R., Chen, A. B., Moore, P., Geier, M. D., and Jones, A. J. S. *J. Immunol. Meth.* 91:213−224 (1986).

14. Towbin, H., Staehelin, T., and Gordon, J. *Proc. Natl. Acad. Sci. USA* 76:4350−4354 (1979).

15. Laemmli, U. K. *Nature* 227:680−685 (1970).

16. Oakley, B. R., Kirsch, D. R., and Morris, N. R. *Anal. Biochem.* 105:361−363 (1980).

17. Jornvall, H., Pohl, G., Bergsdorf, N., and Wallen, P. *FEBS Lett.* 156:47−50 (1983).

18. Pohl, G., Kallstrom, M., Bergsdorf, N., Wallen, P., and Jornvall, H. *Biochemistry* 23:3701−3707 (1984).

19. Rijken, D. C. and Collen, D. *J. Biol. Chem.* 256:7035−7041 (1981).

20. Pennica, D., Holmes, W. E., Kohr, W. J., Harkins, R. N., Vehar, G. A., Ward, C. A., Bennett, W. F., Yelverton, E., Seeburg, P. H., Heyneker, H. L., Goeddel, D. V., and Collen, D. *Nature* 301:214−221 (1983).

21. TIMI Study Group, *N. Engl. J. Med.* 312:932−936 (1985).

22. Hopps, H. E. and Petricciani, J. C. (eds.), *Abnormal Cells, New Products and Risk*. In Vitro Cellular and Developmental Biology Monograph Number 6, Tissue Culture Association, Gaithersburg, MD (1985).

23. Petricciani, J. C. Safety issues relating to the use of mammalian cells as hosts. In *Standardization and Control of Biologicals Produced by Recombinant DNA Technology*, Developments in Biological Standardization, Vol. 59, Karger, Basel, pp. 149−153 (1985).

24. Kafatos, F. C., Jones, C. W., and Efstratiadis, A. *Nucleic Acids Res.* 7:1541–1552 (1979).
25. Edman, P. and Begg, G. *Eur. J. Biochem.* 1:80–91 (1967).
26. Astedt, B., Bladh, B., Christensen, U., and Lecander, I. *Scand. J. Clin. Lab. Invest.* 45:429–435 (1985).
27. Ranby, M., Bergsdorf, N., and Nilsson, T. *Thromb. Res.* 27:175–183 (1982).
28. Verheijen, J. H., Mullaart, E., Chang, G. T. G., Kluft, C., and Wijngaards, G. *Thromb. Haemost.* 48:266–269 (1982).
29. Beebe, D. P. and Aronson, D. L. *Thromb. Res.* 47:123–128 (1987).
30. Gaffney, P. J. and Curtis, A. D. *Thromb. Haemost.* 53:134–136 (1985).
31. Anicetti, V. R. Improvement and experimental validation of protein impurity immunoassays for rDNA products. In *Separations in Analytical Biotechnology, Chemistry and Capillary Electrophoresis* (J. G. Nikelly and C. S. Horvath, eds.), ACS Symposium Series, American Chemical Society, Washington, D.C. (1990) (in press).
32. Chloupek, R. C., Harris, R. J., Leonard, C. K., Keck, R. G., Spellman, M. W., Jones, A. J. S., and Hancock, W. S. *J. Chromatogr.* 463:375–396 (1989).

21

Design and Construction of Manufacturing Facilities for Mammalian Cell-Derived Pharmaceuticals

WILLIAM R. SRIGLEY
Invitron Corporation, St. Louis, Missouri

I. INTRODUCTION

The attractiveness of large-scale mammalian cell culture as a means of producing medically important biomolecules is no longer in question. A large number of potentially important therapeutic products of mammalian cell origin are under development. A few, with the most notable perhaps being Genentech's Activase, have already been approved for marketing in the United States. A relative handful of others are the subject of marketing applications now under consideration by the Food and Drug Administration (FDA). Because the vast majority of these products are in the early stages of development, the amounts of cell-derived products being produced today represent only a small fraction of the quantities likely to be required after 1990. It may well be that a major bottleneck in the conversion of the existing discoveries of biotechnology into commercial products will be the ability of industry to provide manufacturing capacity on a timely basis.

Although there are many manufacturing topics which are deserving of consideration, there are a number of specific issues which are likely to arise during the design and construction of a mammalian cell culture plant for the production of therapeutic products

which are somewhat problematic. In general, they are problematic
for one of two reasons: either the process is a new one which has
not yet been developed and operated successfully at the commercial
scale, or there is no inspectional history to suggest how the FDA
will approach the issue.

Frequently, the major considerations related to the design and
construction of a pharmaceutical plant derive from the basic choices
made by the management of the company regarding its field of interest. These decisions frequently determine the types of products
which the company expects to produce in the plant. Are they, for
instance, intended for human, animal, or agricultural use? The
distinctions are fairly significant since they determine many of the
regulatory requirements which the company will have to meet, as
well as the state or federal agencies responsible for administering
those regulations. The distinctions also define the characteristics
which the products must possess, and those in turn have a direct
bearing on the processes which will be incorporated into the plant.

Clearly, whether the products of interest will be therapeutics
or diagnostics is germane, as is the dosage form. A plant intended
for the production of parenteral therapeutics will differ from one
built for the manufacture of in vitro diagnostics.

Assuming that the focus of the company is on parenterals for
human use, a prime consideration is whether the products will be
classified as drugs or biologicals. While many regulatory requirements and guidelines apply to both, there are some which are unique
to biologicals. The distinctions tend to be unclear in some cases,
however. The fact that some diagnostics are also biologicals and
subject not only to the device regulations (Title 21, Subchapter 18
of the U.S. Code of Federal Regulations), but also to the Biological
Establishment and Product Licensing statutes (Title 21, Part 601 of
the U.S. Code of Federal Regulations), as well as the provisions of
the General Biological Requirements (Title 21, Part 606 of the U.S.
Code of Federal Regulations), can make things somewhat confusing
at times.

Another product consideration which will affect plant design
is the intended form and use of the product. Will the product be
a liquid, powder, or freeze-dried material? Will it be a final finished dosage unit, or is the product to be a partially processed
intermediate intended for further processing by another company?

II. PROCESS CONSIDERATIONS

While regulatory considerations are important in establishing some
of the broader aspects of plant design and construction, the major

influences on many basic performance and construction specifications are often process-driven. In mammalian cell culture, this frequently distills down to a few key factors directly related to the technology, namely, in-process and final product sterility, handling of medium and serum, equipment and facility design, and the potential for biohazards.

A. Cell Culture Technology

One of the most fundamental process considerations is the type of production system to be used in the plant. Cell culture systems take many forms and include a number of diverse techniques (Fig. 1). A basic distinction, however, is whether the system utilizes batch or perfusion culture. Each technology has its advocates and a set of advantages and disadvantages. From a plant design point of view, each also has its own unique set of facility and system requirements.

A manufacturing technology based on batch cell culture typically requires large reactors having a volume of 1000–10,000 liters. Such vessels are normally cleaned and sterilized in place with many, if not all, connections to the reactor already made. Critical parameters such as pH and gas pressures are regulated by the introduction of reagents and gases through preestablished connections. Batch reactors are generally charged at the beginning of the cycle with seed culture and sufficient medium to sustain the anticipated cell mass during the entire production cycle. As the cells increase in number they express product which is secreted into the culture medium. At the end of the production cycle various concentration and purification techniques are employed to separate the desired

- Batch Fermenters
- Roller Bottles
- Multisurface Propagators
- Hollow Fiber Reactors
- Encapsulation Systems
- Perfusion Reactors
- Immobilized Bed Reactors

Figure 1 Equipment typically utilized in the culture of mammalian cells.

product from the mixture of cells, cellular debris, medium, and supplements.

Perfusion culture reactors are typically much smaller in size than batch reactors. In some applications they are cleaned and sterilized in place. Alternatively, however, they may be disassembled for cleaning, reasssembled, and then sterilized in an autoclave. The cell seed culture may be introduced into the reactor at the beginning of the cycle or scaled up in various containers to achieve a cell mass sufficient to inoculate the process reactor. Culture medium is perfused through the reactor at a rate proportional to the number of cells and their metabolic characteristics. Process parameters are maintained by introducing reagents and gases directly into the culture vessel or into the medium as it is being fed to the vessel. Perfusion systems generally require substantially more aseptic connections than batch culture systems. Conditioned medium is removed from contact with the cells on a continuous basis. Concentration and purification typically take place on smaller volumes of conditioned medium than with batch systems.

Without regard to the relative advantages or disadvantages of either of these technologies, they each confer their own specific design requirements on the plant (Fig. 2).

The large reactors used in batch cell culture require commensurately large rooms with high ceilings. Usually the building is constructed around such equipment, which can make installation and startup of that equipment somewhat more difficult. Perfusion systems are typically smaller in size. They can frequently be accommodated in modestly sized rooms with normal ceiling heights. They are normally brought into the facility after construction has been completed, which makes installation easier.

	BATCH	PERFUSION
ROOM SIZE	LARGE HIGH CEILINGS	SMALL LOW CEILINGS
INSTALLATION	DIFFICULT	EASIER
REACTOR/FACILITY INTEGRATION	HIGH	LOW TO MODERATE
MEDIA PREPARATION	NO ADDITIONAL FACILITIES NEEDED	SEPERATE MEDIA MFG. FACILITIES NEEDED
CONCENTRATION /PURIFICATION	LESS FREQUENT LARGE VOLUMES CELL CULTURE DRIVEN	FREQUENT SMALLER VOLUMES NOT CELL CULTURE DRIVEN

Figure 2   Influence of cell culture process on the design of the manufacturing facility.

The level of integration of the reactor with the facility is generally greater with batch culture systems. Once a 5000- or
10,000-liter tank is installed in a building and the steam, condensate, and process piping is connected to it, it is very much a part
of the facility. Removing, replacing, or significantly modifying
such a reactor can be a major undertaking. Perfusion reactors
may also be hardpiped in place, but their smaller size allows them
to be more easily separated from the building. Perfusion systems
utilizing a modular design may be even more easily detached from
the building, thereby providing a great deal of flexibility when
modification or replacement of reactors is called for. In all cell
culture systems a substantial number of electrical and electronic
connections exist between the reactor and the building. It is generally advantageous for these to be designed with quick disconnects
to facilitate the making and breaking of connections.

  Another set of distinctions influenced by the type of cell culture system to be used in the plant relates to the upstream and
downstream processing requirements. In a plant utilizing batch
cell culture, the culture medium is typically made in the reactor
itself. Little is needed in the way of additional facilities for the
large-scale manufacture of culture medium. At the conclusion of
the culture cycle, the entire contents of the reactor vessel are
processed to remove and purify the product of interest. Concentration and purification facilities are therefore usually designed to
handle large volumes of cell-conditioned medium and purification
intermediates on an intermittent basis. This can sometimes be advantageous from the standpoint of process economics.

  In contrast, in a plant utilizing perfusion culture, separate
facilities must be provided for the manufacture of culture medium.
This frequently entails the use of a large number of relatively
small tanks, each of which must be individually handled and tested.
At frequent intervals during the culture cycle, relatively small
batches of cell-conditioned medium become available for concentration and purification. The size and frequency of these batches is
often within the control of the production operators. Perfusion
culture allows the active moiety to be quickly removed from contact with the cells and potentially degrading enzymes which might
be present in the conditioned medium. For some products this is
critical to the preservation of biological activity.

  Invitron's cell culture plant utilizes three types of reactors
ranging in size from 16 to 100 liters. They are modular in construction. This allows them to be completely disassembled for
thorough cleaning between runs. All reactors and associated process
equipment and tanks are sterilized in an autoclave which is unloaded
directly into the contained area (Fig. 3). The plant has three cell
culture suites, each containing six reactor stations (Fig. 4). The
modular approach to reactor design allows any one of the 18 reactor
stations to be utilized for whichever reactor type is best suited

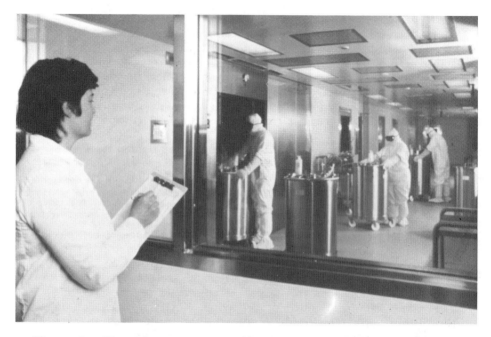

Figure 3   View of the entry corridor in the contained area of Invitron's manufacturing plant. A quality control employee is observing sterilized equipment being unloaded from the clean side of the double-door autoclave.

for a particular cell and product. It also allows individual reactors and the supporting systems to be removed and modified as improvements are made in technology with no impact on the operation of the facility. All of the necessary utility, reagent, process gas, and computer connections are present at each reactor station and quick disconnects are used whenever possible. The rooms are about the size of a long living room and have 8-ft ceilings.

Since the plant utilizes perfusion systems, separate media preparation, filtration, and storage facilities are necessary. Culture medium is made up in batches and filtered into portable stainless steel tanks. Each tank is sampled and then stored at refrigerated temperatures pending completion of quality assurance release testing. Concentration and purification of conditioned medium is typically done shortly after removal from the cell culture reactor. Purification batches are sized to allow the use of intermediate scale equipment and rooms.

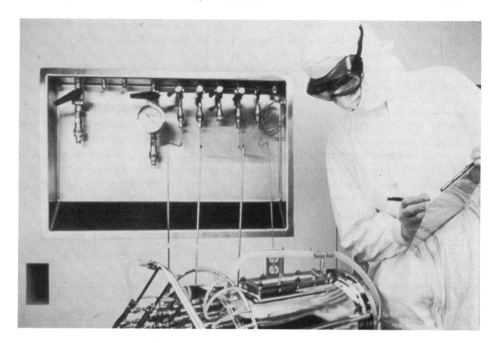

Figure 4 One of eighteen reactor stations in Invitron's cell culture plant. Each station is capable of manufacturing multigram quantities of cell-derived product daily from any one of three different types of reactor.

B. Environmental Control

A second major process-driven issue is the level of environmental control needed in the facility. The fact that environmental control requirements are to a large extent process-driven can be clearly demonstrated by comparing a plant and the production processes used for the large-scale manufacture of recombinant insulin from bacterial cells with those intended for the manufacture of mammalian cell-derived products by perfusion cell culture. There can be an extremely large number of differences between these two types of facilities in terms of room layout, degree of containments, finishes, room air quality and control, level of gowning, and so on. One might justifiably wonder whether these dissimilarities reflect a lack of agreement on precisely what standards apply to production facilities for cell-derived products. I believe, however, that if the manufacturing processes employed in each plant are examined carefully, the conflicts begin to dissolve.

The recombinant insulin process employs E. coli cultures having a doubling time measured in minutes; the batch culture cycle was described as taking less than 24 hr. The product is produced by the cells in a form which requires relatively complex chemical manipulations to properly configure the insulin molecule. The cells themselves produce significant amounts of endotoxin and cell debris which necessitates the use of comparatively harsh separation and purification methods.

In contrast, mammalian cell culture employs cells which have a doubling time measured in hours or days; culture cycles can take from 1 to 2 weeks in a batch culture system to several months in a perfusion system; the products are generally expressed by the cells in structural forms very similar to those found naturally and therefore require very little in the way of chemical modifications. The purification methods employed are typically more gentle and refined.

It might also be noted that insulin is regulated as a drug whereas cell culture products are often regulated as biologicals.

It is, I think, entirely reasonable to conclude that the environmental controls appropriate for the manufacture of pharmaceuticals by these two processes might differ substantially from one another. Similarly, the environmental conditions appropriate for the manufacture of product from mammalian cells in batch cultures utilizing antibiotics might be expected to differ from those necessary when the same mammalian cells are grown in a perfusion system without the use of antibiotics.

A further distinction which could be made relates to the particular manufacturing steps which are to be conducted in the area. For example, rooms in which concentration or purification steps will take place may not need the same level of air quality and control as rooms in which aseptic transfers or sterile filtrations are to be performed.

To a large extent the environmental quality of a processing area is influenced by three factors:

1. Room finishes: primarily a function of the materials of construction, design of surfaces and equipment, and the layout of equipment
2. Room air cleanliness: as determined by

>The number of air changes per hour in the room
>Filtration efficiency of air handling system
>Air pressure differentials between rooms
>Room exhaust air treatment
>Method of controlling air supply and exhaust

3. Personnel contaminant contribution: largely determined by room layout and personnel flow, gowning, and training

In the Invitron plant, environmental control was given a high priority. The fact that antibiotics are not used to preserve the microbiological integrity of the cultures, coupled with the fact that perfusion culture systems are utilized, means not only that a fairly large number of aseptic connections are made but also that those connections must be able to be made with a very high degree of confidence. It was therefore necessary from the outset to utilize design and construction techniques which would provide excellent environmental conditions in the manufacturing areas.

Room finishes, wherever possible, facilitate cleaning. The walls and floors throughout the production areas are seamless vinyl (Fig. 5). The walls were protected during construction by

Figure 5 Photograph taken during construction showing seamless vinyl wall covering on walls of contained area. Utility drops are introduced into rooms in custom-built stainless steel enclosures to facilitate cleaning.

a plastic film. Bumper rails were used liberally throughout the facility to minimize scratching and gouging of the wall surface (Fig. 6). Ceilings are either seamless vinyl or epoxy-painted. All utility services were brought into the production rooms through stainless steel pipes or tubing. All conduits leading to the contained area were sealed to minimize leakage of air. Window frames, doors and door frames, as well as work surfaces and supports were custom designed to reduce cracks and crevices, and constructed of stainless steel to permit frequent cleaning and disinfecting (Fig. 7). Light fixtures and HEPA filters are serviced from an interstitial area (Fig. 8). To the extent possible, equipment is installed so as to permit cleaning behind and underneath individual pieces. Stainless steel shrouding prevents dust from settling on the tops of laminar flow benches (Fig. 9).

Room air quality is a critical element of environmental control. The supply air to the manufacturing areas in the Invitron plant is subjected to HePA filtration. In addition, the number of air changes per hour is maintained at a level sufficient to provide an effective washdown effect so that any particles generated in the

Figure 6 Vinyl wall covering was protected during construction by plastic film.

Figure 7   Custom-built window frames present flush surface,
which minimizes accumulation of particles and facilitates cleaning.

room are promptly removed. The number of air changes ranges
from 30 per hour in less critical rooms to more than 50 per hour
in rooms in which aseptic operations take place. Air returns are
at floor level to facilitate the washdown effect (Fig. 10) and consist of stainless steel ducts located within the walls of the facility
(Fig. 11). All exhaust air, whether returned to the air handlers
or ducted out of the building, is passed through HEPA filters in
stainless steel "bag-in, bag-out" housings located in the interstitial space (Fig. 12). Individual rooms or sets of rooms are
isolated with regard to air flow by the use of 15 individual air
handlers located either in the mechanical room (Fig. 13) or in the
interstitial spaces (Fig. 14) to supply air to the manufacturing
areas. Room air pressures, and therefore the movement of air
from one room to adjacent rooms, as well as the temperature and
humidity of individual rooms are monitored and controlled by a
central, microprocessor-based control system (Fig. 15).

Figure 8   View of interstitial area.

The potential for contaminant contribution by employees was considered during design of the building and was a factor in determining room layout and personnel flow. Although difficult to implement rigidly, from a conceptual point of view it is advantageous to strive for one-way flow through the production facility. Modeling the plant in advance of construction and mapping out the path that you expect people, equipment, and product to take under operating conditions helps to pinpoint and correct design situations which could present logistical problems. Gowning of personnel is an important factor in maintaining low particle levels in critical process areas. In this context, the cleanliness of the gowns themselves as well as the training of employees in their proper use are essential. As in all aspects of pharmaceutical production, the success of these efforts is entirely dependent on the willingness of employees to consistently follow procedures.

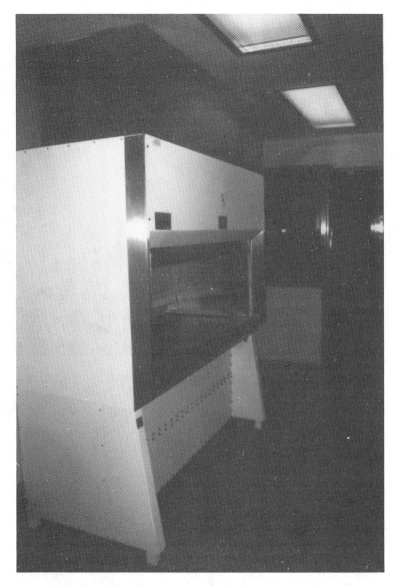

Figure 9 Biosafety hoods are elevated off the floor and positioned away from walls to permit cleaning underneath and behind the units. A custom fabricated shroud on the top of the hood prevents the accumulation of dust.

Figure 10   Seamless vinyl is form-fitted around openings to return air ducts. Note vinyl-covered bumper rail, which prevents equipment from damaging wall surfaces.

C. Production Equipment and Systems

A third process-related issue centers around the production equipment and systems to be used in the plant. While it is true that mammalian cell culture has been used for many years to manufacture pharmaceutical products, the quantities of such materials, and hence the production scale, have been relatively small. The techniques for economically manufacturing diverse products in gram or kilogram quantitites have for the most part been developed only in the past few years. Much of the technology is therefore in a state which might well be characterized as "advanced research." However, this problem is not limited to the specific pieces of equipment used for mammalian cell culture; it extends as well to the sensors used to measure process parameters, to

Figure 11 Stainless steel return air ducts of welded construction are located within the walls. The ducts are joined to form a stainless steel plenum. Each plenum is fitted with a removable hatch on the top surface, which can be removed for inspection and cleaning of the ducts.

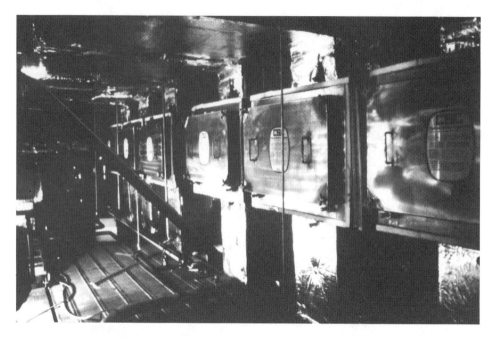

Figure 12   Bag-in, bag-out filter housings for HEPA filters on return air ducts.

the computer algorithms and models used to monitor and control the process, and to the quality of the medium, serum, reagents, and additives used in cell culture and purification.

To some extent these kinds of issues were previously addressed in the manufacture of conventional pharmaceuticals. 3-A standards (formulated by the International Association of Milk, Food and Environmental Sanitarians, published by the U.S. Public Health Service, Dairy Industry Committee), for instance, are relied on by the pharmaceutical and food-processing industries as specifications for the design and construction of sanitary piping systems (Fig. 16). United States Pharmacopoeial (U.S.P.) monographs have been adopted as definitions of acceptability for many raw materials and reagents. At present there are no analogous standards or references for culture media, protease inhibitors, growth factors, purification resins and the component parts of the purification systems, and other recently developed or proprietary processing equipment. Furthermore, many cell culture production systems rely heavily on the use of computers to control critical operating parameters such

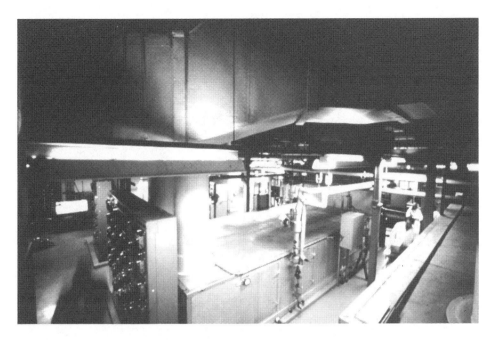

Figure 13 One of fifteen air handlers that supply air to the production areas.

as pH, temperature, and the partial pressures of oxygen and carbon dioxide. In many instances the science behind these control processes is quite imprecise. These are areas of concern which I believe industry must address. In the United States, trade organizations such as the Association of Biotechnology Companies, the International Biotechnology Association, the Pharmaceutical Manufacturers Association, the International Society of Pharmaceutical Engineers, and the Parenteral Drug Association can be effective facilitators of that process.

Some facility design questions derive from regulatory issues which can only be resolved after discussions with the appropriate governmental authority. There are many ways in which the regulations enforced by the FDA have an impact on plant design and construction. It is in a company's best interest to consult with the agency early in a project and at appropriate intervals thereafter to assure that it is proceeding on a sound basis. It may be determined as a result of a meeting with the FDA that they have already adopted a position on the issue at hand, and no matter how

Figure 14   Air handler suspended from trusswork above production area.

convinced the company may be of the appropriateness of its approach, the agency may not agree. On the other hand, sound and well-presented logic is often effective at persuading FDA staff that an alternative proposal is reasonable, thereby allowing a particular facet of design or construction to be employed which provides economic or operational advantages.

Other design issues may be related to process economics. Some fairly basic systems which have been utilized in the pharmaceutical industry for many years are still subject to discussion. The relative merits of water for injection vs. purified water, distillation vs. reverse osmosis, welded stainless steel pipe for process gas lines vs. copper tubing, clean steam generators for autoclaves vs. filtered plant steam are still debatable issues, primarily because they represent significant economic and operational distinctions. Many factors contribute to manufacturing cost sensitivity, and to a large extent they are interrelated. Is it preferable, for instance, to reduce front-end construction costs and perhaps live with the risk of a higher contamination rate, or should one spend more during

Figure 15 A computer system monitors and controls air pressure, temperature, and humidity in individual rooms in the facility.

construction in the expectation of a higher success rate and lower costs in production operations? As we gain experience we may find mammalian cell culture to be more demanding in terms of endotoxin and DNA levels and tolerance to trace chemical contaminants than conventional drug manufacture.

Constructing a plant for the manufacture of pharmaceutical products is a challenging task. Many of the critical pieces of equipment must be custom fabricated and therefore must be ordered many months in advance of installation. When the plant is to utilize biological production processes such as cell culture and protein purification, the job becomes even more formidable. The companies building cell culture production plants are often small biotechnology firms with limited resources and experience. Their continued growth, and sometimes even their ability to continue operations, may be dependent on the success of the plant construction project. In such an instance, there may be little margin for error. For a project of such consequence, management of risk should be a key design consideration. In that context, the following strategies are useful.

Figure 16   Components of the distilled water system are constructed
of stainless steel and fabricated to meet 3-A standards.

First, distinguish those aspects of the facility which are innovative from those which have been done before, To the extent possible,
when there is a successful historical precedent which can be relied
on to guide the design or construction of a plant or process system,
that precedent should be followed.

Second, keep the design of equipment and systems as simple as
the process and good judgment allow. Take into consideration the
likelihood that, particularly during the first few years, it will be desirable to make changes to nearly every one.

Third, strive to do things in a way which takes the uncertainty
out of plant start-up and minimizes manufacturing downtime.

Fourth, recognize that you can't do everything and be everywhere.
Prioritize your own efforts and those of your staff such that critical
events are witnessed, construction errors are minimized, and irre-
trievable data is captured for later analysis.

This strategy was followed during the construction of Invitron's manufacturing plant. Design on the plant commenced in August 1984; construction began in October of that year; the first product was manufactured in the plant 18 months later. The rapid and successful construction and startup were due to two factors: first, a lot of hard work and long hours on the part of many people; and second, the use of design principles and equipment that have a history of success.

In the case of the process water, the combination of a hot, 316-liter stainless steel, electropolished piping loop with a Finn-Aqua still produced a Water-for-Injection (WFI) system which met the U.S.P. requirement for WFI within a week of startup (Fig. 17).

The air distribution system consisted of precleaned ductwork which was kept sealed with plastic sheeting during construction (Fig. 18). The system startup was coordinated with facility cleaning so that dirty air was never drawn into the ductwork and filters. As a result, within 20 min of bringing a room on-line the number of particles greater than 0.5 µm was less than 10,000 per ft^3. The digital control system permits complete rebalancing of the system to be done in less than a day.

Figure 17   Water for injection system. Water is maintained at a temperature in excess of 80°C. Pressure and temperature are monitored continually by the instrumentation at right.

Figure 18   All air duct was cleaned and sealed prior to installation to minimize premature loading of HEPA filters during startup.

D. Quality Control

One of the most critical and yet often neglected aspects of plant
design and construction is quality control. There was a time when
the release of a product was predicated on demonstrating that it
possessed specific characteristics at the end of the production
process. For the most part, meeting final product specifications
meant that the product was acceptable. That is no longer true in
this country and indeed in most countries with high standards of
drug regulation. The measure of success in the manufacture of
pharmaceuticals now encompasses not only the production of an
active material meeting established specifications, but also the si-
multaneous demonstration of the suitability and stability of the
production process. In the United States this demonstration has
taken the form of process validation. The question to be answered
is, "Do the plant systems and processes perform as they were intended to, and does that level of performance consistently result
in acceptable product?" This concept is no longer an issue. What

is an issue, particularly during plant design and construction, is the extent to which the validation of the facility systems, and process equipment find their way into the project at an early stage.

Let us take as an example something which is for the most part an accepted standard for pharmaceutical plant design: "Process sewers and sanitary sewers should not be interconnected." There is only one time when it is possible to verify without a doubt that this requirement has been met; that is during construction, before the floor is poured. This may be the first validation protocol needed in a plant construction project. And yet, all too frequently, perhaps as a result of an observation on an FD-483 form left following an FDA inspection, engineering managers find themselves retrospectively attempting to verify the lack of interconnections between the process and sanitary sewer systems in an operational building.

Realtime verification that the facility is actually constructed in accordance with management's intentions is of immense value. Likewise it is advantageous to be able to monitor the installation and cleaning of critical process equipment and develop plant maintenance and documentation systems during plant construction. This means, however, that the company must invest in the staff necessary to carry out these activities early in the project. While this may be feasible for a large, established pharmaceutical company, it is very difficult for a startup biotechnology company, and many of the companies constructing cell culture facilities today fall into the latter category. Nevertheless, almost any effort which can be made in this regard will be worthwhile. Even if it is not possible to write and assemble all of the protocols in advance of construction, at least know in advance what data you will be need to develop the protocols afterward, and take whatever steps are necessary to preserve and file those records while they exist.

Again using Invitron's plant as an example, systems requiring validation were identified early in the construction project. Shortly after groundbreaking an Invitron trailer was moved onto the job site alongside that of the construction contractor (Fig. 19). This trailer served as the control center for Invitron activities related to the plant construction and startup. Initially, it was staffed by two Invitron employees – one each from Quality Assurance and Engineering. Eventually it became three trailers staffed by over 40 people. Here, drawings and specifications were reviewed to determine which construction details required verification (Fig. 20). Meetings were held on a weekly basis with design and construction engineers to coordinate our activities with theirs (Fig. 21). Materials of construction were inspected and tested as necessary to verify that they met design specifications (Fig. 22). The documentation and procedure center was established here as well.

Figure 19 Owner's staff resided in trailers on plant site during construction to facilitate interaction with design and construction crews.

Figure 20 Details of construction and equipment installation were compared with design drawings. Any deviations from design drawings were evaluated and noted on "as-built" drawings.

Figure 21 Weekly meetings between owner's staff and the design and construction teams ensured close coordination of construction and validation activities.

Figure 22 The documentation trailer served as a command center for the project. Here also, materials of construction were inspected and tested for conformity to design specification.

Figure 23   All underground piping systems were checked to verify proper connections and photographed as a permenent record.

By this means, critical validation information was able to be assembled during plant construction. Using the effluent waste systems as an example, virtually every foot of underground pipe was photographed and labeled in accordance with a coded construction drawing (Fig. 23). Surveying techniques were used to verify that pipes were assembled with the proper slopes (Fig. 24). Videotaping was utilized extensively as a means of documenting construction details (Fig. 25). Using similar techniques the information needed to verify details of construction of other critical aspects of the facility was preserved for later incorporation into formal protocols, even though the staff available at the time was quite limited. What is essential, however, is that the company define these objectives as an early part of the project and hire personnel with the requisite experience and knowledge in a time frame which allows the required activities to be carried out.

It goes without saying that quality control is an essential consideration in the postconstruction stages of plant operation. The ability to utilized fully functional microbiology and chemistry laboratories during startup is immeasurable. If at all possible, these areas should be scheduled for completion well in advance of other

Figure 24 Elevations and slopes and piping systems were verified prior to pouring the floor.

Figure 25   Videotape was used extensively to record the details
of construction. This technique is particularly helpful for areas
that are covered over during construction (i.e., piping behind
walls). It is also an effective training device for maintenance staff.

portions of the plant, not only to serve the needs of facility and
system startup, but also because stocking inventories of raw materials and supplies must be tested for acceptance and release in
advance of need. Ideally, by the time production begins, a comprehensive plan should be in place for environmental analysis as
well as raw material, cell line, in-process, and final product testing. These programs are essential to the successful startup and
operation of a cell culture facility.

III.   REGULATORY CONSIDERATIONS

In conclusion, I would like to address a few additional regulatory
considerations that have an impact on the construction of plants for
the manufacture of mammalian cell-derived products.

A. Current Good Manufacturing Practices (cGMP) Compliance

Because many companies utilizing cell culture are new to the industry, they do not have an extensive base of preexisting validation data and experience on which to draw. Nearly everything must be built from the ground up – a formidable task. To some extent this situation resembles the state in which many pharmaceutical companies found themselves in the mid-1970s when the proposed LVP-GMPs were suddenly adopted by the FDA as guidelines against which they inspected drug plants. There's a lot to do and it is extremely important to understand what is necessary today and what can wait until tomorrow. This is probably one of the most difficult and pressing problems faced by small companies attempting to build and startup a manufacturing plant. It is one which I believe can continue to benefit from dialogue between industry and the FDA. What level of control is necessary for the manufacture of product for preclinical studies or for studies performed in accordance with Good Laboratory Practices (GLPs)? To what extent is it necessary to comply with cGMP – including full validation – in the manufacture of product for use in clinical investigations? While the agency has taken the position that cGMP compliance is necessary for the manufacture of any product intended for administration to humans, there seems to be a recognition that the detail of the programs might be more modest in a facility manufacturing materials for clinical investigation than in a plant manufacturing commercial product. Recently, the Center for Drug Evaluation and Research issued a document entitled "Draft Guidelines on the Preparation of Investigational New Drug Products (Human and Animal)" (February 1988), delineating their expectations with respect to this issue.

B. Drugs vs. Biologicals

It is important to know whether the product you're dealing with will be classified by the FDA as a drug or biological. Not only does it determine how the product approval process will work; it also defines how and when facility inspections will take place, who will conduct them, the type of questions which are likely to be asked, and the emphasis of the inspection. Whether the products produced are drugs or biologicals also determines the means by which changes to the facility will be approved by the agency. It also affects how you go about obtaining FDA comments on your facility design and construction program prior to breaking ground, as well as during construction and startup.

While the FDA has made considerable progress in defining the way in which it will go about deciding how to classify new products, the process is still a dynamic one. If you are contemplating the manufacture of a product which is novel or which has not yet been classified, by all means consult with the FDA as soon as possible.

C. Licensing of Biological Intermediates: Partially Processed Products for Further Manufacture

Until recently, the Office of Biologics followed the practice of not approving a product license application unless the applicant could demonstrate proficiency in all aspects of the manufacture of the particular product. In a practical sense this meant that a company seeking a license for a biological product had to be able to manufacture that product from beginning to end in a facility described in an approved establishment license application. Exceptions to this policy were permitted only under closely defined circumstances. In December 1984, the agency published a notice in the *Federal Register* which stated, in effect, that product licenses would henceforth be granted for monoclonal antibodies intended for further manufacture to licensed biological products. While perhaps not generally recognized as such at the time, this was a significant and far-reaching step, one which has permitted a dramatic change in the structure of the biologicals industry in America. It has, for instance, allowed the development of contract manufacturing of biological products. This concept has since been extended generally to other cell-derived products intended for further manufacture to licensed biologicals.

The best way to assure that the concerns of your company are considered as regulations are developed is to submit well-drafted comments to the proposed regulations when they appear in the *Federal Register* and to be an active participant in industry/FDA discussions on the topic.

The development of regulations is an important process in which industry, in the form of individual companies and trade organizations, must actively participate.

22

Large-Scale Propagation of Insect Cells

JAMES L. VAUGHN
Insect Pathology Laboratory, Plant Sciences Institute, Agricultural Research Service, U.S. Department of Agriculture, Beltsville, Maryland

STEFAN A. WEISS
GIBCO—Life Technologies, Inc., Grand Island, New York

I. INTRODUCTION

Insects hold a unique position in the study of viruses and virus diseases. Viral pathogens of plants and animals, including man, replicate in and are transmitted by insects. Insects are susceptible to several different viruses some which are very similar to the pathogens of plants and other animals and some which are unique to insects. For this reason, insect cell and tissue culture has been of interest to scientists in a wide range of disciplines. Trager demonstrated the replication of a viral pathogen of insects (1) and of encephalomyelitis virus (2) in primary cultures of insect cells. Grace (3,4) and Goodwin et al. (5) demonstrated that insect viruses would replicate in continuously culturable lines of insect cells. This provided the stimulus for extensive research into the development of cell lines and methods for the replication and assay of the wide spectrum of viruses associated with insects.

One goal of research on the viral pathogens of insects was their use to control insect pests. Cell cultures were seen as a reasonable alternative to the production of viruses in mass-reared insects. The latter method is labor-intensive and requires space, equipment, and expertise not commonly available on a commercial scale. Early research established that insect cells could be grown in roller bottles (6) and in spinner cultures (7). Although the roller bottle system is an efficient virus production method (8), insect cells are not substrate-dependent and suspension cultures can be scaled up to commercial levels more easily. Thus, the research concentrated on this method. Hink and Strauss (9), Miltenburger and David (10), and Weiss and his colleagues (11) achieved promising results in a variety of vessels and suspension systems. However, technical problems, including high oxygen demand and the fragility of insect cells, discouraged the commitment of funds and effort for further research. In addition, the high cost of the culture media made virus production too costly to compete successfully against the chemical insecticides. Thus, until recently, progress in large-volume culture was sporadic and slow.

The development of monoclonal antibodies and the use of animal cells to produce metabolites of medical importance described in the preceding chapters of this volume have led to increased research in the culture of animal cells. This included the design of culture systems that resolved some of the technical problems limiting the scale-up on insect cell cultures. Further stimulus for increased research in insect cell culture came with the development of the baculovirus expression vector systems (12,13). In these systems a foreign gene can be inserted into the baculovirus genome in place of the gene for the virus protein polyhedrin. The new gene, under the control of the very active polyhedrin promoter, synthesizes large amounts of the foreign gene product, up to 30% of the total cell protein (13). This expression system has provided the high-dollar products needed to attract research funds and effort into the development of new media and new culture systems for the large-scale propagation of insect cells. In this chapter we will discuss the progress made toward the development of such systems.

II. SELECTION OF CELL LINES OR STRAINS

A. Sources of Cell Lines

The most recent compilation of invertebrate cell lines (14) lists cell lines from 73 insect species from six different orders. These cell lines were originated from all stages of insects, from embryo to adult, and from a variety of tissues, e.g., ovaries, testes, imaginal discs, and hemocytes. There have been nearly 200 insect cell lines developed in the 26 years since the first insect cell lines were reported (15). Examples of cell lines used in large-volume culture and expression vector studies are given in Table 1.

Table 1 Some Insect Cell Lines Susceptible to
Baculovirus Expression Vectors

Cell line	Source	Ref.
IPLB-Sf-21AE	*Spodoptera frugiperda*	36
TN-368	*Trichoplusia ni*	53
IZD-MB-0503	*Mamestra brassicae*	77
BM-5	*Bombyx mori*	78

For most large-volume culture studies an existing cell line can probably be chosen that meets the needs of the project. No insect cells
are known to be substrate-dependent and therefore will grow either
attached to a substrate or in suspension. However, some cells attach so lightly that it may be necessary to develop a strain with
stronger attachment ability than the wild strain (16) if a substrate-
dependent system is desired.

The tissue of origin may be a critical factor in the choice of
cell lines. Many insect viruses are known to have tissue tropisms
in vivo, e.g., some granulosis viruses (GVs) infect only the fat
body, others only the midgut epithelium, and a few the fat body
and the hypodermis. Since only a few cell lines are derived from
these tissues, the replication of GVs has only recently been accomplished in vitro. Insect viruses also have a limited host range
and the cell cultures selected must be derived from the host insect
or a closely related species. This is a good general rule for selecting a cell line; however, it is not an absolute necessity. The
Autographa californica nuclear polyhedrosis virus (AcNPV) will
replicate in a number of cell lines from several insects other than
the original host. Lynn and Hink (17) compared the yields of this
virus in five cell lines from five different insect species and reported greater than 10-fold differences in viral sensitivity based
on plaque numbers. Nearly a fourfold difference in the total number of polyhedra per infected cell was found. In similar studies,
McIntosh et al. (18) found that comparable yields of AcNPV were
obtained in cell lines from three different insect species, none of
which was the species from which the virus was obtained. Their
tests with the *Trichoplusia ni* NPV showed that almost twice the
number of polyhedra were obtained with a cell line of heterologous
origin, *Spodoptera frugiperda*, as with the homolgous TN-368 cell line.

The source of most insect cell lines has been a complex organ
or the whole embryo. Thus, the cell line may be derived from any

number of different stem cells. Goodwin et al. (19) first demonstrated that there was significant variation among cell lines derived
from the same insect organ by the same methods. They found that
among several lines of gypsy moth cells derived from pupal ovaries,
some were susceptible to and replicated the gypsy moth NPV,
whereas others did not. In some susceptible lines only a small number of cells eventually contained polyhedra, evidence for the complete replication of the virus. Their findings were confirmed with
cell lines from *Heliothis* species (20), *Orgyia pseudotsugata* (21),
and *Orgyia leucostigma* (22). When a serological assay was used
to determine infection, the differences among heterologous cell lines
were further defined (23). Volkman and Goldsmith reported, for
example, that a *Bombyx mori* cell line that had previously been
thought to be refractive to infection by AcNPV was infected at a
slightly higher efficiency than a *S. frugiperda* line that produced
more polyhedra. Some heterologous cell lines also produced predominantly single-cell foci of infection even at 88 hr postinfection
and all cell lines produced some level of single-cell infections.

Further evidence of the broad variation in virus susceptibility
was shown in the study of Miltenburger et al (24). A total of 81
primary cell cultures were screened for their response to infection
with a NPV and a GV from the homologous insect the codling moth,
Cydia pomentella. Response to the NPV ranged from no response,
to cytopathic effect (CPE) but no polyhedra, to polyhedra formation. Some of the cell lines that originally replicated the NPV
were unstable and the ability to replicate the virus was lost within
four passages. When challenged with GV, only two primary cell
lines showed CPE visible with the light microscope. An additional
seven showed no CPE but stained positive with monoclonal antibody against the GV antigens. Similar variability was demonstrated
by Granados et al. (25) using *T. ni* cultures and a *T. ni* GV.
They were able to obtain 15 cell lines that had some level of susceptibility to TnGV (1−50% of the cells infected). However,
within 20 passages susceptibility to infection had decreased or
was lost.

These results demonstrate the considerable influence of the
host cell on the regulation of gene expression and virus production. When the baculovirus system is used to produce foreign proteins there are additional effects of the cell-controlled posttranslational modifications of the protein. To date most of the available
data are from the AcNPV vector in cells derived from *S. frugiperda* (13). It is likely that some cell lines from different insects
or from different tissues will produce proteins with different modifications and therefore different biological activity.

Thus, the choice of cell system can have considerable influence
on the end result of a project. Some preliminary testing should be

done before establishing a test or a production system. Once a cell line has been chosen, protocols should be adopted to maintain the purity and the integrity of the cell stock.

B. Characterization

As with microorganisms or other cell lines, insect cell cultures are easily cross-contaminated or misidentified (26). It is therefore important for each cell line to be characterized. A combination of karyology, serology, and enzymology has been used for characterizing most insect cell lines. Cell morphology is not a suitable characteristic since the morphology of cells in a number lines is similar and can vary under changing culture conditions.

Karyotyping has been commonly used for characterizing cell lines from the orders Diptera (27–29), Hymenoptera (30), and Coleoptera (31,32). Karyotyping has not been useful for distinguishing among cell lines from species of Lepidoptera (33,34). Metaphase plates from lepidopteran cells generally contain large numbers of microchromosomes and few typical chromosomes; thus plates from different cell lines do not provide easily recognized differences. Disney and McCarthy (35) reported that more normal, less fragmented chromosomes were obtained when the colchicine treatment was omitted. No significant decrease occurred in the mitotic index and the chromosomes also were less condensed. So perhaps with improved methodology karyotyping will become a useful way to distinguish among cell lines of lepidopteran origin.

Serological methods have had limited application in characterizing insect cell lines (26,36). Aldridge and Knudson (37) evaluated a number of serological methods for distinguishing cell lines. Five cell lines from four different genera in three orders were tested and immunoelectrophoresis proved to be the best of the serological methods. Cell lines from different genera could be distinguished, but not cell lines from different species within the same genus. Another disadvantage is that serological testing requires the additional time and expense to produce and store the antisera.

The most widely used method in addition to karyology is the analyses of cellular isozymes. First used for charcterizing insect cells by Greene and Charney (26), it was investigated extensively by Tabachnick and Knudson (38). Analysis of four isozymes – isocitrate dehydrogenase, malic enzyme, phosphoglucoisomerase, and phosphoglucomutase – was sufficient to distinguish among 16 cell lines studied. Later the study was extended to additional cell lines from Diptera and Acari (39). In both studies the four enzymes were adequate to identify interspecies differences between cell lines but did not distinguish between cell lines from the same species. Harvey and Sohi (40) were able to characterize cell lines from the same species using the enzymes phosphoglucomutase, aspartate aminotransferase, and isocitrate dihydrogenase. Lynn and his coworkers combined phosphoglucoisomerase, phosphoglucomutase, and malic enzyme with the Corning Authentikit (Innovative Chemistry, Marshfield, MA), which contains seven additional enzymes, to characterize a number of new insect cell lines (32,41,42). This kit is commercially available and provides standard materials for comparison.

Enzyme isoelectric focusing was also used to characterize insect cell lines. Maskos and Miltenburger (43) could distinguish between cell lines from four different genera using any of three isozymes: lactate dehydrogenase, glucose-6-phosphate dehydrogenase, or esterases. McIntosh and Ignoffo (44) could distinguish five different cell lines from *Heliothis zea* but not between two lines from *Heliothis virescens* using the single enzyme phosphoglucoisomerase. The cell lines from the two species were clearly distinguished with the single enzyme.

Thus, isozymes appear to provide sufficient markers to identify most insect cell lines, particularly when used with other characteristics. To avoid serious problems, such characterizations should be done as soon as possible for a new cell line. New cell lines should certainly be well characterized before they are released to other laboratories.

C. Cloning of Insect Cell Lines

Isolation of strains from single cells of established cell lines can help to maximize the yields from large-scale processes. A number of methods have been used to clone insect cell lines. Suitor et al. (45) used the capillary tube method to isolate single cells from a line of mosquito cells. Volkman and Summers (46) used a variation of this method in which single cells were picked up in micropipettes with a micromanipulator and transferred to individual chambers containing 0.2–0.3 ml of medium. Single cells can also be obtained by diluting the cell suspension to a concentration calculated to deliver one cell in a fixed volume, e.g., 100 μl. This volume is transferred to individual wells in a 96-well culture plate (47,48). Each well is then examined under the microscope and those containing a single cell are marked. In each of these methods, the culture is observed for growth at regular intervals and either fed by replacing the medium or transferring to larger vessels as the cell growth warrants.

Some insect cells are capable of growth in soft agar and these cell lines can be cloned by methods similar to those used to clone microorganisms (49). In this method a base layer containing 2%

agar in medium was poured into the culture vessel and allowed to solidify. Then a cell suspension containing an appropriate number of cells was combined with a medium containing Special Agar-Noble (final concentration 0.33% agar) and poured over the base layer. Isolated colonies can be removed with a sterile Pasteur pipet and transferred to other culture vessels or diluted and recloned using the same procedure.

Insect cells grow very poorly when inoculated at low densities or as single cells (low colony-forming efficiency). The colony forming efficiency can be increased by using "conditioned" medium. Conditioned medium is usually prepared by removing the medium from a rapidly growing culture (48–72 hr after starting) and re-placing it with fresh medium. The culture is incubated for an additional 24 hr. Then the medium is removed, centrifuged to remove cells, and sterilized by filtration. The conditioned medium can be used alone or mixed with up to equal parts of fresh medium depending on the requirements of each cell line.

D. Cryopreservation

Frozen samples of cloned, characterized cell lines should be stored either as a backup reserve in case of accidental loss of the active culture or as the initial stock to prepare the seed cultures for a production (50). Insect cells may be preserved for months or years without change or loss of viability if properly frozen and stored.

A detailed procedure for freezing cells is described by Weiss and Vaughn (50) and is summarized here. The survival is highest if the cells are frozen in their optimum metabolic state, e.g., during logarithmic growth. The cells are removed from the culture vessel, centrifuged gently, and resuspended in a mixture of one part spent medium and two parts fresh medium containing 7–10% dimethyl sulfoxide (DMSO). This cell suspension is precooled and dispensed into cryotubes which are heat-sealed. The sealed tubes are labeled and tested for the completeness of the heat seal. The cells are cooled slowly (1°/min) until the temperature reaches −40°C. Then the ampules can be transferred directly to liquid nitrogen for storage.

The cells are recovered by transferring an ampule to a 28°C water bath and thawing rapidly. Then the surface of the ampule is sterilized with 95% ethanol and opened. The cell suspension can be transferred to a culture flask or roller bottle and the concentration adjusted by the addition of fresh growth medium to provide a predetermined optimum concentration (not less than 1×10^5 cells/ml for *Spodoptera* cells). The cells are allowed 3–4 hr to attach or are incubated overnight. Then the medium is replaced completely with fresh medium. If the cells are to be used in suspension culture systems they can be inoculated directly into spinner flasks at an optimal density and incubated in a stirred suspension overnight. The next day the concentration of DMSO is reduced to less than 0.05% by the addition of fresh medium and the incubation continued.

The viability of *Spodoptera* cells frozen by this procedure was 85 – 95% (50). The recovered cells had a population doubling time in roller bottles of just under 14 hr and in 500-ml spinner cultures of just under 9 hr when seeded directly into the culture vessel. Thus, it is possible to begin to produce inoculum for a production run directly from frozen stock with no loss in growth capacity or consistency.

III. MEDIA

A. Basal Media

A number of satisfactory culture media have been developed for the growth of insect cells (Table 2). The first successful medium, developed by Wyatt (51) and modified by Grace (15), was based on the known composition of the hemolymph of the commercial silkworm *Bombyx mori*. The modified medium was used by Grace for the culture of the first invertebrate cell lines. It consists of chemically pure amino acids, vitamins, organic acids, and inorganic salts, supplemented with 5% heat-treated insect hemolymph. Most of the media shown in Table 2 are modifications of these original two. The IPL-41 medium was later modified by Weiss et al. (52) by the addition of the trace metals alumimum and zinc as aluminum chloride and zinc sulfate, respectively. This modified medium increased the quantity and the biological activity of a nuclear polyhedrosis virus produced in *Spodoptera* cells.

Table 2   Examples of Insect Media
Useful for the Culture of Insect
Cells in Large Volume

Medium	Ref.
IPL-41	66
Grace *Antheraea*	3
TNM-FH	53
BML-TC/10 (TC-100)	54

One of the most common modifications has been the replacement of purified chemicals with natural products. The two most common substitutes have been yeast extract for B vitamins and lactalbumin hydrolyzate to replace amino acids. Hink's TNM-FH (53) and Gardiner and Stockdale's TC/10 (54) are two such media commonly used in large-volume culture. Goodwin (55) reported that cells from several lines derived from the gypsy moth grew in medium supplemented with one or more peptones, i.e., peptic peptone, liver digest, lactalbumin hydrolyzate, and yeastolate. To obtain maximum growth the peptones were added to an amino acid containing medium such as Goodwin's IPL-52B (56).

Lipids are important growth stimulants for insect cells (for review, see Ref. 57). Goodwin and Adams (56) presented convincing evidence that the nutritional requirements of growing insect cells was significantly different from the requirements for the production of normal virion-containing polyhedra. They found that in serum-free medium typical occluded virus bundle formation required additional fatty acids and sterols. These were added by the preparation of liposomes containing soybean phosphatidylcholine and cholesterol. As new formulations low in protein are developed for use in large-volume culture, this requirement becomes critical and may be growth or yield limiting.

B. Supplements

The most commonly used media supplement is fetal bovine serum (FBS). It has been used at levels ranging from 20% (28) to 4% (58). As is true with the selection of fetal bovine serum for use in mammalian cell culture, there is considerable variation among lots of the FBS. Therefore, pretesting before purchasing a supply is advisable. Comparison of maximum cell numbers achieved with test lots and the lot of serum in use is commonly used. However, Weiss and Vaughn (50) recommended the use of relative plating efficiency for the final evaluation of new lots of sera. In their method five replicate cultures of 500 cells each are prepared in 5 ml of medium containing either the test serum or a previously established control serum. Following incubation, the colonies in each culture are counted and the relative plating efficiency determined as the number of colonies of acceptable size, morphology, and quantity. The test serum should be accepted only when the difference in plating efficiency is not less then 90% that of the control serum.

Other animal sera have been substituted for FBS. Goodwin (59) developed a number of cell lines from insect pupal tissue using a combination of chicken, turkey, and bovine serum at about 3% each. Hink et al. (58) reported that chicken serum supported

growth of insect cells equal to that of FBS. Roder (60) reported
growth of insect cells in medium supplemented with egg yolk emulsion. Calf, horse, lamb, rabbit, guinea pig, and even human sera
have been tested but were usually not suitable for insect cells
(61−63).

Only a few of the new serum replacements or synthetic sera
developed for use in mammalian cell culture have been tested with
insect cells. Lynn and Hung (64) found that Nuserum (Collabo-
rative Research, Lexington, MA) had value in establishing cell lines
from a species of parasitic wasp. Vaughn and Fan (65) reported
that the serum replacement CPSR-3 (Sigma Chemical, St. Louis,
MO) gave adequate cell growth when used at 10% in place of FBS.
These replacements are significantly less expensive and the supply is much more reliable than that of FBS and therefore are more
suited for large-volume culture.

IV. ATTACHED CELL CULTURE

A. General Considerations

1. Selection of a Growth Surface

Insect cells are not substrate-dependent; however, they will attach
to a variety of surfaces and grow to high numbers. Vaughn (6)
obtained growth on glass roller bottles after a pretreatment with
serum-containing medium for 18 hr. Weiss et al. (66) reported
improved growth with increased cell density when polystyrene bottles were substituted for glass. Later Weiss et al. (11) demonstrated improved growth in perfusion systems in bottles containing
a film of Melinex as the growth surface. Goodwin (67) compared
the growth of a number of insect cell lines on standard polystyrene
surfaces and the positively charged Primaria surface. He reported
that the degree of attachment and the ease of cell transfer varied.
The Primaria flasks appear to offer a better surface for some cells
either because attachment was better or because removal and transfer was easier, e.g., the cells from the coleopteran insect *Heteronychus orator*. If an attached system is desired, a variety of surfaces should be tested in order to determine the most suitable one
for each cell line used.

2. Dissociation of Attached Cells

Generally, insect cells attach loosely to a growth surface and can
be detached by a gentle agitation or with a scraper having silicone rubber blades or made from disposable plastic. The cells can
also be removed by treatment with enzymes. Schneider (27) routinely transferred mosquito cells by incubation for 2−3 min with a

trypsin solution containing 0.2% trypsin (1:250 Difco, Difco Laboratories, Detroit, MI) in Rinaldine's salt solution. Weiss and Vaughn (50) used a solution of 0.003% pancreatin plus 0.002% EDTA in a buffered salt solution to remove the cells of the lepidopteran line IPL-SF-21AE from either standard polystyrene or Melinex surfaces.

B. Roller Bottles

The roller bottle system requires little modification from the standard flask culture system. It provides considerable savings in labor, space, and time compared to expansion by increasing the numbers of flasks. Semiautomatic systems for dispensing the medium and the cell inoculum can provide additional savings. Yields of 20 times the inoculum can be obtained (50,68). In the above study roller bottles were inoculated with cell densities ranging from 1.27×10^7 cells in 100 ml growth medium for 490-cm^2 bottles and 3.17×10^9 cells in 250 ml medium for 1750-cm^2 bottles. Because insect cells generally attach much less firmly to the growth surface than do mammalian cells, the bottles are roller slower than for mammalian cells, i.e., one revolution every 8.5 min (66) to 10 min (6).

C. Perfusion Cultures

Further increases in cell density can be obtained using perfusion systems. In such systems the level of nutrients and metabolic waste products can be maintained at constant levels by the continuous replacement of spent culture medium with fresh medium. Such a system also provides a method for harvesting labile cell products which are excreted from the cell before they are destroyed or altered.

One such system has been described for insect cells by Weiss and Vaughn (50). The vessel used was a modified roller bottle containing a spiral of Melinex and three access ports (68). Three of these vessels were connected in series to a steady-state, continuous-perfusion system containing a fresh medium reservoir, a spent medium reservoir, pH-DO monitoring systems, a solid state channelyzer, and a recorder. Medium was perfused through the system with a medical perfusion pump (Model No. 921, Criticon Inc., Miami, FL).

The culture was established by inoculating each unit with 5×10^8 cells in 1600 ml of growth medium and rotating the units at 3.4 revolutions per hour for 48 hr. After the cell attachment was complete the units were connected to the perfusion system and fresh medium perfused through the bottles at 25 ml/hr. This

perfusion rate was sufficient to maintain a constant pH and DO
level in the system. IPL-SF-21AE cells could be maintained in
this system for 30 days without cell degeneration or loss of viability. If labile cell products are produced, the spent medium reservoir can be refrigerated and emptied as needed to protect the
product.

V. SUSPENSION SYSTEMS

A. General Considerations

Although the roller bottle methods permit a significant scale-up for
production of viruses or cellular metabolites, they do not provide
sufficient capacity for producing many products of low economic
value. Since the growth of insect cells is not dependent on attachment to a substrate, suspension cultures are the most efficient systems for large-volume culture. Such cultures are similar to the deep-tank fermentations developed for producing microbial products such as antibiotics, enzymes, and growth factors.
As described elsewhere in this book, the experience with microbial fermentations has been utilized as much as possible to develop vessels, instrumentation, and methods for the culture of mammalian cells in large-volume suspension systems.

This experience in turn provided the basis for the development of the culture of insect cells in suspension. Although suspension cultures in a variety of vessels at the laboratory scale
have been successful, two characteristics of insect cells hampered
the development of suspension culture systems at the pilot plant
scale or larger: (1) the high oxygen demand of insect cells, particularly when virus-infected, and (2) the fragility of the cells.
Recent research on large-volume culture systems for other animal
cells has provided some possible solutions to these problems. The
development of the baculovirus expression vectors to produce products for the clinical market has provided an economic stimulus
for new research to adapt these large-volume methods to the culture of insect cells.

B. Control of Shear Stress in Suspension Cultures

The shear sensitivity of insect cells has presented the major ob-
stacle to the scale-up of insect cells in suspension. Tramper
et al. (69), using a Haake rotaviscometer to quantify the shear
stress, reported that IPLB-Sf-21 cells lost viability at a shear
stress of $1.5 \, N/m^2$. Hink (70) had earlier shown that TN-368
cells ceased to grow in medium containing 0.1% methyl cellulose
50 cP when impeller speeds in the fermenter reach 220 rpm but

would grow if the level of methyl cellulose was increased to 0.3%. Tramper et al. (69) calculated that this reflected a critical shear stress of the order of 1.5 N/m^2. Other workers have used a variety of viscosity agents to protect cells in suspension cultures. Vaughn (7) added 0.15% 400 cP methyl cellulose to medium for the growth of the Grace *Antheraea* cells in spinner flasks. Weiss et al. (11) added 0.25% 15 cP methyl cellulose and 0.02% Darvan N2 to medium for the growth of IPL-Sf-21AE cells in either spinner flasks or spin-filter flasks (Virtis Co. Biospin filter). Weiss et al. (71) used this combination of viscosity agents as protectants in airlift fermenter cultures in volumes up to 40 liters.

C. Dissolved Oxygen Control

Complicating the problem of shear stress in suspension culture is the requirement for oxygen by insect cells. Vaughn and Goodwin (72) demonstrated that suspension cultures of gypsy moth cells required sparging to maintain satisfactory levels of dissolved oxygen in 250 ml cultures in 500 ml-capacity spinner flasks stirred at 200 rpm. Hink and Strauss (73) reported that the level of dissolved oxygen in 2-liter fermenter cultures dropped to 2% within 24 hr if the culture was not aerated. Maximum cell density could be obtained in their fermenter if the level of dissolved oxygen was maintained between 15 and 50%. This level could be maintained by sparging with air at flow rates that did not cause excessive foaming or vacuolation of cells. Nevertheless they added 0.02% Silicone DC antifoam emulsion (Dow Chemical) to the medium. Foaming has been shown to have a very deleterious affect on insect cells (69). Tramper et al. reported that bursting foam bubbles at the surface of the suspension was a more serious cause of cell death than the velocity at which the bubble rose through the medium and that cell survival in foam was very low. The additional stress due to sparging was avoided by Miltenburger and David (10) by the use of a modified fermenter, Model BIOSTAT-S (B. Braun AG, Melsungen, FRG). The 12-liter vessel was fitted with 5 m of silicone rubber tubine (1 mm wall thickness, 5 mm inside diameter) that was coiled around the metal tube of the heating system and provided about 2800 cm^2 surface area through which oxygen could diffuse into the medium. Filtered air was blown through the silicone tube at a pressure of 0.5−1.5 bar depending on the cell density in the culture. With this arrangement they were able to achieve cell growth in volumes up to 9.5 liters.

Weiss et al. (71,74) achieved good cell growth in a modified spinner in volumes up to 8 liters using serum-free medium. In this system two additional bars were added to the vessel's magnetic spinning bar and the air sparge ring placed directly under

the magnetic stirring bars. Satisfactory growth was obtained when
the impeller speed was reduced to 90 rpm and air was added at a
rate of 75 ml/min.

These studies represented the largest volumes of insect cell
cultures until the recent work of Weiss and his colleagues (75)
using airlift systems. In this system the bioreactor is a long,
tube-shaped vessel containing an inner draught tube. Gas or air
is introduced at the bottom of the open draught tube. As the air
bubbles rise up the draught tube medium is drawn up with them
creating a medium flow up the center draught tube and down the
perimeter of the bioreactor. No additional impellers are needed to
maintain the cells in a homogeneous mixture. The sparge gases
used were N_2, air, CO_2 and 10% PO_2 initially. The PO_2 was increased to 20% on day 4. Table 3 shows the sparge rate and the
maximum cell yields obtained in a series of fermentors up to 40
liters.

D. Optimization of Other Conditions

Another advantage of suspension culture systems is that other
parameters can be monitored easily and controlled to maximize the
yield of a desired product. One such parameter is pH. Insect
cells, particularly lepidopteran cells, grow best at a slightly acid
pH. Hink and Strauss (73) found that the yield of TN-368 cells
declined when the pH was outside a range of 5.8 − 6.7 and the

Table 3   Growth of Insect Cells in Airlift Fermenters

Volume (liters)	Sparge rate (ml/min)[a]	Final cell density $(\times 10^6)$[b]	Days in tank
5	120	6.40	8
10	200 initial	2.96	30
	600 final		
40	400 initial	2.40	11
	800 final		

[a]Sparge gases were N_2, air, CO_2, and 10% PO_2 initially, increased
to 20% PO_2 on day 4.

[b]All fermenters were seeded with 2×10^5 cells/ml.

maximum cell yield was obtained when the pH was maintained between 6.0 and 6.25 by the addition of 1.0M H_3PO_4 by automatic
pH controllers. Weiss and his colleagues found that the IPLB-
Sf-21AE cells in bulk cultures produce lactic acid causing the pH
to fall rapidly from an initial pH of 6.2 and the control must be
obtained by the addition of alkalai (11). Miltenburger and David
(10) automatically maintained a pH of 6.6 in their fermenter studies.

Studies with other animal cells have shown that levels of lactic acid and ammonia can be controlled by adjustments in the composition of nutrients or their concentration (76). The substitution of galactose for glucose as the carbohydrate source can result in the decrease of lactic acid formation. Similarly, the batch
feeding of glucose to keep the concentration below a certain minimum will reduce the amount of lactic acid produced. The levels
of ammonia produced can be reduced by maintaining a lower level
of glutamine by continuous feeding rather than high initial levels.
None of these have been tried with insect cells. Studies by Weiss
et al. (71) showed that in a number of suspension systems 60–70%
of the original glutamine and glucose were metabolized by IPLB-Sf-
21AE cells within 8 days. The major waste product produced, i.e.,
lactic acid or ammonia, appears to vary with the cell line as deduced by the resulting shifts in pH to either more acidic or more
basic. However, no determinations of the actual waste products
have been made.

Perfusion systems allow for the maintenance of constant levels
of both nutrients and metabolic wastes and at least in theory provide a more normal environment for cell growth. In addition, pH
and dissolved oxygen levels can be monitored and adjusted outside of the growth vessel if desirable. Weiss and Vaughn (50)
tested a perfusion system in which complete medium was used to
replace the spent medium in a bulk culture roller bottle designed
by House et al. (68). The cells were grown attached to a Melinex substrate and the vessel had a working volume of 1.7 liters.
A perfusion rate of 25 ml/hr was used and the concentration of
glucose was maintained constant for 9 days following a slight initial drop. In batch systems the level of glucose decreased from
325 mg/100 ml to 50 mg/100 ml. There was also a significant drop
in pH presumably due to the accumulation of lactic acid.

VI. SUMMARY

Cultured insect cells have many uses in agriculture and medicine.
They can be used in the diagnosis and isolation of a number of
viruses infecting both animals and plants and for the laboratory
study of these viruses. Large-volume culture of insect cells has

been envisioned as a way of producing viruses for use in controlling insect pests and for the production of viral antigens for vaccine preparations. Recently they have become a potentially valuable way of producing a variety of proteins for human and veterinary medicine using the genetically engineered baculovirus expression vectors.

The development of satisfactory cell lines and culture methods has proceeded at a slow, irregular pace, inhibited by the lack of knowledge of the physiology of the insect, its small size, and often by the lack of consistent, adequate support for the necessary developmental research. However, now that the basic culture systems are available, cell lines have been developed or can easily be developed for most needs. Suitable media are available and recent developments in refining existing media formations have resulted in low-cost media containing little protein to interfere with downstream processing of cellular metabolites. Future developments are likely to further improve the media formulations and lower the cost.

Technical problems relating to oxygen demand and cell fragility that inhibited the continued development of large-volume culture systems beyond the laboratory a few years ago now appear to be solved or at least are solvable. The successful culture of the *Spodoptera* cells in bioreactors of 40-liter capacity indicates that means of producing insect cells or their metabolic products on a commercial scale can be made economically feasible.

REFERENCES

1. Trager, W. Cultivation of the virus of grassirie in silkworm tissue cultures. *J. Exp. Med.* 61:501−505 (1935).
2. Trager, W. Multiplication of the virus of equine encephalomyelitis in serviving mosquito tissues. *Am. J. Trop. Med.* 18:387−393 (1938).
3. Grace, T. D. C. The development of acytoplasmic polyhedrosis in insect cells grown in vitro. *Virology* 18:33−42 (1962).
4. Grace, T. D. C. Insect cell culture and virus research. *In Vitro* 3:104−117 (1967).
5. Goodwin, R. H., Vaughn, J. L., Adams, J. R., and Louloudes, S. J. Replication of a nuclear polyhedrosis virus in an established insect cell line. *J. Invertebr. Pathol.* 16:284−288 (1970).
6. Vaughn, J. L. The production of nuclear polyhedrosis viruses in large-volume cell cultures. *J. Invertebr. Pathol.* 28:233−237 (1976).

7. Vaughn, J. L. Growth of insect cell lines in suspension culture. In *Proc. 2nd Intern. Colloq. Invertebr. Tissue Cult. Tremezzo, Italy* 1967 (A. Baselli, ed.), Istit. Lombardo, Parisa, pp. 119–125 (1968).

8. Weiss, S. A., Smith, G. C., Kalter, S. S., Vaughn, J. L., and Dougherty, E. Improved replication of *Autographa californica* nuclear polyhedrosis virus in roller bottles: Characterization of the progeny virus. *Intervirology* 15:213–222 (1981).

9. Hink, W. F. and Strauss, E. Growth of the *Trichoplusia ni* (TN-368) cell line in suspension culture. In *Invertebrate Tissue Culture Applications in Medicine, Biology, and Agriculture* (E. Kurstak and K. Maramorosch, eds.), Academic Press, New York, pp. 297–300 (1976).

10. Miltenburger, H. G. and David, P. Mass production of insect cells in suspension. *Develop. Biol. Standard.* 46:183–186 (1980).

11. Weiss, S. A., Peplow, D., Smith, G. C., Vaughn, J. L., and Dougherty, E. Biotechnical aspects of a large-scale process for insect cells and baculovirus. In *Techniques in the Life Sciences*, Vol. C1, *Techniques in Setting Up and Maintenance of Tissue and Cell Cultures*, (E. Kurstak, ed.), Elsevier, New York, pp. C110/1–C110/16 (1985).

12. Smith, G. E., Summers, M. D., and Fraser, M. J. Production of human beta interferon in insect cells infected with a baculovirus expression vector. *Mol. Cell Biol.* 3:2156–2165 (1983).

13. Luckow, V. A. and Summers, M. D. Trends in the development of baculovirus expression vectors. *Biotechnology* 6:47–55 (1988).

14. Hink, W. F. and Bezanson, D. R. Invertebrate cell culture media and cell lines. In *Techniques in the Life Sciences*, Vol. C1, *Techniques in Setting Up and Maintenance of Tissue and Cell Cultures* (E. Kurstak, ed.), Elsevier, New York, pp. C111/1–C111/30 (1985).

15. Grace, T. D. C. Establishment of four strains of cells from insect tissues grown in vitro. *Nature* 195:788–789 (1962).

16. Bilimoria, S. L. and Sohi, S. S. Development of an attached strain from a continuous insect cell line. *In Vitro* 13:461–466 (1977).

17. Lynn, D. E. and Hink, W. F. Comparison of nuclear polyhedrosis virus replication in five lepidopteran cell lines. *J. Invertebr. Pathol.* 35:234–240 (1980).

18. McIntosh, A. H., Ignoffo, C. M., and Andrews, P. L. In vitro host range of five baculoviruses in lepidoteran cell lines. *Intervirology* 23:150–156 (1985).

19. Goodwin, R. H., Tompkins, G. J., and McCawley, P. Gypsy moth cell lines divergent in viral susceptibility. I. Culture and Identification. *In Vitro* 14:485–494 (1978).

20. McIntosh, A. H. and Ignoffo, C. M. Replication and infectivity of the single-embedded nuclear polyhedrosis virus, Baculovirus heliothis, in homologous cell lines. *J. Invertebr. Pathol.* 37:258–264 (1981).

21. Sohi, S. S., Percy, J., Cunningham, J. C., and Arif, B. M. Replication and serial passage of a multicapsid nuclear polyhedrosis virus of *Orgyia pseudotsugata* (Lepidoptera: Lymantriidae) in continuous insect cell lines. *Can. J. Microbiol.* 27:1133–1139 (1981).

22. Sohi, S. S., Percy, J., Arif, B. M., and Cunningham, J. C. Replication and serial passage of a single enveloped baculovirus of *Orgyia leucostigma* in homolgous cell lines. *Intervirology* 21:50–60 (1984).

23. Volkman, L. E. and Goldsmith, P. A. Generalized immunoassay for *Autographa californica* nuclear polyhedrosis virus infectivity in vitro. *Appl. Environm. Microbiol.* 44:227–233 (1982).

24. Miltenburger, H. G., Naser, W. L., and Harvey, J. P. The cellular substrate: A very important requirement for baculovirus in vitro replication. *Z. Naturforsch.* 39c:993–1002 (1984).

25. Granados, R. R., Derkson, A. C. G., and Dwyer, K. D. Replication of the Trichoplusia ni granulosis and nuclear polyhedrosis viruses in cell cultures. *Virology* 152:472–476 (1986).

26. Greene, A. E. and Charney, J. Characterization and identification of insect cell cultures. *Curr. Topics Microbiol. Immunol.* 55:51–61 (1971).

27. Schneider, I. Establishment of three diploid cell lines of Anopeles stephensi (Diptera:Culicidae). *J. Cell. Biol.* 42:603–606 (1969).

28. Pudney, M. and Varma, M. G. R. Anopheles stephensi var. mysorensis: establishment of a larval cell line (Mos. 43). *Exp. Parasitol.* 29:7–12 (1971).

29. Dolphini, S. Karyotype polymorphism in a cell population of *Drosophila melanogaster* cultured in vitro. *Chromosoma* (Berl.) 33:196–208 (1971).

30. Sohi, S. S. and Ennis, T. J. Chromosomal characterization of cell lines of *Neodiprion lecontei* (Hymenoptera: Diprionidae). *Proc. Entomol. Soc. Ont.* 112:45–48 (1981).

31. Crawford, A. M., Parslow, M., and Sheehan, C. Changes in the karyotype of the cell line, DSIR-HA-1179, and a

comparison with that of its parent insect, *Heteronychus arotor* (F.) (Coleoptera:Scarabaeidae). *N. Zeal. J. Zool.* 10:405−408 (1983).

32. Lynn, D. E. and Stoppleworth, A. Established cell lines from the beetle, *Diabrotica undecimpunctata* (Coleoptea: Chrysomelidae). *In Vitro* 20:365−368 (1984).

33. Nickols, W. W., Bradt, C., and Boune, W. Cytogenetic studies on cells in culture from the class Insecta. *Curr. Topics Microbiol. Immunol.* 55:61−69 (1971).

34. Schneider, F. I. Characteristics of insect cells. In *Tissue Culture Methods and Applications* (P. F. Kruse, Jr., and M. K. Patterson, Jr., eds.), Academic Press, New York, pp. 788−790 (1973).

35. Disney, J. E. and McCarthy, W. J. A modified technique for the improved characterization of lepidopteran chromosomes from cells in culture. *In Vitro Cell. Dev. Biol.* 21:563−568 (1985).

36. Vaughn J. L., Goodwin, R. H., Tompkins, G. J., and McCawley, P. The establishment of two cell lines from the insect *Spodoptera frugiperda* (Lepidoptera: Noctuidae). *In Vitro* 13:213−217 (1977).

37. Aldridge, C. A. and Knudson, D. L. Characterization of invertebrate cell lines. I. Serologic studies of selected lepidopteran lines. *In Vitro* 16:384−391 (1980).

38. Tabachnick, W. J. and Knudson, D. L. Characterization of invertebrate cell lines. II. Isozyme analyses employing starch gel electrophoresis. *In Vitro* 16:392−398 (1980).

39. Brown, S. E. and Knudson, D. L. Characterization of invertebrate cell lines. IV. Isozyme analysis of dipteran and acarine cell lines. *In Vitro* 18:347−350 (1982).

40. Harvey, G. T. and Sohi, S. S. Isozyme characterization of 28 cell lines from five insect species. *Can. J. Zool.* 63:2270−2276 (1985).

41. Rochford, R. Dougherty, E. M., and Lynn, D. E. Establishment of a cell line from embryos of the cabbage looper, *Trichoplusia ni* (Hubner). *In Vitro* 20:823−825 (1984).

42. Lynn, D. E. Dougherty, E. M., McClintock, J. T., and Loeb, M. Development of cell lines from various tissues of Lepidoptera. In *Invertebrate and Fish Tissue Culture* (Y. Kuroda, E. Kurstak, and K. Maramorosch, eds.), Springer-Verlag, Berlin, pp. 239−242 (1988).

43. Maskos, C. B. and Miltenburger, H. G. Isoenzyme-Analyse von Insektenzellinien durch isoelektrische Fokussierung in Dünnschicht-Polyacrylamid-Gelen. *Mitt. dtsch. Ges. allg. angew. Ent.* 4:44−47 (1983).

44. McIntosch, A. H. and Ignoffo, C. M. Characterization of five cell lines established from species of Heliothis. *Appl. Ent. Zool.* 18:262–269 (1983).
45. Suitor, E. C., Jr., Chang, L. L., and Liu, H. A. Establishment and characterization of a clone from Grace's in vitro cultured mosquito (*Aedes aegypti* L.) cells. *Exp. Cell Res.* 44:572–578 (1966).
46. Volkman, L. E. and Summers, M. D. Comparative studies with clones derived from a cabbage looper ovarian cell line, TN-368. In *Invertebrate Tissue and Culture Applications in Medicine, Biology, and Agriculture* (E. Kurstak and K. Maramorosch, eds.), Academic Press, New York, pp. pp. 289–296 (1976).
47. Lynn, D. E. and Oberlander. H. The effect of cytoskeletal disrupting agents on the morphological response of a cloned Manduca sexta cell line to 20-hydroxyecdysone. *Wilhelm Roux Arch.* 190:150–155 (1981).
48. Corsaro, B. G. and Fraser, M. J. Characterization of cloned populations of Heliothis zea cell line IPLB-HZ 1075. *In Vitro Cell. Dev. Biol.* 23:855–862 (1987).
49. McIntosh, A. H. and Rechtoris, C. Insect cells: Colony formation and cloning in agar medium. *In Vitro* 10:1–5 (1974).
50. Weiss, S. A. and Vaughn, J. L. Cell culture methods for large-scale propagation of baculoviruses. In *The Biology of Baculoviruses, Vol. 2 Practical Application for Insect Control* (R. R. Granados and B. A. Federici, eds.), CRC Press, Boca Raton, pp. 64–87 (1986).
51. Wyatt, S. S. Culture in vitro of tissue from the silkworm, *Bombyx mori* L. *J. Gen. Physiol.* 39:841–857 (1956).
52. Weiss, S. A., Smith, G. C., Vaughn, J. L., Dougherty, E. M., and Tompkins, G. Effect of aluminum chloride and zinc sulfate on Autographa californica nuclear polyhedrosis virus (AcNPV) replication in cell culture. *In Vitro* 18:937–944 (1982).
53. Hink, W. F. Established insect cell line from the cabbage looper, *Trichoplusia ni. Nature* 226:466–467 (1970).
54. Gardiner, G. R. and Stockdale, H. Two tissue culture media for production of lepidopteran cells and nuclear polyhedrosis viruses. *J. Invertebr. Pathol.* 25:363–370 (1975).
55. Goodwin, R. H. Insect cell growth on serum free media. *In Vitro* 12:303–304 (1976).
56. Goodwin, R. H. and Adams, J. R. Nutrient factors influencing viral replication in serum-free insect cell line cultures. In *Invertebrate Systems In Vitro* (E. Kurstak, K. Maramorosch, and A Dübendorfer, eds.), Elsevier, Amsterdam, pp. 493–509 (1980).

57. Vaughn, J. L. and Louloudes, S. J. Lipid nutrition and metabolism of cultured insect cells. In *Metabolic Aspects of Lipid Nutrition in Insects* (T. E. Mittler and R. H. Dadd, eds.), Westview Press, Boulder, CO, pp. 223−244 (1983).

58. Hink, W. F., Strauss, E., and Mears, J. L. Effects of media constituents on growth of insect cells in stationary and suspension cultures. *In Vitro* 9:371 (1974).

59. Goodwin, R. H. Insect cell culture: Improved media and methods for initiating attached cell lines from the lepidoptera. *In Vitro* 11:369−378 (1975).

60. Roder, A. Development of a serum-free medium for cultivation of insect cells. *Naturwissenschaften* 69:92−93 (1982).

61. Martignoni, M. E. and Scallion, R. J. Preparation and uses of insect hemocyte monolayers "in vitro." *Biol. Bull.* 121:507−620 (1961).

62. Mitsuhashi, J. and Maramorosch, K. Leafhopper tissue culture: Embryonic, nymphal, and imaginal tissue from aseptic insects. *Contrib. Boyce Thompson Inst.* 22:435−460 (1964).

63. Ignoffo, C. M., Shapiro, M., and Dunkel, D. Effects of insect, mammalian, and avian sera on in vitro growth of Heliothis zea cells. *Ann. Entomol. Soc. Am.* 66:170−172 (1973).

64. Lynn, D. E. and Hung, A. C. F. Development of a continuous cell line from the insect egg parasitoid, *Trichogramma pretiosum* (Hymenoptera: Trichogramenatidae). *In Vitro Cell. Dev. Biol.* 22:440−442 (1986).

65. Vaughn, J. L. and Fan, F. Use of commercial serum replacements for the culture of insect cells. *In Vitro Cell. Dev. Biol.* 25:143−145 (1989).

66. Weiss, S. A., Smith, G. C., Kalter, S. S., and Vaughn, J. L. Improved method for the production of insect cell cultures in large-volume. *In Vitro* 17:495−502 (1981).

67. Goodwin, R. H. Growth of insect cells in serum-free media, In *Techniques in the Life Sciences, Vol. C1, Cell Biology Techniques in Setting Up and Maintenance of Tissue and Cell Cultures* (E. Kurstak, ed.), Elsevier, New York, pp. C109/1−C109/28 (1985).

68. House, W., Shearer, M., and Maroudas, N. G. Method of bulk culture of animal cells on plastic film. *Exp. Cell. Res.* 71:293−296 (1972).

69. Tramper, J., Williams, J. B., and Joustra, D. Shear sensitivity of insect cells in suspension. *Enzyme Microb. Technol.* 8:33−36 (1986).

70. Hink, W. F. Production of Autographa californica nuclear polyhedrosis virus in cells from large-scale suspension

cultures. In *Microbial and Viral Pesticides* (E. Kurstak, ed.), Marcel Dekker, New York, pp. 493–506 (1982).

71. Weiss, S. A., Belisle, B. W., DeGiovanni, A., Godwin, G., Kohler, J., and Summers, M. D. Insect cells as substrates for biologicals. *Dev. Biol. Standard* 70:271–279 (1989).

72. Vaughn J. L. and Goodwin, R. H. Large-scale culture of insect cells for virus production. In *Beltsville Symposia on Argic. Res. I. Virology in Agriculture* (J. A. Romberger, ed.), Allenheld Osmun, Montclair, NJ pp. 109–116 (1977).

73. Hink, W. F. and Strauss, E. M. Semi-continuous culture of the TN-368 cell line in fermentors with virus production in harvested cells. In *Invertebrate Systems In Vitro* (E. Kurstak, K. Maramorosch, and A. Dübendorfer, eds.), Elsevier, Amsterdam, pp. 27–33 (1980).

74. Weiss, S. A., DeGiovanni, A. M., Godwin, G. P., and Kohler, J. P. Large scale cultivation of insect cells. In *Conference on Biotechnology Biological Pesticides and Novel Plant-Pest Resistance for Insect Pest Management* (D. W. Roberts and R. R. Granados, eds.), Boyce Thompson Institute for Plant, Research, Cornell University, July 18–20, 1988, Cornell University Press, Ithaca, NY, pp. 22–30 (1989).

75. Weiss, S. A., Belisle, B., De Giovanni, H., and Godwin, G. Use of insect cells in biotechnology. *In Vitro Cell. Dev. Biol.* 24:53 (1988).

76. Glacken, M. W., Fleischaker, R. J., and Sinskey, A. J. Large-scale production of mammalian cells and their products: Engineering principles and barriers to scale-up. *Ann. NY Acad. Sci.* 413:355–372 (1983).

77. Miltenburger, H. G., David P., Mahr, U., and Zipp, W. Ueber die Erstellung von Lepidopteren-Dauerzellinien und die in vitro Replikation von insektenpqthogenen Viren. I Mamestra brassicae (Kohleule)-Zellinien und NPV-Replikation. *Z. ang. Ent.* 82:306–323 (1977).

78. Grace, T. D. C. Establishment of a line of cells from the silkworm Bombyx mori. *Nature* 216:613 (1967).

Index